Smart Innovation, Systems and Technologies

Volume 389

The Smart Innovation, Systems and Technologies book series encompasses the topics of knowledge, intelligence, innovation and sustainability. The aim of the series is to make available a platform for the publication of books on all aspects of single and multi-disciplinary research on these themes in order to make the latest results available in a readily-accessible form. Volumes on interdisciplinary research combining two or more of these areas is particularly sought.

The series covers systems and paradigms that employ knowledge and intelligence in a broad sense. Its scope is systems having embedded knowledge and intelligence, which may be applied to the solution of world problems in industry, the environment and the community. It also focusses on the knowledge-transfer methodologies and innovation strategies employed to make this happen effectively. The combination of intelligent systems tools and a broad range of applications introduces a need for a synergy of disciplines from science, technology, business and the humanities. The series will include conference proceedings, edited collections, monographs, handbooks, reference books, and other relevant types of book in areas of science and technology where smart systems and technologies can offer innovative solutions.

High quality content is an essential feature for all book proposals accepted for the series. It is expected that editors of all accepted volumes will ensure that contributions are subjected to an appropriate level of reviewing process and adhere to KES quality principles.

Indexed by SCOPUS, EI Compendex, INSPEC, WTI Frankfurt eG, zbMATH, Japanese Science and Technology Agency (JST), SCImago, DBLP.

All books published in the series are submitted for consideration in Web of Science.

Sankar K. Pal · Sabu M. Thampi · Ajith Abraham
Editors

Intelligent Informatics

Proceedings of Eighth International Symposium on Intelligent Informatics (ISI 2023)

Springer

Editors
Sankar K. Pal
Center for Soft Computing Research
Indian Statistical Institute
Kolkata, West Bengal, India

Ajith Abraham
Bennett University
Greater Noida, India

Sabu M. Thampi
School of Computer Science
and Engineering
Kerala University of Digital Sciences
Innovation and Technology (Digital
University Kerala)
Thiruvananthapuram, Kerala, India

ISSN 2190-3018 ISSN 2190-3026 (electronic)
Smart Innovation, Systems and Technologies
ISBN 978-981-97-2146-7 ISBN 978-981-97-2147-4 (eBook)
https://doi.org/10.1007/978-981-97-2147-4

This Springer imprint is published by the registered company Springer Nature Singapore Pte Ltd.
The registered company address is: 152 Beach Road, #21-01/04 Gateway East, Singapore 189721, Singapore

Organized by

PES University, Bengaluru, India

Conference Organization

Chief Patron

M. R. Doreswamy, Chancellor, PES University

Patrons

D. Jawahar, Pro Chancellor, PES University
Ajoy Kumar, COO, PES Institutions
J. Surya Prasad, Vice Chancellor, PES University
K. S. Sridhar, Registrar, PES University

General Chairs

Sankar K. Pal, Center for Soft Computing Research, Indian Statistical Institute, Kolkata, India
Sabu M. Thampi, Kerala University of Digital Sciences, Innovation and Technology, India
Ajith Abraham, FLAME University, India

TPC Chairs

Swagatam Das, Indian Statistical Institute, India
Sougata Mukherjea, Indian Institute of Technology Delhi, India

General Executive Chair

Shikha Tripathi, PES University, Bangalore

Steering Committee

Sudarshan T. S. B., Dean of Research, PES University (Chair)

Organizing Chair

B. N. Krupa, PES University

Organizing Secretaries

M. S. Sunitha, PES University
Event Management Chair
M. Rajasekar, PES University

TPC Members

Vo Nguyen Quoc Bao, Posts and Telecommunications Institute of Technology, Vietnam
Phan Cong-Vinh, NTT University, Vietnam
Thanh D. Nguyen, Banking University of Ho Chi Minh City, Vietnam
Tri-Thanh Nguyen, Vietnam National University, Hanoi, Vietnam
Afrand Agah, West Chester University of Pennsylvania, USA
Lie Lu, Dolby, USA
Haijun Pan, New Jersey Institute of Technology, USA
Arijit Bhattacharya, University of East Anglia, UK
Thomas Chen, City University London, UK
Ali Hessami, Vega Systems, UK
Mohammed Mujahid Ulla Faiz, University of Westminster, UK
Quoc-Tuan Vien, Middlesex University, UK
Hanen Idoudi, University of Manouba, Tunisia
Permanand Mohan, The University of The West Indies, Trinidad and Tobago
Justin Dauwels, Delft University of Technology, The Netherlands

Nattee Pinthong, Rajabhat Rajanagarindra University, Thailand
Grienggrai Rajchakit, Maejo University, Thailand
Yue-Shan Chang, National Taipei University, Taiwan
Uei-Ren Chen, Hsiuping University of Science and Technology, Taiwan
Chien-Fu Cheng, National Taiwan Ocean University, Taiwan
Ying-Ren Chien, National I-Lan University, Taiwan
Tzung-Pei Hong, National University of Kaohsiung, Taiwan
Wei-Chiang Hong, Asia Eastern University of Science and Technology, Taiwan
Gwo-Jiun Horng, Southern Taiwan University of Science and Technology, Taiwan
Wen-Liang Hwang, Institute of Information Science, Academia Sinica, Taiwan
Wen-Yang Lin, National University of Kaohsiung, Taiwan
Ming-Chi Liu, Feng Chia University, Taiwan
Jeng-Shyang Pan, National Kaohsiung University of Applied Sciences, Taiwan
Ming-Fong Tsai, National United University, Taiwan
Sheng-Shih Wang, Lunghwa University of Science and Technology, Taiwan
You-Chiun Wang, National Sun Yat-Sen University, Taiwan
Christian Buddendick, ZEB, Switzerland
Athanasios V. Vasilakos, Lulea University of Technology, Sweden
Vijayaratnam Ganeshkumar, Just In Time Group, Sri Lanka
Rafael Asorey-Cacheda, Technical University of Cartagena, Spain
Carlos Fernandez-Llatas, Universitat Politècnica de València, Spain
Felix J. Garcia Clemente, University of Murcia, Spain
Javier Gozalvez, Universidad Miguel Hernandez de Elche, Spain
Antonio LaTorre, Universidad Politécnica de Madrid, Spain
Miguel Sepulcre, Universidad Miguel Hernandez de Elche, Spain
Engin Zeydan, Centre Tecnològic de Telecomunicacions de Catalunya (CTTC), Spain
Roman Jarina, University of Zilina, Slovakia
El-Sayed El-Alfy, King Fahd University of Petroleum and Minerals, Saudi Arabia
Dmitry Korzun, Petrozavodsk State University, Russia
Sergey Mosin, Kazan Federal University, Russia
Felix Albu, Valahia University of Targoviste, Romania
Monica Chis, Freelancer Information Technology and Services, Romania
Anca Daniela Ionita, University Politehnica of Bucharest, Romania
Ramiro Barbosa, Institute of Engineering of Porto, Portugal
Luis Barreto, Instituto Politécnico de Viana do Castelo, Portugal
Eugénia Bernardino, Polytechnic Institute of Leiria, Portugal
Isabel Jesus, Institute of Engineering of Porto—ISEP, Portugal
Carlos Vaz de Carvalho, Instituto Superior de Engenharia do Porto, Portugal
Dariusz Barbucha, Gdynia Maritime University, Poland
Dariusz Gasior, Wroclaw University of Technology, Poland
Marek Wegrzyn, Space Research Centre of the Polish Academy of Sciences, Poland
Piotr Zwierzykowski, Poznan University of Technology, Poland
Hussain Saleem, University of Karachi, Pakistan
Kenneth Nwizege, Ken Saro-Wiwa Polytechnic, Bori, Nigeria

Cheng-Lian Liu, Pacific University, Nicaragua
Paulus Sheetekela, The International University of Management, Namibia
Mohamed Moussaoui, Abdelmalek Esaadi UniversitY, Morocco
Mohd Ashraf Ahmad, Universiti Malaysia Pahang, Malaysia
Boon Chong Ang, Intel, Malaysia
Rozmie Razif Othman, Universiti Malaysia Perlis, Malaysia
Huong Yong Alan Ting, University of Technology Sarawak, Malaysia
Farrah Wong, Universiti Malaysia Sabah, Malaysia
Jin-Han Park, Pukyong National University, Korea (South)
Osama Abu-Sharkh, Princess Sumaya University for Technology, Jordan
Hugang Han, Prefectural University of Hiroshima, Japan
Hiroshi Sakai, Kyushu Insitute of Technology, Japan
Antonio Cimmino, Lasting Dynamics, Italy
Paolo Crippa, Marche Polytechnic University, Italy
Arianna D'Ulizia, CNR, Italy
Angelo Trotta, University of Bologna, Italy
Vuong Ngo, Technological University Dublin, Ireland
Aws Yonis, Ninevah University, Iraq
Kambiz Badie, Iran Telecom Research Center, Iran
Saeed Olyaee, Shahid Rajaee Teacher Training University, Iran
Hamed Vahdat-Nejad, University of Birjand, Iran
Ida Giriantari, Udayana University, Bali, Indonesia
Tutut Herawan, Ambarrukmo Tourism Institute, Indonesia
Naveen Aggarwal, Panjab University, India
Sachin Agrawal Sony, Sony AI, India
Manjunath Aradhya, Sri Jayachamarajendra College of Engineering, India
Keerthi Balasundaram, Researchers Academy, India
Usha Banerjee, College of Engineering Roorkee, India
D. Shanmugapriya, Avinashilingam Institute, India
Radhakrishnan Delhibabu, VIT Vellore, India
Durairaj Devaraj, Kalasalingam University, India
Anirban Dutta Choudhury, Tata Consultancy Services, India
Omid Mahdi Ebadati E., Hamdard University, India
Bibhas Ghosal, IIIT Allahabad, India
Avik Ghose, Tata Consultancy Services, India
Ankur Gupta, Model Institute of Engineering and Technology, India
Sandhya Harikumar, Amrita Vishwa Vidyapeetham, India
J. Amudha, Amrita Vishwa Vidyapeetham, India
Ramkumar Jaganathan, Sri Krishna Arts and Science College, India
Avinash Jha, OppCorp Learning and Development Private Limited, India
K. C. Raveendranathan, College of Engineering Thiruvananthapuram, India
Sanjay Kimbahune, Tata Consultancy Services Ltd., India
K. V. Krishna Kishore, Vignan University, India
Sunil Kumar Kopparapu, Tata Consultancy Services, India
K. S. Hareesha, Manipal Institute of Technology, India

Adesh Kumar, UPES, India
Naresh Kumar, GGSIPU, India
Ashwani Kush, IIT knapur and KUK India, India
M. Suresh, Amrita Vishwa Vidyapeetham, India
Noor Mahammad Sk, IIIT Design and Manufacturing Kancheepuram, India
Ravibabu Mulaveesala, Indian Institute of Technology Ropar, India
Sakthi Muthiah, LNMIIT, India
Nithin Nagaraj, National Institute of Advanced Studies, India
Subrata Nandi, National Institute of Technology, Durgapur, India
Kanubhai Patel, Charotar University of Science and Technology (CHARUSAT), India
Jaynendra Kumar Rai, Amity University Uttar Pradesh, India
Hanumantha Raju, BMS Institute of Technology and Management, India
G. Ramachandra Reddy, Vellore Institute of Technology, India
Jaydip Sen, Praxis Business School, India
Aditi Sharma, Parul University, Vadodara, India
Durga Prasad Sharma, AMUIT, MOSHE FDRE under UNDP and Adviser (IT) ILO-UN, India
Ajay Singh, NIIT University-Neemarana India, India
Ravi Subban, Pondicherry University, Pondicherry, India
Syed Zafaruddin, BITS Pilani, India
Kalman Palagyi, University of Szeged, Hungary
Jozsef Vasarhelyi, University of Miskolc, Hungary
Katerina Kabassi, Ionian University, Greece
Sotiris Kotsiantis, University of Patras, Greece
Dimitrios Koukopoulos, University of Patras, Greece
Michael Vrahatis, University of Patras, Greece
Feng Cheng, University of Potsdam, Germany
Christian Veenhuis, CARIAD SE (VW Group), Germany
Ramin Yahyapour, GWDG—University Göttingen, Germany
Mohamed Ba khouya, University of Technology of Belfort Montbeliard, France
Mohammed Chadli, University of Paris Saclay, France
Mounir Kellil, CEA LIST, France
Pascal Lorenz, University of Haute Alsace, France
Amir Nakib, University Paris East, France
Patrick Siarry, University of Paris XII, France
Roberto Carlos Herrera Lara, Electricity Company of Quito, Ecuador
Frantisek Zboril, Brno University of Technology, Czech Republic
George Dekoulis, Aerospace Engineering Institute (AEI), Cyprus
Philip Moore, Lanzhou University, China
Hongbo Ni, Northwestern Polytechnical University, China
Peiyan Yuan, Henan Normal University, China
Michael McGuire, University of Victoria, Canada
Marie-Jose Montpetit, Concordia University, Canada
Ali Rafiei, General Motors, Canada

Arshin Rezazadeh, University of Western Ontario, Canada
Elizabeth Goldbarg, Federal University of Rio Grande do Norte, Brazil
Lisandro Lovisolo, State University of Rio de Janeiro, Brazil
Júlio Nievola, Pontificia Universidade Catolica do Paraná—PUCPR, Brazil
Otavio Teixeira, Universidade Federal Do Pará (UFPA), Brazil
Carlos Becker Westphall, Federal University of Santa Catarina, Brazil
Dimitri Papadimitriou, University of Antwerp—imec, Belgium
Gancho Vachkov, Baku Higher Oil School (BHOS), Baku, Azerbaijan
Lloyd Wood, Ericsson, Australia
Fatiha Merazka, LISIC Laboratory. USTHB University, Algeria
Hamouma Moumen, University of Batna 2, Algeria

Preface

The 8th International Symposium on Intelligent Informatics (ISI'23) was held in Bengaluru (Bangalore), India, from December 18 to 20, 2023. ISI'23 provided a platform to share and discuss theoretical and practical developments in intelligent informatics. It was co-located with the International Conference on Applied Soft Computing and Communication Networks (ACN'23). The conference included keynote addresses, contributed papers, workshops, and tutorials. The event was organized by PES University, Bengaluru, and received technical support from the IEEE Signal Processing Society Bangalore Chapter and the IEEE Communications Society Bangalore Chapter.

This volume comprises 30 papers presented at the symposium and is organized into different sections, such as Computer Vision, Image Processing, Signal Processing, Machine Learning and Deep Learning Applications, Healthcare and Medical Diagnostics, Biotechnology and Environmental Applications, IoT Security and Data Encryption, and Quantum Computing and Intelligent Systems.

All submissions underwent evaluation based on their significance, novelty, and technical quality. A double-blind review process was conducted to ensure that the author names and affiliations were unknown to the Technical Program Committee (TPC).

We extend our gratitude to all the authors who contributed their papers to the success of ISI'23. We acknowledge the pivotal role played by PES University, Bengaluru, as the organizing institution, and express our thanks to the IEEE Signal Processing Society Bangalore Chapter and IEEE Communications Society Bangalore Chapter for their technical support. The dedication of the Local Organizing Committee members is commendable, as is the selfless contribution of time by the faculty, staff, and student volunteers who played vital roles in ensuring the success of ACN'23.

Lastly, we express our appreciation for the collaboration with our publisher, Springer, and extend our sincere thanks to Senior Editor Aninda Bose for their invaluable support.

Kolkata, India Sankar K. Pal
Thiruvananthapuram, India Sabu M. Thampi
Greater Noida, India Ajith Abraham
December 2023

Contents

Part II Machine Learning and Deep Learning Applications

Part III Healthcare and Medical Diagnostics

About the Editors

Sankar K. Pal (Life Fellow, IEEE) received the first Ph.D. degree in radio physics and electronics from the University of Calcutta, Kolkata, India, in 1979, and the second Ph.D. degree in electrical engineering along with DIC from Imperial College, University of London, London, UK, in 1982. He is currently National Science Chair, Government of India, and President of the Indian Statistical Institute (ISI). He is also Distinguished Scientist and Former Director of ISI, Former Distinguished Professor of the Indian National Science Academy, and Former Chair Professor of the Indian National Academy of Engineering. He founded the Machine Intelligence Unit and the Center for Soft Computing Research: a national facility in the institute in Calcutta. In 1975, he joined ISI as CSIR Senior Research Fellow where he became Full Professor in 1987, Distinguished Scientist in 1998, Director in 2005–2010, and President in 2022–2024.

Sabu M. Thampi is a Professor at the School of Computer Science and Engineering, Digital University Kerala, Trivandrum, India. His current research interests include the Internet of Things (IoT), cognitive security, social networks, endpoint security, and smart cyber-physical systems. Sabu is also coordinating the Connected Systems and Intelligence (CSI) Lab at the University. He holds a Ph.D. in Computer Engineering from the National Institute of Technology Karnataka. Dr. Sabu has been actively involved in funded research projects and published papers in book chapters, journals, and conference proceedings. He has authored and edited a few books, as well as edited 45+ conference proceedings published by Springer in various series, as well as a few others published by IEEE, ACM, and Elsevier.

Ajith Abraham received his Ph.D. in Computer Science from Monash University, Melbourne, Australia. He has a Master of Science in Control and Automation from Nanyang Technological University, Singapore. He holds a bachelor's degree in electrical and electronic engineering from the University of Calicut, Kerala, India. He has over 32 years of industry and academic experience. His primary research is on developing advanced machine intelligence using hybridization of function approximation methods, approximate reasoning and global optimization methods focused on big data analytics, understanding networks, information security, Web intelligence, decision support systems, the Internet of things, etc. He is Founding Director of Machine Intelligence Research Labs, a not-for-profit Scientific Network for Innovation and Research Excellence connecting industry and academia.

Part I
Computer Vision, Image Processing. Signal Processing

Chapter 1
Identification of Taurine Cattle Breed Based on Convolutional Neural Network

Fulbert Bembamba, Ozias Bombiri, Albert Soudré, Frédéric Ouedraogo, and Sadouanouan Malo

Abstract Identifying West African taurine cattle breed has become vital since uncontrolled crossing with zebus is jeopardizing their genetic heritage and trypanoresistance capacity. In this study, a computer vision solution is proposed for lobi taurine cattle classification. We implemented a customized Convolutional Neural Network (CNN) on a dataset containing 2379 images taken from three angles: front, side, and rear. The CNN was trained on four subsets of the image data according to the angle of shooting. The training is accelerated by a GeForce RTX3080 laptop GPU. The model yields only 78% precision for the mixed image dataset. Precision rises when the images are split by angle of view: 91% for side images and up to 99% for rear view images. Transfer learning has also been applied for comparison between our model and pretrained models. VGG-16 improved the results for all subsets.

1.1 Introduction

Computer vision intends to set up computer systems and applications that can replicate human vision capacity. This includes the task of classifying an image into the right class category. Since the advent of Convolutional Neural Networks (CNN), this technology has made a clear difference in the field of image classification. In the meantime, the need for image analysis and the use of computer for image recognition is in constant development and is generating many benefits in many areas of knowledge [1]. Computer vision based on convolutional neural networks has enabled people to accomplish tasks that had been considered impossible in the past few centuries [2]. For example, with the growth of the demand for efficient traceability and identification systems for livestock [3], CNN has outperformed over traditional Machine Learning (ML) techniques, providing excellent and robust solutions.

F. Bembamba (✉) · A. Soudré · F. Ouedraogo · S. Malo
Université Norbert Zongo, Koudougou, Burkina Faso
e-mail: bembaplus@gmail.com

O. Bombiri
Université Nazi BONI, Houet, Burkina Faso

S. K. Pal et al. (eds.), *Intelligent Informatics*, Smart Innovation, Systems and Technologies 389, https://doi.org/10.1007/978-981-97-2147-4_1

Taurines are a specific breed of cattle that can resist an animal disease called trypanosomosis that is endemic in tropical West African countries, in general, and Burkina Faso, in particular. Taurine breed has different denominations depending upon their regions. For example, they are called *baoulé* in Côte d'Ivoire and *lobi* in Burkina Faso. The present study is done using data collected on the lobi-type taurines that live mainly in the south-west region of Burkina Faso. The Taurines' innate resistance to trypanosomosis makes them robust and hardy animals, and perfectly integrated into their wet and tse-tse-infected environment, albeit smaller than the zebus. However, because of uncontrolled crossing with the zebus species, it has become difficult to distinguish a purebred lobi from a crossbred, unless through blood analysis. But using DNA for routine identification is not cost-effective as it is a very time-consuming process to get unique DNA identifiers [4]. Therefore, lobi cattle genetic heritage and their trypanoresistance ability are in danger if a cheaper and easy-to-use solution is not found. In paper [5], a Machine Learning (ML) model is applied with good results. But this solution requires morphological measures to be collected from the animal.

In this paper, we propose a method inspired by computer vision and doesn't involve contact with the cow. A customized three-layer convolutional neural network is trained on a dataset containing 2379 images of cattle. The images were taken from three different shooting angles: side, front, and rear. The results show low performance (78%) on the entire mixed dataset but excellent performance in dataset from same shooting side, hitting 99% accuracy. We also applied transfer learning to compare our own model's outcomes with two pretrained models, namely, VGG-16 and Inception-V3. The first model improved the performance on all subsets. Whereas Inception-V3 performed less than the customized model on head and rear images.

The remainder of the paper is organized as follows: in Sect. 1.2, a brief literature review is made, in Sect. 1.3, we present materials and methodology. We show results in Sect. 1.4 and conduct discussions in Sect. 1.5, whereas Sect. 1.6 concludes the paper and presents future work.

1.2 Related Work

Computer vision is becoming an increasingly trendy word in the area of image processing. In the era of precision livestock, there is a growing demand for computer-based system of identification and tracking of animals. A number of solutions have been unveiled in scientific literature with the aim of replacing traditional identification methods like tattooing body, tagging of ear, microchip implants and branding, radio-frequency identification (RFID), etc.

In [9], a CNN is modeled on a 1000 image dataset containing 10 different species of cows. The trained model reached 89.95% accuracy. The authors of [6] suggested a cattle face recognition method based on a two branch CNN. The purpose of the research was to build a cattle face recognition network model to efficiently and quickly identify individual cattle without contact.

Silva [1] presented a hybrid method using CNN and support vector machine (SVM) to detect brandings of cattle on a total of 39 brands. CNN is used for segmentation and feature extraction while SVM is used for classification. Two different experiments were carried out, the first on 1950 images and the second on 2730 images. Overall accuracy reached 93% for Experiment I. In Experiment II, the method reached an accuracy of 95% and an algorithm processing time of 42 s for the same brands. The authors also made a correlation between increase of accuracy and number of sample images. This is confirmed by [8] who declares that the breakthrough of CNN in computer vision is largely due to the introduction of large amounts of data and readily available hardware.

However, some techniques can be used to make the most of small datasets. Laith Alzubaidi et al. [18] brought forward a few methods to address data scarcity including, but not all, transfer learning (TL), self-supervised learning (SSL), generative adversarial networks (GANs), model architecture (MA), etc. To achieve state-of-the-art results with as few as 150 images of 26 cattle breeds, Manoj [7] removed noise and background from images and converted them into grayscale. A convolutional neural network is then applied.

Raduly et al. [10] set up a system for determining the breed of dogs. In his approach, two different networks were trained on the Standford Dog dataset (12000 images of 120 different breeds). A client–server system was also implemented for friendly usage of the trained CNN.

Very few authors worked on lobi cattle classification. Bembamba [5] implemented a machine learning solution using a dataset of 1968 cattle. Six morphological features were measured on purebred and crossbred taurines. The resulted data were used to train 5 ML algorithms. Random Forest yielded the best accuracy score (86%). This solution needs parameters to be measured directly on the animals.

In this study, we propose a non-invasive image-based model, obtained from a customized convolutional neural network trained on data subsets. The results are compared with pretrained models performance leveraging transfer learning technology.

1.3 Materials and Methods

1.3.1 Materials

In this study, we dispose of image data collected in 2007 by a PhD student during his fieldwork [11]. The collect was done in three regions of Burkina Faso with different composition of breeds: northern region, cascades region, and south-west region. The pure lobi taurines were mostly located in the south-west region (Noumbiel, Poni, Ioba, see Fig. 1.1). The images are in JPG format with a resolution of 2560 by 1920 pixels. They were taken in day time, without flash, by an amateur KODAK CX7530

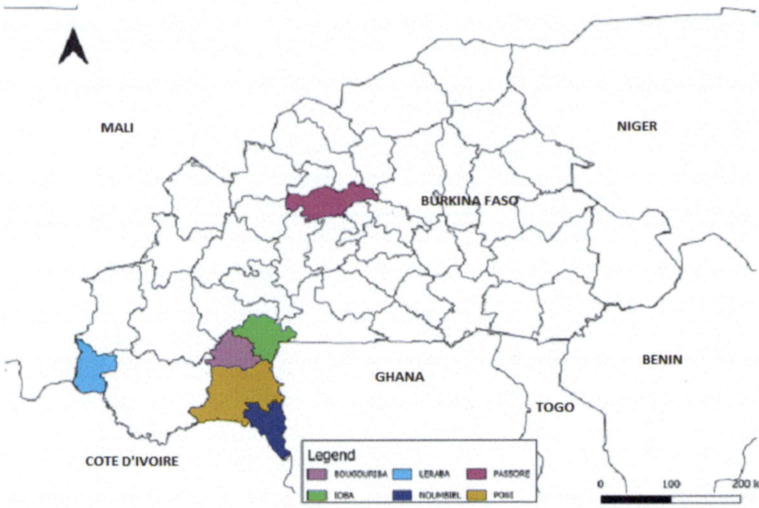

Fig. 1.1 Sampling area

Fig. 1.2 Sample of images

camera. Basically, the images were captured from three shooting angles: front, side, and rear. See Fig. 1.2 for samples.

On the majority of pictures the animals were immobilized by breeders or the operator team mates, which caused unwanted objects to appear on the images like pen bars, humans, vegetation, shadows, other animals, etc. So, for an initial total number of 4792 pictures, more than 600 have been discarded because they were too polluted with noise, 1400 had not been labeled by domain experts yet. Finally, for the study, we retained 2379 images comprising 44% of lobi and 56% of non-lobi cattle (crossbred and zebus).

For the implementation, we used a personal computer from MSI brand, equipped with a NVIDIA GeForce RTX3080 GPU and 11th Gen Intel Core i9- 2.50 GHz CPU. Coding was done in Python language with Tensorflow-Keras libraries.

1.3.2 Methods

1.3.2.1 Overview on Convolutional Neural Network

A Convolutional Neural Network (CNN) is a deep learning algorithm that can take in an input image, assign importance (learnable weights and biases) to various aspects/objects in the image, and be able to differentiate one from the other [16]. A CNN is basically composed of two components: a feature extraction part and a classification part. The feature extraction component is made up of one or more convolution layers which contain filters that scan the input image for prominent features. This process also occurs as the image passes through the filters in the different convolution layers [13]. An activation function provides for non-linear characteristics and a Pooling layer reduces the intermediate image dimensions. Features extracted from the convolution layers are fed into a fully connected neural network for training and prediction through a dedicated activation function. The number of layers, the number of filters, and the dense network design are hyperparameters that depend on the network architecture.

In general, solving complex problems using CNN requires a lot of data that not every study can afford to collect. Utilizing a large dataset helps to enhance the model's ability to learn patterns, while diversity in the dataset ensures that the model can generalize to new and unseen instances [18]. Transfer learning is one of the techniques applied to address data scarcity. It consists in reusing a pretrained model knowledge for another task [15]. In practical terms, it aims to extract knowledge from one or more source tasks and apply to the target domain [17]; hence, helping to fast prototype a new model, using pretrained models tuned on millions of data, that normally would have required lot of time and computation resources.

1.3.2.2 Methods

In the present study, convolutional neural network is used to classify cattle images into two classes: *Lobi* and *non-Lobi*. We built a customized CNN from scratch. Figure 1.3 displays the architecture of our own designed network.

The first part is made of three convolution layers followed by a Rectified Linear Unit (ReLU) activation function. A (2×2) MaxPooling layer is inserted between every two convolution layers. After the convolution phase, convolved features are flattened before being fed into two dense layers. A sigmoid function at the end of the

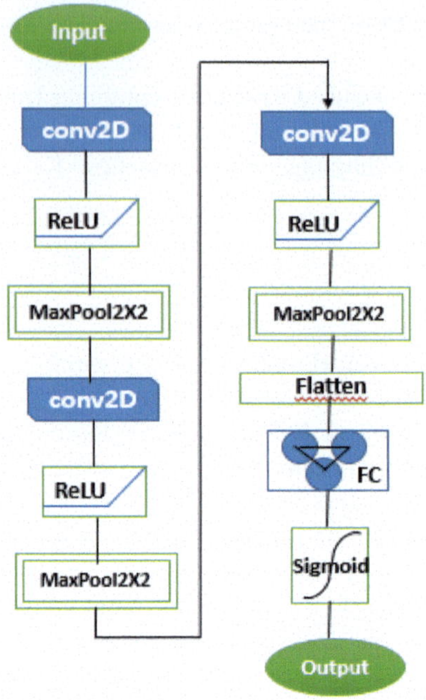

Fig. 1.3 Architecture of customized CNN

chain is responsible for classification. To tackle overfitting issues, a Dropout layer was inserted right before the dense classification layer.

In the preprocessing phase, we cropped and zoomed some images to delete noise since, as pointed out by [3], Deep Learning model results are not satisfactory when the full body or any part of the body overlaps with another animal. Images are then rescaled into 256 by 256 pixel format before going through the convolution process. The dataset was divided into four subsets on which models were trained and evaluated separately. The first three subsets contain images from the three shooting angles: side, rear, and front. The fourth contains the whole dataset, see Table 1.1.

Table 1.1 Performance for own customized CNN

	Number of images	Precision	Recall	Accuracy
All views	2379 (100%)	0.78	0.68	0.70
Side view	1080 (45%)	0.91	0.89	0.88
Head view	790 (33%)	0.98	98	1
Rear view	514 (22%)	0.99	1	1

We applied two state-of-the-art models (VGG-16 and Inception-V3), pretrained on ImageNet, a dataset containing 1.4 million annotated images. This dataset was made available to the public, in order to benchmark object category detection and classification as part of large-scale visual recognition challenge. Feature extractors of the pretrained models were frozen during training and a classification layer was added on top. For the Inception model, a fully connected 1048 neurone layer was also inserted for training.

1.4 Results

Our customized convolutional neural network was trained over 20 epochs on four subsets, namely: all views dataset, side view, head, and tail view subsets. Results are shown on Table 1.1. The all views dataset offers 78% for precision and 70% for overall accuracy. Side-view images provide better performance with 91% precision. The last two subsets are close to or hit 100% for precision and overall accuracy.

In general, pretrained models outperfomed our own customized model (Fig. 1.6). However, for small datasets (head and rear view images) our model equaled VGG-16 and even outperfomed Inception-V3. Learning curves have been plotted to assess the algorithm learning trend. For mixed image dataset, we can see that validation accuracy stabilizes around 78% while training accuracy keeps rising steadily until 99.5%. See Fig. 1.4.

As far as side image set is concerned (Fig. 1.7), validation curves increase up to 90%. For the two last subsets, head and tail images, training and validation accuracies increase together steadily, reaching 99% (Fig. 1.5). These subsets contain relatively small numbers of data: 790 head pictures and 514 rear pictures. Pretrained VGG-16 and Inception-V3 models provide similar learning curves.

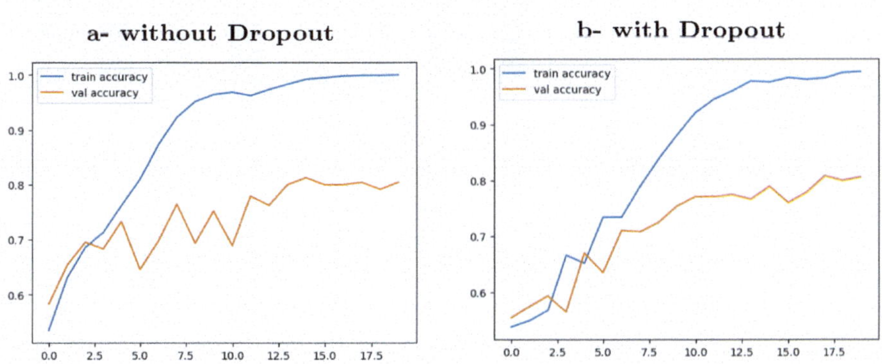

Fig. 1.4 Learning curves for all views dataset

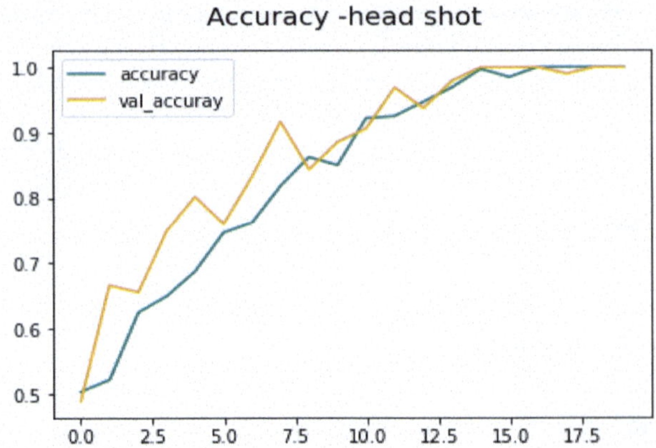

Fig. 1.5 Learning curves for head images

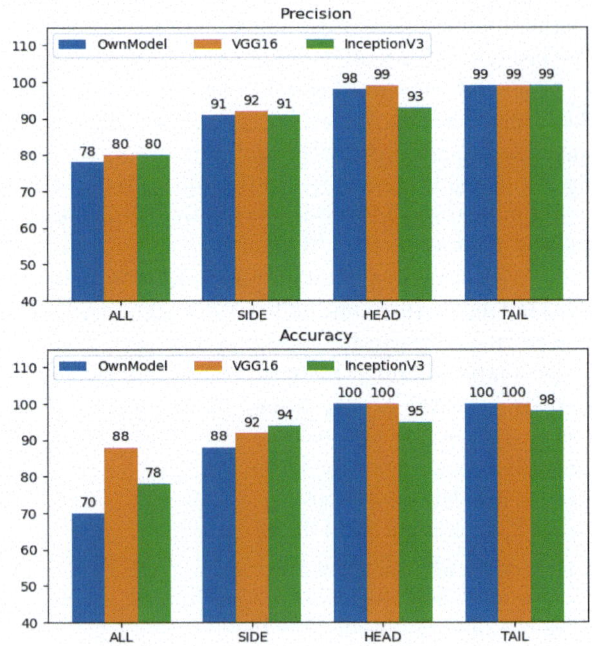

Fig. 1.6 Performance comparison for three models over four data subsets

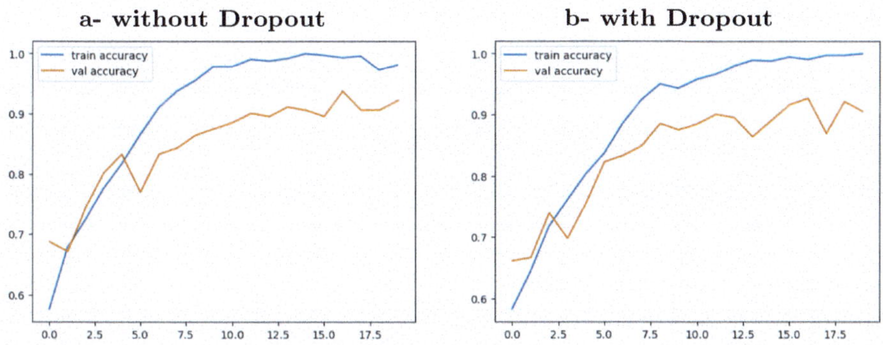

Fig. 1.7 Learning curves for side view images

1.5 Discussion

On all learning curves plotted above, we note that training accuracy reaches 99%. This high value means that the model has learned the features of cattle images in our datasets well. However, for the all views subset, the shape or the validation curve clearly indicates overfitting starting as early as the fifth epoch. Despite dropout regularization was applied to address the overfitting issue, we only see small improvement in the validation accuracy (Fig. 1.4b). This dataset mingles all images regardless of the angle of shooting. The mixture makes it difficult for the model, trained on less than 3,000 heterogenous data, to figure out the discriminating features and generalize accurately. The model then captures noise and remains dataset dependent.

This is confirmed by rise of validation accuracy when we split data in separate bins according to shooting angles. Data then become more coherent. Side shot images achieve a 90% validation accuracy. But there is still a piece of generalization issue because of noise. As stated by [3], one challenge for cattle identification using Deep Learning models is that the model results are not satisfactory when the full body or any part of the body overlaps with another animal. The system also works better if the image is a complete single animal [7]. In order to capture the whole animal body side, the camera operator needed a wider range, which leads to appearance of more undesired objects on our side view pictures. Yet, training a CNN on noisy picture data increases the misclassification error [14].

Head and rear view datasets perform nearly 100% accuracy and quite smooth learning curves. Nevertheless, the quantity of data is too small to draw a general conclusion since it is known that a small number of data can bias the model [13]. Indeed, deep learning models require a vast amount of data during the training phase. Thus, they are usually outperformed over traditional Machine Learning models in the area of text, speech, image, video, and audio processing where data are generally large [12].

1.6 Conclusion

In this research, we presented a computer vision method for identifying purebred lobi taurine cattle. A convolutional neural network was trained on four subsets of cattle images taken from different angles of view. The model obtained from the mingled data offered poor performance with a high trend to overfitting. The ones from separate shooting angles with homogenous data achieved good accuracy. Transfer learning improved the performances. In particular, results are excellent for VGG-16 on head and tail datasets with 100% accuracy.

However, we need to enhance model robustness and stability by collecting and labeling more data. Later work could expand preprocessing phase by systematizing image cropping, zooming, etc., especially for wide range shooting in order to get rid of maximum noise. Neural network design could be improved as authors in [8] stated that the choice of hyperparameters has a significant impact on the CNN performance. A friendly mobile application for taurine cattle image identification could also be implemented to apply the model.

Acknowledgements We wish to express our sincere gratitude to all the partners who provided us with data for this study. In particular, we would like to thank Ministère de l'Enseignement Supérieur de la Recherche Scientifique et de l'Innovation (MESRSI) of Burkina Faso and Austrian Partnership Programme in Higher Education and Research for Development (APPEAR), especially APPEAR project 120: Local cattle breed of Burkina Faso-characterization and sustainable utilization.

References

1. Silva, C., Weber, J., Belloni, B., et al.: Segmentation and detection of cattle branding images using CNN and SVM classification (2019)
2. Li, Z., Liu, F., Yang, W., Peng, S., Zhou, J.: A survey of convolutional neural networks: analysis, applications, and prospects. IEEE Trans. Neural Netw. Learn. Syst. (2021)
3. Hossain, M.E., Kabir, M.A., Zheng, L., Swain, D.L., McGrath, S., Medway, J.: A systematic review of machine learning techniques for cattle identification: datasets, methods and future directions. Artif. Intell. Agric. (2022)
4. Santosh Kumar and Sanjay Kumar Singh: Cattle recognition: a new frontier in visual animal biometrics research. Proc. Natl. Acad. Sci. India Sect. A: Phys. Sci. **90**(4), 689–708 (2020)
5. Bembamba, F., Ouédraogo, F.T., Albert, S., Traoré, A.: Toward an intelligent system for taurine cattle recognition. J. Intell. Learn. Syst. Appl. **14**(1), 1–13 (2022)
6. Weng, Z., Meng, F., Liu, S., Zhang, Y., Zheng, Z., Gong, C.: Cattle face recognition based on a two-branch convolutional neural network. Comput. Electron. Agric. **196**, 106871 (2022)
7. Manoj, S., Rakshith, S., Kanchana, V.: Identification of cattle breed using the convolutional neural network. In: 2021 3rd International Conference on Signal Processing and Communication (ICPSC), pp. 503–507. IEEE (2021)
8. Bhatt, D., Patel, C., Talsania, H., Patel, J., Vaghela, R., Pandya, S., Modi, K., Ghayvat, H.: CNN variants for computer vision: history, architecture, application, challenges and future scope. Electronics **10**(20), 2470 (2021)
9. Bello, R.-W., Talib, A.Z., Mohamed, A.S.A., Olubummo, D.A., Otobo, F.N.: Image-based individual cow recognition using body patterns. Image **11**(3) (2020)

10. Ráduly, Z., Sulyok, C., Vadászi, Z., Zölde, A.: Dog breed identification using deep learning. In: 2018 IEEE 16th International Symposium on Intelligent Systems and Informatics (SISY), pp. 000271–000276. IEEE (2018)
11. Soudre, A.: Trypanosomosis, genetic diversity and admixture in cattle breeds of Burkina Faso. PhD thesis, University of Natural Ressources and Life Sciences, Vienna (2011)
12. LeCun, Y., Bengio, Y., Hinton, G.: Deep learning. Nature **521**(7553), 436–444 (2015)
13. Diarra, A., Tegawende Bissyandé, F., Poda, P.: A deep learning app for counterfeit banknote detection in the WAEMU. In: JRI 2022: Proceedings of the 5th edition of the Computer Science Research Days, JRI 2022, 24-26 November 2022, Ouagadougou, Burkina Faso, p. 40 (2023)
14. Szegedy, C., Liu, W., Jia, Y., Sermanet, P., Reed, S., Anguelov, D., Erhan, D., Vanhoucke, V., Rabinovich, A.: Going deeper with convolutions. In: Proceedings of the IEEE Conference on Computer Vision and Pattern Recognition, pp. 1–9 (2015)
15. Tammina, S.: Transfer learning using vgg-16 with deep convolutional neural network for classifying images. Int. J. Sci. Res. Publ. (IJSRP) **9**(10), 143–150 (2019)
16. Khan, A., Sohail, A., Zahoora, U., Qureshi, A.S.: A survey of the recent architectures of deep convolutional neural networks. Artif. Intell. Rev. **53**, 5455–5516 (2020)
17. Ribani, R., Marengoni, M.: A survey of transfer learning for convolutional neural networks. In: 2019 32nd SIBGRAPI Conference on Graphics, Patterns and Images Tutorials (SIBGRAPI-T), pp. 47–57. IEEE (2019)
18. Alzubaidi, L., Bai, J., Al-Sabaawi, A., Santamaría, J., Albahri, A.S., Al-dabbagh, B.S.N., Fadhel, M.A., Manoufali, M., Zhang, J., Al-Timemy, A.H., et al.: A survey on deep learning tools dealing with data scarcity: definitions, challenges, solutions, tips, and applications. J. Big Data **10**(1), 46 (2023)

Chapter 2
Engineering a Mecanum Wheel Mobile Robot with Raspberry Pi for SLAM

Prajakta Salunkhe, Harsh Kshatriya, and Mahesh Shirole

Abstract Mobile robots have become increasingly significant in numerous fields due to their mobility, autonomy, and capacity to carry out tasks in dynamic conditions. However, as we delve deeper into the field of robotics, we face a recurring challenge: maneuvering mobile robots to accurately determine their own position while simultaneously creating a map of their surroundings. This critical process is defined as Simultaneous Localization and Mapping (SLAM). A major challenge SLAM faces is the strategic selection of the optimal set of sensors to ensure the acquisition of precise and reliable results. This research paper presents a comprehensive exploration of the design and implementation of a mobile robotic platform meticulously engineered for SLAM applications. By integrating sensors like LiDAR (light detection and ranging), IMU (inertial measurement unit), and wheel encoders into a mecanum wheel robot, it perceived the environment more accurately and Raspberry Pi eliminated the need for Arduino for data processing. It is observed that the developed robot navigates efficiently and maps the environment precisely. The selection of mecanum wheel configuration helped in precise positioning, for enhanced maneuverability and versatility. However, the LiDAR's restricted range hindered mapping in cluttered areas.

2.1 Introduction

Mobile robots come in various forms, each designed to excel in specific tasks and environments. Examples include wheeled robots, tracked robots, legged robots, aerial drones, and underwater vehicles, each tailored for diverse applications [1]. SLAM (simultaneous localization and mapping) is a crucial technology that empowers mobile robots with autonomy, allowing them to navigate, understand, and adapt to their surroundings in real time. It enables robots to estimate their own pose (position and orientation) while simultaneously creating a map of the environment they are exploring. This simultaneous process of localization and mapping has far-reaching

P. Salunkhe (✉) · H. Kshatriya · M. Shirole
Veermata Jijabai Technological Institute (VJTI), Mumbai, India
e-mail: sprajakta1012@gmail.com

S. K. Pal et al. (eds.), *Intelligent Informatics*, Smart Innovation, Systems and Technologies 389, https://doi.org/10.1007/978-981-97-2147-4_2

implications for the field of robotics [2, 3]. For the successful implementation of SLAM, a mobile robot or platform equipped with suitable sensors such as cameras, LiDAR (light detection and ranging), IMU (inertial measurement unit), and wheel encoders for odometry is essential [4]. Robots need these sensors to observe their environment and estimate their position.

While most prior works emphasize standard wheeled robots or differential drive robots for SLAM applications, these designs have a primary limitation in achieving precise turns and accurate odometry. Whereas mecanum wheels offer unique omnidirectional movement capabilities by enabling efficient navigation and precise control. In contrast, differential drive robots possess two independently driven wheels, providing straightforward forward and backward motion as well as turning in place. This simplicity in control makes differential drive robots less maneuverable but often easier to manage. The mecanum wheel exhibits a range of basic motions, including forward, reverse, leftward, rightward, left diagonal forward, left diagonal reverse, right diagonal forward, and right diagonal reverse [5]. This advantage proves particularly beneficial for mapping and localization tasks in complex and dynamic environments, making mecanum wheels a promising alternative for enhancing the accuracy and maneuverability of robots in SLAM applications [6, 7]. In this context, the mobile robot prototype development for the SLAM application, utilizing mecanum wheels, is worth trying. By incorporating mecanum wheels, our research aims to bridge this gap and contribute insights into the potential benefits and challenges associated with their use in SLAM scenarios.

Figure 2.1 shows the primary motions of the mecanum wheels.

This research paper offers an extensive investigation into the creation and deployment of a mobile robot platform intricately crafted for the specific purpose of facilitating SLAM applications. At its core, this platform boasts a unique mechanical design, featuring the mecanum wheel configuration. The developed platform has onboard state-of-the-art sensors, including LiDAR, wheel encoders, and IMU, that allow for accurate perception of the surroundings. The Raspberry Pi is an important part of

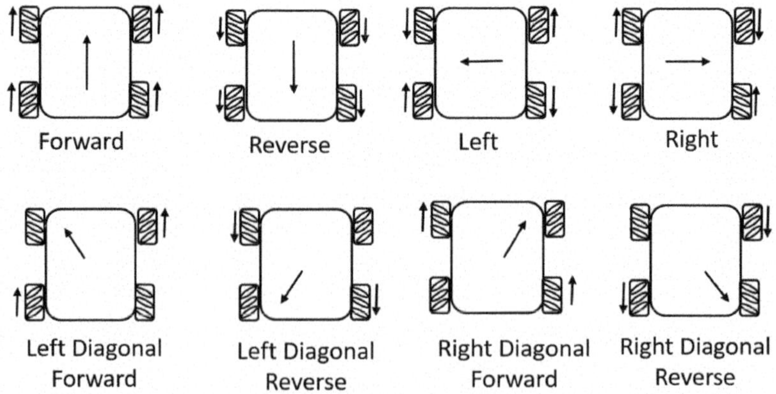

Fig. 2.1 The primary motions of the mecanum wheel

this study because it is the main component that does all the combination of sensor data. Putting together sensor data and computing power has a big effect on how well the SLAM tool works. By presenting the inner workings of this robotic system, we aim to shed light on the crucial role of design and integration in the realization of a highly capable and adaptable robotic platform optimized for the rigors of SLAM.

The main contributions of this research paper are

(i) We introduced a mobile robotic platform designed specifically for SLAM applications. This platform is a valuable contribution as it caters to the growing demand for robots capable of autonomously mapping and localizing themselves within dynamic environments.
(ii) By utilizing mecanum wheels, the device improved maneuverability, providing a significant advantage in mapping and localization tasks in an indoor environment.
(iii) Eliminated use of Arduino board by replacing it with Raspberry Pi board. Also, a specific set of sensors enhanced the robot's perception capabilities, allowing it to create detailed maps.
(iv) The experimental results demonstrated the platform's ability to map and navigate its surroundings. These results serve as empirical evidence of the robot's capabilities, underscoring the practicality and effectiveness of the system in real-world scenarios.

The paper is structured into several sections. Section 2.2 provides an overview of the related work in the field. Section 2.3 details the materials and kinematics employed in the design of the mecanum wheel mobile robot. In Sect. 2.4, the hardware and software design considerations that form the foundation of the platform's performance are presented. Section 2.5 demonstrates the platform's effectiveness in mapping and navigating custom environments. Finally, Sect. 2.6 presents the conclusion of this study.

2.2 Related Work

In a comprehensive study, considerations for selecting the most suitable wheel type and configuration were discussed, differentiating between holonomic and non-holonomic robots based on their wheels [8]. The paper compared various wheel types, including conventional, omnidirectional, mecanum, caster, and standard steering wheels, analyzing their optimal applications, degrees of freedom, manufacturing complexities, and sensitivities to surface conditions and objects. It serves as a practical guide for beginners in mobile robot platform design. Another work provided an evolutionary perspective on wheeled mobile robotics, emphasizing omnidirectional navigation [9]. Covering different wheel categories and control aspects, it addressed challenges faced by omnidirectional drive systems, such as slippage, vibration problems, singularities, and robustness. This review offers valuable insights into the advancements and challenges in omnidirectional mobile robotics.

The paper [10] discussed the analysis of mobile robot indoor mapping using IRobot Create equipped with only one sensor which is a Hokuyo Laser Range Finder. The research aimed to understand the effects of different parameters, such as robot speed, map update interval, and particle filter, on the mapping quality of SLAM using the GMapping algorithm. Given the robot's ability to travel in a spiral manner, it was determined that the mapping path would be executed in a spiral pattern. This approach may require increased computational resources and processing power, which could impact the real-time processing capabilities of the SLAM algorithm. While the paper investigates the effects of different parameters on mapping quality, it does not provide an in-depth analysis or explanation of why these parameters impact the accuracy and time taken for mapping. Another work [11] talks about real-life experiments that were done in a lab setting using their crawler robot, which was equipped with RPLidar A1, IMU sensors, and odometry. They used Gazebo, a ROS (robotic operating system) simulation tool, to test the algorithms in an urban search and rescue (USAR) environment. Nehate et al. [12] developed a robot with omnidirectional wheels and a set of sensors like 2D LiDAR, wheel encoders, IMU, and a depth camera for the evaluation of SLAM algorithms. Alatise et al. [13] highlighted the hurdles associated with autonomous mobile robots. These issues, including navigation and localization, are the primary factors constraining the robot's performance. Also suggested, the utilization of sensors affixed to the mobile robot to enhance its overall performance. Relying solely on a single sensor to ascertain an object's position may not yield dependable and precise results; thus, the adoption of multiple sensors is advocated. Mutalib et al. [14] developed a basic mecanum wheel robot prototype to overcome the slippage issues in the existing system. Housein et al. [15] used a *turtolbot* which is a differential drive robot equipped with sensors such as a camera, 3D depth sensor, rplidar, and wheel encoders to map the indoor environment. They utilized GMapping algorithm for SLAM. The system utilized ROS package based on Extended Kalman Filter (EKF) for mapping and Adaptive Monte Carlo localization (AMCL) package for localization. The paper [16] discussed a mobile robot navigation control system that integrates laser SLAM localization and real-time obstacle avoidance. The mobile robot equipped with two laser scanners, one on the front and another at the rear, was utilized for SLAM. These laser scanners were used for robot localization, obstacle detection, and obstacle avoidance. However, the type of LiDAR and robot used in this paper is not explicitly mentioned.

In the review of the relevant literature, it can be asserted that the utilization of the mecanum wheel robot system for SLAM applications is limited and hence holds significant merit and warrants attention. Furthermore, it is worth considering the design of a robotic system designed for SLAM from the base up. Conducting comprehensive trials in the field is necessary to test every set of sensors used in combination with a mecanum wheel robot for SLAM purposes. Also, existing systems underscore a need for more in-depth investigations into the combined use of LiDAR, IMU, and wheel encoders in SLAM applications. While individual studies often focus on one or two of these sensors, a comprehensive exploration of their synergistic integration is lacking. Our research recognizes this gap and strives to address it by providing a holistic understanding of how these sensors when combined enhance the accuracy and robustness of SLAM algorithms.

2.3 Materials and Methods

Mecanum wheels, a remarkable innovation in the field of robotics and mobility, are renowned for their design that enables omnidirectional movement with three-degrees of freedom (3-DOF). Comprising multiple rollers mounted at an angle around the wheel's circumference, mecanum wheels allow a mobile platform to move forward, backward, and sideward with precision and agility, eliminating the need for a standard steering mechanism. By manipulating the rotation speed and direction of these wheels, a mecanum wheel-equipped robot can effortlessly execute intricate movements along the X-, Y-, and rotational Z-axes [7]. However, slippage is a common problem with mecanum wheels due to their design, which includes a single roller that is in constant contact with the ground [9]. Figure 2.2 shows the image of the mecanum wheel utilized in our robotic chassis.

2.3.1 Kinematics of Mecanum Wheel

Kinematics, as a fundamental and classical subject in robotics, pertains to the examination of the correlation between a robot's joint coordinates and its spatial arrangement. The application of kinematics can provide precise calculations in several scenarios, such as determining the relative position of a gripper, devising a mechanism for moving a tool from one location to another, or forecasting potential collisions between a robot's movements and obstacles [17]. Figure 2.3 shows the general wheel configuration for the four-wheel mecanum drive robotic chassis.

Fig. 2.2 Design of mecanum wheel

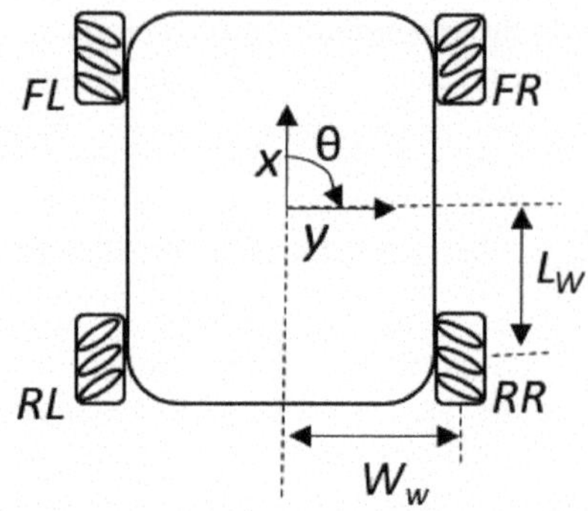

Fig. 2.3 Wheel Configuration

To understand the kinematics, we can start with some basic principles:

- Wheel configuration: A typical four-wheeled mecanum robot has its wheels arranged in a square configuration. The wheels are labeled as front-right (FR), front-left (FL), rear-right (RR), and rear-left (RL).
- Wheel motion: Each wheel can be independently controlled in terms of both speed and direction. This is usually done using motor controllers.
- Wheel axes: Each mecanum wheel is mounted at an angle of 45 °C relative to the robot's forward direction. When you look at the robot from above, each wheel's rollers form an "X" pattern.
- Robot motion: The combination of the wheel motions results in the robot's overall linear and angular velocity.

The kinematic equations for a four-wheeled mecanum robot can be derived using the following formulas:

$$\text{Linear Velocity (Vx):} \quad Vx = \frac{1}{4}(V_{FR} + V_{FL} + V_{RR} + V_{RL}) \tag{2.1}$$

$$\text{Lateral Velocity (Vy):} \quad Vy = \frac{1}{4}(-V_{FR} + V_{FL} + V_{RR} - V_{RL}) \tag{2.2}$$

$$\text{Angular Velocity}(\omega): \quad \omega = \frac{1}{d}(V_{FR} - V_{FL} - V_{RR} + V_{RL}) \tag{2.3}$$

where
Vx : Linear velocity of the robot in the X-axis.
Vy : Linear velocity of the robot in the Y-axis.
ω : Angular velocity (rate of rotation).
V_{FR} : Linear velocity of the front-right wheel.
V_{FL} : Linear velocity of the front-left wheel.
V_{RR} : Linear velocity of the rear-right wheel.
V_{RL} : Linear velocity of the rear-left wheel.
d : Distance between the mecanum wheels.
(the distance from the center of the robot to the center of each wheel).

2.3.2 *Inverse Kinematics*

The inverse kinematics equation for a four-wheel mecanum drive robot determines the individual wheel velocities required to achieve a desired robot motion (linear and angular velocities) [17, 18]. The equation is as follows:

$$\begin{bmatrix} \omega_{FL} \\ \omega_{FR} \\ \omega_{RL} \\ \omega_{RR} \end{bmatrix} = \frac{1}{r} \begin{bmatrix} 1 & -1 & -(L+W) \\ 1 & 1 & (L+W) \\ 1 & 1 & -(L+W) \\ 1 & -1 & (L+W) \end{bmatrix} \begin{bmatrix} v_x \\ v_y \\ \omega_z \end{bmatrix} \tag{2.4}$$

where

$\omega_{FR}, \omega_{FL}, \omega_{RR},$ and ω_{RL} : angular velocities of the Front-Right, Front-Left, Rear-Right, and Rear-Left wheels, respectively.
V_x : the desired linear velocity in the X-axis.
V_y : the desired linear velocity in the Y-axis.
ω_z : the desired angular velocity (rate of rotation)
L : the distance between the front and rear wheels.
W : the distance between the left and right wheels.

2.3.3 *Forward Kinematics*

The forward kinematics equation for a four-wheel mecanum drive robot calculates the robot's position and orientation (pose) in a 2D plane based on the individual wheel velocities [17, 18]. The equation is as follows:

$$\begin{bmatrix} v_x \\ v_y \\ \omega_z \end{bmatrix} = \frac{r}{4} \begin{bmatrix} 1 & 1 & 1 & 1 \\ -1 & 1 & 1 & -1 \\ -\frac{1}{L+W} & \frac{1}{L+W} & -\frac{1}{L+W} & \frac{1}{L+W} \end{bmatrix} \begin{bmatrix} \omega_{FL} \\ \omega_{FR} \\ \omega_{RL} \\ \omega_{RR} \end{bmatrix} \tag{2.5}$$

2.4 System Setup

Every moving agent starts with a mobile platform. The developed prototype is equipped with a versatile set of hardware and software components to facilitate its functionalities. It uses a four-wheeled mecanum drive chassis with built-in encoders and high-torque DC motors. It enhances the robot's agility and maneuverability. Motor control is facilitated by Cytron dual-channel DC motor drivers, providing efficient and precise control over the robot's movements. At its computational core, the platform features a Raspberry Pi 4B with 8GB of RAM, serving as a single-board controller capable of handling complex tasks. The robot's perception capabilities are augmented by an RPLiDAR A1M8 360-degree laser range scanner, enabling accurate environmental mapping. Built-in incremental rotary encoders contribute to odometry information, aiding in precise localization. Furthermore, the robot is equipped with an MPU 9250 (IMU), a 9-degree-of-freedom (DOF) real-time motion tracking device, enhancing the platform's spatial awareness. This comprehensive assembly of components makes the mobile platform well suited for a range of applications, from autonomous navigation to complex robotics tasks. In order to achieve functionality, additional hardware components are also incorporated. Tables 2.1 and 2.2 present the hardware and software specifications of the developed mobile platform and remote PC, respectively.

The system setup has been carefully designed to ensure smooth operation. Both the motor drivers and the Raspberry Pi (RPi) are powered by a lithium polymer (LiPO) battery and a power bank, respectively. To initialize the RPi, the imager app is used to flash a server image file of Ubuntu OS version 20.04. Subsequently, the Noetic version of ROS is installed on top of Ubuntu OS. In the laboratory setup, ROS is configured to operate across multiple machines to prevent the RPi from experiencing an overload due to multiple SSH connections. Simultaneously, on the remote PC, the Noetic distribution of ROS version 1.X is installed. This computer essentially runs the Ubuntu 20.04 operating system, which is a prerequisite for the Noetic distribution of ROS. The choice of the ROS distribution was made after a thorough examination of the package dependencies required for running SLAM algorithms. It was discovered that these packages were compatible exclusively with ROS 1.X distributions, specifically Melodic and Noetic.

Table 2.1 Hardware and software specifications of mobile platform

S. no.	Component	Configuration
1.	Chassis	Four-wheeled mecanum drive with built-in encoders
2.	Motor Drivers	Cytron dual-channel DC motor drivers
3.	Single board controller	Raspberry Pi 4B 8GB RAM
4.	Laser range finder	RPLiDAR A1M8 360°C laser range scanner
5.	Encoders	Built-in incremental rotary encoders
6.	OS	Ubuntu 20.04
7.	ROS version/distro	ROS 1/Noetic
8.	IMU	MPU 9250: 9 DOF real-time motion tracking device

Table 2.2 Hardware and software specifications of Remote PC

S. no.	Component	Configuration
1.	Processor	Intel i7
2.	GPU	NVIDIA Quadro P620
3.	RAM	16 GB
4.	OS	Ubuntu 20.04
5.	ROS version/distro	ROS 1/Noetic

As for the hardware components, the LiDAR, wheel encoders, and IMU can all be added to RPi without any problems. The Cytron MDD10A dual-channel DC motor driver is used to connect Moebius motors to RPi. The motor driver can only control two DC motors at once. Because of this, two motor drivers are used to control all four motors at the same time. Also, a 3:9 terminal block is used so that power from the battery can be sent in parallel. Since the only way to communicate to ROS on RPi (which is a headless server) is through SSH, X11 forwarding is enabled on RPi so that RViz and resource occupation programs can run so that the map can be seen and plots can be used to track how much resource is being used. Figure 2.4 shows the components interfacing of the developed mobile platform.

SSH is used to connect the remote PC to the virtual server (RPi) that runs Ubuntu OS. Using pins SDA and SCL and the IIC serial communication protocol, this RPi is linked to an IMU. The RPi also has the IIC protocol turned on so that it can communicate with the sensor and share info. The power bank is a steady source that sends power to the RPi all the time. For that, a type C USB cable is plugged into a power bank. This turns on the RPi so that ROS can be used. The LiDAR is connected to the RPi through one of the USB 2.0 ports using a type A cable. The GPIO library is installed and set up on the RPi so that it can connect to two motor drivers and send PWM and direction signals to make the wheels connected to the motors controlled by the motor driver move forward or backward. A handy battery

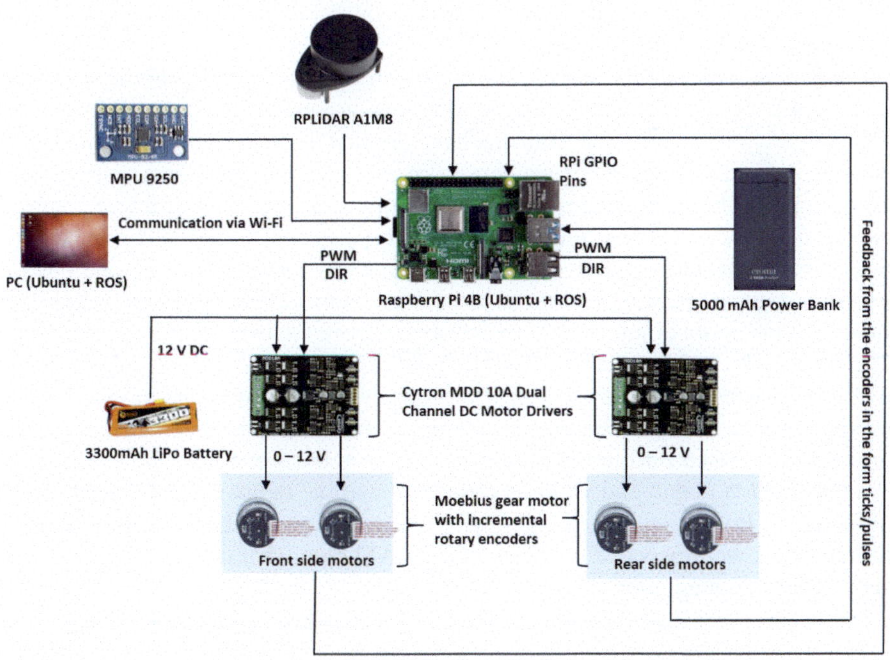

Fig. 2.4 Components interfacing of the mobile platform

Fig. 2.5 Developed mobile robot prototype

powered two motor drivers so that the wheel could turn at the right PWM. Encoders on the motors connect to the RPi through GPIO pins. Interrupts are turned on for these pins so that the edges that rise and fall can be used to figure out the ticks. This makes a feedback loop, in which the RPi tells the motor driver to use new PWM and DIR numbers to move the way it wants.

Then we proceeded to configure transforms in ROS using the *tf* library to align the onboard sensors effectively. Transformations are essential when dealing with different parts of a robot or sensors that have their own coordinate frames, and you need to convert data between them. This is crucial for robot perception, control, and navigation, as it helps in understanding the relationships between different components of a robot. To achieve this, we employed quaternions with the assistance of the *tf* library. Quaternions are often used to represent orientations or rotations in three-dimensional space. Quaternions are a mathematical concept used to describe orientation in a more efficient and robust way compared to using Euler angles. They have several advantages, such as avoiding gimbal lock and providing smooth interpolation between orientations. The next part was to set up an odometry system for the developed robot. For that, the wheel separation width and wheel separation length are calculated based on the measurements of the robot's chassis. The average of wheel separation length and width is calculated and the radius is determined using the diameter. Following the completion of the system assembly and the required programming, we proceeded to execute the SLAM algorithm in order to assess the adequacy of the integration.

Figure 2.5 shows the real image of the developed mobile robot prototype.

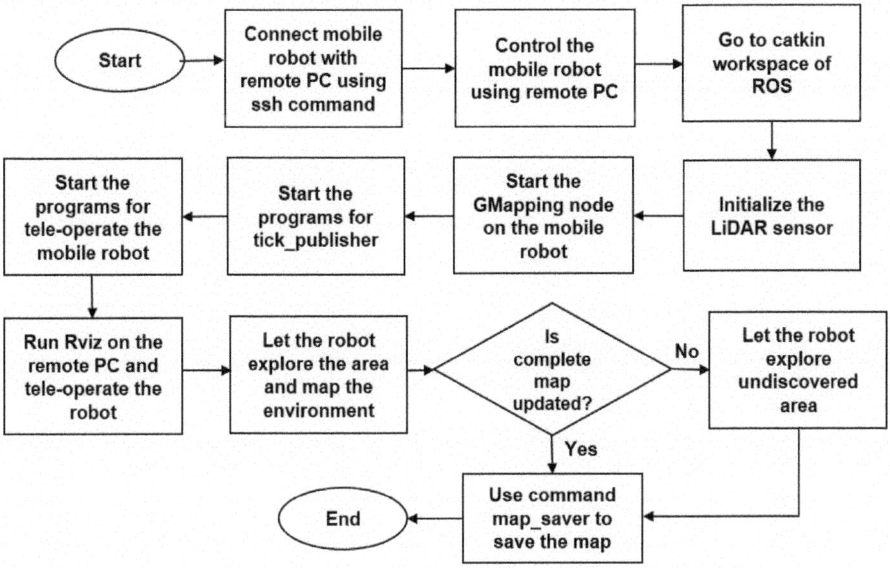

Fig. 2.6 SLAM process flow utilizing developed system

2.4.1 Algorithm Used for SLAM

In our robotic system employing a mecanum wheel chassis, we leverage the integration of LiDAR, IMU, and wheel encoders for SLAM, specifically using the GMapping algorithm. The LiDAR sensor captures detailed point cloud data, allowing for precise mapping of the robot's environment. Wheel encoders contribute odometry information, aiding in estimating the robot's position based on its movement. The IMU enhances pose estimation by providing orientation and acceleration data. The *tf* library in ROS facilitates the alignment of coordinate frames, crucial for integrating diverse sensor inputs. GMapping, a popular SLAM algorithm, utilizes this data to iteratively update the map and refine the robot's pose estimation. The algorithm works by incrementally building a map while simultaneously estimating the robot's location within that map. Figure 2.6 shows the flow of the mapping process using our developed mobile robot.

GMapping also known as grid-based mapping is an open-source SLAM algorithm that relies on particle filter algorithms. GMapping is commonly used with 2D LiDAR data but can be less suitable for larger scenes and lacks loop closure detection [11]. It is a common SLAM method used in robotics and autonomous systems, particularly in the context of ROS. GMapping is a probabilistic technique that constructs a 2D occupancy grid map of an environment while estimating the robot's pose inside that map. It accomplishes this by taking sensor input from a laser range finder or 2D LiDAR sensor and applying a Bayesian filter to update the map and localize the robot. GMapping is frequently utilized for applications such as autonomous navigation, exploration, and mapping of indoor spaces, and it provides a stable and fast method for mapping new settings. It is a useful technique in robotics for producing precise maps and enabling autonomous actions.

2.5 Results and Discussion

Values for the parameters used by the GMapping LiDAR-based SLAM method, such as number of particles, map update interval, max range of LiDAR, and linear and angular updates, were determined and fixed prior to the construction of the map. Figure 2.7 shows the lab area to be mapped (a) and the map started to update as the robot was moving ahead (b).

The system employed a range of onboard sensors, including LiDAR, wheel encoders, and an IMU. In this study, the RPLiDAR A1M8 was employed as it is a cost-effective and lightweight device. Nevertheless, the device's range is rather restricted, posing a challenge in accurately mapping the densely populated laboratory setting. Consequently, its suitability for applications necessitating long-range scanning or outdoor usage in intense sunlight may be limited. Additionally, the reduced data density had an effect on the level of detail observed in the created map. The reduced data density may not be optimal for mapping jobs that need high levels of precision. The scanning speed of the RPLiDAR A1M8 is comparatively lower. The aforementioned constraint had an impact on the accuracy of the map production process. Then to check the capacity of the developed system, a tailored experimental configuration was devised. Figure 2.8 vividly illustrates the robot's efficient updating process on the map.

In the given configuration, we erected provisional partitions and initiated the robotic system to perform environmental mapping. The map in question revealed

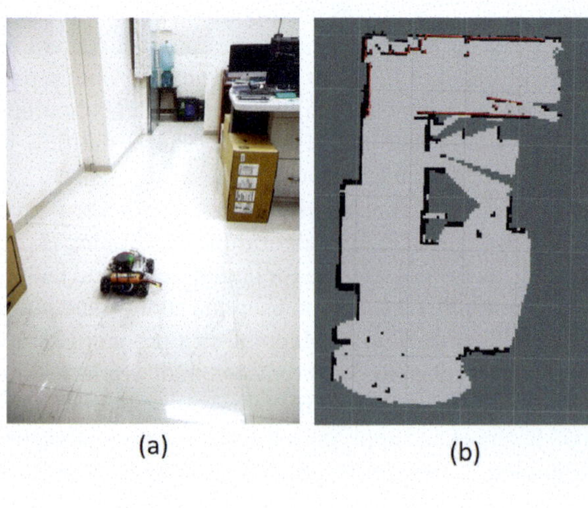

Fig. 2.7 **a** The area to be mapped and **b** The area mapped using Gmapping

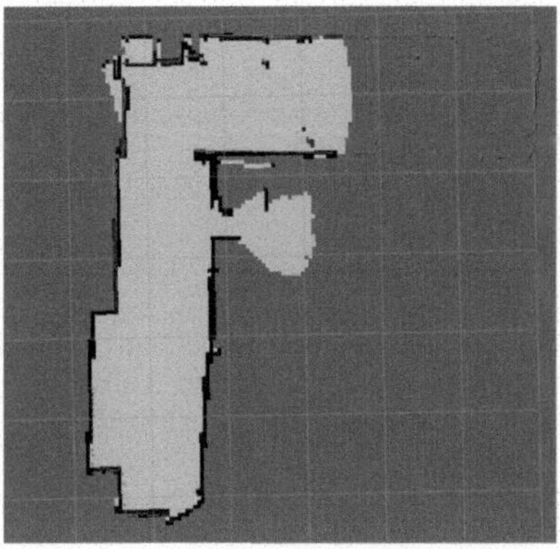

Fig. 2.8 Map of the customized workspace obtained using GMapping

the presence of a glass wall on one side. It is crucial to acknowledge that LiDAR technology operates using infrared wavelengths, allowing it to penetrate translucent materials like glass. Consequently, it was unable to detect those elements. The robot, however, was unable to complete the map of the area. Nevertheless, the laser range data was successfully displayed on RViz.

2.6 Conclusion

This research investigated the growing importance of mobile robotic systems in a variety of disciplines, attributing it to their mobility, autonomy, and adaptability in dynamic environments. Our contribution consists of the complete design, construction, and integration of a mobile robot platform optimized for SLAM. The integration of LiDAR, IMU, and wheel encoders into our mecanum wheel robot configuration resulted in a wonderfully versatile and adaptable robotic platform with remarkable SLAM capabilities. We validated the system's proficiency in mapping and navigating workspaces through rigorous empirical trials, confirming its practical utility. Notably, the RPLidar A1M8 developed as a low-cost and easily accessible LiDAR system that is well suited for a wide range of robotic applications, particularly in indoor and close-range outdoor conditions. However, it is critical to understand its limits, which include a limited range, lesser data density, and vulnerability to environmental variables. These factors must be carefully considered while picking the best LiDAR for the specific needs of a given project. The future scope includes using more SLAM methods to map the environment and evaluating these methods in order to identify the best one for the environment. In addition, camera integration with the present set of sensors could be added in the future.

References

1. Ben-Ari, M., Mondada, F.: Robots and Their Applications, pp. 1–20. Springer International Publishing (2018)
2. Jaulin, L.: Mobile Robotics, Edition 2, revised. Wiley (2019)
3. Taheri, H., Xia, Z.C.: Slam; definition and evolution. Eng. Appl. Artif. Intell. **97**, 1 (2021)
4. Huang, J., Junginger, S., Liu, H., Thurow, K.: Indoor positioning systems of mobile robots: a review 4 (2023)
5. Thongpance, N., Chotikunnan, P.: "Design and construction of electric wheelchair with mecanum wheel. J. Robot. Control. (JRC) **4**, 71–82, 1 (2023)
6. Kanjanawanishkul, K.: Omnidirectional wheeled mobile robots: wheel types and practical applications. Int. J. Adv. Mechatron. Syst. **6**(6), 289–302 (2015)
7. Zeidis, I., Zimmermann, K.: Dynamics of a four-wheeled mobile robot with mecanum wheels 12 (2019)
8. Shabalina, K., Sagitov, A., Magid, E.: Comparative analysis of mobile robot wheels design, pp. 175–179. Institute of Electrical and Electronics Engineers Inc., 7 (2018)
9. Taheri, H., Zhao, C.X.: Omnidirectional mobile robots, mechanisms and navigation approaches 11 (2020)
10. Norzam, W.A., Hawari, H.F., Kamarudin, K.: Analysis of Mobile Robot Indoor Mapping Using Gmapping Based SLAM with Different Parameter, vol. 705, IOP Publishing Ltd., 12 (2019)
11. Xuexi, Z., Guokun, L., Genping, F., Dongliang, X., Shiliu, L.: SLAM algorithm analysis of mobile robot based on lidar. In: 2019 Chinese Control Conference (CCC), pp. 4739–4745 (2019)
12. Nehate, C., Shinde, R., Naik, S., Aradwad, M., Bhurke, A., Kazi, F.: Implementation and Evaluation of SLAM Systems for a Mobile Robot. Institute of Electrical and Electronics Engineers Inc. (2021)
13. Alatise, M.B., Hancke, G.P.: A review on challenges of autonomous mobile robot and sensor fusion methods (2020)
14. Abd Mutalib, M.A., Azlan, N.Z.: Prototype development of mecanum wheels mobile robot: a review. Appl. Res. Smart Technol. (ARSTech) **1**(2), 71–82 (2020)
15. Housein, A.A., Xingyu, G.: Simultaneous localization and mapping using differential drive mobile robot under ros. In: Journal of Physics: Conference Series, vol. 1820, p. 012015. IOP Publishing (2021)
16. Song, K.-T., Chiu, Y.-H., Kang, L.-R., Song, S.-H., Yang, C.-A., Lu, P.-C., Ou, S.-Q.: Navigation control design of a mobile robot by integrating obstacle avoidance and lidar SLAM. In: 2018 IEEE International Conference on Systems, Man, and Cybernetics (SMC), pp. 1833–1838. IEEE (2018)
17. Li, Y., Ge, S., Dai, S., Zhao, L., Yan, X., Zheng, Y., Shi, Y.: Kinematic modeling of a combined system of multiple mecanum-wheeled robots with velocity compensation. Sensors **20**(1), 75 (2019)
18. Röhrig, C., Heß, D., Künemund, F.: Motion controller design for a mecanum wheeled mobile manipulator. In: 2017 IEEE Conference on Control Technology and Applications (CCTA), pp. 444–449. IEEE (2017)

Chapter 3
Multi-filter-Based Image Pre-processing on Face Mask Detection Using Custom CNN Architecture

Devrim Kayali and Kamil Dimililer

Abstract After the COVID-19 pandemic, the effectiveness of face mask usage in such an environment has been seen. This situation triggered much research and experiments on this subject. Most of this research was done by using artificial intelligence. This paper focuses on a filter-based approach using a convolutional neural network for classification of face mask usage, which are mask correct, mask wrong, and no mask. Seventeen chosen filters are applied to the input images as an image pre-processing phase which results in 17 input images per image. Our customized network takes these 17 images as input and eliminates them into a single image before flattening and feeding to the dense layers. For training, 5 learning rates were tried starting from 10^{-7} and increasing 10 times for each try stopping at 10^{-3}. Results showed that increasing the learning rate shortened the training time and increased the accuracy. The highest test accuracy was obtained at 10^{-3} learning rate with 98.86%. At 10^{-4} learning rate, the second good result of 98.16% was obtained.

3.1 Introduction

When COVID-19 came up and spread rapidly and became a pandemic that affected the whole world, we had to take precautions to take the spread under control and protect ourselves individually. One of the effective methods for this was to use face masks. When it comes to public places, it is important to check the face mask usage since improper and no usage can increase the spread and affect other's lives. When the situation is serious and constantly running applications are needed with minimum

D. Kayali (✉)
Electrical and Electronic Engineering Research Center for Science, Technology and Engineering (BILTEM), Near East University, Via Mersin 10, Nicosia, N. Cyprus, Turkey
e-mail: devrim.kayali@ktemo.org

K. Dimililer
Electrical and Electronic Engineering, Applied Artificial Intelligence Research Centre (AAIRC), Research Center for Science, Technology and Engineering (BILTEM), Near East University, Via Mersin 10, Nicosia, N. Cyprus, Turkey
e-mail: kamil.dimililer@neu.edu.tr

S. K. Pal et al. (eds.), *Intelligent Informatics*, Smart Innovation, Systems and Technologies 389, https://doi.org/10.1007/978-981-97-2147-4_3

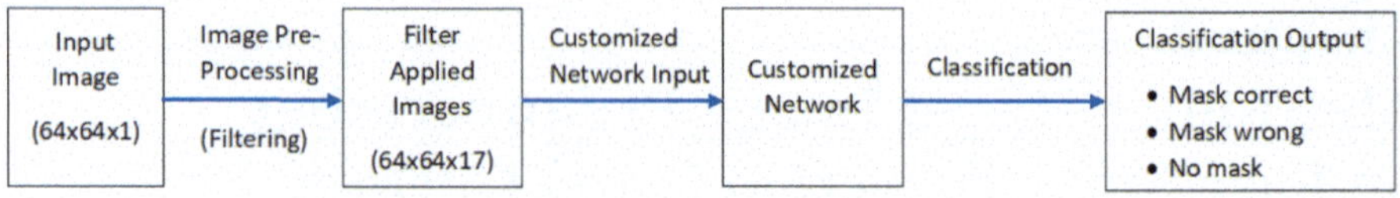

Fig. 3.1 Block diagram of the proposed process

errors, automated detection systems become very useful in many different areas. Image processing and artificial intelligence-related applications are used and proven to be effective in many different areas. These areas include education [1], image compression [2, 3], and classification and detection systems [4–9]. Because of this, research about face mask detection and classification has been made in recent years. Eladham et al. used machine learning and deep learning for face mask detection. According to their results, all NB, KNN, and CNN methods achieved an accuracy of over 80%, with CNN having the best accuracy [10]. Prasad et al. proposed a CNN-based gate control which is opened or closed according to face mask detection and body temperature. They used Raspberry pi for face mask monitoring with CNN, and Arduino for body temperature monitoring to open or close the gate using motor drivers [11]. Kowalczyk et al. used a thermal image dataset annotated with the type of face mask and the location of it within the face. They obtained 95% precision for face mask detection with Yolov5 "nano", and 91% for classification of mask type using a pre-trained convolutional neural network [12].

The inspiration behind this research was to have a process that acts as a pre-processing step and extracts features from the images to feed to a customized convolutional neural network at the same time. The idea of using multiple filters was to obtain multiple images that contain different directional or regional features of the image and feed them to a convolutional neural network to eliminate them and make classification. To achieve these objectives, this paper focuses on three-class face mask detection with a customized network that makes classification using the multiple input images which are obtained by applying various filters to the input image as a pre-process. The filter-based approach in this paper is applied to the face mask dataset, but it is not just limited to this application and can be used as an image pre-processing step in any application. Block diagram of the whole process in this paper is shown in Fig. 3.1.

3.2 Literature Review

Filters are widely used in image processing and by using them it is possible to obtain good results. Chanoui et al. used the median filter as a part of a CubeSat onboard image processing architecture. They mentioned that satellite images can be disturbed and to remove the impulse noise from the images median filter is generally used [13]. Li et al. conducted a study on color image edge detection. They used multi-scale Gabor filter and according to their results, the proposed method has high edge detection accuracy and is good at maintaining noise robustness [14]. Dong et al. studied generating raw monochrome images from raw colored images, and low-light image enhancement. For the first task, they used deep neural networks based on De-Bayer filter simulator. For the low-light image enhancement, they proposed a fully convolutional network. For training, they proposed a dataset called Mono-Colored Raw paired dataset (MCR) which is collected by a color camera with Bayer filter and a monochrome camera without Bayer filter [15]. Lv et al. proposed an adaptive bilateral filter that combines bilateral filtering and the edge detection operator to enhance infrared images. This method acts as an improved convolutional kernel for bilateral filtering and enhances the details in infrared images while suppressing the noise [16]. For edge detection, Siddharth et al. mentioned that using Kalman filter with ANN has lower calculation rates and quicker merging. They used ANN for object localization and Kalman filtering on obtained object coordinates which lowered the localization error distances and improved localization accuracy [17]. Schmalfuss et al. studied blind image inpainting using directional filters. They used a dictionary of filters and combined them with the trainable weights of a lightweight network. Their approach had faster network convergence and improved inpainting quality [18]. Dasari and Reddy applied Gabor filter to the public ILD (interstitial lung diseases) dataset to obtain texture-enhanced input images. Then they gave these input images to a Multi-scale Convolution Neural Network (M-CNN) and obtained 90.67% accuracy on lung tissue classification [19]. Putra et al. studied a face mask detection system on CCTV that sends notifications on a mobile application. They used MobileNetV2 and stated that using a distance of one meter gave good results in their experiments [20]. Sheikh and Zafar proposed a system called RRFMDS (rapid real-time face mask detection system). They fine-tuned MobileNetV2 for classification while using single-shot multi-box detector for face detection. The system detects three classes which are incorrect mask, with mask, and without mask. According to their results, training accuracy was 99.15% and the test accuracy was 97.81% [21].

3.3 Material and Method

3.3.1 Dataset

The dataset used in this research is a three-class synthesized dataset obtained by using the Labeled Faces in the Wild dataset (LFW) [22]. The dataset was obtained by adding face masks to these images to correct and wrong face mask-wearing positions, which resulted in three different classes, namely, mask correct, mask wrong, and no mask. This dataset was obtained prior the this research and was also used in previous researches [23–25].

3.3.2 Image Pre-processing

Before feeding the images to the convolutional neural network, image pre-processing was applied. This process was done by applying various 3×3 image filters to the images and feeding all of the output images to the network at once. To extract various features from the image, 17 different filters were chosen and applied. The idea behind choosing these filters was to have multiple filters that put forward different directional and regional details. By leaving out the center pixel, surrounding values were adjusted to keep up with different details from the image. Then the center pixel value was set so that all the numbers in the filter sum up to 0, otherwise loss of details would occur. The chosen filters are shown in Fig. 3.2. The steps included in the whole image pre-processing phase are resizing, converting images to grayscale, and custom filtering step. In the means of input shape, after the first step, the input shape is $64 \times 64 \times 3$. Then in the second step, the input shape becomes $64 \times 64 \times 1$. And after the last step custom filtering, the resulting input shape becomes $64 \times 64 \times 17$. This means there are 17 64×64 images which are obtained by applying 17 different 3×3 filters to the images.

3.3.3 Network and Training

By applying pre-processing to the images, the input shape becomes $64 \times 64 \times 17$, so the input layer of the customized convolutional neural network is specified to accept this input shape. After the input layer, a convolutional layer is added to reduce the input to $64 \times 64 \times 1$, which is a single-channel image of 64×64 size. So up to this point, the network actually eliminates the features from the whole 17 images and obtains a single resulting image. After this layer, a flattening layer is applied before continuing with the final dense layers. After two fully connected dense layers with "relu" activation function, the last layer is a dense layer with "softmax" function for the classification result. The custom network architecture is shown in Fig. 3.3.

For training, the dataset is split into three which are train, validation, and test datasets. 20% of the dataset is reserved for test. From the remaining 80%, again 20%

Fig. 3.2 Used 17 filters

$$\begin{bmatrix}0&1&0\\0&-2&0\\0&1&0\end{bmatrix}\begin{bmatrix}0&0&0\\1&-2&1\\0&0&0\end{bmatrix}\begin{bmatrix}0&0&1\\0&-2&0\\1&0&0\end{bmatrix}\begin{bmatrix}1&0&0\\0&-2&0\\0&0&1\end{bmatrix}$$

$$\begin{bmatrix}1&1&1\\0&-3&0\\0&0&0\end{bmatrix}\begin{bmatrix}0&0&0\\0&-3&0\\1&1&1\end{bmatrix}\begin{bmatrix}0&0&1\\0&-3&1\\0&0&1\end{bmatrix}\begin{bmatrix}1&0&0\\1&-3&0\\1&0&0\end{bmatrix}$$

$$\begin{bmatrix}1&1&0\\1&-3&0\\0&0&0\end{bmatrix}\begin{bmatrix}0&1&1\\0&-3&1\\0&0&0\end{bmatrix}\begin{bmatrix}0&0&0\\0&-3&1\\0&1&1\end{bmatrix}\begin{bmatrix}0&0&0\\1&-3&0\\1&1&0\end{bmatrix}$$

$$\begin{bmatrix}1&-1&1\\-1&0&-1\\1&-1&1\end{bmatrix}\begin{bmatrix}-1&1&-1\\1&0&1\\-1&1&-1\end{bmatrix}\begin{bmatrix}1&0&1\\0&-4&0\\1&0&1\end{bmatrix}\begin{bmatrix}0&1&0\\1&-4&1\\0&1&0\end{bmatrix}\begin{bmatrix}1&1&1\\1&-8&1\\1&1&1\end{bmatrix}$$

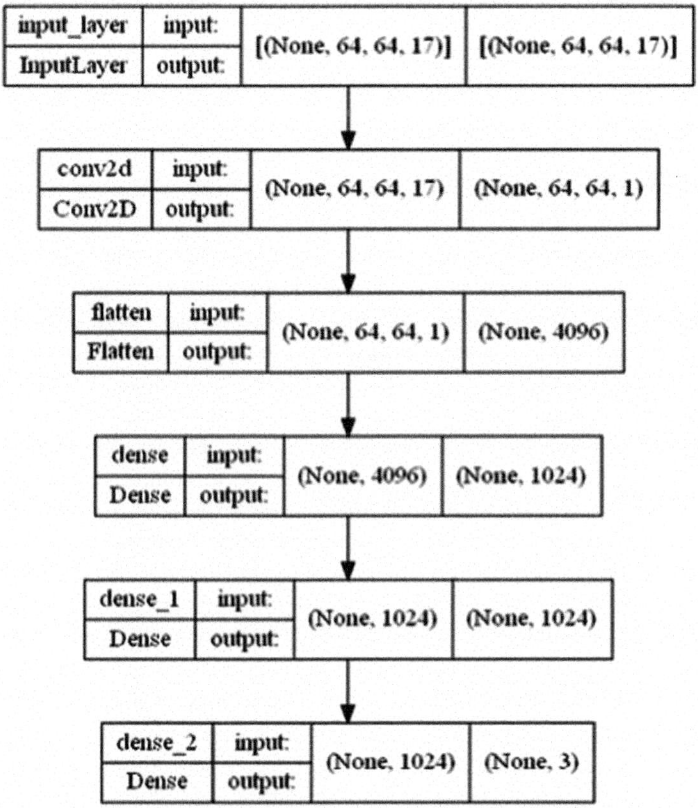

Fig. 3.3 Custom network architecture

is used for validation 80% is used for training. This results in 64% for training, 16% for validation, and 20% for test. Adam optimizer was used and different learning rates were tried for comparison. These learning rates are 10^{-3}, 10^{-4}, 10^{-5}, 10^{-6}, and 10^{-7}. These learning rates were selected to have a good comparison of their effect on accuracy and training time but without being very low or very high. Very low learning rates can cause the network to have a very long training time without having a good fit on the data and remain at low accuracy. On the other hand, using a high learning rate can cause the network to overfit on the training data which will then cause generalization problems and accuracy will be low on any other data than the training data. During the training process, accuracy, precision, recall, TP, TN, FP, and FN metrics are monitored and the loss function is set as "categorical cross entropy" which is also monitored. Patience-based training is used which traces validation loss and stops the training process when it does not decrease by a given threshold value in the number of patience epochs which is 100 in this research. When the training process is stopped, the weights of the best resulting epoch are restored to the model.

All of the training process is done with a computer with 32GB system memory, 10th generation Intel Core i7-10750H processor, and NVIDIA GeForce RTX2070 SUPER graphics card with 8GB memory.

3.4 Results and Discussion

Results were obtained for five different learning rates. At the start, the learning rate was set to 10^{-7} and increased by 10 times for each try to compare the results. It is observed that the increase in the learning rate resulted in faster training and improved accuracy. A test dataset was used which was not involved in the training process to also check the test accuracy if any overfitting problem is caused. According to the obtained results, no overfitting problem was observed. At the 10^{-7} learning rate, the training process runs for 4545 epochs for 180 min. The validation accuracy was 96.2% while the test accuracy was 95.69%. When the learning rate was increased to 10^{-6}, the training process was about 4 times faster with 44 min with 1010 epochs. The validation accuracy was slightly better with 96.4% but the test accuracy was 95.57%. At the 10^{-5} learning rate, the training time was improved and the results were obtained in 470 epochs in 20 min. Both validation and test accuracies were better than the previous ones with 97.4% and 96.33%, respectively. When the learning rate was increased to 10^{-4} a good performance increase occurred both for time and accuracy. The epochs run was 115 in 5.5 min. The obtained validation accuracy was 99.3% and the test accuracy was 98.16% which was a good increase. Then the last learning rate of 10^{-3} was tried and the results were close to the results of 10^{-4}. 114 epochs were run in 5.4 min. The validation accuracy was 98.8% which was lower, but the test accuracy was 98.86% which was better and the highest result obtained in the research. Obtained results are summarized in the Table 3.1. Since our dataset is obtained by image processing by adding face masks to the images, these accuracies may vary if images with low quality, bad lighting conditions, or face mask types that were not included in the dataset are used. Of course, these kinds of minor problems can be solved with image processing or fine-tuning the network. It is also important to mention that the final size of the customized network in this research is about 60.4MB which is a good size to be implemented in an application and does not require devices with high-performance needs.

Table 3.1 Results obtained for each learning rate

Learning rate	Validation accuracy (%)	Test accuracy (%)	Time(s)	Epochs
10^{-3}	98.8	98.86	323	114
10^{-4}	99.3	98.16	326	115
10^{-5}	97.4	96.33	1170	470
10^{-6}	96.4	95.57	2649	1010
10^{-7}	96.2	95.69	10791	4545

3.5 Conclusion

This research aimed to obtain an accurate classification of face mask usage using a filter-based approach as a pre-process for feature extraction to give them to a customized neural network. For this purpose, 17 filters were chosen and applied to input images. Then the input of the network was adjusted accordingly to accept these 17 images obtained by the filters. At the training stage of the network, different learning rates were tried to find the optimal learning rate. Results showed that at 10^{-3} and 10^{-4} learning rates, both training time and accuracy were better. The highest accuracy was obtained at 10^{-3} learning rate by 98.8% which was followed by 98.16% at 10^{-4}. The filter-based approach is not limited to the face mask dataset used in this research, it is possible to use it for feature extraction for any application. For future work, this process will be embedded in the convolutional neural network which will then be a part of the network model. In this way, it is possible to obtain different versions of the network by using different filters and different combinations of filters. After achieving this, the pre-process involving neural networks will be used on different types of datasets to check performance variation between various datasets. Also, performance gains in the means of accuracy and training time are aimed by modification of parameters during training.

References

1. Sekeroglu, B., Dimililer, K., Tuncal, K.: Artificial intelligence in education: application in student performance evaluation. Dilemas Contemporáneos: Educación, Política y Valores **7**(1) (2019)
2. Amirjanov, A., Dimililer, K.: Image compression system with an optimisation of compression ratio. IET Image Process. **13**(11), 1960–1969 (2019)
3. Dimililer, K.: Neural network implementation for image compression of x-rays. Electron. World **118**(1911), 26–29 (2012)
4. Khan, S., Khan, A.: Ffirenet: deep learning based forest fire classification and detection in smart cities. Symmetry **14**, 2155 (2022)
5. Rahhal, D., Alhamouri, R., Albataineh, I., Duwairi, R.: Detection and classification of diabetic retinopathy using artificial intelligence algorithms. In: 2022 13th International Conference on Information and Communication Systems (ICICS), pp. 15–21 (2022)
6. Rossi, J.G., Rojas-Perilla, N., Krois, J., Schwendicke, F.: Cost-effectiveness of artificial intelligence as a decision-support system applied to the detection and grading of melanoma, dental caries, and diabetic retinopathy. JAMA Netw. Open **5** (2022)
7. Nafisah, S.I., Muhammad, G.: Tuberculosis detection in chest radiograph using convolutional neural network architecture and explainable artificial intelligence. Neural Comput. Appl. 1–21 (2022)
8. Dimililer, K., Ever, Y.K., Ratemi, H.: Intelligent eye tumour detection system. Procedia Comput. Sci. **102**, 325–332 (2016)
9. Dimililer, K., Ever, Y.K., Ugur, B.: ILTDS: intelligent lung tumor detection system on ct images. In: Intelligent Systems Technologies and Applications 2016, pp. 225–235. Springer (2016)
10. Eladham, M.W., Nassif, A.B., AlShabi, M.: Face mask detection using machine learning. In: Real-Time Image Processing and Deep Learning 2023, vol. 12528, pp. 103–109. SPIE (2023)

11. Prasad, T.G., Turukmane, A.V., Kumar, M.S., Madhavi, N.B., Sushama, C., Neelima, P.: Cnn based pathway control to prevent covid spread using face mask and body temperature detection. J. Pharm. Negat, Results (2022)
12. Kowalczyk, N., Sobotka, M., Rumiński, J.: Mask detection and classification in thermal face images. IEEE Access **11**, 43349–43359 (2023)
13. Chanoui, M.A., Bouganssa, I., Sbihi, M., Alaoui Ismaili, Z.E.A., Salbi, A.: Design and simulation of a median filter for a cubesat image processing application using an FPGA architecture. In: ITM Web of Conferences (2022)
14. Li, Y., Bi, Y., Zhang, W., Ren, J., Chen, J.: Color image edge detection using multi-scale and multi-directional gabor filter (2022). abs/2208.07503
15. Dong, X., Xu, W., Miao, Z., Ma, L., Zhang, C., Yang, J., Jin, Z., Teoh, A., Shen, J.: Abandoning the bayer-filter to see in the dark. In: 2022 IEEE/CVF Conference on Computer Vision and Pattern Recognition (CVPR), pp. 17410–17419 (2022)
16. Lv, H., Shan, P., Shi, H., Zhao, L.: An adaptive bilateral filtering method based on improved convolution kernel used for infrared image enhancement. Signal Image Video Process. **16**, 2231–2237 (2022)
17. Siddharth, D., Saini, D.K.J., Singh, P.: An efficient approach for edge detection technique using kalman filter with artificial neural network. Int. J, Eng (2021)
18. Schmalfuss, J., Scheurer, E., Zhao, H., Karantzas, N., Bruhn, A., Labate, D.: Blind image inpainting with sparse directional filter dictionaries for lightweight CNNs. J. Math. Imaging Vis. **65**, 323–339 (2022)
19. Dasari, N.B., Reddy, B.V.R.: Multi-scale lung tissue classification for interstitial lung diseases using learned gabor filters. Microsyst. Technol. **29**, 599–607 (2023)
20. Putra, R.M., Yossy, E.H., Suharjito, Saputro, I.P., Pratama, D., Prasandy, T.: Face mask detection using convolutional neural network. In: 2023 8th International Conference on Business and Industrial Research (ICBIR), pp. 133–138 (2023)
21. Sheikh, B.U.H., Zafar, A.: RRFMDS: rapid real-time face mask detection system for effective covid-19 monitoring. Sn Comput. Sci. **4** (2023)
22. Huang, G.B., Ramesh, M., Berg, T., Learned-Miller, E.: Labeled faces in the wild: a database for studying face recognition in unconstrained environments. Technical Report 07-49, University of Massachusetts, Amherst, October 2007
23. Kayali, D., Dimililer, K., Sekeroglu, B.: Face mask detection and classification for covid-19 using deep learning. In: 2021 International Conference on Innovations in Intelligent SysTems and Applications (INISTA), pp. 1–6. IEEE (2021)
24. Kayali, D., Olawale, P., Kirsal-Ever, Y., Dimililer, K.: The effect of compressor-decompressor networks with different image sizes on mask detection using convolutional neural networks-VGG-16. In: 2022 Innovations in Intelligent Systems and Applications Conference (ASYU), pp. 1–5. IEEE (2022)
25. Dimililer, K., Kayali, D.: Mask detection and categorization during the covid-19 pandemic using deep convolutional neural network. Ingeniería e Investigación **43**(3), e101817 (2023)

Chapter 4
A Handy Simulated Radar Interface for Black Flight Identification System

Arwin Datumaya Wahyudi Sumari, Rosa Andrie Asmara, Helda Risman, Ika Noer Syamsiana, Dimas Rossiawan Hendra Putra, and Astika Ayuningtyas

Abstract Recognizing and identifying unidentified aerial objects or black flights has been a challenge for the National Air Operations Command Radar Unit. For this reason, a recognition and identification system that utilizes Artificial Intelligence (AI) technology is needed. But in the absence of such a system interface with the current Radar system, research has built an interface that connects the Radar system with an AI-based recognition and identification system. The main requirement is that this interface is simple and easy to operate considering the dynamics in airspace surveillance is very fast. By understanding the working mechanism in recognition and identification, a User-Case Diagram has been successfully designed that shows the working mechanism of the interface between these systems which is implemented to support the exchange of information between the Radar system and the AI-based recognition and identification system.

4.1 Introduction

The high-risk challenge in maintaining national sovereignty over airspace is to prevent unidentified aerial objects intending to carry out illegal acts such as intrusion, close detection, and attack of objects important national bodies. Surveillance of the airspace is an important task of the radar units of the National Air Combat Command [1], performing continuous airspace observation activities around the clock without stopping. This operation is supported by two types of radar, namely primary surveillance radar (PSR) and secondary surveillance radar (SSR).

A. D. W. Sumari (✉) · R. A. Asmara · I. N. Syamsiana · D. R. H. Putra
State Polytechnic of Malang, Malang, Indonesia
e-mail: arwin.sumari@polinema.ac.id

A. D. W. Sumari · A. Ayuningtyas
Adisutjipto Institute of Aerospace Technology, Bantul, Indonesia

H. Risman
Republic of Indonesia Defense University, Bogor, Indonesia

S. K. Pal et al. (eds.), *Intelligent Informatics*, Smart Innovation, Systems and Technologies 389, https://doi.org/10.1007/978-981-97-2147-4_4

The first type of radar is used to detect the presence of aerial objects up to 240 miles or about 400 km. From the detected objects will be obtained data in the form of flight speed, flight altitude, flight direction, and one important arrival is the cross-section of the object body called Radar Cross Section (RCS) [2–4]. To be able to recognize and identify airborne objects that have been detected, the second type of radar, namely SSR, will carry out its task by sending an identification signal called the Identification Friend or Foe (IFF) signal [5]. This signal consists of several modes that can be applied to both commercial aircraft and military aircraft [6].

The problem is that aircraft or aerial objects that have been designed to perform illegal acts will turn off IFF transponders to mask their identity [7]. This is a critical situation, but the Commander-in-Chief of the National Air Operation Command must be very careful when acting. Without knowing the aircraft or aerial objects' identification, will cause very hard decisions, but the time is ticking, and the necessary decision cannot be waited too long. No action can be taken but dispatching observer fighters to observe closely the detected aircraft or aerial object. Then, the nearest Air Squadron will be instructed to dispatch a fighter flight to the coordinate point where the aerial unidentified object is detected by the nearest Radar Unit to make visual observations that can be followed by an interception and forced to land. The second problem is that if the dispatched fighter aircraft turns out to have a lower combat capability than an aerial unidentified object, then the act of intercepting and forcing landing becomes very unlikely because it presents a high risk to both the crew and the fighter itself. This situation will directly damage the sovereignty of Indonesia's airspace (Fig. 4.1).

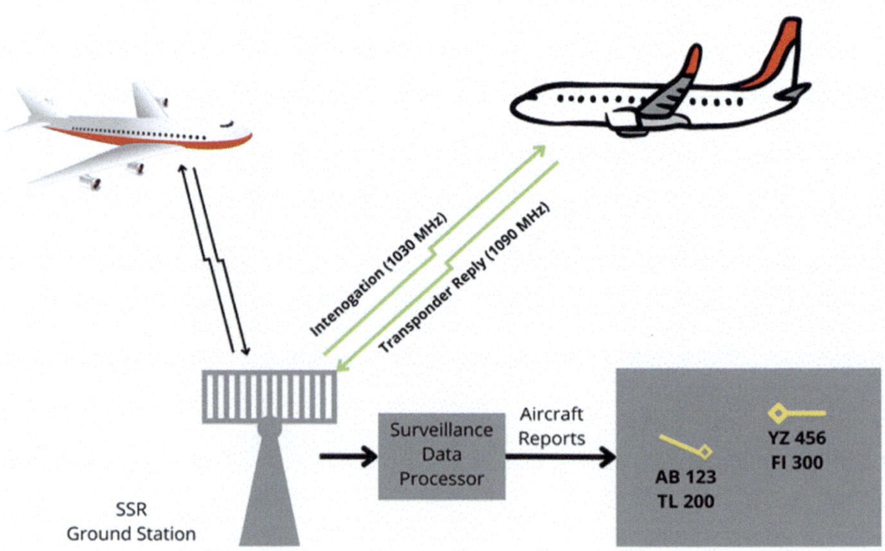

Fig. 4.1 Air object identification with SSR

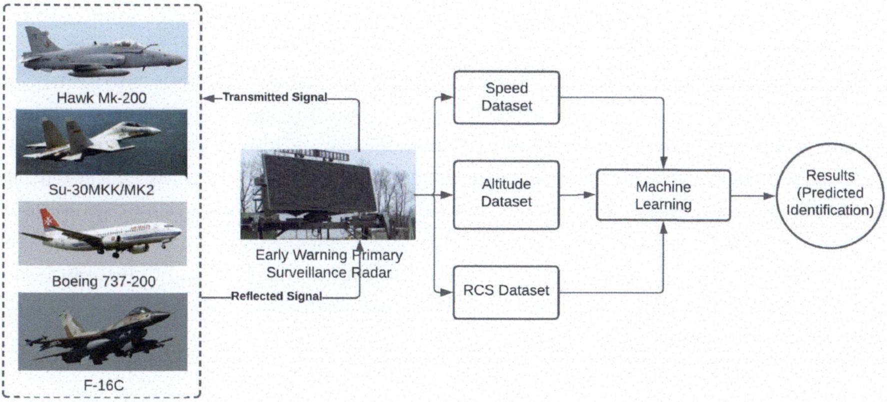

Fig. 4.2 Our new intelligent system for aircraft recognition and identification uses machine learning [8]

Faced with the absence of IFF data, the main challenge faced by the Radar Unit is how to recognize and identify the detected airborne object by utilizing only data from the PSR. One of the unique PSR data that provides a way to recognize and identify any one object is RCS. In research that has been done previously [8], the use of a combination of PSR data, namely flight speed and RCS, can be used to recognize and identify an aerial object by utilizing Artificial Intelligence (AI) technology, which in this case is machine learning. The next challenge is to build an interface that connects the Radar system with the AI application.

There have been many studies on the use of RCS for the identification of aerial objects both for aircraft [9, 10] and for unmanned aerial vehicles [11–13] using AI methods [14–17] or not [7, 18]. However, recognition and identification using a combination of PSR data in the form of speed and RCS has not been done. For this reason, in this study, a special interface was built to make it easier for Radar Unit personnel to recognize and identify detected aerial objects but turn off the IFF system. As a first step before being implemented into real conditions, a simulated interface has been built between the Radar system and machine learning applications to show how this interface works and its ease of operation or handy (Fig. 4.2).

4.2 Methods

4.2.1 Unified Modeling Language (UML)

UML is a general-purpose visual modeling language used to specify, visualize, construct, and document software system artifacts in an efficient manner [19]. It captures the decision and understanding of the systems that need to be built. It involves the use of understanding, designing, browsing, configuring, maintaining, and controlling information on such a system.

The development of UML has several goals but the most important is to define an understandable and easy-to-use general-purpose modeling language that all modelers can use. UML diagrams are designed for businesses and developers. Users are the public, and anyone interested in understanding the system. A system can be a software system or a non-software system. It is therefore clear that UML is not a development methodology but a set of processes for building a successful system. The goal of UML can be defined as a general software modeling technique [20] that models all possible practical systems in today's complex environment especially Information Technology (IT)-based systems [21].

4.2.2 Use-Case Diagram

Use-Case Diagram (UCD) provides simple and fast facilities to describe the working mechanism of the system from the user side and its interaction with the system [22, 23]. System users can also be referred to by the term actor. UCD shows who the actors are and can interact with the system [24]. For example: the Commander-in-Chief of the National Air Operation Command actor is the system user with the highest level of authentication and can access all facilities provided by the system. On the other side, field administration actors should only interact at lower authorization levels because their job is only to enter data into the system.

UCD can also show scenarios of system usage by all actors connected to the system created. In UCD, actors can be added or subtracted according to the purpose for which the system is built. The more actors, the more complex UCDs will be. For this reason, building UCD must first determine the actors who are entitled to have access to the system and the level of authorization.

4.2.3 Activity Diagram

An activity diagram is a visual representation of the sequence of operations or control flow in a system such as a flowchart or data flow diagram to show the logic of an algorithm. The activity diagram shows the steps in a UCD and how the systems and objects behave [25]. Activities modeled can be sequential and concurrent. In both cases, the activity diagram has a beginning (initial state) and an end (final state). An activity diagram is another important feature in UML that describes the dynamic aspects of a system and a chart that represents the flow from one activity to another or its algorithm [26].

Operations in a system can be called operations and flow control from one operation to another. This flow can be derived in phase or parallel. Action diagrams deal with any type of power flow using various elements such as fork links. An activity is a specific process of a system. Action diagrams are used not only to visualize a dynamic system but also to operate the system using forward and reverse engineering techniques. The only thing missing from the activity icon is the message part. Message flow from one activity to another is not shown. The activity of the diagram can be considered as fluorescence. Although the shape of the table looks like a flower it is not a flower. It indicates various currents like simple parallel branches, etc.

4.2.4 Application Programming Interface (API)

The Radar system application is one of the subsystems of a larger system so to be able to communicate with the main application and other applications requires an interface. For this reason, an Application Programming Interface (API) is built that is tasked with bridging communication between systems with certain rules. API can also be referred to as interfaces in the form of software to give access to those who want to use certain resources [27].

4.2.5 Radar System

Air defense systems consist of several systems integrated and generally consist of radar systems, guided missile systems, air defense artillery systems, and air defense command and control centers.

Primary Surveillance Radar (PSR). A system that is operated to detect the presence of one object or some objects at a certain distance and direction. The result of detection is the position, speed, direction, and shape of the object. Radar operates on electromagnetic waves, namely, on microwaves and radio with wavelengths of about millimeters to approximately one meter. In the military field, defense radar functions to detect the presence of the enemy and guide aircraft or weapons to intercept, anticipate possible attacks, and guide in attacks. In general, defense radar serves as defensive eyes and ears. In the context of air defense, PSR is equipped with Early Warning (EW) and Ground Control Interception (GCI) capabilities. GCI capabilities are required to guide the fighter when making direct visual observations to detect aerial objects to confirm them as friend, or foe. One type of radar used by the air defense system is the Thomson TRS 2215D radar as shown in Fig. 4.3, which can detect 3D targets (bearing, range, altitude/direction of speed, distance, and altitude) as well is equipped with an Early Warning System (EWS) and Ground Control Interceptor (GCI) capabilities. This radar has technical specifications among others are operational frequency between 2.9 and 3.1 GHz, a distance range between 10 and 470 km, an antenna rotation of 6 rpm $\pm$ 10%, and a motor speed of 700 rpm.

The radar antenna will rotate 360° continuously to scan all air areas within its coverage. For that, a display is needed that can show activities and objects scanned by radar antennas through the Plan Position Indicator (PPI) as shown in Fig. 4.4. In

Fig. 4.3 Thomson TRS 2215D radar antenna [28]

Fig. 4.4 Plan Position Indicator (PPI) example [29]

this view, all detected objects, both stationary and moving objects, will appear. For each moving object, the radar system calculates altitude, speed, bearing, and RCS.

Secondary Surveillance Radar (SSR), SSR as shown in Fig. 4.5, is a surveillance radar system that uses transmitters/receivers (interrogators) and transponders. SSR has several modes for interrogation of a single airborne object detected by the PSR and the data obtained is highly dependent on the mode it operates. The interrogator signal is used to identify the air object with an important note that the object has the IFF transponder turned on. In this case, a combination of data from PSR and SSR will be used to identify and identify. If the combination of data has not been able to be recognized and identified, then the operating procedure that must be done is to send a fighter flight to make direct visual observation. In the case of black flight, an aerial object will always turn off the transponder to mask its identity. On the other hand, the black flight pilot will close communication with the Air Traffic Controller (ATC) officer in the airspace he is flying.

SSR in military terms is referred to as IFF, a tool used to identify an aerial object whether friend or foe. IFF supporting technologies include [30]:

Interrogators. Interrogators are electronic devices that emit an "interrogating" radio signal at one frequency, prompting an IFF Transponder to emit a reply signal at a

Fig. 4.5 SSR in operation

different frequency, indicating that an approaching aircraft is "friendly." If there is no reply to the interrogating signal, it indicates that it is "foe".

Transponders. Transponders are transmitters/responders mounted in aircraft, naval vessels, and some ground vehicles to identify the craft as "friendly" by responding to "interrogation" signals emitted by IFF system Interrogators.

Combined Interrogator Transponders (CITs). CITs consolidate the functions of dedicated IFF Interrogators and Transponders into a single unit, specifically for use with modern mobile platforms to identify "friendlies" and save lives within secure zones.

Cryptographic Computers (CC). CC encrypts signals being sent by IFF Interrogators and encodes replies from corresponding Transponders. This makes it difficult for adversaries to steal identification codes to imitate friendly aircraft for executing surprise attacks.

Emulators. Emulators provide testing of IFF Interrogators and Transponders, including cryptographic capabilities, in multiple Modes appropriate to each system.

Antennas. Antennas enhance both Interrogator and Transponder signals and work with all standard IFF Modes.

4.2.6 Radar Cross Section (RCS)

Radar Cross Section (RCS) is the ability of a target to reflect radar signals toward the source from the radar transmitter. RCS has units of square meters because it shows the signature of a target. It can be said that RCS shows the projected area of a metal sphere which will emit a certain amount of radiated power toward a target. An aircraft's RCS is a measure of its stealth capability, and it is not affected by the distance between the radar and the detected target [31]. A low radar signature means better stealth with a shorter detection range and shorter reaction time. The RCS of the target is the approximate area of the object that disperses the same energy in the same direction as the metal ball. It is measured in m^2 or dBsm and refers to the comparison of the power density of the wave incident on the target and the power density of the scattered wave received at the radar [32].

The larger the RCS the easier it is for radar to detect objects both in the air and at sea. Radar typically works by emitting electromagnetic waves. When these electromagnetic waves hit an object, the object reflects the waves to the radar which allows the control center to detect and identify the object. More precisely RCS can be interpreted as the ease of object detection [33, 34]. In other words, the smaller the RCS the harder it is for the radar to detect the object. There is a term called stealth aircraft. Stealth means that the plane has much less RCS than the one with the same size. The American-made B-2 Spirit bomber has a wingspan of 524 m, the length from nose to tail is 21 m, but its RCS is only 0.1.

4.3 Implementing the Interface

4.3.1 The Development Methodology

The interface of the built radar system is part of a major application but plays an important role in aerial surveillance systems. The steps to building this interface include creating a block diagram of the relationships between parts of the application and the interface, creating a UCD that shows the relationship between all users and the application, creating an activity diagram that shows the flow of interface operations, and integrating the APIs available with the interface. In addition, trials are also carried out by prospective users to obtain inputs for the improvement of the interface that has been made.

4.3.2 Use-Case Diagram Development

Several things must be prepared before building a UCD, namely the actors and the number, the facilities supported by the system, and the inputs and outputs needed for the system to work as designed. UCD Radar system interface with AI-based black flight identification system is shown in Fig. 4.6. The UCD showed that the actors who had access to the system were high-ranking officials in the National Air Operations Command. Access to the system will give them information about the name and location of the Radar Unit shown in the form of icons on the Earth map. In addition, the system also provides facilities to find out the identity of air objects that pass

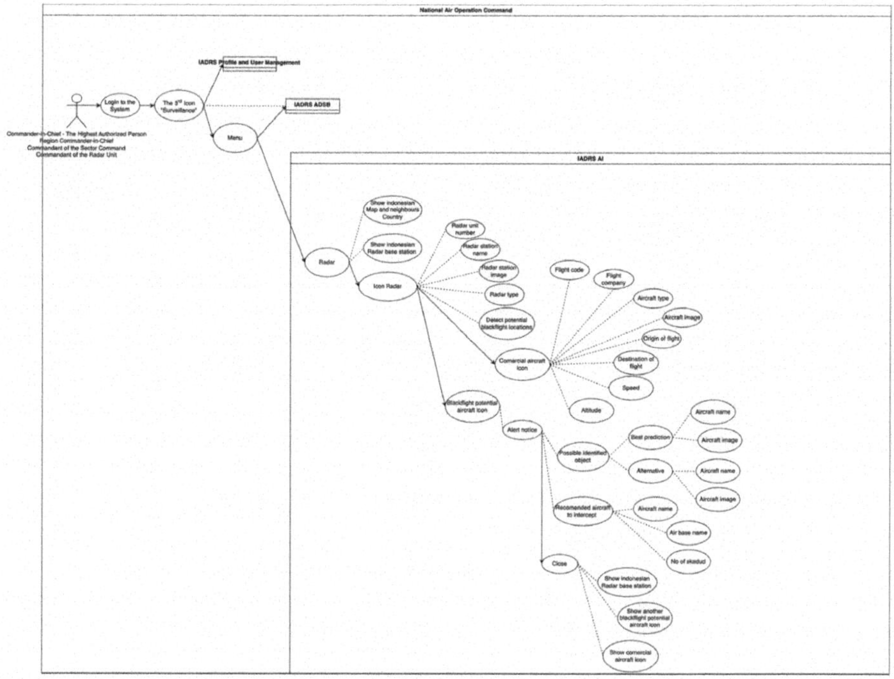

Fig. 4.6 UCD interface radar system and AI-based identification system

through national airspaces such as aircraft codes, departure, and flying destinations, as well as flight duration and coordinates of the position of air objects in real time.

In the context that the detected aerial object does not turn on the transponder and is then declared as black flight, the system will perform recognition and identification by utilizing data from the PSR to predict the identity of the object. From the prediction, results will inform the prediction of the type and name of the air object and its capabilities. From the results of this prediction, the National Air Operations Command will fly one fighter flight to conduct an intercept in the air. The actors who use the system are the Commander-in-Chief of the National Air Operation Command, the Commander-in-Chief of the Air Operation in each operation region, the Commandant of the Sector Command that commands a group of Radar Units, and the Radar Units Commandant. Concisely, the Radar Unit that detects unidentified aircraft or aerial objects will report to the Sector Command which then reports to the Commander-in-Chief of the National Air Operation Command. He then makes decisions and instructs the related Region Air Operation Command to dispatch a fighter flight consisting of three fighter aircraft, that can intercept or force down the detected unidentified aircraft or aerial objects.

Simply, the National Air Operation Command's authorized user opens the IADRS system and logs in using the user and password set by the administrator. If not registered, the user must register with the system administrator to gain access to the system. Errors in filling in the account username or password will produce an error notification, and the user will be asked to enter the correct data. After successfully entering the system, the National Air Operation Command's Commander-in-Chief or other authorized users will enter the system's main display, which shows a map of Indonesia as the default, which can be shifted in all directions, zoom-in and zoom-out. The maps also support worldwide map viewing. To enter the main display of the radar system interface, select the icon "**Surveillance**". At the top center of the main system display, there are two main system features, which are represented in the form of icons for accessing the ADSB facility and the Radar facility.

By default, ADSB will be active automatically and display all aerial objects on the highlighted map. All identified aerial objects captured by ADSB will appear on the main display. The aerial objects' data displayed includes the type of aircraft, the origin airport, the destination airport, flight speed, and flight altitude as well as the direction from the Radar station. If the Radar feature is selected, all aerial objects within the range will be displayed in real time on the main display screen and the PPI. Aerial objects that turn off IFF and ADSB are called black flights or unidentified aerial objects. The black flight will be identified using AI technology by utilizing speed and altitude data equipped with important data called RCS. If the black flight is a threat, the system will display threat notification to provide alertness to the National Air Operation Commander-in-Chief as the decision maker to make future decisions and actions.

4.3.3 Activity Diagram Development

Activity diagrams must be used parallel (horizontally) with other modeling techniques, such as UCDs and State diagrams. This diagram depicts a complex sequential algorithm and modeling with parallel processes. As shown in Fig. 4.7, the process starts from the start point to indicate the initial status, initial action, or starting point of activity in the Radar system interface with an AI-based black flight identification system.

The National Air Operations Command carries out an authorization process on the system first before accessing the surveillance menu. In this menu, there are two information menus, namely ADSB Information and Radar Information. The radar menu displays all radar data, while the ADSB menu displays information on objects in the air that have been captured. There are two pieces of information, namely, registered objects that will display some information (code, airline name, city of origin, destination city, altitude, speed) and objects that are not registered and caught by radar will be identified using the AI method. The identification process will display information output on two object predictions and recommended aircraft to intercept. So that this information output can help the National Air Operations Command in deciding whether to fly a flight of fighter aircraft to conduct an airborne intercept.

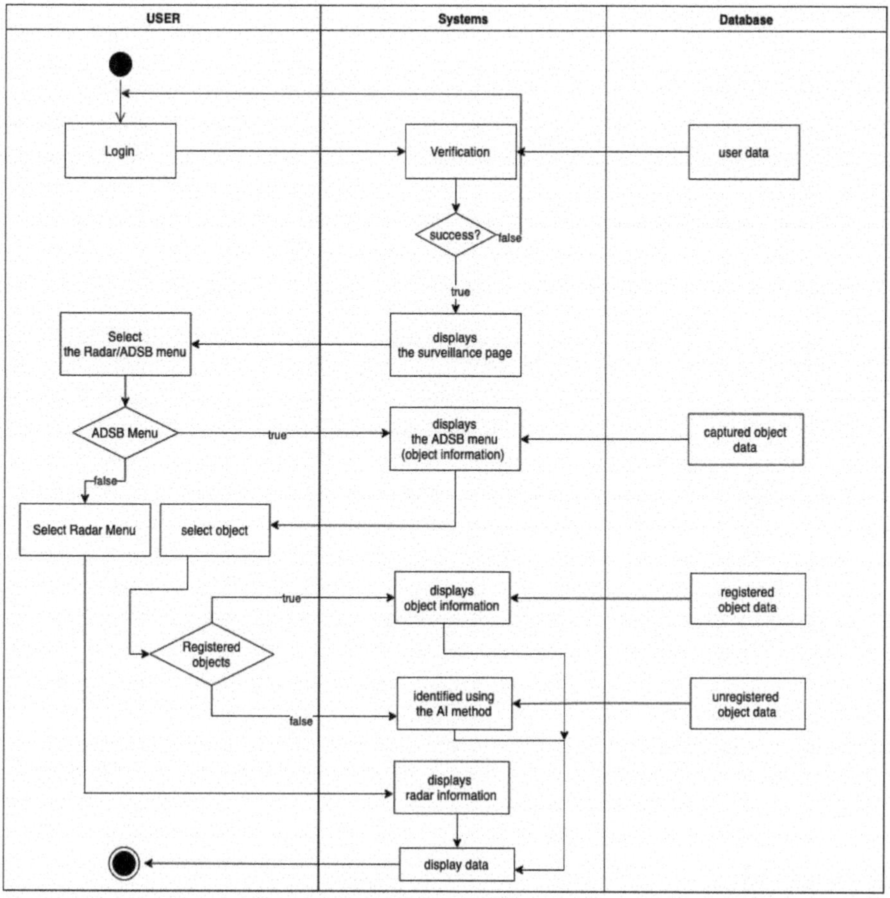

Fig. 4.7 Activity diagram radar system and AI-based identification system

4.3.4 The Simulation Environment

The radar system interface consists of a Front End (FE) built using React JS on the vue.js framework, and a Back End (BE) built using the Python programming language on the Flask framework on the Windows operating system. ADSB data is obtained by combining it with the Flightradar24 application API on a paid basis. This data from API is only for commercial aircraft that are required to turn on their IFF transponders. For the server, IBM Z16 was used to host all software including all interfaces required to run the complete application.

4.3.5 The Application

Integrating AI with the Simulated Radar System. The focus of this research is to simulate a radar system with an AI-based recognition and identification system at the software level before being implemented on a real radar system at a later stage of research.

Interface Implementation. The UCD that has been conveyed in the previous section, in this section several parts of the interface that have been successfully implemented are conveyed, namely:

- A radar system interface that displays the radar coverage area along with its PPI simulation. Figure 4.8 shows the coverage area of Radar Unit 211 Tanjung Kait which covers the capital city of Indonesia to the southern part of Sumatra Island and the Indian Ocean. The picture also shows aircraft flying in the air space under its guard. Radar signal coverage represents actual coverage of 240 miles or 400 km. All aircraft detected in the air space within the radar signal coverage will be displayed, including aerial objects that do not reveal their identity.
- Radar system interface showing the identity of aircraft passing through airspace under guard of 211th Radar Unit as shown in Fig. 4.9. This identity will be displayed by the system when the aircraft icon is clicked. For example, the icon clicked on this display displays information that the aerial object is a Qantas commercial aircraft with flight code QFA82 and is serving flights from Singapore to Sydney, Australia.
- Radar system interface that shows the identity of black flight successfully predicted by AI-based recognition and identification systems as depicted in Fig. 4.10. For example, the display shows the results of predictions on the air

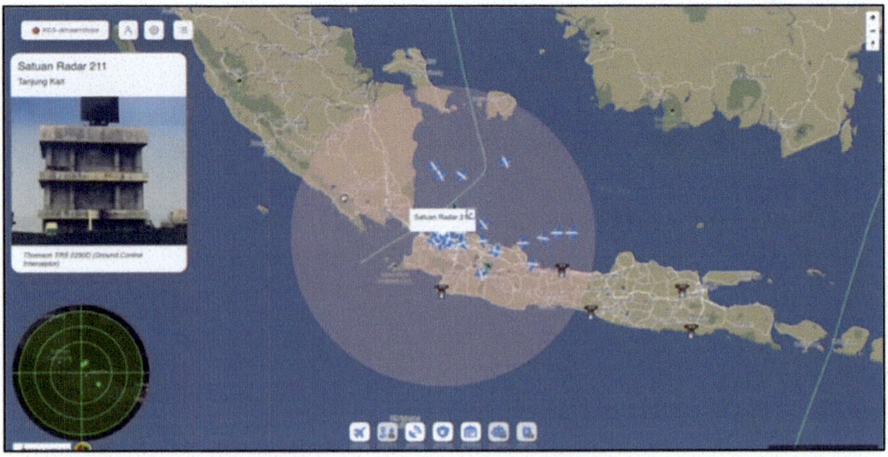

Fig. 4.8 A view of the 211th Radar unit and its guarded airspace

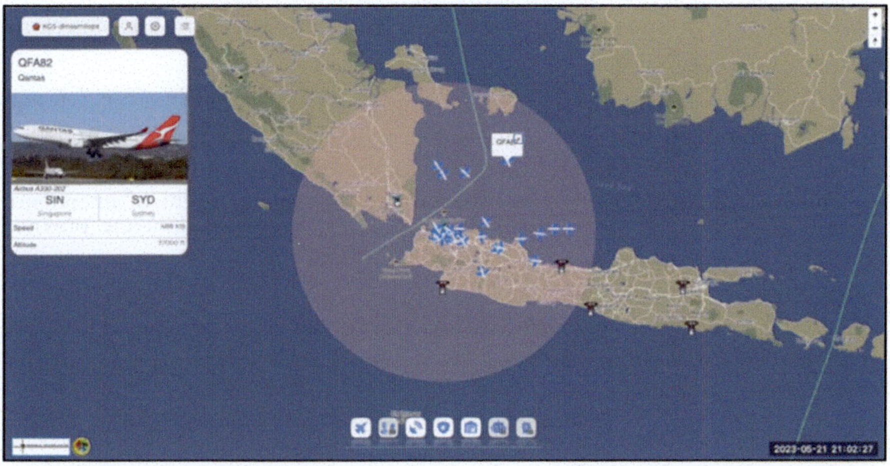

Fig. 4.9 Identity of aircraft passing through airspace under the guard of the 211th Radar unit

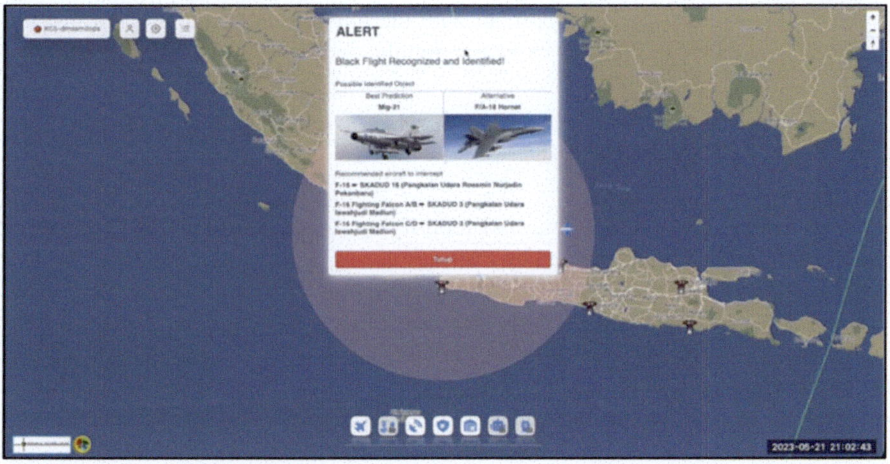

Fig. 4.10 Predicted results of black flight recognition and identification

object detected by the PSR as a MiG-21 intercept fighter or F-18 Hornet. In response, the system recommends a F-16 fighter equivalent to the black flight.

4.3.6 User Evaluation

To ensure that the radar system interface that has been created provides benefits to users in the field as well as related industries, questionnaires have been sent to two senior Radar Officers and a representative from the industry to operate this interface. Their reviews are as follows.

- What stands out in the IADRS application is the ability of the application to provide information on black flight targets and provide information on possible types of black flight aircraft to make it easier for the National Air Operation Command Commander-in-Chief to determine the capable flight of interceptor fighter aircraft to intercept the black flight.
- Interception activities against suspect or black flight are greatly helped by this application, where the type of black flight can be known in detail along with the types of aircraft and from which Air Squadron locations can be deployed to carry out interception. Besides that, this application is web-based, making it easier for operator personnel to monitor flights passing through its area.

On the other hand, the reviewers also provide inputs for system improvements as follows.

- The IADRS application can be developed so that it can be used as an interface for the integration of several types of radar used by the Indonesian Air Force. There also needs to be development with the integration of active radar capture and passive radar so that it is expected that the display raised is more optimal.

4.4 Concluding Remarks

The big challenge in recognizing and identifying an unidentified aerial object or black flight that has been detected by PSR has been given a solution in the form of a combination of PSR data, namely flight speed and object cross-section or RCS in the form of body size, combined with AI technology. In the existing radar system besides not yet supported by an interface that connects with the AI-based recognition and identification system, no mechanism relates between the flight speed of air objects and RCS. In this study, an interface has been successfully built between the radar system and a simple and easy-to-use AI-based recognition and identification system to assist Radar Unit personnel in carrying out their duties. The next step is to build an interface between AI-based recognition and identification systems and ADSB systems to support more accurate recognition and identification of airborne objects. Important thing. Another thing that will be done is to study and understand the mechanism of black flight identification which is the concern of the Sector Command. The speed of recognition and identification is the key to maintaining the sovereignty of national airspace.

Acknowledgements We want to thank the Academic Directorate of Vocational Higher Education, Directorate General of Vocational Education of the Ministry of Education, Culture, Research, and Technology of the Republic of Indonesia for support in the form of Vocational Product Research grants in 2023 number 173/SPK/D.D4/PPK.01.APTV/VI/2023, and the State Polytechnic of Malang number 12901/PL2/TU/2023 for the opportunity to conduct research in the field of Defense and

Security. Appreciation is also conveyed to PT. Republik Defense for its in-kind support of this research.

References

1. National Air Defense Command: Decree of the Commander of the National Air Defense Command Number Kep/79/XII/2017 regarding Permanent Procedures for Air Defense Operations. Komando Pertahanan Udara Nasional, Jakarta (2017).
2. Nohara, T.J., Beason, R.C., Weber, P.: Using radar cross-section to enhance situational awareness tools for airport avian radars. Human-Wildlife Interactions **5**, 210–217 (2011). https://doi.org/10.26077/sgas-w455
3. Charris, V.D., Gómez Torres, J. M.: Analysis of radar cross section assessment methods and parameters affecting it for surface ships. Ship Sci. Technol. **6**, 91–106 (2012)
4. de Andrade, L.A., dos Santos, L.S.C., Gama, A.M.: Analysis of radar cross section reduction of fighter aircraft by means of Computer Simulation. J. Aerosp. Technol. Manag. **6**, 177–182 (2014). https://doi.org/10.5028/jatm.v6i2.259
5. IFF—Identification—Friend or Foe. http://www.tscm.com/iff.pdf.
6. Miller, J.: IFF and Mode 5: Past Present and Future. https://www.ieee.li/pdf/viewgraphs/iff_past_present_future.pdf (2002)
7. Sumari, A.D.W., Nugraheni, A.M., Yunhasnawa, Y.: A novel approach for recognition and identification of low-level flight military aircraft using Naive Bayes classifier and information fusion. Int. J. Artif. Intell. Res. **6,** (2022). https://doi.org/10.29099/ijair.v6i1.248
8. Sumari, A.D.W., Asmara, R.A., Risman, H., Syamsiana, I.N., Handayani, A.N., Arai, K.: Black Flight identification using radar cross section (RCS), speed, and altitude from RADAR data using supervised machine learning. In: Proceedings—IEIT 2022: 2022 International Conference on Electrical and Information Technology. pp. 377–382. IEEE (2022). https://doi.org/10.1109/IEIT56384.2022.9967914
9. Chan, Y.T., Ho, K.C., Wong, S.K.: Aircraft identification from RCS measurement using an orthogonal transform. In: IEE Proceedings—Radar, Sonar and Navigation, pp. 93–102. IET Digital Library (2000)
10. Jeong, Y.-R., Park, C.-S., Ko, Y.-K., Yook, J.-G.: Analysis of RCS of low observable aircraft in VHF band. Int. J. Antennas Propag. (2018). https://doi.org/10.1155/2018/5435837
11. Rosamilia, M., Aubry, A., Balleri, A., Carotenuto, V., De Maio, A.: RCS Measurements of UAVs and Their Statistical Analysis. In: 2022 IEEE 9th International Workshop on Metrology for AeroSpace (MetroAeroSpace), pp. 179–184. IEEE (2022). https://doi.org/10.1109/MetroAeroSpace54187.2022.9856394
12. Ritchie, M., Fioranelli, F., Griffiths, H., Torvik, B.: Micro-drone RCS analysis. In: 2015 IEEE Radar Conference. pp. 452–456. IEEE (2015). https://doi.org/10.1109/RadarConf.2015.7411926
13. Patel, J.S., Fioranelli, F., Anderson, D.: Review of radar classification and RCS characterisation techniques for small UAVs or drones. IET Radar Sonar Navig. **12**, 911–919 (2018). https://doi.org/10.1049/iet-rsn.2018.0020
14. Zhu, S., Peng, Y., Alexandropoulos, G.C.: RCS-based flight target recognition using deep networks with convolutional and bidirectional GRU layer. In: The 2020 the 4th International Conference on Innovation in Artificial Intelligence, pp. 137–141 (2020). https://doi.org/10.1145/3390557.3394300.
15. Sehgal, B., Shekhawat, H.S., Jana, S.K.: Automatic radar target identification using radar cross section fluctuations and recurrent neural networks. In: IEEE Region 10 Annual International Conference, Proceedings/TENCON. 2019-Oct, pp. 2490–2495 (2019). https://doi.org/10.1109/TENCON.2019.8929635

16. Darusman, M., Sumari, A.D.W., Wuryandari, A.I.: Sistem Identifikasi Pesawat Menggunakan Kecepatan dan Radar Cross Section Pesawat Berbasis Jaringan Syaraf Tiruan Backpropagation. In: Seminar Radar Nasional III, pp. 11–15 (2008)
17. Emre Aydemir, M., Gose, E.: Radar cross section identification of air targets using the cosine transform and neural networks. Recent Patents Eng. **6**, 65–69 (2012). https://doi.org/10.2174/187221212799436745
18. Sumari, A.D.W., Pranata, A.S., Mashudi, I.A., Syamsiana, I.N., Sereati, C.O.: Automatic target recognition and identification for military ground-to-air observation tasks using support vector machine and information fusion. In: 9th International Conference on ICT for Smart Society: Recover Together, Recover Stronger and Smarter Smartization, Governance and Collaboration, ICISS 2022—Proceedings, pp. 1–8 (2022). https://doi.org/10.1109/ICISS55894.2022.9915256
19. Gosala, B., Chowdhuri, S.R., Singh, J., Gupta, M., Mishra, A.: Automatic classification of UML class diagrams using deep learning technique: convolutional neural network. Appl. Sci. **11**, 4267 (2021). https://doi.org/10.3390/app11094267
20. Rocha, M., Simão, A., Sousa, T.: Model-based test case generation from UML sequence diagrams using extended finite state machines. Softw. Qual. J. **29**, 597–627 (2021). https://doi.org/10.1007/s11219-020-09531-0
21. Górski, T.: UML profile for messaging patterns in service-oriented architecture, microservices, and internet of things. Appl. Sci. **12**, 12790 (2022). https://doi.org/10.3390/app122412790
22. Fauzan, R., Siahaan, D., Rochimah, S., Triandini, E.: A different approach on automated use case diagram semantic assessment. Int. J. Intell. Eng. Syst. **14**, 496–505 (2021). https://doi.org/10.22266/ijies2021.0228.46
23. Li, S., Rajapuri, A.S., Felix Gomez, G.G., Schleyer, T., Mendonca, E.A., Thyvalikakath, T.P.: How do dental clinicians obtain up-to-date patient medical histories? Modeling strengths, drawbacks, and proposals for improvements. Front Dig. Health **4** (2022). https://doi.org/10.3389/fdgth.2022.847080
24. Amponsah, A.A., Adekoya, A.F., Weyori, B.A.: Improving the financial security of national health insurance using cloud-based blockchain technology application. Int. J. Inf. Manag. Data Insights **2**, 100081 (2022). https://doi.org/10.1016/j.jjimei.2022.100081
25. Abbas, M., Rioboo, R., Ben-Yelles, C.-B., Snook, C.F.: Formal modeling and verification of UML Activity Diagrams (UAD) with FoCaLiZe. J. Syst. Architect. **114**, 101911 (2021). https://doi.org/10.1016/j.sysarc.2020.101911
26. Řezník, T., Herman, L., Klocová, M., Leitner, F., Pavelka, T., Leitgeb, Š, Trojanová, K., Štampach, R., Moshou, D., Mouazen, A.M., Alexandridis, T.K., Hrádek, J., Lukas, V., Širůček, P.: Towards the development and verification of a 3D-based advanced optimized farm machinery trajectory algorithm. Sensors **21**, 2980 (2021). https://doi.org/10.3390/s21092980
27. Kumar, S., Suleski, M., Craig, J.M., Kasprowicz, A.E., Sanderford, M., Li, M., Stecher, G., Hedges, S.B.: TimeTree 5: An expanded resource for species divergence times. Mol. Biol. Evol. **39** (2022). https://doi.org/10.1093/molbev/msac174
28. Mathurin, R. Des: Long Range 3-D Mobile Fixed Radar Thomson TRS 2215. Thomson CSF (1985)
29. Middleton, W.E.K., Mai, A.: Radar Technical Overview
30. BAE Systems: What are IFF Technologies?
31. Herbette, Q., Darces, M., Bourey, N., Saillant, S., Jangal, F., Hélier, M.: Redefining of the radar cross section and the antenna gain to make them suitable for surface wave propagation. Progr. Electromagn. Res. C **119**, 1–16 (2022). https://doi.org/10.2528/PIERC21111204
32. Taj, Z.U.D., Bilal, A., Awais, M., Salamat, S., Abbas, M., Maqsood, A.: Design exploration and optimization of aerodynamics and radar cross section for a fighter aircraft. Aerosp. Sci. Technol. **133** (2023). https://doi.org/10.1016/j.ast.2023.108114
33. Singh, H.: Radar cross section minimization analysis for different target shapes. Mater Today Proc. (2022). https://doi.org/10.1016/j.matpr.2022.10.306
34. Pieraccini, M., Miccinesi, L., Rojhani, N.: RCS measurements and ISAR images of small UAVs (2017). https://doi.org/10.1109/MAES.2017.160167

Chapter 5
Spatial Pyramid Image Representation with DCT Features for Offline Signature Verification

Bharathi Pilar, B. H. Shekar, Wincy Abraham, and D. S. Sunil Kumar

Abstract This paper presents a Spatial Pyramid image representation-based technique with global and local features captured through DCT coefficient at various levels for offline signature verification. The spatial pyramid feature vector is an extension of an orderless bag of features. We employed a spatial pyramid with 4 levels and took the entire signature image at the first level. The rest of the 3 levels in the pyramid consist of images with 4 partitions, 16 partitions, and 8×8 non-overlapping blocks. Our approach captures both local and global DCT features from the image and its sub-blocks at various levels. The AC DCT coefficients obtained are represented in the form of a matrix. The variation in the statistical properties of AC coefficients of the image is made use of in detecting the forgery. For that, the standard deviation and count of non-zero DCT coefficients corresponding to each row in the DCT matrix are found. The standard deviation of DCT coefficients is a good measure of representing the spread of values in an image. The extracted feature vector consists of the standard deviation and the number of non-zero values in the DCT matrix. The Support Vector Machine (SVM) is used for the classification. The classification accuracy of our approach on standard datasets is calculated and the results are compared with a few well-known approaches. This shows that the performance of the proposed approach is better than the other approaches in the state-of-the-art literature.

B. H. Shekar · D. S. S. Kumar
Department of Computer Science, Mangalore University, Konaje, Karnataka, India

W. Abraham (✉)
Department of Computer Science, Assumption College, Changanassery, Kerala, India
e-mail: wincy@assumptioncollege.edu.in

B. Pilar
Department of Computer Science, University College Mangalore, Mangaluru, Kerala, India

S. K. Pal et al. (eds.), *Intelligent Informatics*, Smart Innovation, Systems and Technologies 389, https://doi.org/10.1007/978-981-97-2147-4_5

5.1 Introduction

Although there are many biometrics for person authentication and identification, a signature is considered a means of identification in many financial and legal systems. There are many crimes related to signature forgery reported from time to time. Most of the forged signatures are not blind signatures which can be detected easily as they have less similarity with the original signature. Here the forger does not have any idea about the original signature. Trace-over forgeries created by tracing over the original signature are difficult to detect manually. Skilled forgeries are difficult to detect as they resemble the original signature and are almost undetectable by the human eye. Manual signature forgery detection is thus found to be error-prone with false rejections and false acceptance. So it is high time to develop methods for signature forgery detection which is reliable and dependable. In our approach, we have explored the Spatial Pyramid image representation-based technique with global and local DCT coefficients. The standard deviation and the count of non-zero DCT coefficients corresponding to each AC frequency component of the image at 4 levels of the spatial pyramid are combined to form the feature vector. Many of the methods to reveal the signature forgery exist in the literature. The following section gives a summary of some of them. However, designing an accurate algorithm for offline signature verification is still a challenging task and hence, motivates researchers across the globe to come up with better approaches. The following section presents a brief review of related works in the state-of-the-art literature.

5.2 Literature Review

In this section, we have presented a brief overview of the various methods that exist in the state-of-the-art literature for offline signature verification. The Blockwise Binary Pattern (BBP) [13] takes 3×3 neighborhood of every point of the signature image and uses histogram representation for representing the local BBP features. Hafemann et al. [5] present formulations learning features for offline signature verification. Learning features are used to train writer-dependent classifiers using deep convolutional neural networks. Shekar et al. [12] proposed a morphological pattern spectrum, and Bhattacharya et al. [2] have proposed a pixel matching technique (PMT). Yasmine et al. (Guerbai et al. 2015) [4] propose a design of handwritten signature verification by using a one-class support vector machine, i.e. OC-SVM. This method creates a genuine signature model. We know that many signature-based person identification systems demand only genuine samples for verification as in bank account creation. In such cases, only genuine signatures are stored in the database. One class SVM will effectively classify genuine signatures from forge based on how effectively the discriminating features are extracted and provided to the classifier. The bag-of-visual words (BoVW) [9] uses forensic document examiners' (FDEs) cognitive process for feature extraction by extracting KAZE features. Jahandad et al.

(Jahandad 2019) [10] proposed Deep Convolutional Neural Network-based offline signature verification. This method uses the architectures GoogLeNet Inception-v1 and Inception-v3. GoogLeNet is a deep CNN architecture that won the completion ImageNet Large Scale Visual Recognition Competition (ILSVRC) in 2014. Compared to VGGNET it has a lower error rate and compared to AlexNet it has fewer filters. Jagtap et al. [6] proposed offline handwritten signature verification using Convolutional Neural Network. The authors proposed a deep convolutional neural network architecture with Support Vector Machine(SVM) for classification. Shekar et al. [14] too use CNN and SVM in their methodology for offline signature verification. CNN is used for feature extraction and SVM for classification. In the next section, we describe the proposed approach. The document is outlined as follows. Section 5.3 gives the proposed methodology, Sect. 5.4 details the Experimentation and discussion followed by the Conclusion in Sect. 5.5.

5.3 Proposed Approach

The proposed methodology uses the spatial pyramid image [3] technique on the two-dimensional signature images. The global and local DCT features are extracted from the signature image and the standard deviation and count of non-zero DCT coefficients corresponding to each AC frequency are computed forming the feature vector. We have made use of the fact that authentic signatures appear accurate and fluent the lack of which in the signature implies forgery. The standard deviation of DCT coefficients is a good measure of representing the spread of intensities in an image. The spatial pyramid is an extension of the bag of features that consists of a collection of orderless features. Figure 5.1 illustrates the spatial pyramid applied on a signature image.

In the proposed approach, a given image say I is partitioned into sub-regions recursively. The extraction of local features in the form of dominant DCT coefficients from I and sub-blocks $S = \{s_1, s_2, \ldots, s_k\}$ of I is performed recursively for L levels, where $L = \{l_0, l_1, \ldots, l_x\}$ and x=4 in our experimentation. In the level, l_0, the entire image is taken for the feature extraction process in order to capture global DCT features. The image I is applied with DCT to obtain DCT matrix D. The top-left nXn sub matrix is obtained from D containing $C = nXn$ number of coefficients. Out of C coefficients, one is DC and the rest are all AC coefficients. We have discarded the DC coefficient and taken only B=$C - 1$ AC coefficients as before. The level l_1 feature vector can be represented as follows.

$$F_0 = \begin{bmatrix} f_1 \\ f_2 \\ \vdots \\ f_B \end{bmatrix}$$

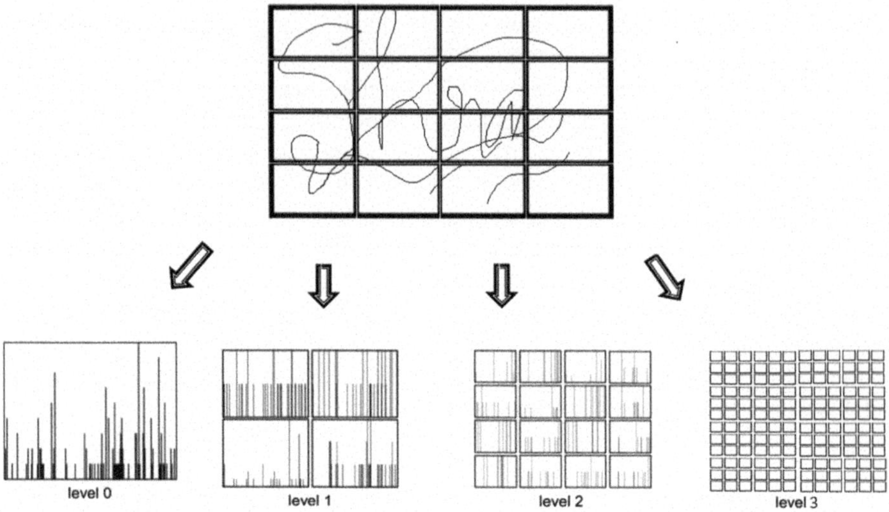

Fig. 5.1 Spatial pyramid with 4 levels, Level 0: Signature image, Level 1: Image is partitioned into 4 blocks, Level 2: Image is partitioned into 16 sub-blocks, and Level 3: Image is partitioned into 8 × 8 non-overlapping sub-blocks; Image courtesy [7]

In level 1 of Spatial Pyramid, the image I is partitioned into $I = \{p_1, p_2, \ldots, p_j\}$, where $j = 4$ in our experiments. For each p_i, where $1 \le i \le 4$, DCT is applied resulting in $D = \{d_1, d_2, \ldots, d_i\}$ corresponding DCT matrices. The top-left nXn sub-matrices are extracted for each d_i. Out of $C = nXn$ DCT coefficients in each di, we have discarded the DC component and retained only $B{=}C - 1$ AC coefficients. So in the level l_1 of spatial pyramid, we obtain $B * 4$ coefficients in the form of 4 vectors $V{=}\{v_1, v_2, \ldots, v_j\}$, where $j = 4$ for level l_1, with each vector having B coefficients. The level l_1 feature vector can be represented as a matrix as follows:

$$F_1 = \begin{bmatrix} f_{11} & f_{12} & \cdots & f_{1j} \\ f_{21} & f_{22} & \cdots & f_{2j} \\ \vdots & & & \vdots \\ f_{B1} & f_{B2} & \cdots & f_{Bj} \end{bmatrix}$$

where $j{=}4$, $n{=}8$, and $B{=}63$ in our experimentation. In level 2 of Spatial Pyramid, the image I is partitioned into $I = \{p_1, p_2, \ldots, p_j\}$, where $j = 16$ in our experiments. For each p_i, where $1 \le i \le 16$, the above procedure is repeated to obtain the feature vectors. So in the level l_2 of spatial pyramid, we obtain $B * 16$ coefficients in the form of 16 vectors $V{=}\{v_1, v_2, \ldots, v_j\}$, where $j = 16$ for level l_2, with each vector having B coefficients. The level l_2 feature vector can be represented as a matrix as follows:

$$F_2 = \begin{bmatrix} f_{11} & f_{12} & \cdots & f_{1j} \\ f_{21} & f_{22} & \cdots & f_{2j} \\ \vdots & & & \vdots \\ f_{B1} & f_{B2} & \cdots & f_{Bj} \end{bmatrix}$$

where j= 16, n=8, and B=63 in our experimentation. In the topmost level l_3, the image I is partitioned into nXn non-overlapping blocks, say, $I = \{p_1, p_2, \ldots, p_q\}$, where q = the number of nXn blocks that depends on the size of the image, and n=8 in our experimentation. For each p_i, where $1 \leq i \leq q$, the above procedure is repeated to obtain the feature vectors. So in the level l_3 of spatial pyramid, we obtain $B * q$ coefficients in the form of q number of vectors V=$\{v_1, v_2, \ldots, v_q\}$ for level l_3, with each vector having B coefficients. The level l_3 feature vector can be represented as a matrix as follows.

$$F_3 = \begin{bmatrix} f_{11} & f_{12} & \cdots & f_{1q} \\ f_{21} & f_{22} & \cdots & f_{2q} \\ \vdots & & & \vdots \\ f_{B1} & f_{B2} & \cdots & f_{Bq} \end{bmatrix}$$

where q is the number of nXn blocks in the image, n=8, and B=63 in our experimentation. The resultant feature matrix F is computed as follows by concatenating the feature vectors in each level. $F = \{F_0, F_1, F_2, F_3\}$ The resultant feature matrix is of dimension $B * E$, where E is the total number of columns in F_i with $0 \leq i \leq 3$. The row-wise standard deviation is obtained from F. As a result we obtain a feature vector $T_1 = \{t_1, t_2, \ldots, t_B\}$. Further, for each row of matrix F, the number of non-zero coefficients is computed forming another vector, $T_2 = \{t_1, t_2, \ldots, t_B$. The above procedure is repeated for the cropped version of the signature image. The cropped version of the image is obtained by discarding the first four rows and columns of the image. After performing the same procedure, we will get two more vectors $T_3$and T_4 similar to $T_1$and T_2, with the same dimension corresponding to the cropped version of the image. These T_is, where $1 \leq i \leq 4$, are concatenated forming a single vector with size $B * 4$. $T = \{T_1, T_2, T_3, T_4\}$ This is the final feature vector of the image I. The steps involved in the feature extraction process are given below. The above process is repeated for all the images in the training set and test set, and features are extracted and fed to the Support Vector Machine classifier. For a given training set, it creates a linear decision boundary. It finds a decision plane that maximizes the margin. For n training samples in the set $\{(x_i; y_i)\}_{i=1}^{n}$ where each $x_i \in R^L$ is taken from the domain X and each of the label y_i is an integer from the set $Y = \{0, 1\}$. For a given data, label pair (x_i, y_i) , $i = 1, \ldots, l$ where $x_i \in R^n$ and $y \in \{0, 1\}^l$, the SVM has to find the solution for the optimization problem [12]:

$$min_{(W,b,\xi)} \frac{1}{2} W^T W + C \sum_{i=1}^{l} \xi_i \tag{5.1}$$

1 Steps involved in feature extraction and forgery detection

- Step 1: Form the Spatial pyramid with 4 levels.
- Step 2: if it is level-0 then
  obtain the global DCT features as follows
  Apply DCT on entire image
  Crop top-left nXn sub matrix, say M.
  Discard DC coefficient at M(0,0).
  Form AC DCT coefficient feature vector F_1
  end
- Step 3: For the image at levels l= 1,2 do
  Divide the image into $N = 4^l$ non overlapping sub-blocks
- Step 4 : For each block b_i do
  Apply DCT to get block DCT matrix
  Crop top-left nXn sub matrix, say M.
  Form AC DCT coefficient feature vector F_l
  Combine column wise the F_is of all the blocks to form
  DCT coefficient Matrix
  end
- Step 5 : Divide the image in to N non overlapping n x n blocks
- Step 6 :For each n x n block do
  Apply DCT to get Block DCT
  Extract the last $(n * n - 1)$ AC coefficients from Block DCT as a
  column vector V
  Place Vs of all the blocks in a $(n * n - 1)XN$ matrix
  Combine all the feature vectors at various levels.
  Compute row-wise standard deviation $T1$ and number of non-zero
  coefficients T_2 in each row of the $(n * n - 1)XN$ matrix to get
  feature vectors of size $(n * n - 1)X1$ each.
  end
- Step 7: Discard the first four rows and columns of the image
- Step 8: Repeat the above procedure to get feature matrix T_3 and T_4
  of size $(n * n - 1)X1$ each
- Step 9: Compute the final feature vector as $T = T_1, T_2, T_3, T_4$ of size $4 * (n * n - 1)$

with the condition that

$$y_i \left(W^T \phi(x_i) + b\right) \geq 1 - \xi_i \tag{5.2}$$

and

$$\xi_i \geq 0. \tag{5.3}$$

ϕ is the function used to map x_i,the training features to a higher-dimensional space. The SVM then finds a linear hyperplane with the maximal margin in the higher-dimensional space to separate the data into classes. C is the penalty parameter. To transform lower-dimensional data into higher-dimensional one, a kernel trick is made use of. N SVM classifiers are needed to classify signatures of N writers since SVM is a binary classifier. So, the number of SVM classifiers used by the proposed work is equal to the number of writers [12].

5.4 Experimental Results and Discussions

Offline signature datasets like CEDAR [1] and MUKOS (Mangalore University Kannada Off-line Signature) are used for experimentation. The CEDAR dataset contains the signatures of 55 individuals with 24 genuine and 24 skilled forged images for each signer. The MUKOS dataset contains the signatures of 30 signers with 30 genuine and 15 skilled forged Signature images each. Each dataset is divided into training and test sets in four different ways as follows.

- Set-1: The training set is made of the first 10 genuine and 10 forged images and the rest belong to the test sample set.
- Set-2: The training set is made of the first 15 genuine and 15 forged images and the rest belong to the test sample set.
- Set-3: The training set is made of 10 genuine and 10 randomly chosen forged images and the rest belong to the test sample set.
- Set-4: The training set is made of 15 genuine and 15 randomly chosen forged images and the rest belong to the test sample set.

CEDAR dataset: Experimental Results
Experiments on the four sets of datasets as explained in the previous section were carried out. Table 5.1 presents the results.

Table 5.2 shows the results of the comparison with some other methods in the literature.

MUKOS dataset: Experimental Results
Experiments on the four sets of datasets as explained in the previous section were carried out. Table 5.3 presents the results. Table 5.4 shows the results of the comparison with some other methods in the literature.

Table 5.1 Results for CEDAR Dataset

Dataset no.	Accuracy	FRR	FAR
Set-1	92.12	7.8	7.9
Set-2	94.65	5.28	4.91
Set-3	92.82	6.8	7.4
Set-4	94.88	5.37	5.21

Table 5.2 Results for CEDAR dataset—a comparison

Method	Classifier used	Accuracy	FAR	FRR
PS (Morphological Spectrum) [12]	EMD	91.06	10.63	9.4
InterPoint Envelop [8]	SVM	92.73	6.36	8.18
Proposed approach	SVM	94.88	5.37	5.21

Table 5.3 Results for MUKOS dataset

Dataset	Accuracy	FRR	FAR
Set-1	99.2	0.0	1.6
Set-2	98.26	1.0	2.4
Set-3	96.2	2.8	4.8
Set-4	96.4	2.0	5.4

Table 5.4 Results for MUKOS dataset—a comparative analysis

Methodology	Classifier	Accuracy	FAR	FRR
Eigen signature [11]	Euclidean distance	93.00	11.07	6.40
Morphological spectrum [12]	Earth mover distance	97.39	5.6	8.2
Proposed approach	SVM	99.2	0.0	1.6

5.5 Conclusion

In this paper, a Spatial Pyramid image representation-based technique using DCT is used for offline signature verification. The extracted features are found to provide valid clues regarding the genuineness of the signatures. This claim is supported by the results of the experimentation. The Support Vector Machine (SVM) is used for the classification. This method works better than many of the state-of-the-art methods for signature verification.

5.6 Competing Interests and Data Availability

The authors declare no competing interests.
Data used in the experimentation CEDAR [1] are publicly available at the link https://cedar.buffalo.edu/NIJ/data/. MUKOS is a regional language dataset, namely, the Mangalore University Kannada Off-line Signature (MUKOS) dataset available on request.

References

1. object dataset. https://cedar.buffalo.edu/NIJ/data/
2. Bhattacharya, I., Ghosh, P., Biswas, S.: Offline signature verification using pixel matching technique. Procedia Technol. **10**, 970–977 (2013)
3. Grauman, K.: Pyramid match kernels: discriminative classification with sets of image features. In: Grauman, K., Darrell, T. (eds.), Computer Science and Artificial Intelligence Laboratory Technical Report, Technical Report (2005)
4. Guerbai, Y., Chibani, Y., Hadjadji, B.: The effective use of the one-class SVM classifier for handwritten signature verification based on writer-independent parameters. Pattern Recognit. **48**(1), 103–113 (2015)
5. Hafemann, L.G., Sabourin, R., Oliveira, L.S.: Learning features for offline handwritten signature verification using deep convolutional neural networks. Pattern Recognit. **70**, 163–176 (2017)
6. Jagtap, A.B., Hegadi, R.S., Santosh, K.: Feature learning for offline handwritten signature verification using convolutional neural network. Int. J. Technol. Hum. Interact. (IJTHI) **15**(4), 54–62 (2019)
7. Lazebnik, S., Schmid, C., Ponce, J.: Beyond bags of features: spatial pyramid matching for recognizing natural scene categories. In: 2006 IEEE Computer Society Conference on Computer Vision and Pattern Recognition (CVPR'06), vol. 2, pp. 2169–2178. IEEE (2006)
8. Manoj, K.M., NB, P.: Inter-point envelope based distance moments for offline signature verification. In: 2014 International Conference on Signal Processing and Communications (SPCOM), pp. 1–6. IEEE (2014)
9. Okawa, M.: Offline signature verification based on bag-of-visual words model using KAZE features and weighting schemes. In: Proceedings of the IEEE Conference on Computer Vision and Pattern Recognition Workshops, pp. 184–190 (2016)
10. Sam, S.M., Kamardin, K., Sjarif, N.N.A., Mohamed, N., et al.: Offline signature verification using deep learning convolutional neural network (CNN) architectures GoogLeNet inception-v1 and inception-v3. Procedia Comput. Sci. **161**, 475–483 (2019)
11. Shekar, B.H., Bharathi, R.K.: Eigen-signature: a robust and an efficient offline signature verification algorithm. In: 2011 International Conference on Recent Trends in Information Technology (ICRTIT), pp. 134 –138 (2011)
12. Shekar, B.H., Bharathi, R.K., Pilar, B.: Local morphological pattern spectrum based approach for off-line signature verification. In: International Conference on Pattern Recognition and Machine Intelligence, pp. 335–342. Springer (2013)
13. Shekar, B.H., Pilar, B., Sunil, K.D.S.: Blockwise binary pattern: a robust and an efficient approach for offline signature verification. Int. Arch. Photogramm., Remote. Sens. Spat. Inf. Sci. **42**, 227 (2017)
14. Shekar, B., Abraham, W., Pilar, B.: Offline signature verification using CNN and SVM classifier. In: 2022 IEEE 7th International Conference on Recent Advances and Innovations in Engineering (ICRAIE), vol. 7, pp. 304–307 (2022). https://doi.org/10.1109/ICRAIE56454.2022.10054336

Chapter 6
Detection of AI Manipulated Videos Using Modern Deep Learning Algorithms

Satendra Gupta, Tapas Saini, and Anoop Kumar

Abstract Video manipulation is the method to edit the video for various purposes. The deepfake is the manipulated media created using modern deep learning techniques which are the new AI-based methods for video manipulation, Deepfakes are generated for various good and bad purposes. The bad use of Deepfakes has been rising prominently. Researchers worldwide have been developing methods for Deepfake creation and Deepfake detection. The rate of Deepfake Creation has been rising owing to the opportunity for adversaries for financial fraud. This work is toward the detection of Deepfakes using various AI-based methods. The AI-based methods are majorly based on Deep learning models which are trained for Deepfake Detection through the form of AI Inferencing. Over the last decade, many AI models have been designed for various real-world computer vision problems starting from AlexNet followed by new popular models for instance VGGNet, ResNet, InceptionNet, and many more. EfficientNets and Transformers are the recent state of the art in Deep learning models. For the purpose of Deepfake Detection, we have used these newer architectures. We have trained the model for Deepfake Detection using the publicly available Dataset of DFDC Faceforensics++ and CelebDFV2. We have achieved comparable results of 84% accuracy on the popular Deepfake Detection challenge test set.

S. Gupta (✉) · T. Saini · A. Kumar
Centre for Development of Advanced Computing, Hyderabad, Telangana, India
e-mail: satendrag@cdac.in

T. Saini
e-mail: stapas@cdac.in

A. Kumar
e-mail: anoopkr@cdac.in

S. K. Pal et al. (eds.), *Intelligent Informatics*, Smart Innovation, Systems and Technologies 389, https://doi.org/10.1007/978-981-97-2147-4_6

6.1 Introduction

The Deepfakes are the new source of deceiving people away from the truth. Deepfakes term is made by the combination of Deep learning and Fake. Deep learning is a subfield of Machine learning, and which is further a subfield of Artificial Intelligence. Deep learning has been successful in many fields, for instance Computer vision, Natural Language Processing, Medical diagnostics, etc. As there is a dark side to any field, Deep learning has also a dark side. It is getting used to spread misinformation in the form of Deepfakes. Deepfakes can be termed as subfield of generative AI. Deepfakes are majorly getting used for spreading misinformation and executing financial fraud. Alternatively, Deepfakes are also used for positive purposes in the form of re-creating the videos of deceased celebrities for example dalli, creating movie scenes using deepfakes of actor, and creating avatars of people enhancing privacy specially in social media platforms. The positive use of Deepfake is generally observed as per ethical standards. But the negative use of Deepfakes is executed without the user's consent and leads to a high possibility of financial fraud or defaming people. The nonconsensual deepfakes of celebrities are getting viral in social media. It is increasingly becoming difficult to judge between Deepfake and real videos as the deepfake creation technology has highly improved in terms of quality. There is a high need to develop Anti-Deepfake technology which can assist humans to judge between real and deepfake content.

To perform the detection of deepfakes, algorithms can be implemented from simple heuristics to state-of-the-art Deep learning algorithms. Though heuristics algorithms need less data to develop, they may require high domain expertise and may not generalize well to novel creation techniques. On the other hand, having abundant data in the form of availability of big Deep fake datasets enables the Deep learning technology natural candidate for Deepfake detection applications. Hence, to counter the spread of Deepfakes through detection, effective Deep learning AI models needs to be developed and trained. Over the last decade, many Deep learning models have been created for various real-world computer vision problems starting from AlexNet followed by popular models for instance VGGNet, ResNet, InceptionNet, and many more. EfficientNets and Transformers are newer architectures among Deep learning models. For the purpose of Deepfake Detection, we have used these new architectures, as these learn highly complex semantic features for solving classification problems.

6.2 Related Works

The Deepfakes detection aims to solve the problem of Deepfakes creation. Any questionable media in the form of video needs to be analyzed to report if it is pristine or manipulated. It is therefore needed to discuss the Deepfake media generation methodologies before deepfake detection methods.

6.2.1 Deepfake Media Generation

Modern Deepfakes are generated using Deep learning methods. One of the new advancements is being incorporated in the deep learning architectures in the form of Variational AutoEncoders (VAEs) [1] and Generative Adversarial Networks (GANs) [2]. The majority of Deepfake creation techniques are based on these GANs and VAEs. GAN architectures have two distinct networks, generator and discriminator. The discriminator is the deep network that is expected to classify whether the query media is fake or real, and the generator is the network that manipulates the face features in videos in a realistic manner in order to fool its counterpart. Deepfake media which are being generated by using GANs architecture are very realistic and accurate, and over time, numerous architectures have been designed in the form of StarGAN [3] and DiscoGAN [4]. StyleGAN-V2 [5] is one of the best-suited networks to generate the high-quality deepfake videos. On the other hand, VAE-based architectures are combinations of two encoder-decoder pairs, both network pairs are trained to deconstruct and reconstruct their respective input faces. After training of pairs, the decoding part is switched, and this enables the reconstruction of the target person's face. The popular implementations of this technique were DeepFaceLab [6], DFaker, and DeepFaketf.

6.2.2 Deepfake Media Detection

To address the problem of deepfake video detection, many deep learning architectures are proposed over the years. Good results are obtained using EfficientNetB7 ensemble techniques in [7]. Other methods include spatiotemporal artifacts combined CNN with Gated Recurrent Unit (GRU) as covered in [8]. Some promising works to detect spatiotemporal inconsistencies were made using 3DCNN networks as per [9]. Amerini et al. [10] presented an architecture that majorly exploits optical flow to detect artifacts in videos. More recently, after the success of Transformer in natural language processing, a combined architecture Convolution Neural Network (CNN) with Transformers is proposed, which seems more generalized architecture in [11] and gives good result on cross dataset tests.

6.3 Proposed Approach

The scope of the activity included only detection of single person Deepfake detection. Accordingly, the proposed methods first extract the face of person from the video followed by analysis of the extracted faces from the source video to determine whenever they have been manipulated. As a first step toward Deepfake detection, faces are preextracted using a state-of-the-art face detector, MTCNN, and blazeface

[12]. After the faces are extracted, they are provided as input into the Deep Learning models which process the faces and output the result as Fake or Real with confidence scores. Many Deep learning model approaches were attempted using different architectures which are as follows:

6.3.1 Deepfake Detection Approach

Deepfake Detection approaches may include simple heuristics-based methods to machine learning approaches. Due to the increasing availability of datasets, the Deep Learning approaches proved to outperform heuristics and traditional machine learning techniques. The success of AlexNet for Computer Vision domain in 2012 and Transformers in Natural Language Processing domain motivated the use of the Deep learning architectures for Deepfake detection problem.

Deepfake Detection is majorly a Computer Vision problem especially for the detection of Image and Video Deepfakes. Convolutional Neural Networks (CNNs) are widely used for solving Computer Vision problems. Vision Transformers (ViTs) have given some competition to CNNs but the recent state-of-the-art architectures of CNN for example EfficientNetV2 proved that CNNs are still useful in comparison to transformers along with magnitudes smaller in size than Transformers. For the purpose of Deepfake Detection training, both Transformers and CNN-inspired architectures are chosen as per detailed below.

EfficientNet

EffcientNets [13] are the family of Deeplearning models created from low resource systems to high resource systems. EfficientNet was designed with the objective of balanced scaling of deep learning architectures. The authors of EfficientNets systematically experimented model scaling and observed that carefully balancing network depth, width, and resolution can produce models which have better performance for model training. They proposed a new scaling method that uniformly scales all dimensions of width/depth/resolution using an effective compound coefficient. Further, they demonstrated the effectiveness of compound scaling on MobileNets and ResNets. Using the technique of neural architecture search the authors designed a new baseline network and scaled it up to obtain a family of models. In particular, they produced EfficientNet models in range EfficientNetB0 to EfficientNetB7. The EfficientNetB7 achieved state-of-the-art 84.3% top-1 accuracy on ImageNet, along with 8.4 $\times$ smaller and 6.1$\times$ faster on inference than Gpipe [14] which was the best existing ConvNet during the time of paper publication. It was also stated that EfficientNets are transferred well and achieved high accuracies on computer vision datasets in the form of CIFAR-100 (accuracy: 91.7%), Flowers (98.8%), and other datasets with significantly fewer parameters. The basic architecture of EfficientNetB0 is shown in Fig. 6.1.

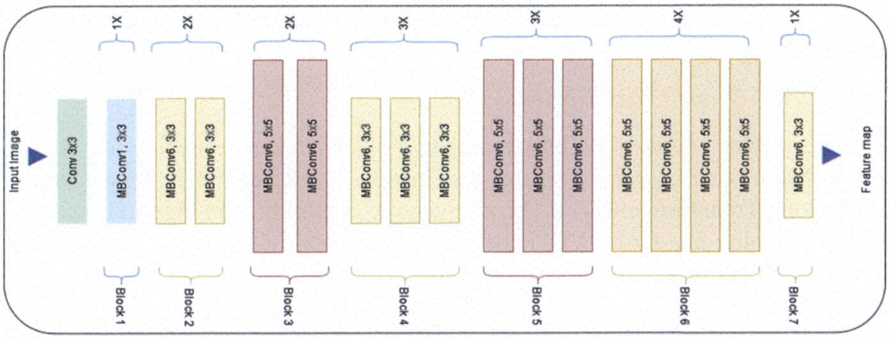

Fig. 6.1 The basic diagram of EfficientNetB0

Due to the success of EfficientNets, we used EfficientNets for training of Deepfake Detection models. Due to the balanced nature of complexity and performance, the EfficientNetB4 was chosen as a Deepfake detection model training.

EfficientNetAutoAttB4

Deep learning models are promising to solve variety of real-world problems provided enough data. A high majority of Deep learning models provide satisfactory results in terms of accuracies and other evaluation metrics. But these models lack explanation in their results. Owing to the success of transformers and residual attention networks in providing details in which part of the input is more relevant for arriving at the result for the task. For the purpose of video deepfake detection, it may be useful to get the information on which part of input gave the deep network for arriving at the result of deepfake detection. Based on the inspiration from [15], we utilized the attention mechanism and incorporated the EfficientNet in the form of Self-attention. Specifically, the attention is inserted after third MBConv block of EfficientNetB4. This modified EfficientNetB4 with self-attention mechanism embedded is named EfficientNetAutoAttB4. This network helped to provide the information on which portions of face image were significant for arriving the decision on whether the face image in video is Fake or Real.

Convolutional Neural Networks combined with Transformers

In addition to EfficientNet, we used architecture which is the combination of both convolutional neural network and transformer. The input of architecture is the extracted face in batch and the output will be the probability of face manipulation in frames. This is a hybrid architecture trained as a binary classifier in a supervised manner to classify the respective videos based on introduced artifacts in the face area.

The proposed models are trained on a face basis which are extracted from videos. To draw the conclusion at the time of inference, all faces are extracted and aggregated the processed output both across multiple faces and in time. Specifically, it is proposed based on the Convolutional Cross ViT-based architectures explained in the following paragraphs.

The Efficient ViT and Cross Efficient ViT

The EfficientNet ViT is a combination of two blocks, where convolutional blocks work as feature extractor and Transformer Encoder block function is almost similar to the Vision Transformer (ViT). The architecture is presented in Fig. 6.2.

The Convolutional Cross ViT is designed using both the Efficient ViT and the multi-scale Transformer architecture by [16].

Efficient ViT

As EfficientNet gives promising results, EfficientNetB0 the smallest in the EfficientNet network family is used as a convolutional feature extractor for cropping faces from the videos. The EfficientNet creates the chunk of 7 × 7 for the local receptive field of the input face. These chunks are projected linearly which is processed by the Vision Transformer. There is a CLS token which is utilized for the binary classification score. The EfficientNetB0 is initialized with the pretrained weights and fine-tuned which works as a feature extractor. The extracted features by the

Fig. 6.2 Convolutional ViT

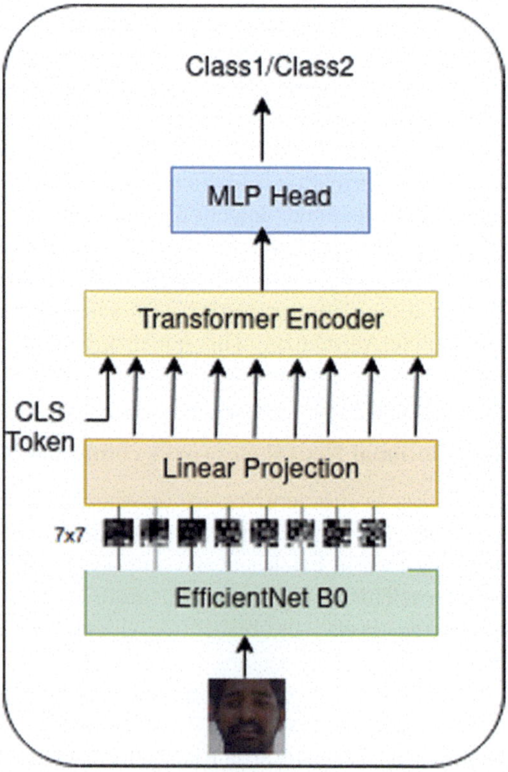

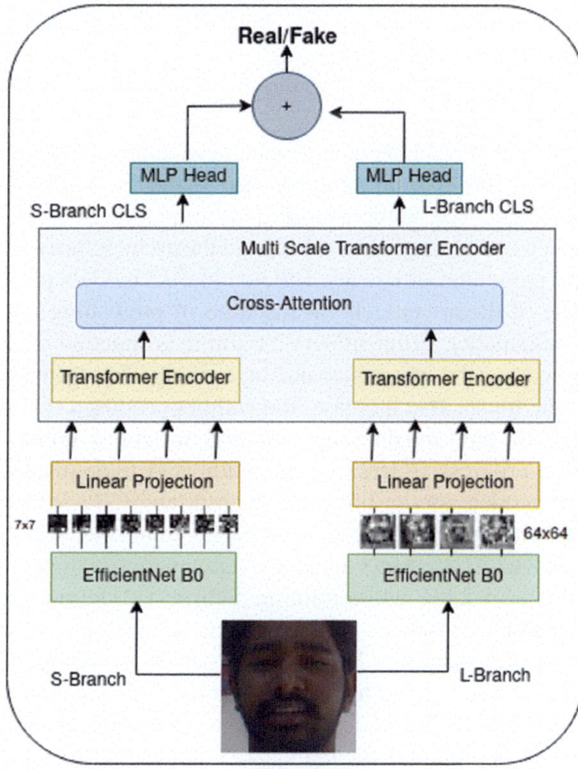

Fig. 6.3 Cross efficient ViT

EfficientNetB0 network provided the feature chunks for the training of the Vision Transformer, as these features contain important localized embedded features from the input face image. Refer to Fig. 6.3 for graphical details.

Cross Efficient ViT

As in EfficientNet ViT architecture only local features are used for training the network may have low efficiency to distinguish between real and deepfake images in case where artifacts introduced globally. The introduced artifacts by deepfake generation methods could be both locally and globally. So that to tackle both at same time cross architecture suited the best as it will learn from both local and global input face feature. The EfficientNet creates the chunk of 7×7 for the local receptive field and 64×64 for the global receptive field of input face. These chunks will be projected linearly which is processed by the Vision Transformer. There is a CLS token which is utilized for the binary classification score. The Convolutional Cross ViT architecture has two distinctive branches: the S-branch, which deals with local receptive field patches, and the L-branch, which works on global receptive field patches to contain a wider receptive field. The output of the Transformer Encoder from both S-branch and L-branch are connected with a cross attention, which allows both branches to interact directly. In last for these two branches, corresponding separate logits are produced by CLS token. The final probabilities will be produced by taking the sigmoid of the sum of these logits. Refer to Fig. 6.3 for graphical details.

EfficientNetV2

EfficientNetV2 [17] comprises a new family of newer convolutional networks that promise faster training speed and better parameter efficiency than classical EfficientNets. The authors used a combination of training-aware neural architecture search and scaling, hence designed the architecture to jointly optimize training speed and parameter efficiency. EfficientNetV2 models promise to train much faster than state-of-the-art models (at the time of publishing) at the same time are up to 6.8 times smaller. EfficientNetV2 training is stated to further speed up by progressively increasing the image size during training. To compensate for the accuracy drop due to the image size increase, the authors proposed to adaptively adjust regularization (e.g., dropout and data augmentation) to achieve both fast training and good accuracy. With progressive learning, the authors' EfficientNetV2 outperformed the previous best models on the ImageNet dataset and CIFAR/Cars/Flowers datasets. By doing pretraining on the ImageNet21k, on ImageNet ILSVRC2012, the EfficientNetV2 achieved 87.3% top-1 accuracy, outperforming the previous best ViT by an accuracy margin of 2.0% while training 5 times to 11 times faster on the same computing resources.

As EfficientNetV2 provided at par or sometimes higher accuracy over classical ViTs, it was decided to train EfficientNetV2-based architectures for Deepfake Detection. EfficientNetV2 is designed in 3 variants: Small (S), Medium (M), and Large (L). In particular, EfficientNetV2S was chosen for training the Deepfake Detection model training. Refer to diagram of EfficientNetV2S in Fig. 6.4.

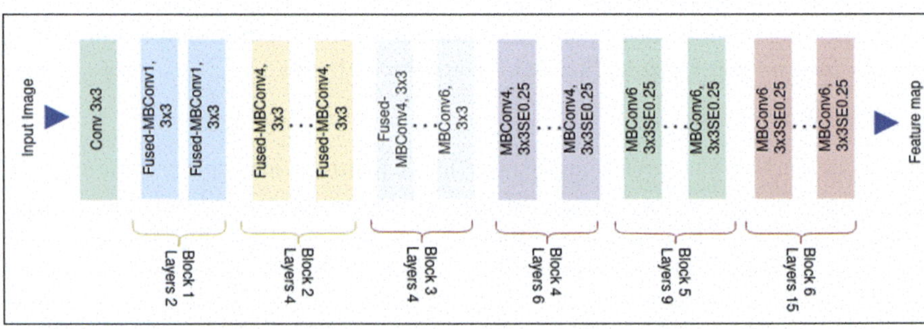

Fig. 6.4 The basic diagram of EfficientNetV2S

6.4 Datasets and Experimental Details

6.4.1 Datasets

For the purpose of Deepfake detection, many datasets are available in the public domain. The popular datasets are Face Forensics++, Deeper Forensics, Celeb-DF, DFDC, and more.

To address the problem of deepfakes detection in videos, many datasets have been created over the years. These datasets are categorized into three or more generations, the first generation includes datasets of DF-TIMIT, UADFC, and FaceForensics++ [18], the second generation datasets are Google Deepfake Detection Dataset, Celeb-DF [19], and the third generation datasets include DFDC dataset [20] and Deep-Forensics. The further the generations go, the larger these datasets are, and the more frames they contain. Every subsequent generation includes larger datasets, increased number of frames, and higher quality of deepfakes than previous generations.

FaceForensics++

The FaceForensics++ dataset [18] is composed of 1000 originals and fake videos generated through 4 different face manipulation techniques: FaceSwap, Deepfakes, Face2Face, and NeuralTextures. FaceSwap is a graphics processing-based approach to transfer the face region from a source video to a target video. NeuralTextures uses GAN to alter the facial expression or face in an Image or video.

Deep Fake Detection Challenge (DFDC) Dataset

The DFDC dataset [20] is one among the largest and publicly available face swap video dataset, with more than 100,000 total video clips created by collection from 3,426 paid volunteers, created using several Deepfake, GAN-based, and non-deep learning methods. Specifically, DFDC constraints 104,500 unique fake videos out of 128,154 videos. There was a total of 960 people who agreed to the usage of their images and videos for Deepfake creation.

Celeb-DFv2 Dataset

Celeb-DF (v2) [19] is one of the high quality dataset. In Celeb-DF (v2), a total number of original videos were collected from YouTube having subjects of different ages, environments, and genders, and by using these 590 original videos 5639 corresponding deepfake videos were created.

Team Created Dataset

Our team has collected 100 videos from different ages, ethnicities, and gender and created 1500+ corresponding deepfake videos using many open-source deep fake creation tools like DeepfaceLab, SwimSwap, etc. Refer to Fig. 6.5 for sample datasets images.

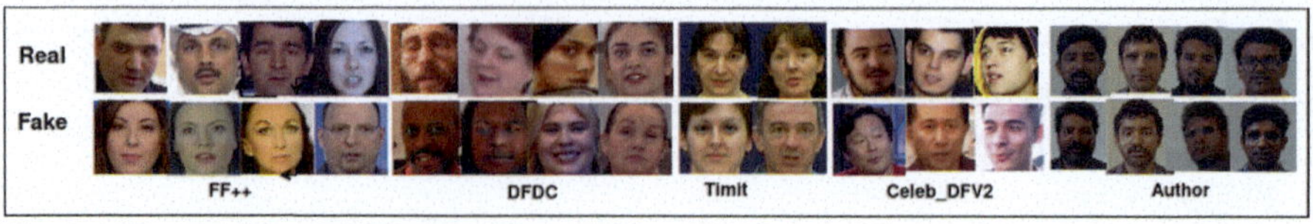

Fig. 6.5 Deepfake detection datasets

6.4.2 Experimental Details

Training of EfficientNet and EfficientNetV2

The models of EfficientNetB4, EfficientNetAutoAttB4, and EfficientNetV2S are trained on the DFDC Training set. DFDC training set is huge as it contains approximately 1 lakh Face swap Videos. The Transfer learning technique was used for model training. The weights of the models were initialized using ImageNet pretrained weights. The total iterations for training were kept to a maximum of 30,000. The learning rate was initialized to 1e-5. Typical batch size was taken as 32. The input face was kept to the size of 224 × 224. For the purpose of optimization, Adam Optimizer was used. Regarding hardware resources, training was done on NVIDIA GeForce 1080 and NVIDIA Quadro P5000.

We have used the DFDC test set which contains 5000 videos. The model is trained on the entire training set of DFDC. During training, we extracted the faces from the videos using an MTCNN, and we performed data augmentations. We extracted and resized the faces in squared aspect ratio form and without padding. The input face images obtained are only used during the training and discarded the remaining part of the frames. We applied common transformations such as the introduction of blurness, Gaussian noise, transpositions, rotations, and various isotropic resizes during training.

Training of Convolutional Cross ViT

We have used the DFDC training set and FaceForensics++ training set to train the cross ViT network from which we have extracted 212,360 face images, we used a validation dataset created from the DFDC validation set videos. The training set was constructed which maintained a balance between the real class and the fake class. The real class consists of 111,236 images and fakes have 101,124 images.

We extract the features using pretrained EfficientNetB0 for feature extraction. We fine-tuned all layers in EfficientNetB0 feature extractor network instead of using the pretrained weights which gives better results. We have used the standard binary cross-entropy loss function for the updating weights during training. We also used a Stochastic Gradient Decent optimizer to optimize our network end-to-end with a learning rate of 0.001.

Experimental Results

Multiple models were trained for Deepfake Detection problem. The DFDC Training dataset was used for Training for multiple Deep learning Architectures.

Figures 6.6 and 6.7 show the training details in graphical form on DFDC training set for EfficientNetAutoAttB4 architecture.

After training, the DFDC test set was used to check the accuracy and F1 score as per below Table 6.1.

The results in Table 6.1 indicate that the detection performance of the models is satisfactory but still needs further improvements in terms of accuracy and F1 score results. Any model needs to have more than 95% accuracy in order to be used as effective utility by users.

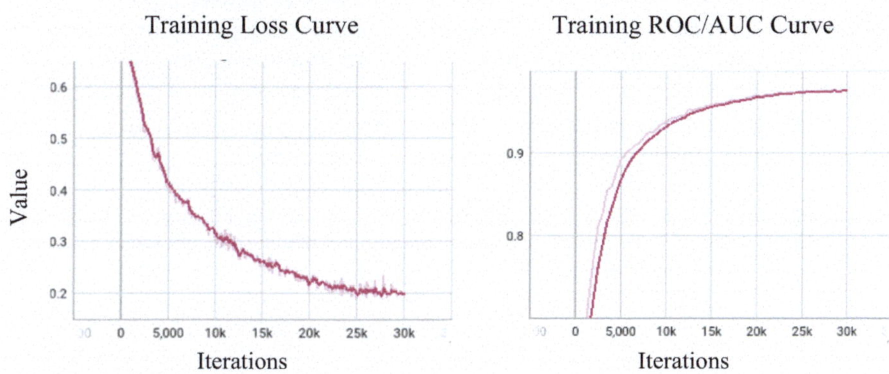

Fig. 6.6 Training accuracy and loss details of EfficientNetAutoAttB4 on DFDC training dataset

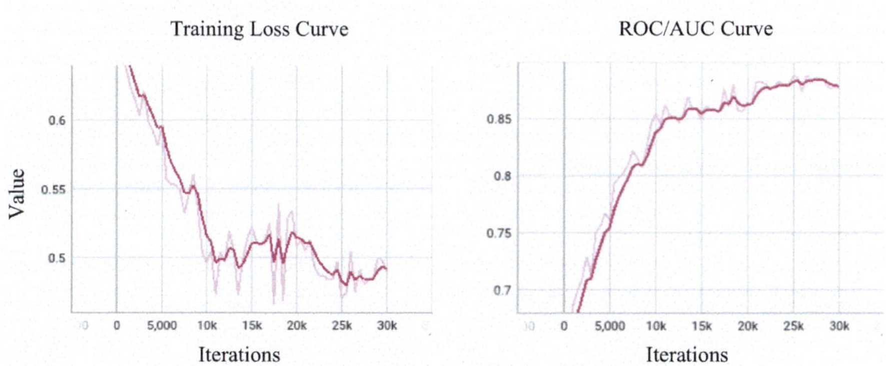

Fig. 6.7 Validation accuracy and loss details of EfficientNetAutoAttB4 on DFDC dataset

Table 6.1 Deep learning model metrics on DFDC test set

S. No	Model architecture	Accuracy	F1 score
1	Efficient ViT	83.43	83.76
2	Cross Efficient ViT	85.06	87.00
3	EfficientNetAutoAttB4	80.98	84.02
4	EfficientNetB4	79.46	82.96
5	EfficientNetV2S	84.16	86.30

6.5 Conclusion and Future Work

Deepfake detection is a challenging task due to the rising quality of Deepfakes. We have attempted to perform deepfake detection through various deep learning models and achieved acceptable accuracies. There still remains lot of scope for improvements in generalization capability of detection models. To improve accuracy further, it is planned to train models with additional datasets and also consider ensembling of multiple models. Further temporal level processing along with audio processing is also planned for improvements in Deepfake Detection accuracy.

Acknowledgements The authors thank Ministry of Electronics and Information Technology (MeitY), Government of India, for providing opportunity and funding through sanction number 4(15)/2021-ITEA to conduct this research.

References

1. Kingma, D.P., Welling, M.: Auto-encoding variational bayes. arXiv preprint arXiv:1312.6114 (2013)
2. Goodfellow, I.J., Pouget-Abadie, J., Mirza, M., Xu, B., Warde-Farley, D., Ozair, S., Courville, A., Bengio, Y.: Generative adversarial networks. In: Advances in Neural Information Processing Systems, vol. 27 (2014)
3. Choi, Y., et al.: Stargan: unified generative adversarial networks for multi-domain image-to-image translation. In: Proceed
4. Kim, T., Cha, M., Kim, H., Lee, J.K., Kim, J.: Learning to discover cross-domain relations with generative adversarial networks. In: International Conference on Machine Learning, pp. 1857–1865. PMLR (2017)
5. Karras, T., Laine, S., Aittala, M., Hellsten, J., Lehtinen, J., Aila, T.: Analyzing and improving the image quality of stylegan. In: Proceedings of the IEEE/CVF Conference on Computer Vision and Pattern Recognition, pp. 8110–8119 (2020)
6. Perov, I., Gao, D., Chervoniy, N., Liu, K., Marangonda, S., Umé, C., Dpfks, M., Facenheim, C.S., RP, L., Jiang, J., et al.: Deepfacelab: a simple, flexible and extensible face swapping framework. arXiv preprint arXiv:2005.05535 (2020)
7. Seferbekov, S.: DFDC 1st place solution (2020). https://github.com/selimsef/dfdc_deepfake_challenge
8. Montserrat, D.M., Hao, H., Yarlagadda, S.K., Baireddy, S., Shao, R., Horváth, J., Bartusiak, E., Yang, J., Guera, D., Zhu, F., et al.: Deepfakes detection with auto matic face weighting. In: Proceedings of the IEEE/CVF Conference on Computer Vision and Pattern Recognition Workshops, pp. 668–669 (2020)

9. de Lima, O., Franklin, S., Basu, S., Karwoski, B., George, A.: Deepfake detection using spatiotemporal convolutional networks. arXiv preprint arXiv:2006.14749 (2020)
10. Amerini, I., Galteri, L., Caldelli, R., Del Bimbo, A.: Deepfake video detection through optical flow based cnn. In: Proceedings of the IEEE/CVF International Conference on Computer Vision Workshops (2019)
11. Wodajo, D., Atnafu, S.: Deepfake video detection using convolutional vision trans former. arXiv preprint arXiv:2102.11126 (2021)
12. Bazarevsky, V., Kartynnik, Y., Vakunov, A., Raveendran, K., Grundmann, M.: Blazeface: sub-millisecond neural face detection on mobile gpus. CoRR, vol. abs/1907.05047 (2019). http://arxiv.org/abs/1907.05047
13. Tan, M., Le, Q.V.: Efficientnet: Rethinking model scaling for convolutional neural networks. In: International Conference on Machine Learning, (ICML) 2019, ser. Proceedings of Machine Learning Research, vol. 97. PMLR, pp. 6105–6114 (2019)
14. Huang, Y., Cheng, Y., Chen, D., Lee, H., Ngiam, J., Le, Q. V., Chen, Z.: Gpipe: efficient training of giant neural networks using pipeline parallelism. arXiv preprint arXiv:1808.07233 (2018)
15. Bonettini, N., Cannas, E.D., Mandelli, S., Bondi, L., Bestagini, P., Tubaro, S.: Video face manipulation detection through ensemble of CNNs. In: 2020 25th International Conference on Pattern Recognition (ICPR), pp. 5012–5019 (2021). https://doi.org/10.1109/ICPR48806.2021.9412711
16. Chen, C.F., Fan, Q., Panda, R.: Crossvit: cross-attention multi-scale vision trans former for image classification. arXiv preprint arXiv:2103.14899 (2021)
17. Tan, M., Le, Q.V.: EfficientNetV2: Smaller Models and Faster Training (2021)
18. R¨ossler, A., Cozzolino, D., Verdoliva, L., Riess, C., Thies, J., Nießner, M.: FaceForensics++: learning to detect manipulated facial images. In: International Conference on Computer Vision (ICCV) (2019)
19. Li, Y., Yang, X., Sun, P., Qi, H., Lyu, S.: Celeb-DF: a large-scale challenging dataset for deepfake forensics. In: Proceedings of the IEEE/CVF Conference on Computer Vision and Pattern Recognition, pp. 3207–3216 (2020)
20. Deepfake Detection Challenge (DFDC). https://deepfakedetectionchallenge.ai/ (2019)

Part II
Machine Learning and Deep Learning Applications

Chapter 7
Comprehensive Exploration of Deepfake Detection Using Deep Learning

Pratham Agrawal, Anchalaa Jha, and Avinash Bhute

Abstract The proliferation of deepfake technology poses a significant threat to the veracity of multimedia content, necessitating robust countermeasures for detection. This research explores the efficacy of three distinct deep learning architectures—Long Short-Term Memory (LSTM), Gated Recurrent Unit (GRU), and Convolutional Neural Network (CNN)—in discerning deepfake manipulations within video data. The study encompasses a diverse dataset encompassing real and manipulated videos, ensuring comprehensive evaluation. Each model is meticulously trained and rigorously tested, with performance assessed in terms of accuracy, precision, recall, and F1-score. The results underscore the CNN's exceptional prowess in spatial feature extraction, yielding superior accuracy compared to its temporal counterparts. A comprehensive discussion delves into the strengths and limitations of each model, providing valuable insights for future research in this critical domain. This research not only contributes to the arsenal of techniques for deepfake detection but also underscores the dynamic landscape of multimedia forensics, necessitating continuous innovation to safeguard the integrity of visual content in an era of increasingly sophisticated manipulations. The findings herein offer a foundational framework for further advancements in deepfake detection methodologies.

7.1 Introduction

The rapid expansion of multimedia technologies in today's digitally networked society has resulted in both unprecedented opportunities and unprecedented challenges. The proliferation of deepfake content is one of the most intriguing and concerning challenges. Deepfakes, which are hyper-realistic synthetic media made with artificial intelligence (AI) techniques, have gotten a lot of attention because of their ability to

P. Agrawal (✉) · A. Jha
SCTR's Pune Institute of Computer Technology, Pune, 411046, India
e-mail: pratham05072002@gmail.com

A. Bhute
Pimpri Chinchwad College of Engineering, Pimpri-Chinchwad 411044, India
e-mail: avinash.bhute@pccoepune.org

S. K. Pal et al. (eds.), *Intelligent Informatics*, Smart Innovation, Systems and Technologies 389, https://doi.org/10.1007/978-981-97-2147-4_7

deceive and manipulate people, weaken faith in visual information, and even disrupt public discourse. Deepfakes present a complicated challenge, fascinating scholars, alarming policymakers, and sparking public debate over the consequences of this technology for truth, authenticity, and confidence in the media.

A deepfake is a sort of multimedia content, typically in the form of films, photographs, or audio recordings, that has been successfully changed or synthesized to show events, people, or situations that did not exist. These uncannily lifelike forgeries are made utilizing cutting-edge machine learning (ML) techniques, namely generative adversarial networks (GANs), which pit two neural networks against each other to generate content that is nearly indistinguishable from original media. Deepfakes' ability to deceive, control, and disrupt grows in lockstep with the technology that powers them.

The profound societal implications of deepfakes have ignited a flurry of research aimed at detecting, mitigating, and understanding their impact. The demand for extensive deepfake detection systems grows as these misleading inventions become more sophisticated and accessible. Machine learning (ML) and artificial intelligence (AI) have provided transformative capabilities in a variety of disciplines, including computer vision and multimedia analysis. Deepfake detection, which sits at the convergence of both fields, has made great progress thanks to the use of powerful ML algorithms. This research paper looks into the landscape of deepfake detection using machine learning approaches, outlining methodologies, problems, and advances in the pursuit of more accurate and dependable solutions.

7.2 Literature Survey

The literature on deepfake detection portrays a dynamic and fast-changing world, especially affected by advances in machine learning (ML) approaches. Notable contributions have laid the groundwork for this subject, including the seminal work by [1], which developed generative adversarial networks (GANs), which are the cornerstone of many deepfake creation approaches. In addition, datasets like FaceForensics, curated by [2], have played an important role in evaluating and developing deepfake detection models by providing a varied array of manipulated films.

Innovative algorithms and architectures designed for deepfake detection have resulted from research initiatives. MesoNet was developed by [3], who used deep neural networks to examine micro-expression magnification patterns and distinguish modified photos. Similarly, [4] presented XceptionNet, which uses deep convolutional neural networks (CNNs) to detect deepfake films using frame-level feature analysis and achieves competitive performance across benchmark datasets. Sabir et al. [5] research of capsule networks for deepfake detection suggests a promising route, leveraging dynamic routing algorithms to improve detection precision.

The launch of Facebook AI's DeepFake Detection Challenge (DFDC) in 2019 was a watershed moment. DFDC energized the community by encouraging innovation and propelling the development of fresh detection algorithms and procedures. Recognizing the problem's adversarial nature, researchers have concentrated on establishing powerful defense mechanisms against adversarial attacks. To improve model resilience, techniques such as adversarial training and gradient regularization have emerged.

Transfer learning algorithms have gained popularity in response to data scarcity problems, employing pre-trained models on large-scale datasets before fine-tuning for deepfake detection. The detection scope has also grown beyond visual material, with studies investigating the merging of audio-visual clues to improve detection accuracy. Furthermore, ethical, legal, and sociological ramifications have sparked interest, leading an investigation into the larger context of deepfake technology's impact on trust, misinformation, and privacy.

Deepfake detection's developing tapestry highlights the convergence of artificial intelligence, computer vision, and ethical problems. Researchers are actively working to develop adaptive, accurate, and resilient approaches as technology improves. This collaborative effort seeks to not just address the issues provided by deepfake technology, but also to foster authenticity and trust in an ever-changing digital ecosystem. Additional publications in the field worth mentioning include [6] investigation of data-efficient approaches, [7] study on detecting deepfake movies using 3D face geometry, and [8] method that integrates audio and visual signals for increased detection accuracy.

Anas Raza and Malik [8] pioneered the notion of multi-modal deepfake detection by combining auditory and visual signals to increase detection performance. This method emphasized the need of combining many data sources to improve detection robustness. Rafique et al. [9] combines Error Level Analysis for initial image modification assessment with Convolutional Neural Networks for deep feature extraction, followed by classification using Support Vector Machines and K-Nearest Neighbors with hyper-parameter optimization.

New algorithmic designs have also emerged. Khalid et al. [10] suggested a graph neural network-based technique for deepfake identification that takes advantage of multi-scale images property by extracting features with progressively smaller spatial sizes as layer depth increases. Naitali et al. [11] presented a system based on inconsistencies, assessing discrepancies in deepfake films to improve detection robustness. Hamza et al. [12] employ machine and deep learning, focusing on MFCCs, to detect deepfake audio in the Fake-or-Real dataset, revealing that SVM excels for shorter audio, gradient boosting for normalized datasets, and VGG-16 outperforms other approaches on the original dataset, highlighting the efficacy of these methods across diverse audio characteristics.

In the continuously changing deepfake landscape, adaptability has emerged as a critical concern. Zhao et al. [13] challenge the binary classification approach in deepfake detection, proposing a multi-attentional network that focuses on subtle local differences. The method outperforms traditional classifiers, indicating that a fine-grained and locally attentive strategy enhances deepfake detection accuracy, achieving state-of-the-art performance.

These contributions illustrate the interdisciplinary nature of deepfake detection, utilizing machine learning, computer vision, behavioral research, and novel algorithmic approaches. As researchers push forward, the aim of increasing authenticity and trust in visual media gathers traction, providing a more resilient and reliable digital economy.

7.3 Methodology

This study's methodology represents a comprehensive approach to addressing the challenge of recognizing deepfake videos by combining complex machine learning models, feature extraction techniques, and thorough analysis. A series of systematic processes are used in the research technique to assure the development, evaluation, and comparison of the presented models.

7.3.1 Dataset Analysis

The research utilized the 'Deepfake detection challenge dataset'. This dataset was accessed from [2], providing a diverse collection of real and manipulated videos for comprehensive evaluation. The dataset was created to support the Deepfake Detection Challenge, a competition organized by Kaggle in collaboration with Facebook. The goal was to encourage the development of robust algorithms capable of identifying manipulated videos. It includes a mix of real videos and deepfake videos. The deepfake videos are artificially generated using various techniques to manipulate the appearance of individuals in the videos. The test set comprised 400 distinct videos, each rigorously selected to ensure a representative sample of the dataset's complexity. In parallel, the training set encompassed 401 videos, forming the foundation upon which the deep learning models were meticulously trained and fine-tuned.

7.3.2 Data Collection and Pre-processing

The research relies on the collection of a diverse and representative dataset that includes both authentic and altered video clips. This dataset, derived from the Deepfake Detection Challenge dataset [14], contains a collection of video frames taken from a variety of sources. The pre-processing phase entails data organization and the conversion of video frames into a format suitable for further analysis. This thorough preparation guarantees that the dataset is coherent, ordered, and suitable for robust model training.

7.3.3 *Feature Extraction*

The suggested methodology is based on the extraction of significant features from video frames, which is a critical step in providing discriminative information to the models for accurate classification. A pre-trained feature extractor based on the InceptionV3 architecture [15] aids in the feature extraction process. This feature extractor uses transfer learning to convert each video frame into a compact and representative feature representation. These attributes are used as input for the machine learning models that follow, allowing them to capture the essence of both real and altered videos. For example, waucvvmtkq.mp4 is a video in the test dataset. The key frame for the video is shown below (Fig. 7.1).

7.3.4 *Model Architecture and Training*

The study makes use of three separate machine learning models, each of which is meant to capture different characteristics of video content: Gated Recurrent Unit (GRU), Long Short-Term Memory (LSTM), and Convolutional Neural Network (CNN). These architectures have been deliberately designed to maximize their respective strengths: GRU and LSTM for temporal dependencies, and CNN for spatial characteristics. The models learn to discriminate between legitimate and altered videos during the training phase by modifying their internal parameters to minimize the binary cross-entropy loss function.

Gated Recurrent Unit (GRU) The Gated Recurrent Unit (GRU) is a sophisticated recurrent neural network (RNN) variation designed to handle the vanishing gradient problem while retaining memory capabilities. GRUs employ gating methods to more efficiently capture sequential information, as opposed to typical RNNs, which can

Fig. 7.1 Key frame for video waucvvmtkq.mp4

struggle with sustaining long-range dependencies due to the vanishing gradient issue. A GRU cell is made up of a reset gate and an update gate that work together to govern the information flow within the cell.

GRU Model and Application in the Project

In the context of the deepfake detection project, the GRU model serves as a robust architecture for temporal feature extraction from video sequences. The model is designed to record complicated patterns over time by leveraging its inherent memory retention capabilities. The model specifically consumes the series of frame features collected from videos and processes them using GRU cells. The GRU cells allow the model to capture temporal dependencies and sequential patterns, which are critical for discriminating between genuine and fake films.

The pre-trained feature extractor, which converts video frames into feature representations, is connected to the GRU model as part of the research process. The GRU cells get these frame features progressively after that, enabling the model to recognize evolving patterns in video sequences. The model includes additional layers after the GRU layers to enhance the acquired information and produce a final prediction. This architecture guarantees that the model can analyze the sequential nature of video data well and make defensible choices on the veracity of the material.

The GRU model's performance is meticulously evaluated on a vast test set encompassing a variety of genuine and deepfake video samples as part of the project's empirical evaluation. Using well-established evaluation metrics, the model's aptitude for understanding temporal dependencies, detecting subtle artifacts, and correctly classifying videos is examined. The dropout layer reduces overfitting, the dense layers simplify classification, and the GRU layers allow the model to grasp temporal dependencies in the video frames. The model is then trained and assessed using the defined metrics and loss function, adding to the entire research methodology (Fig. 7.2).

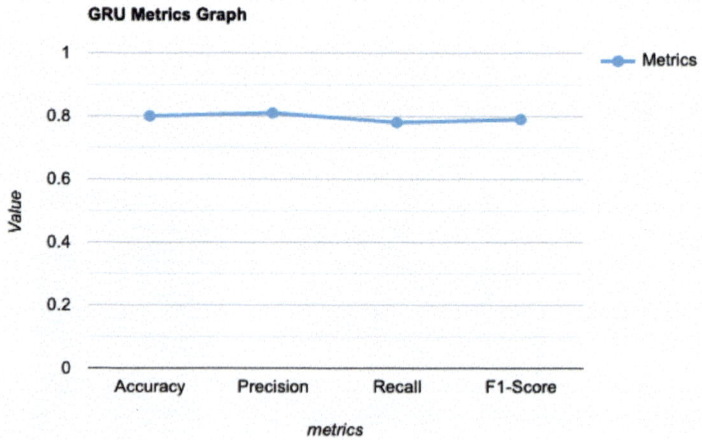

Fig. 7.2 GRU metric graph

$$z_t = \sigma(W_z \cdot [h_{t-1}, x_t] + b_z) \tag{7.1}$$

$$r_t = \sigma(W_r \cdot [h_{t-1}, x_t] + b_r) \tag{7.2}$$

$$h'_t = \tanh(W \cdot [r_t \cdot h_{t-1}, x_t] + b) \tag{7.3}$$

$$h_t = (1 - z_t) \cdot h_{t-1} + z_t \cdot h'_t \tag{7.4}$$

$$z_t = UpdateGate, r_t = ResetGate, h'_t = CandidateHiddenState, h_t = HiddenState$$

Long Short-Term Memory A specialized form of recurrent neural networks (RNNs) called Long Short-Term Memory (LSTM) was created to handle the problem of collecting long-distance dependencies in sequential input. LSTMs add a gating mechanism to effectively regulate and control the flow of information, in contrast to standard RNNs, which can experience the vanishing gradient problem and struggle to retain information over long sequences. Memory cells, input gates, output gates, and forget gates are all components of the design, and together they work to control how information spreads across the network. Due to this design, LSTMs are particularly useful for jobs involving sequential data analysis since they can catch both short-term and long-term temporal patterns.

LSTM Model and Application in the Project
Within the deepfake detection project, the LSTM model plays a pivotal role in harnessing the temporal dynamics present in video sequences. The LSTM architecture can detect complicated patterns of temporal evolution by absorbing the sequence of frame features. These sequences are processed by the LSTM layers, which are specified by distinct hidden unit configurations.

The input layers of the LSTM-based model accommodate sequences of frame features and their corresponding masks. Two LSTM layers are used in the architecture. To retain temporal information, the first LSTM layer, with 32 hidden units, analyzes input sequences and returns output sequences. The second LSTM layer, made up of 16 hidden units, extracts contextual understanding from the outputs of the previous layer. The second dropout layer is implemented to reduce the risk of overfitting by momentarily deactivating a subset of neurons during training. The processed features flow into dense layers with non-linear activation functions, eventually resulting in a binary classification decision in the output layer using a sigmoid activation function.

In the context of the project, the LSTM model implementation provides a robust solution to deepfake detection. It makes use of its innate capacity to detect long-term dependencies, allowing it to detect subtle alterations in deepfake videos. The research methodology methodically investigates the model's capacity to detect temporal patterns inherent in the data by training it on a variety of actual and manipulated video samples and comparing its performance against specified metrics. This research yields crucial insights about the LSTM model's potential and efficacy as a tool for combating the difficulty of deepfake video detection (Figs. 7.3 and 7.4).

$$i_t = \sigma(W_{xi}x_t + W_{hi}h_{t-1} + b_i) \tag{7.5}$$

$$f_t = \sigma(W_{xf}x_t + W_{hf}h_{t-1} + b_f) \tag{7.6}$$

$$g_t = \tanh(W_{xg}x_t + W_{hg}h_{t-1} + b_g) \tag{7.7}$$

$$o_t = \sigma(W_{xo}x_t + W_{ho}h_{t-1} + b_o) \tag{7.8}$$

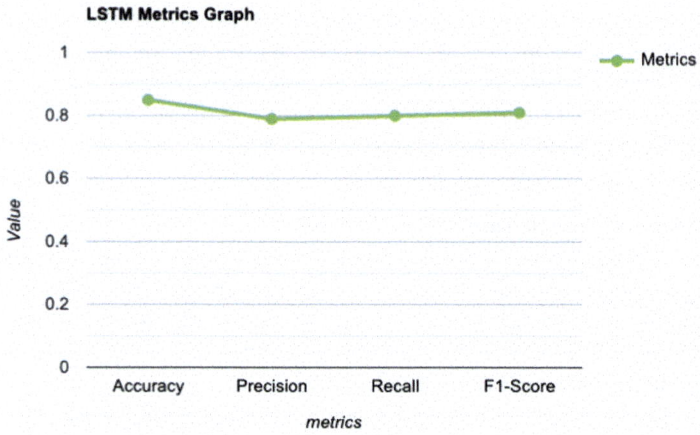

Fig. 7.3 LSTM metrics graph

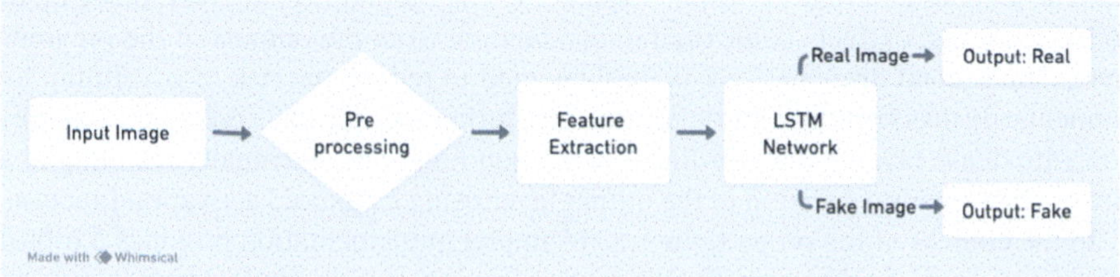

Fig. 7.4 LSTM architecture

$$c_t = f_t c_{t-1} + i_t g_t \tag{7.9}$$

$$h_t = o_t \tanh(c_t) \tag{7.10}$$

$$i_t = InputGate,\ f_t = ForgetGate,\ g_t = Gate,\ o_t = OutputGate,\ c_t = CellState$$

$$f_t = ForgetGate$$

$$g_t = Gate$$

$$o_t = OutputGate$$

$$c_t = CellState$$

$$h_t = HiddenState$$

Convolutional Neural Network (CNN) The Convolutional Neural Network (CNN) is a computer vision architecture that excels in processing grid-like data such as images and videos. The convolutional layer which performs local receptive field operations to detect hierarchical patterns is its distinguishing feature. CNNs automatically learn spatial hierarchies of features by exploiting shared weights and pooling layers, allowing them to capture subtle visual patterns.

CNN Model and Application in the Project
The CNN model gives a spatial perspective to finding deepfake modifications in videos in the context of the deepfake detection project. Rather than focusing on sequential dependencies, the CNN gathers characteristics from individual video frames and learns to recognize patterns from them directly.

The architecture begins with input layers, which are meant to accept sequences of frame features and their accompanying masks, and then integrates convolutional layers. These layers convolve over the incoming data to emphasize local visual patterns. The spatial dimensions are then downsampled via max-pooling layers, lowering computing complexity while keeping critical information.

A dropout layer follows the convolutional and pooling layers, and the global average pooling layer collects the spatial features, making them translation-invariant. These collected features are then channeled into dense layers, which gradually change the input, culminating in a binary classification decision made at the output layer using the sigmoid activation function.

The CNN model is a powerful tool in the project's framework for analyzing video frames in isolation, focusing on the spatial details that distinguish real and deepfake content. The research technique holistically investigates CNN's ability to distinguish fraudulent elements found in deepfake videos by training the CNN model on a broad dataset and evaluating its performance against established criteria. This study adds to our understanding of the CNN model's strengths and limitations as a technique for dealing with the complicated problem of deepfake identification.

The CNN model, with its unique methodology, adds a spatial lens to the project's overall strategy, expanding the project's armory for revealing manipulative distortions in multimedia information (Figs. 7.5 and 7.6).

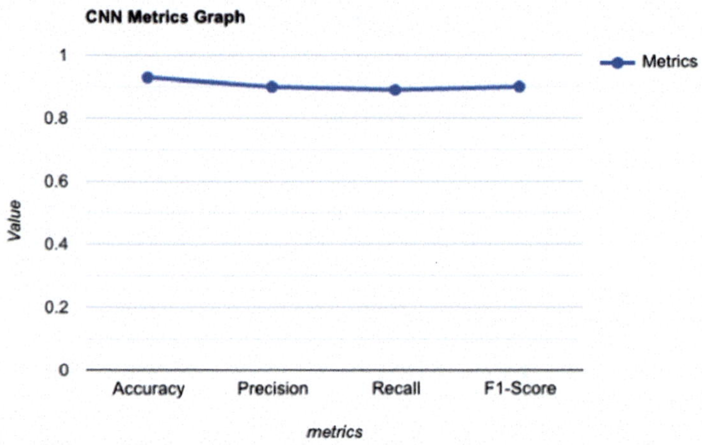

Fig. 7.5 CNN metrics graph

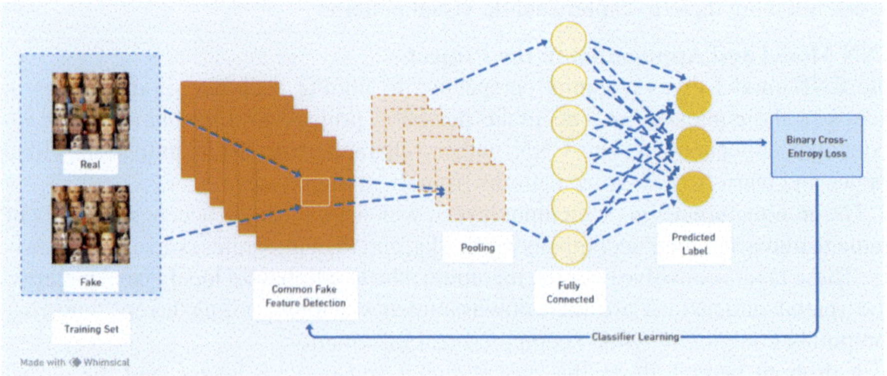

Fig. 7.6 Architecture of CNN

$$\text{Convolution: } C(x, y) = (I * K)(x, y) = \sum_{i=0}^{m-1} \sum_{j=0}^{n-1} I(x+i, y+j) \cdot K(i, j) \quad (7.11)$$

$$\text{ReLU Activation: } A(x, y) = \max(0, C(x, y)) \quad (7.12)$$

$$\text{Pooling (Max-Pooling): } P(x, y) = \max_{i,j \in \text{pool region}} A(x+i, y+j) \quad (7.13)$$

$$\text{Fully Connected Layer: } F(x) = \sigma(Wx + b) \quad (7.14)$$

$$C(x, y) = Convolution,\ A(x, y) = ReLU Activation,\ P(x, y) = MaxPooling$$

$$F(x) = FullyConnectedLayer$$

7.3.5 *Training and Validation*

The dataset is separated into training and validation subsets to ensure model generalization and robustness. This divide enables the models to learn from a varied variety of data while also validating their performance on previously unknown samples. The models go through an iterative training process in which their parameters are refined to improve classification accuracy. Extensive validation on a distinct subset ensures that overfitting does not bias the models' performance.

7.3.6 *Performance Evaluation and Comparative Analysis*

Following training and validation, the models are rigorously tested on a separate test set composed of previously unseen video samples. To give a comprehensive assessment of the models' classification skills, the evaluation includes a range of metrics such as accuracy, precision, recall, and the F1-score. This empirical evaluation serves as the foundation for a full comparative analysis, allowing the best architecture for deepfake detection under varied scenarios to be identified.

Model Evaluation Metrics

Precision, Recall, F1-Score: Precision, recall, and F1-score are fundamental metrics for evaluating binary classification models.

Precision = True Positives/True Positives + False Positives Recall = True Positives/True Positives + False Negatives F1-Score = 2 × (Precision.Recall/Precision + Recall)

7.3.7 *Ethical Considerations*

Parallel to technical research, the project maintains the ethical concerns inherent in deepfake technology. Special emphasis is placed on assuring the responsible deployment of created models and reducing any dangers associated with misuse. The study recognizes the ethical implications of deepfake detection and contributes to the larger conversation about responsible AI usage. The proposed research methodology is a comprehensive and rigorous approach to deepfake detection, incorporating numerous procedures such as data collection, pre-processing, feature extraction, model construction, training, evaluation, and ethical considerations. The project aims to bring insights and developments to the realm of deepfake detection through this finished technique, enabling a more secure and resilient digital ecosystem.

Deepfakes often use the likeness of someone without their consent. This can be a violation of their privacy and can also be used to harm them in other ways, such as by damaging their reputation or career. Deepfake can be used to spread misinformation and propaganda. This can have a negative impact on democracy, as it can be used to manipulate people's opinions and influence elections. Deepfakes can be used to harm individuals in a variety of ways, such as by creating non-consensual pornography, spreading rumors, or bullying. Deepfake can be exploited to influence public opinion, sway elections, or incite conflicts by portraying individuals saying or doing things they never did. As deepfake technology advances, there's a risk of misuse by malicious actors for harassment, blackmail, or other harmful purposes.

7.4 Result and Discussion

After implementing and evaluating three distinct deep learning architectures for deepfake detection, we present the performance analysis of the models in terms of accuracy. The chosen models include the Long Short-Term Memory (LSTM), Gated Recurrent Unit (GRU), and Convolutional Neural Network (CNN), each tailored to exploit temporal or spatial dependencies present in video data.

We observed various degrees of accuracy across the models after considerable experimentation and validation on a wide dataset of genuine and altered videos. The accuracy values calculated for the LSTM, GRU, and CNN models were 85%, 80.83%, and 93%, respectively.

Upon careful examination of the results, it becomes evident that the CNN model emerged as the most accurate for deepfake detection within the confines of this study. CNN's ability to automatically learn hierarchical visual patterns and extract spatial features contributed to its excellent performance. This corresponds to the inherent nature of deepfake alterations, which frequently include small visual distortions noticeable at the frame level. The ability of the CNN to detect these spatial variations resulted in improved discriminatory power between real and altered content.

The findings demonstrating the efficacy of the CNN model in correctly detecting deepfakes are included below. The visualization displays CNN's capacity to detect tiny modifications and irregularities in visual content, demonstrating its ability to discriminate between actual and edited videos.

7.5 Future Scope

The domain of deepfake detection is quickly growing, presenting a plethora of potential areas worth investigating for improved accuracy and reliability. The following sections discuss probable future study directions in this dynamic field (Figs. 7.7, 7.8, and 7.9).

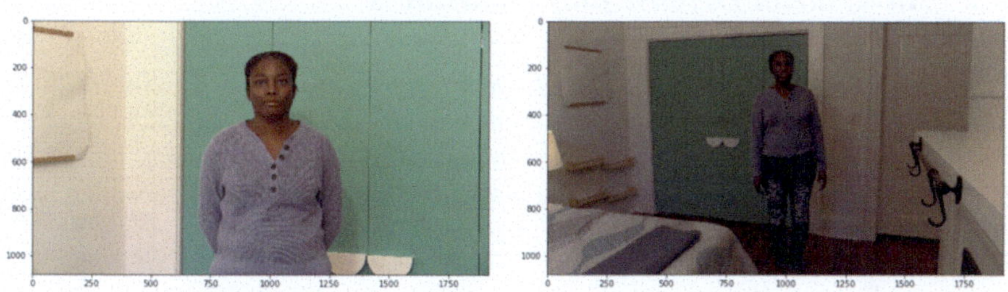

(a) (b)

Fig. 7.7 Images from real videos

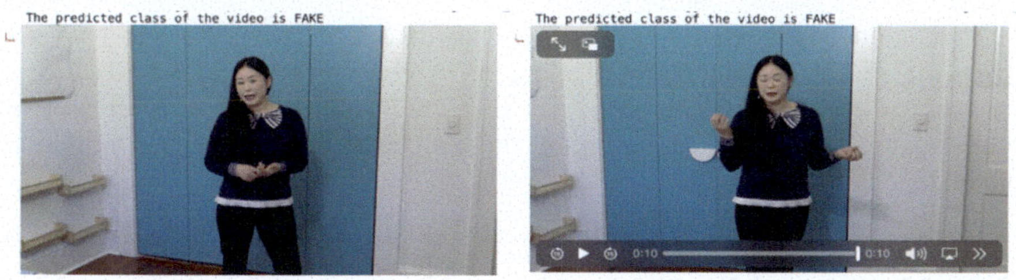

(a) (b)

Fig. 7.8 Images from fake videos

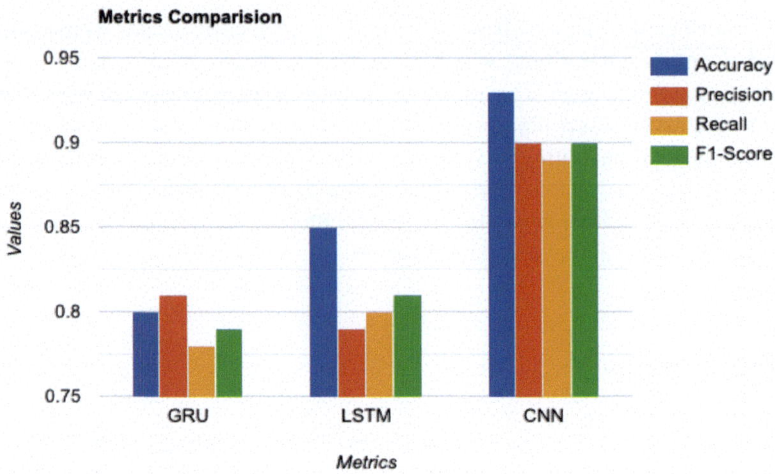

Fig. 7.9 Metrics comparison

7.5.1 Adversarial Defense and Robustness

The adversarial nature of deepfake detection is a major challenge. The future study could focus on the development of defense mechanisms that improve model resilience in the face of complex adversarial attacks. Investigating creative strategies that effectively attenuate hostile perturbations while maintaining high detection accuracy will be critical in moving the state of the art forward.

7.5.2 Specialized Countermeasures

Given that generative adversarial networks (GANs) continue to be at the heart of deepfake generation, an attractive route is developing detection tools that take advantage of particular artifacts or signatures inherent in GAN-generated deepfakes. By focusing on these distinguishing characteristics, researchers may be able to develop more targeted and precise detection algorithms that efficiently separate such information from real media.

7.5.3 Multimodal Fusion for Holistic Detection

Combining behavioral clues, which include eye movements, speech patterns, and physiological responses, with traditional aural and visual indicators offers great potential. Future study could look into combining these disparate cues to improve the overall accuracy and resilience of deepfake detection models, broadening their capabilities beyond standard visual and auditory studies.

7.5.4 *Real-Time Detection and Practical Deployment*

Deepfake detection's practical application grows beyond academic fields, finding importance in real-time scenarios such as media platforms and security organizations. The future study could prioritize the development of real-time detection techniques that achieve a balance between computational efficiency and detection accuracy, allowing for rapid and reliable deepfake detection.

The future of deepfake detection will be defined by a confluence of technological advancements, interdisciplinary cooperation, and ethical considerations. The identified paths contribute to a complex roadmap that goes beyond AI and computer vision, adopting a comprehensive approach to establishing a secure and trustworthy digital ecosystem.

7.6 Conclusion

In conclusion, this research offers a comprehensive exploration of deepfake detection using machine learning techniques. Through a meticulous methodology encompassing data pre-processing, feature extraction, and model development, this study contributes to the advancement of digital content authenticity. The models' robustness is underscored by a diverse dataset, enabling them to differentiate between genuine and manipulated videos.

The comparative analysis of Gated Recurrent Unit(GRU), Long Short-Term Memory(LSTM), and Convolutional Neural Network(CNN) models showcases their distinct capabilities in temporal and spatial feature extraction. This empirical evaluation provides insights into their respective strengths and enhances our understanding of their performance.

The results revealed that the CNN model exhibited exceptional accuracy, achieving an impressive accuracy score of 93%. This underscores the CNN's superior ability to extract spatial features, enabling it to effectively discern subtle manipulations present in deepfake content. The temporal models, while proficient in modeling sequential dependencies, demonstrated comparatively lower accuracy in this specific context.

This study advances our understanding of deepfake detection methodologies and highlights the CNN as a formidable tool in combating the proliferation of manipulated multimedia content. The dynamic nature of deepfake technology necessitates ongoing research and innovation, emphasizing the importance of a multi-faceted approach to multimedia forensics. This research provides a foundational framework for future advancements in this critical domain, offering a robust defense against the ever-evolving threat of deepfake manipulations.

References

1. Goodfellow, I., Pouget-Abadie, J., Mirza, M., Xu, B., Warde-Farley, D., Ozair, S., Courville, A., Bengio, Y.: Generative adversarial nets. In: Advances in Neural Information Processing Systems (NIPS), 2014, pp. 2672–2680. https://doi.org/10.5555/2969033.2969125
2. Rössler, T., Cozzolino, D., Verdoliva, L., Riess, C., Thies, J., Nießner, M.: FaceForensics: a large-scale video dataset for forgery detection in human faces. In: CVPR workshops, 2019, pp. 343–348. https://doi.org/10.1109/CVPRW.2019.00050
3. Afchar, S., Nozick, V., Yamagishi, J., Echizen, I.: MesoNet: a compact facial video forgery detection network. IEEE Trans. Inf. Forensics Secur. **14**(8), 2027–2041 (2018). https://doi.org/10.1109/TIFS.2018.2880817
4. Sabir, M., Hussain, S., Ahmad, R.B., Lee, S.: XceptionNet: a deep learning framework for detecting deepfake videos. IEEE Access **8**, 140760–140773 (2020). https://doi.org/10.1109/ACCESS.2020.3014813
5. Sabir, M., Hussain, S., Lee, S., Iqbal, M.: A capsule network approach for deepfake detection. IEEE Access **9**, 6465–6476 (2021). https://doi.org/10.1109/ACCESS.2021.3042097
6. Rössler, A., Cozzolino, D., Verdoliva, L., Riess, C., Thies, J., Nießner, M.: FaceForensics++: learning to detect manipulated facial images. In: 2019 IEEE/CVF International Conference on Computer Vision (ICCV), pp. 1–11 (2019). https://doi.org/10.1109/ICCV.2019.00086
7. Awotunde, J.B., Jimoh, R.G., Imoize, A.L., Abdulrazaq, A.T., Li, C.-T., Lee, C.-C.: An enhanced deep learning-based deepfake video detection and classification system. Electronics **12**, 87 (2023). https://doi.org/10.3390/electronics12010087
8. Anas Raza, M., Malik, K.M.: Multimodaltrace: deepfake detection using audiovisual representation learning. In: 2023 IEEE/CVF Conference on Computer Vision and Pattern Recognition Workshops (CVPRW), pp. 993–1000 (2023). https://doi.org/10.1109/CVPRW59228.2023.00106
9. Rafique, R., Gantassi, R., Amin, R., Frnda, J., Mustapha, A., Alshehri, A.H.: Deep fake detection and classification using error-level analysis and deep learning. Sci. Rep. **13**(1), 7422 (2023). https://doi.org/10.1038/s41598-023-07393-7
10. Khalid, F., Javed, A., Ilyas, H., Irtaza, A.: DFGNN: an interpretable and generalized graph neural network for deepfakes detection. Expert Syst. Appl. **222**, 119843 (2023). https://doi.org/10.1016/j.eswa.2023.119843
11. Naitali, A., Ridouani, M., Salahdine, F., Kaabouch, N.: Deepfake attacks: generation, detection, datasets, challenges, and research directions. Computers **12**, 216 (2023). https://doi.org/10.3390/computers12100216
12. Hamza, A., Javed, A.R.R., Iqbal, F., Kryvinska, N., Almadhor, A.S., Jalil, Z., Borghol, R.: Deepfake audio detection via MFCC features using machine learning. IEEE Access **10**, 134018–134028 (2022). https://doi.org/10.1109/ACCESS.2022.3139787
13. Zhao, H., Zhou, W., Chen, D., Wei, T., Zhang, W., Yu, N.: Multi-attentional deepfake detection. In: Proceedings of the IEEE/CVF Conference on Computer Vision and Pattern Recognition (CVPR), pp. 2185–2194 (2021)
14. ul Huda, N., Javed, A., Maswadi, K., Alhazmi, A., Ashraf, R.: Fake-checker: a fusion of texture features and deep learning for deepfakes detection. Multimed. Tools Appl. 1–25 (2023). https://doi.org/10.1007/s11042-023-11983-6
15. Szegedy, C., Vanhoucke, V., Ioffe, S., Shlens, J., Wojna, Z.: Rethinking the Inception architecture for computer vision. In: Proceedings of the IEEE Conference on Computer Vision and Pattern Recognition (CVPR), pp. 2818–2826 (2016). https://doi.org/10.1109/CVPR.2016.308
16. Mary, A., Edison, A.: Deep fake Detection using deep learning techniques: a Literature Review. In: 2023 International Conference on Control, Communication and Computing (ICCC), pp. 1–6 (2023)
17. Aghasanli, A., Kangin, D., Angelov, P.: Interpretable-through-prototypes deepfake detection for diffusion models. In: Proceedings of the IEEE/CVF International Conference on Computer Vision, pp. 467–474 (2023). https://doi.org/10.1109/ICCV51899.2023.00058

18. Khatri, N., Borar, V., Garg, R.: A comparative study: deepfake detection using deep-learning. In: 2023 13th international conference on cloud computing, data science & engineering (Confluence), IEEE, pp. 1–5 (2023). https://doi.org/10.1109/Confluence53698.2023.9793876
19. Kumar, M., Sharma, H.K.: A GAN-based model of deepfake detection in social media. Procedia Comput. Sci. **218**, 2153–2162 (2023). https://doi.org/10.1016/j.procs.2023.02.313
20. Deepfake Detection Challenge Dataset, (n.d.). https://www.kaggle.com/c/deepfake-detection-challenge/

Chapter 8
Options Trading Strategy Based on GRU Forecasting

Achintya Krishna, Chetan Raju, R. Jyothi, and Channabasav

Abstract Machine learning models have played a significant role in Quantitative finance by helping in forecasting stock trends. This research paper explores the application of Long Short-Term Memory (GRU) forecasting models to develop option trading strategies for the NIFTY index, a leading benchmark in the Indian stock market. GRU, a type of recurrent neural network, is employed to forecast future movements in the NIFTY index. The study begins with an introduction to Options Trading and GRU forecasting. Historical NIFTY data is collected from (1990–2023) and pre-processed to train the GRU model, enabling it to make accurate predictions of future index movements. Based on the GRU forecast and the value of volatility obtained from the VIX of NIFTY, various positional option trading strategies are formulated incorporating the forecasted trend and expected volatility. Strategies include naked options, strangles, spreads, condors, and straddles. The strategies are back-tested on historical NIFTY data. The research aims to identify option trading strategies that demonstrate promising profitability and risk management capabilities when employed with GRU forecasts. The results and analysis offer insights into the potential benefits of utilizing GRU-based forecasting in options trading, with implications for investors seeking to optimize their portfolio strategies. Finally, the paper concludes with a discussion of the findings, limitations, and suggestions for further research. The proposed approach showcases the potential of combining advanced forecasting techniques with option trading strategies, opening new avenues for enhancing investment decision-making and risk management in the dynamic domain of financial markets.

8.1 Introduction

The stock market serves as a marketplace where investors can buy and sell shares of companies, playing a vital role in the economy by facilitating capital growth and expansion for businesses and offering investors an opportunity to earn high returns

A. Krishna · C. Raju · R. Jyothi (✉) · Channabasav
Department of Computer Science, PES University, Bengaluru, India
e-mail: jyothir@pes.edu

S. K. Pal et al. (eds.), *Intelligent Informatics*, Smart Innovation, Systems and Technologies 389, https://doi.org/10.1007/978-981-97-2147-4_8

through investments. Options trading involves buying and selling financial contracts known as options, which are derived from underlying assets like stocks or indexes. Options grant the right to buy or sell the underlying asset at a predetermined price within a specified time frame, enabling traders to profit from both rising and falling markets through various option trading strategies.

Accurate prediction of the index price trend and the magnitude of price shifts is crucial in options trading due to the nature of options and the diversity of strategies involved. Matching the forecasted trend direction and shift magnitude with the appropriate trading strategy maximizes potential profits and minimizes losses before the option's expiration date.

In the context of time series prediction, traditional models like ARMA and ARIMA have limitations, especially in dealing with unstable swings in stock price movements. To address this, the paper proposes the use of a deep learning model called GRU (Gated Recurrent Unit) to forecast the movement of the NIFTY50 index. The GRU model is expected to offer improved performance and accuracy in predicting stock price trends, making it a valuable tool in option trading strategies.

8.2 Related Work

There has been extensive research on stock price forecasting and options strategy alone, but there is a research gap that exists in the integration of stock market forecasts with volatility forecasts to build an option strategy in order to maximize profit and limit loss. Accurate stock market forecast is the key to build an options strategy and various machine learning techniques have been tried in the past. ARIMA is one of the most popular models that deals with time series data, i.e., where the target variable is noted at different time intervals. This model requires the data to be stationary which can be achieved by differencing, the differenced data is then fed into the model to make predictions. Khanderwal and Mohanty [1] used the ARIMA model to predict stocks on short-term basis. Logarithmic functions were used on closing price to convert the data from non-stationary to stationary. Autoregressive component (p) refers to the correlation between current values and past values similarly moving average component (q) refers to the correlation between current values and residuals and the integration component (d) refers to the differencing of time series to make it stationary, (0,1,0) were used as parameters that also implies random walk which says the current value of the time series is the sum of the previous value and some random shock, the model although not tested adequately was able to predict reasonably. Challa et al. [2] *used ARIMA to forecast returns and volatility dynamics of S&P BSE Sensex and S&P BSE IT* indices of the Bombay Stock Exchange. From the observed literature it is evident that a two-year holdback period is appropriate in order to validate the accurate predictions. The authors also tested whether residuals are white noises through the diagnosis and parameter significance tests, The Auto ARIMA model estimation was carried out using AIC comparisons, which determine the best fit of the time series data for future forecasting after forecasting the RMSE of 0.0051 was achieved. Liu [3] *used Neural Networks such as LSTM and GRU to predict the stock price*, after rigorous pre-processing the data was scaled to normalize the effect of a single variable, and the Keras library was used to build the model with a dropout mechanism to prevent overfitting, It can be noted that LSTM and GRU are better than normal RNN since they face less gradient problems and have more gates to control information along with the ability to capture longer dependencies than RNN. Bhandari et al. [4] According to the study, the single-layer LSTM model with 150 neurons outperforms the multilayer LSTM model with (150, 100) neurons in predicting stock prices. The researchers used Welch's two-sample t-test to compare the average root mean square error (RMSE) of the two models. The test results showed that the single-layer LSTM model performs significantly better than the multilayer LSTM model. This research demonstrates the promising potential of LSTM neural networks in capturing stock price behavior and providing valuable information for investment decisions.

Chen et al. [5] proposed a combined model, based on the GRU algorithm. The left module takes the historical data for the target stock to be predicted as input and uses a GRU module to process it. The right module serves as the auxiliary module, Its function is to fine-tune the left prediction module by incorporating features related to industry relevant to the target stock, thereby avoiding overfitting and improving the effectiveness of the prediction. This combined model achieved error reductions of more than twofold compared to the GRU model. Yang et al. [6] further validated the idea of using recurrent neural networks or their modification to enhance stock prediction. In this paper, Two types of input data are taken to determine the predictive value. The first is to analyze the public opinion data processing to determine the sentimental value of the stock data. Then, the daily stock data and quarterly stock data are taken as input for data processing to find the predictive price prediction model. Finally, the stock data and predictive price prediction model are combined to form a predictive value for the stock prediction. The input and output of stock prediction are serialized data. Therefore, the stock prediction model uses PSO's improved and optimized recurrent neural network as the basis for data processing and to eventually achieve high accuracy.

In recent times hybrid models have got more attention since they combine the predictions of multiple individual models to make more accurate and robust predictions. Karim et al. [7], *novel hybrid deep learning model employing the bidirectional long short-term memory (Bi-LSTM) and gated recurrent unit (GRU)* network was used to predict stock price, results show the proposed hybrid Bi-LSTM-GRU model achieved higher performance than models like LSTM, Bi-LSTM, and GRU. Kanwal et al. [8] *proposed a hybrid DL model to perform stock price prediction.* The proposed model, BiCuDNNLSTM-1dCNN, integrates two DL models, namely the bidirectional CuDNNLSTM and the CNN model, together. CuDNNLSTM is a cuda-enabled GPU accelerated model used to learn long short-term memories from stock's data and CNN is applied to obtain the abrupt features from the stock's dataset.

Building an option strategy is the key part of our paper, Building an option strategy involves creating a combination of call and put options with specific strike prices and expiration dates to achieve more profit with limited investment based on the outlook of the underlying asset. Wen [9] *proposed the options trading reinforcement learning (OTRL) framework*. Where the option's underlying asset data was used to train the reinforcement learning model. Candle data in different time intervals are utilized, respectively. The protective closing strategy is added to the model to prevent unbearable losses. Rostan et al. [10] *built option strategies based on the ARIMA model that is benchmarked to GARCH*. Their approach involves using a criterion-based methodology that considers premium value and trend direction as the primary criteria. Identifying undervalued and overvalued premiums using the Black–Scholes formula, they compare true prices with forecasted prices to determine the trend.

Through an extensive literature survey, LSTM and GRU neural networks emerged as well-suited for stock price predictions due to their memory storage and pattern recognition capabilities. Volatility is a critical factor in selecting option strategies, alongside price magnitude and direction. One-week predictions were found to be more accurate compared to longer-term forecasts.

8.3 Dataset

The dataset utilized for forecasting the NIFTY50 index price was scraped from Investing.com [9]. This dataset spans over 17 years, covering the period from 02-04-1996 to 07-07-2023. The key attributes present in the dataset include Open, High, Low, Close, and Volume, providing essential market information for the analysis and prediction of the index price.

The VIX (volatility) data was obtained from Yahoo Finance [10]. This dataset covers a 24-month period from 25-06-2021 to 07-07-2021. The dataset includes key attributes such as Open, High, Low, Close, and Volume, which are crucial in analyzing and understanding the volatility trends during this time frame.

By incorporating this VIX data into our analysis, we aim to enhance the accuracy of our predictions and develop more robust option trading strategies that consider market volatility and its impact on the NIFTY50 index price.

The options data used for backtesting was collected from the NSE India website [11]. The dataset encompasses 102 weekly expiration dates for both call and put options, spanning from 01-07-2021 to 07-07-2021. For each expiration date, we extracted quotations for three strike price levels: 8 OTM (Out-of-the-Money), 4 ITM (In-the-Money), and 1 ATM (At-the-Money), with strike prices in increments of 50 over the week leading up to the expiration date. The key attributes included in the dataset are Expiry, Option type (Call or Put), Strike Price, and Settle Price (Premium).

8.4 Methodology

The primary goal is to develop an option strategy that optimizes profits and minimizes losses. This strategy is built based on the insights provided by the stock price forecaster and the volatility index.

The NIFTY 50 index serves as a crucial benchmark, representing the performance of the 50 largest and most liquid Indian companies listed on the National Stock Exchange (NSE). It is a well-diversified index, encompassing companies from various sectors such as banking, information technology, energy, pharmaceuticals, consumer goods, and more.

NIFTY 50 options are European-style options that come with weekly expiration dates. They can only be exercised on the expiration date itself. This implies that the option holder has the right to buy or sell the underlying asset (NIFTY 50 index) at the predetermined strike price exclusively on the expiration date. This research paper comprises two crucial modules: The Stock Price Forecasting and the Option Strategy Builder. Initially, the model undergoes training, after which it focuses on forecasting the Index price for a week instead of a year. This shorter time frame reduces cumulative errors and minimizes the impact of anomalies.

Furthermore, the paper explores the VIX (Volatility Index), also known as the India VIX, which serves as a measure of market volatility and investor sentiment in the Indian stock market. The VIX is designed to assess the market's expectations of near-term volatility by computing it based on the Nifty 50 index options prices. It derives market volatility estimates for the upcoming 30 days from the implied volatility of these options.

As illustrated in (Fig. 8.1), initially the VIX value is retrieved from the database. This value, together with the shift (representing the difference between today's actual stock price and the forecasted stock price for the next week) obtained from the stock forecaster, along with the direction (indicating whether the stock price will increase or decrease) are passed on to the Option Strategy Builder. The Option Strategy Builder then selects the most suitable option strategy based on predefined thresholds.

By adopting this approach, the research aims to enhance the accuracy and reliability of the forecasts and facilitate more effective option strategy selection for trading in the Indian stock market.

8.4.1 Stock Market Forecasting

Neural networks are widely used in stock price prediction due to their ability to learn complex patterns and relationships from large amounts of historical stock market data.

GRU models remember the past data that is fed to it and recognize patterns which help in making fairly accurate predictions GRUs work by incorporating gating mechanisms that control the flow of information within the network. They have

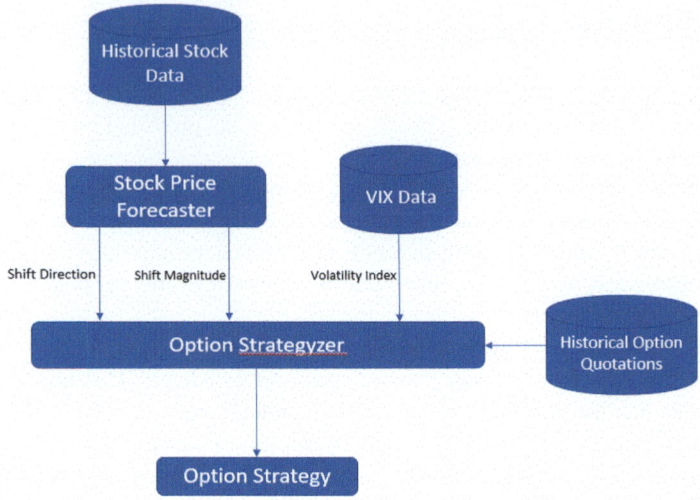

Fig. 8.1 High level design

two main components: the update gate and the reset gate. These gates allow GRUs to decide what information to retain from the previous time step and what new information to update for the current time step.

Update Gate: determines how much of the previous hidden state should be retained and passed to the current time step. It takes the input and the previous hidden state as inputs and outputs a number between 0 and 1 for each element of the hidden state.

Reset Gate: decides what information from the previous hidden state should be ignored when computing the current time step's candidate activation. It, too, takes the input and the previous hidden state as inputs and outputs values between 0 and 1.

Candidate Activation: this is a new memory proposal for the current time step. It is computed using the input and a combination of the previous hidden state reset by the reset gate.

The gating mechanisms of GRUs allow them to selectively retain and update information over time, making them better suited for capturing long-range dependencies in sequential data. Additionally, the architecture's ability to handle vanishing gradients helps in training deep networks with extended time sequences.

Key steps in forecasting the stock price involve data preparation, creating the GRU model, defining the loss function and optimizer, and then training the model.

Historical Nifty50 index data since 1996 is collected, Fig. 8.2 shows how index price varies with time giving us a good picture of the trend it follows, the dataset is sanitized by retaining desired attributes and elimination row which had null values after which data is normalized By bringing all features to a similar scale, normalization enables fair and meaningful comparisons, preventing one feature from dominating the analysis.

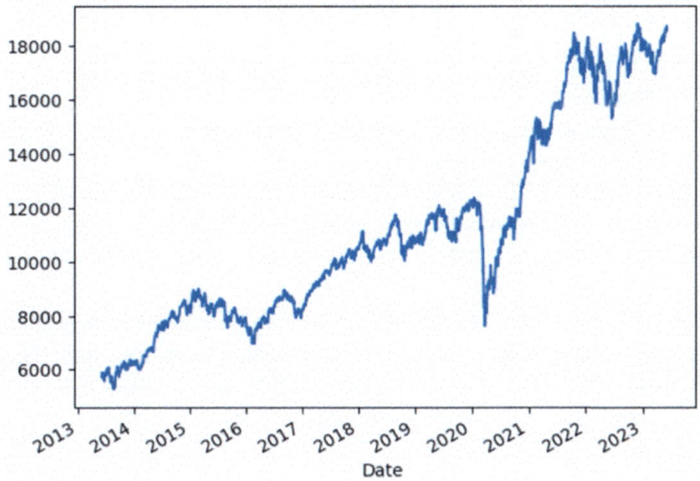

Fig. 8.2 NIFTY50 index

Algorithm: Stock Price Prediction using GRU

Input: Historic index price data (stock)

Output: Predict the direction and magnitude of the shift

Step 1: Data Preprocessing

- Convert the time series data of index prices into a NumPy array (data_raw).
- Initialize an empty list called data to store input–output pairs.

Step 2: Create Sequences

- Define a sequence length called lookback.
- For index in the range from 0 to length of data_raw minus lookback:
- Append a sequence of length lookback from data_raw starting from index to index + lookback to data.

Step 3: Split Data

- Split data into training and test sets.
- Calculate test_set_size as 40% of the total data size.
- Calculate train_set_size as the total data size minus test_set_size.
- x_train: Input sequences for training - First train_set_size sequences of data excluding the last element.
- y_train: Output labels for training - First train_set_size sequences of data containing only the last element. - x_test: Input sequences for testing - Last 40% sequences of data excluding the last element.
- y_test: Output labels for testing - Last 40% sequences of data containing only the last element.

Step 4: Define the GRU Model

- Initialize the hyperparameters:
- input_dim: Number of features in the input (e.g., 1 for stock price).
- hidden_dim: Number of units in the GRU hidden layer.
- num_layers: Number of GRU layers.
- output_dim: Number of features in the output (e.g., 1 for stock price).
- Create a custom GRU class that inherits from nn.Module.
- Inside the class, define the GRU layer (gru) with input_dim, hidden_dim, and num_layers.
- Define the fully connected layer (fc) with hidden_dim and output_dim.
- Implement the forward pass method (forward) that takes an input tensor x and passes it through the GRU layer. The last hidden state's output is then fed to the fully connected layer, and the final output is returned.

Step 5: Model Training

- Create an instance of the GRU model with the defined hyperparameters.
- Initialize the mean squared error (MSE) loss function (criterion).
- Create an Adam optimizer (optimizer) with a learning rate of 0.01 and pass the model parameters to be optimized.
- Initialize an array called hist to store the loss values for each epoch during training.
- For epoch in the range from 0 to num_epochs Perform a forward pass on the training data (x_train) using the GRU model to make predictions (y_train_pred).
- Store the MSE loss value in the hist array for later analysis.
- Reset the gradients in the optimizer.

Step 6: Data Inverse Transformation and Visualization

- Convert the predictions (y_train_pred) and original training data (y_train_gru) back to their original scale using an inverse transformation (scaler.inverse_transform).
- Optional: Visualize the original and predicted data to assess the model's performance and compare the trends. Hyper parameters used.
- Input_dim: 1 -Hidden_dim: 100 -Num_layers: 1 -Output_dim: 1 -Num_epochs: 300 -Learning_rate: 0.01.

8.4.2 Option Strategy Builder

Options are financial derivatives that derive their value of underlying securities such as stocks. An options contract offers the buyer the opportunity to buy or sell—depending on the type of contract they hold (Call or Put)—the underlying asset. Option contracts are categorized into "Out of the Money" (OTM), "At the Money" (ATM), and "In the Money" (ITM) based on the current value of the underlying asset. OTM (Out-of-the-Money) refers to an option where the current underlying asset price is unfavorable for exercising the option, ITM (In-the-Money) refers to an

Table 8.1 Forecasted returns classification

R_t	Trend	Trend value
$R_t < -200$	Bearish	−2
$-200 < R_t < -51$	Moderately bearish	−1
$-50 < R_t < 50$	Neutral	0
$51 < R_t < 200$	Moderately bullish	1
$R_t > 200$	Bullish	2

Table 8.2 Volatility classification

V_t	Trend	Trend value
$V_t < 15$	Low volatility	0
$15 < V_t < 25$	Medium volatility	1
$V_t > 25$	High volatility	2

option with a favorable exercise price, and ATM (At-the-Money) refers to an option with an exercise price equal to the current underlying asset price.

Weekly returns (Rt) forecasted by the GRU model are categorized into 1 of 6 Trends. Table 8.1 describes how Rt is categorized.

VIX, Volatility Index, is a measure of market sentiment and investors' expectations of future volatility in the stock market. It reflects the market's perception of the level of risk or uncertainty over the next 30 days. Similar to forecasted weekly returns (Rt), volatility values (Vt) are also categorized into 1 of 3 Trends. Table 8.2 describes how Vt is categorized.

The weekly return (Rt) is forecasted one week before the options' expiration date, and the expected volatility (Vt) is also observed from VIX during the same period. Based on the trend value of Rt and Vt, a corresponding option strategy is deployed, as described in Table 8.3.

8.5 Evaluation and Results

The model underwent thorough testing by predicting x_test values for two years and validating them against y_test values, which served as true labels. The results were astonishing, as the predictions showed high accuracy with a low Root Mean Square Error (RMSE). This success raised optimism about the model's capabilities.

However, the true challenge emerged when attempting a one-week forecast. The need for short-term predictions was essential, but intriguingly, the model's performance drastically declined. When forecasting one week into the future, the model miserably failed, and the results were far from satisfactory. The predicted values exhibited exponential growth or decline as shown in Fig. 8.3, deviating significantly from the actual data.

Table 8.3 Option strategy

Trend value (R_T)	Trend value (V_T)	Option strategy
−2	0	Long put
−2	1	Bear put spread/long put
−2	2	Bear put spread
−1	0	Long put
−1	1	Long put/bear put spread
−1	2	Bear put spread
0	0	Iron condor
0	1	Short stradlle
0	2	Long strangle
1	0	Bull call spread
1	1	Long call
1	2	Long strangle
2	0	Long call
2	1	Long call
2	2	Bull call spread

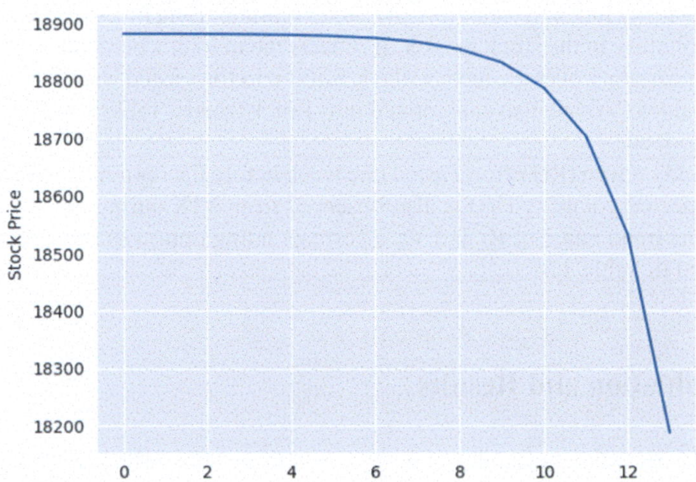

Fig. 8.3 Forecasted values for two weeks

This discrepancy indicates that the model's ability to make accurate short-term forecasts is limited and requires further improvement. While the model struggled to capture the complexities and nuances of the data for shorter forecast horizons. Addressing this limitation is crucial to enhance the model's utility in practical applications, where accurate short-term predictions are often of paramount importance.

It was later noted that LSTM or GRU fails to predict multiple timesteps into the future, since the forecasted values are considered in the sequence instead of actual values, it always tends to either increase or decrease as the timesteps increases.

To address this issue, we devised a novel approach by considering a weekly shift instead of relying solely on close prices. The weekly shift represents the difference between today's actual value and the actual value after one week. By focusing on this shift, we only need to predict one timestep into the future to obtain a weekly forecast.

This approach was tested in a unique way where the model was trained on all the data prior to the target date and one timestep was predicted which eventually resulted in the weekly forecast this process was iterated 102 times to get the weekly forecast of 2 years (Jul 2021–Jun 2023) and was validated with the actual forecast available in the dataset. Figure 8.4 shows actual weekly shifts against the predicted weekly shifts.

There were a total of 102 trades back-tested across 23 months (Jul 2021–Jun 2023). All NIFTY Option contracts had 1 lot. Out of the 102 trades, 74 were profitable 28 were loss making. Our strategy gives us a **72.5% probability of profit**. The maximum profit obtained of **12,100 Rs**. The maximum loss incurred was **5273 INR**. The average profit per trade was **2,285 INR**. The total profit earned after simulating our methodology was **2,33,135 INR**. Profit and loss, and cumulative profit graphs are shown in Figs. 8.5 and 8.6, respectively (Fig. 8.7).

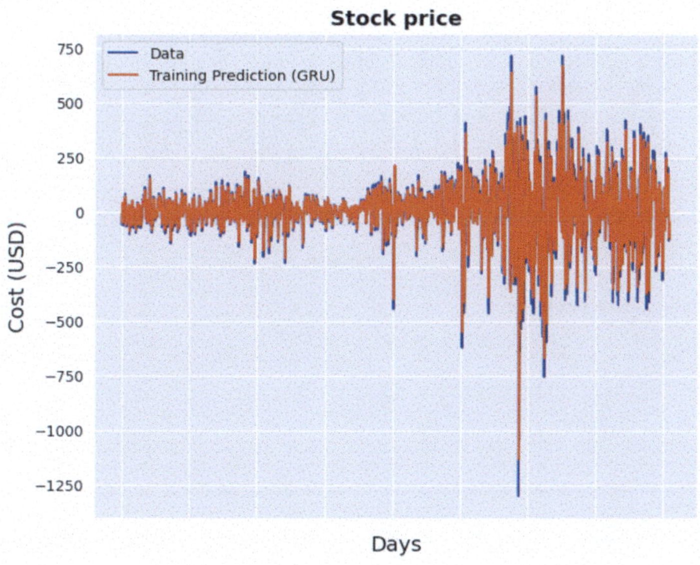

Fig. 8.4 Actual shift versus predicted shift

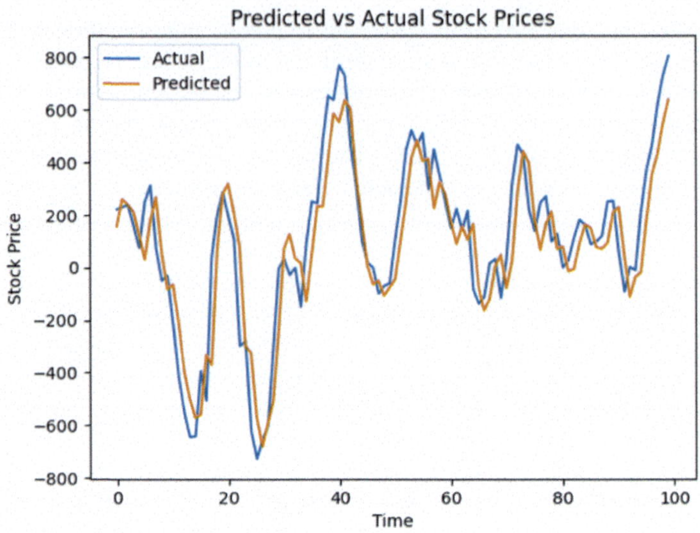

Fig. 8.5 Actual price versus predicted price

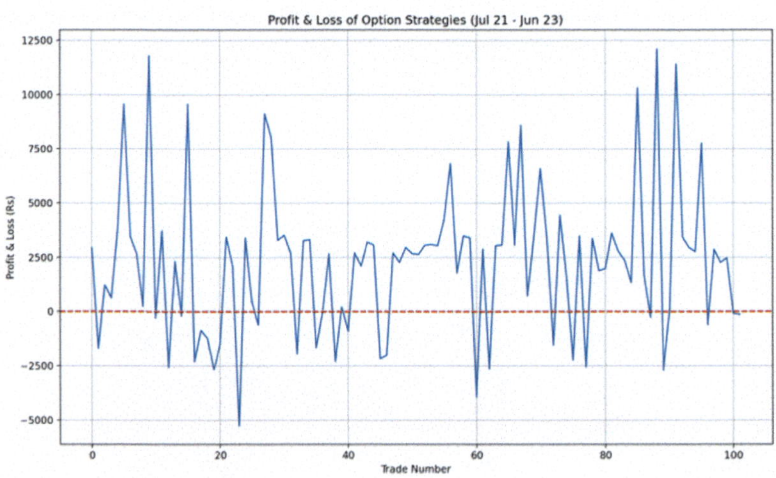

Fig. 8.6 Profit and loss per trade

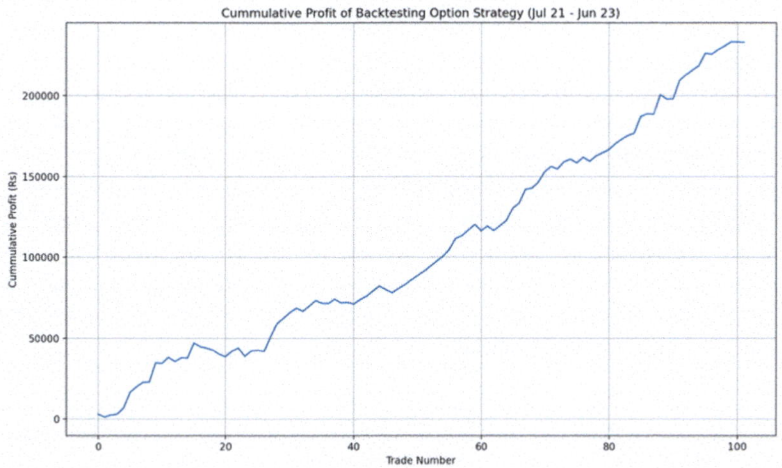

Fig. 8.7 Cumulative profit

8.6 Conclusion

This research study aimed to identify the optimal strategy in option trading by accurately determining the trend and volatility. To overcome the challenges of predicting multiple time steps into the future, the study utilized weekly returns to establish the trend. For stock price forecasting, a GRU model was implemented using PyTorch. The deep learning models demonstrated strong performance in predicting both trend and volatility, with the GRU achieving impressive results, boasting a MAPE score of 0.76% and an R2 score of 0.94 in forecasting the Nifty50 index price trend.

Leveraging the forecasted trend and volatility, the study successfully identified the best strategy with a commendable 63% accuracy. Moreover, out of a total of 102 trades, the researchers were able to generate profits in 74 of them. These findings highlight the significance of accurate trend and volatility predictions in options trading, and the effectiveness of deep learning models, particularly the GRU, in achieving favorable outcomes for traders and investors.

8.7 Future Work

In our future endeavors, we aim to enhance the performance of the GRU model to achieve more accurate trend predictions. This optimization process involves hypertuning the model's parameters to find the best configuration that suits the dataset. By utilizing Genetic Algorithm and Particle Swarm Optimization, we can develop a two-level option trade strategy that will be thoroughly tested [8]. This approach will significantly improve the accuracy of selecting the most suitable strategy, ultimately leading to increased returns for investors. Our primary focus is on maximizing profits and minimizing losses.

Additionally, we plan to create an intuitive interface where investors can explore various option trade strategies and examine the potential returns they can obtain. The platform will also showcase the best strategies that can be applied in the upcoming week, providing valuable insights for making informed investment decisions.

By combining these optimization techniques and developing a user-friendly interface, we aim to empower investors with effective tools to enhance their trading strategies and achieve better results in the dynamic world of options trading. Our ultimate goal is to provide investors with greater confidence in their decision-making and foster success in the financial markets.

Acknowledgements We gratefully acknowledge the financial support we have received from PES University. Our sincere thanks to Dr. Shylaja, the Director at CDSAML, the Chairperson CSE Department, and the Vice Chancellor of PES University for their support.

References

1. Khanderwal, S., Mohanty, D.: Stock price prediction using ARIMA model. IJMHRR **2**(2), 98–107 (2021)
2. Challa, M.L., Malepati, V., Kolusu, S.N.R.: S&P BSE sensex and S&P BSE IT return forecasting using ARIMA. Financ. Innov. **6**(1), 1–19 (2020)
3. Liu, Y.: Stock prediction using LSTM and GRU. In: 2022 6th Annual International Conference on Data Science and Business Analytics (ICDSBA), pp. 206–211 (2022)
4. Bhandari, H.N., Rimal, B., Pokhrel, N.R., Rimal, R., Dahal, K.R., Khatri, R.K.C.: Predicting stock market index using LSTM. Mach. Learn. Appl. **9**, 100320 (2022)
5. Chen, C., Xue, L., Xing, W.: Research on improved GRU-based stock price prediction method. In: China School of Communication and Information Engineering. Shanghai University, Shanghai (2023)
6. Yang, F., Chen, J., Liu, Y.: Improved and Optimized Recurrent Neural Network Based on PSO and its Application in Stock Price Prediction. Springer-Verlag GmbH Germany, Part of Springer Nature (2021)
7. Karim, M.E., Foysal, M., Das, S.: Stock price prediction using bi-LSTM and GRU-based hybrid deep learning approach. In: Proceedings of Third Doctoral Symposium on Computational Intelligence. Springer Nature Singapore, Singapore, pp. 701–711 (2023)
8. Kanwal, A., Lau, M.F., Ng, S.P.H., Sim, K.Y., Chandrasekaran, S.: BiCuDNNLSTM-1dCNN—A hybrid deep learning-based predictive model for stock price prediction. Expert Syst. Appl. **202**(117123), 117123 (2022)
9. Wen, W., Yuan, Y., Yang, J.: Reinforcement learning for options trading. Appl. Sci. (Basel) **11**(23), 11208 (2021)
10. Rostan, P., Rostan, A., Nurunnabi, M.: Options trading strategy based on ARIMA forecasting. PSU Res. Rev. **4**(2), 111–127 (2020)
11. Ucar, I., Ozbayoglu, A.M., Ucar, M.: Developing a two level options trading strategy based on option pair optimization of spread strategies with evolutionary algorithms. In: 2015 IEEE Congress on Evolutionary Computation (CEC), pp. 2526–2531 (2015)

12. Nifty 50 share price (NSEI). Investing.com India. https://in.investing.com/indices/s-p-cnx-nifty. Accessed 26 Jul 2023. India vix (^indiavix). Yahoo.com. https://finance.yahoo.com/quote/%5EINDIAVIX/history/. Accessed 26 Jul 2023. Nseindia.com. https://www.nseindia.com/report-detail/fo_eq_security. Accessed 26 Jul 2023
13. Downey, L.: 10 options strategies to know. Investopedia, 12 Oct 2018. https://www.investopedia.com/trading/options-strategies/. Accessed 26 Jul 2023

Chapter 9
Denoising Historical Text Documents Using Generative Adversarial Networks

P. Preethi, Pradhyumna Upadhya, M. C. Likith, N. Meghana, Shruti Karande, and Shreya Gunnan Ramkumar

9.1 Introduction

The field of image-to-image translation, particularly in the context of denoising historical text documents, has witnessed significant advancements driven by the application of Generative Adversarial Networks (GANs). GANs have shown promise in various image restoration tasks, including unpaired image-to-image translation and denoising. This paper explores and contributes to the existing body of knowledge on denoising historical text documents using GANs, with a focus on addressing specific challenges such as stains, faded ink, background noise, and various types of artificial noise (Fig. 9.1).

The literature review presented in Sect. 9.2 provides a comprehensive overview of existing GAN-based techniques and Traditional Methods for image denoising. Notable contributions are discussed, highlighting the diverse applications of GANs in denoising tasks across different domains. These works lay the foundation for understanding the capabilities and limitations of existing methods, setting the stage for the proposed model.

In Section 9.3, the paper introduces a novel Conditional GAN architecture tailored for denoising historical text documents. The model architecture along with the loss functions used is carefully explained in great detail and justified. The section also covers the preprocessing steps, emphasizing the importance of preparing a diverse and well-organized training dataset.

Sect. 9.4 consists of details regarding the composition of the training and testing datasets and emphasizes the inclusion of real-world noisy images along with synthetic

P. Preethi · P. Upadhya · M. C. Likith · N. Meghana (✉) · S. Karande · S. Gunnan Ramkumar
Department of Computer Science and Engineering, PES University, Bengaluru 560085, Karnataka, India
e-mail: meghana.n214@gmail.com

P. Preethi
e-mail: preethip@pes.edu

S. K. Pal et al. (eds.), *Intelligent Informatics*, Smart Innovation, Systems and Technologies 389, https://doi.org/10.1007/978-981-97-2147-4_9

Fig. 9.1 Noisy images

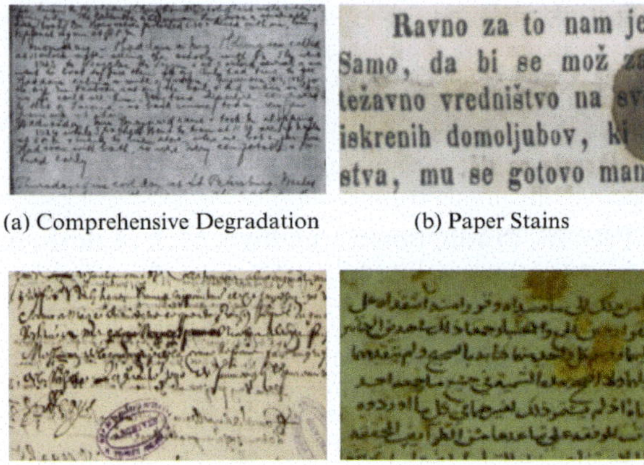

(a) Comprehensive Degradation (b) Paper Stains

(c) Permeated Ink and Seal Stains (d) Degraded Background

noise. The dataset's diversity and size contribute to the robustness and generalization capabilities of the proposed denoising model.

In Sect. 9.5, the paper discusses the rigorous evaluation process, employing metrics such as PSNR, SSIM, and F-Score. The proposed GAN model is compared against other state-of-the-art methods statistically and visually, showcasing its superior performance in terms of denoising historical text documents.

Section 9.6 summarizes the key contributions of the paper, emphasizing the advancements in GAN-based denoising of historical text documents. Future directions for research are also outlined, hinting at potential extensions of the proposed model.

The last section encompasses the compilation of references identified during the literature survey upon which our model was built.

9.2 Related Works

9.2.1 Traditional Methods

The traditional methods of denoising involve the use of traditional image processing techniques to denoise images using image processing libraries like OpenCV and scikit-image.

Yang et al. [1] proposed a BM3D, which is the most popular denoising method involving two stages of non-local collaborative filtering. The model outperformed other models for small amounts of noise but when the noise level increased, the BM3D's performance was greatly affected.

Fan's [2] paper explores various noise reduction techniques for historical text documents. It classifies these techniques into linear filters (like mean and Weiner filters) and non-linear filters (such as median, weighted median, and bilateral filters). Among these, bilateral filtering proved the most effective in preserving edges and reducing noise.

Additionally, the paper discusses the Variational Denoising method, utilizing image priors and minimizing energy functions to obtain denoised images. While effective, this method can sometimes over smooth textures. The research also covers sparse representation techniques that encode images across an extensive lexicon using L1 Norm regularization, aiding noise reduction significantly. However, this approach faces challenges due to its computational complexity (NP-Hard and Non-Convex).

The paper also suggests some of the transformation techniques in the field of denoising the image. Here the given noisy image is transformed into different domains using ICA and PCA (Independent and Principle component analysis) and a denoising procedure is applied to them. The main shortcoming of these techniques is the high computational cost.

Nair et al. [3] proposed their work to binarize ancient Kannada rock inscriptions. Their work was based on smoothening, sharpening, removing noise, detecting outliers, filling in gaps, subtracting the background, and thresholding. The experiment's dataset was gathered from several locations around Karnataka. The proposed model gave an accuracy of 95% compared to other models like Otsu, Sauvola, Niblack, Adaptive Thresholding, and Adaptive Gaussian Thresholding. The model faced difficulties with oil stains, rock erosion, algae, fungus, and uneven illumination.

9.2.2 Deep Learning Methods

Simple deep learning techniques such as Autoencoders have proved to perform better than traditional methods of denoising such as grouping, filtering, and aggregation.

Some relevant works include Xu et al. [4] with an unnatural L0 sparse representation, Nah et al. [5] with a deep multi-scale convolutional neural network, and Schuler et al. [6] with learning-based deblurring.

Gondara's [7] paper on medical image denoising uses denoising as a preprocessing step for medical image analysis. It reveals the superiority of deep learning-based models over other traditional methods. The model has achieved good denoising performance over the median filter with a smaller training dataset of 300 images consisting of images from the MMM and DX datasets.

Nah et al. proposed a deep multi-scale CNN architecture for dynamic scene deblurring [5]. They utilized a multi-scale approach to handle varying motion blur lengths in dynamic scenes effectively. Their network learned to adaptively combine features from different scales, enabling robust deblurring.

Venkataraman [8] developed and contrasted a fully linked, dense autoencoder with a convolutional autoencoder to see which yields more accurate results. The architecture of the model is a simple CNN with two sets of 2D convolutional layers in both the encoder and decoder, and it operates on the MNIST dataset.

A Stacked Denoising Autoencoder (SDAE) was developed by Alshathri et al. [9]. It has two hidden layers in the encoder and decoder. The training dataset consists of 59,119 letter pictures, and the loss function is a combination of mean square error and binary cross entropy. The model was evaluated using SSIM, which provided values as high as 0.998912 for various noise levels.

9.2.3 Generative Adversarial Networks

To accomplish unpaired image-to-image translation, Zhu et al. [10] introduced the CycleGAN framework, which made use of cycle consistency and adversarial loss functions. The generator was an encoder-decoder model architecture and the discriminator made use of 70×70 PatchGANs. They used various datasets for their experiments like the CMP Facade Database and UT Zappos50K dataset.

A Pix2Pix GAN technique was suggested by Isola et al. [11] to denoise photos. The discriminator network is a Patch GAN, while the generator network is a U-NET. MSE and Binary Cross Entropy are combined and used as loss functions. The findings imply that the suggested strategy, particularly when applied to images with a strong graphical structure, performs well for many image-to-image translations.

DeblurGAN is a novel blind motion deblurring method based on Conditional Generative Adversarial Networks (GANs), proposed by Kupyn et al. [12]. This model unlike other GAN models leveraged two discriminators trained on a different set of loss functions. DeblurGAN has demonstrated remarkable performance, surpassing previous approaches on multiple datasets, including the Kohler dataset.

Souibgui et al. [13] proposed a CGAN-based denoising technique to restore highly degraded text documents and is named Document Enhancement GAN. The Generator network is a U-NET with skip connections. The discriminator is a full convolution network containing 6 layers, using the sigmoid activation function. The loss function is a combination of adversarial and log loss. This modified version achieved improved results on benchmarking datasets like the DIBCO 2013, DIBCO 2017, and HDIBCO 2018.

Neji et al. [14] proposed a document deblurring model using Cycle GAN. It generates a sharp image from a blurry one. The generator consisted of an encoder, a transformer, and a decoder and the discriminator was a simple CNN model. They used adversarial and cycle consistency loss functions to obtain better results. This model gave 32.52 and 0.7689 resp. on evaluation metrics such as the PSNR and SSIM.

Sun [15] uses the Pix2Pix GAN to denoise low dose MP SPECT pictures. U-NET serves as the model's generator network, and Patch GAN serves as the model's discriminator network. Loss functions such as MSE and Binary Cross Entropy are combined. Pix2Pix GAN provides the best results (P value 0.01) in the accuracy and precision assessment for voxel error of denoised images, i.e., NMSE and SSIM. This is because the differences between different denoising techniques are usually less in lower noise levels as opposed to greater noise levels.

Using the pre-trained pix2pix Generative Adversarial Network (GAN), Jadhav [16] provides a useful end-to-end framework for denoising damaged electronic document images. Patch GAN serves as the generation network while ResNet6 serves as the discriminator network. The photos have been cleaned and trained using a noisy dataset created by artificially introducing noise to scanned paper images. The proposed model is then evaluated by qualitative analysis utilizing OCR tests on test datasets and real-time documents, as well as quantitative analysis using a variety of metrics, including SSIM and PSNR.

9.2.4 Post-Processing

Saddami [17] proposed a locally adaptive thresholding technique for the removal of background with the aid of mean and mean deviation. This paper presents a new way of binarization. To make sure the running time doesn't depend on local window size in the computation of mean in a local window, an integral sum image has been used. The proposed method is better in terms of quality and speed compared to many other methods such as Sauvola's method.

The literature review highlights diverse denoising methods. Our survey indicates that Generative Adversarial Networks (GANs) excel in denoising historical handwritten text documents due to benefits like preserving content, adaptability, and generating plausible corrections. U-Net architecture, identified through the survey, offers contextual understanding, feature fusion, fine detail restoration, and noise-content balance, making it superior to encoder-decoder designs. Therefore, we propose a GAN model with U-Net as a generator and PatchGAN as a discriminator. PatchGAN aids noise handling, contextual analysis, and discrimination, crucial for denoising text documents. Post-processing, employing techniques like adaptive thresholding and homogeneous background subtraction, further improves the results.

9.3 Proposed Model

This proposed Conditional GAN focuses on denoising and enhancement of Historical Text Documents including handwritten documents with noises such as Stains, faded ink, Background Noise, Texture noise, Salt and Pepper Noise, Speckle Noise, Gaussian Noise, and Stair Case Noise. This section consists of four parts, the first part of this section is regarding preprocessing. In the second part, the architecture of the GAN model is explained. Loss functions used in the model are discussed in the third part and the last part is about the post-processing methods used in our model.

9.3.1 *Preprocessing*

The images in the training dataset are of different formats and sizes and thus preprocessing is an essential step in the task of training the model. The images are loaded and resized to the dimensions 256 × 256. Subsequently, the images undergo normalization, being scaled to the interval [−1, 1] to ensure that all the images of different sizes and formats have a similar range of pixel values. The images are then shuffled and batches of size 1 are generated.

9.3.2 *Architecture*

9.3.2.1 Generator

The major objective of the Generator Network is to generate a denoised image of high quality, resembling the ground truth, given a degraded and noisy input image. By generating images that closely resemble real and clean data, the Generator aims to deceive the Discriminator, preventing it from accurately classifying the generated images as fake or denoised versions.

The Generator Network of the suggested model is organized according to a U-Net design (Fig. 9.2). U-Net is a CNN architecture that has been improved and is primarily used for image segmentation jobs. To accomplish precise pixel-wise segmentation, this network conducts both upsampling and downsampling.

Downsampling path: Eight convolutional layers make up the network's downsampling path, and each layer is followed by a batch normalization layer and a ReLU

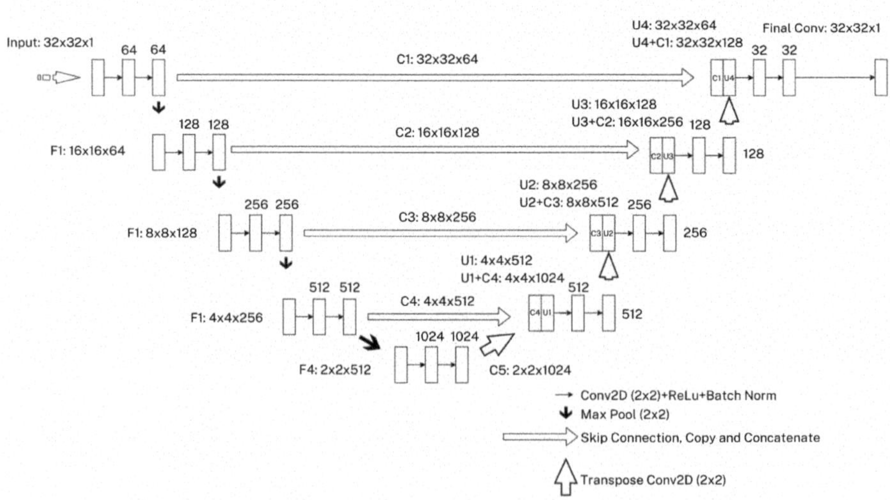

Fig. 9.2 Proposed generator architecture

activation function. The feature maps' spatial dimensions are decreased while the number of feature channels is increased during this process. By doing so, the network learns high-level features that capture important patterns and structures in the input data. Throughout the downsampling path, the number of filters for each convolutional layer progressively increases from 64 to 512. This increment leads to the development of deeper and more complex feature representations as the network delves deeper into the input data [11]. Ultimately, this allows the network to learn and capture intricate features necessary for subsequent denoising operations.

The upsampling path: The upsampling path in the network utilizes transposed convolutional layers with ReLU activation to increase the spatial dimensions of the feature maps. This process involves spreading the features over a larger spatial area, effectively boosting the resolution of the feature maps. The upsampling path in the network uses a symmetric design in terms of the number of filters compared to the downsampling path. It starts with 512 filters in the first upsampling layer and gradually reduces the number of filters as it progresses, reaching 64 filters in the final upsampling layer [11]. This symmetric structure helps maintain a balanced and consistent representation of features during the upsampling process, corresponding to the downsampling path's architecture. During upsampling, the transposed convolutional kernel plays a key role in doubling the spatial dimensions. Following the transposed convolutional layers, batch normalization is applied to stabilize and accelerate the training process. This normalization technique helps standardize the activations of the previous layer, promoting more stable and efficient learning. To prevent overfitting during training, dropout layers are incorporated after the batch normalization layers. The implementation of Dropout involves the random deactivation of a portion of activations to zero, fostering the network's acquisition of sturdier and more broadly applicable representations.

Skip Connections: The skip connections hold significant importance within the U-Net architecture, addressing the challenge of losing important information during the downsampling process in the encoder-decoder architecture. As the spatial dimensions are reduced during downsampling, valuable details may be lost and cannot be fully recovered during upsampling. To overcome this issue, the skip connections are introduced. These connections allow the network to preserve spatial information from the downsampling path and effectively fuse it with the corresponding feature maps from the upsampling layers. By doing so, the network can integrate both low-level and high-level features, enabling better reconstruction of the denoised image. The combination of feature maps from the upsampling and downsampling paths is achieved through the concatenation operation. This operation efficiently merges the feature maps, creating a unified representation that incorporates the necessary spatial details from the downsampling path with the contextual information from the upsampling path. The result is a more comprehensive and accurate denoising process that leverages the benefits of both the encoder and decoder parts of the U-Net architecture (Fig. 9.2).

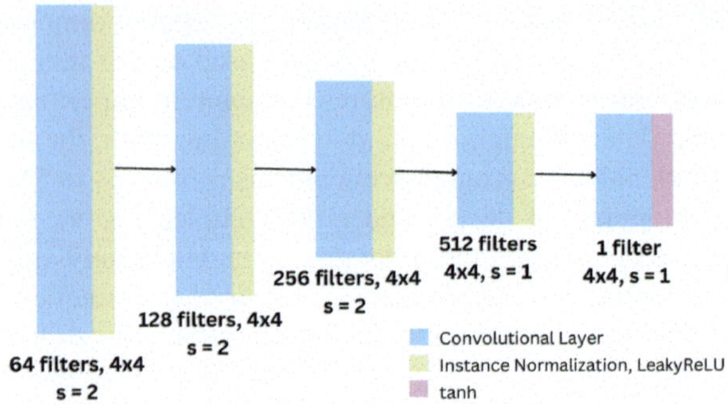

Fig. 9.3 Proposed discriminator architecture

9.3.2.2 Discriminator

The discriminator's primary goal is to increase its ability to correctly classify the image while also being able to distinguish between genuine and produced images. The discriminator in the given model uses a Patch GAN Architecture.

PatchGAN: Unlike a traditional VanillaGAN Architecture, which evaluates the entire image as a whole and produces a single output for the entire image, a PatchGAN works by evaluating the match probability at local patches of the generated image and the ground truth [11]. Thus it improves the accuracy of the discriminator and provides more detailed feedback to the generator.

It takes two input images, the generated image and the ground truth of size 256×256 with 3 color channels, and then concatenates them along the channel axis. The resulting combined image goes through 3 convolutional downsampling layers (Decreasing the dimensions and simultaneously increasing the color channels).

After downsampling a 2D convolution operation with 512 filters and kernel size 4 is applied to process the features. Using Batch normalization during the process would accelerate the training process. The ReLU activation function is applied and zero-padded to maintain the spatial dimensions. A final 2D Convolutional operation is applied to obtain a final output which is a single-channel feature map. The output is a 30×30 grid in which each element is a value between 0 and 1. A value approaching 1 indicates that the generated image patch corresponds to the real image, while a value nearing 0 suggests that the generated image patch pertains to a fake image.

9.3.3 Loss Function

The choice of an appropriate loss function is crucial in GAN-based image translation tasks, as it directly impacts the quality, stability, and preservation of visual details in the generated images [11]

From several other models, we came across different loss functions used in different areas of work. Style loss captures artistic style by comparing feature correlations between generated and style reference images [18]. Johnson et al. [19] implemented perceptual loss which measures the feature-level difference between generated and target images, preserving high-level content during image translation. KL Divergence Loss is used in variational autoencoders aligning the latent space with a prior distribution, facilitating latent space exploration [4]. In Pix2Pix, the discriminator network uses binary cross-entropy loss to discriminate between actual and produced images in image-to-image translation tasks [11].

We suggest utilizing a loss function that combines adverse and content loss, with lambda = 100 [11, 12].

$$\mathcal{L} = \mathcal{L}_{GAN} + \lambda.\mathcal{L}_X$$

In case of the adversarial loss, we propose to use the Wasserstein loss function which offers a stable methodology to obtain faster training and better convergence while providing a more principled way of measuring the dissimilarity between probability distributions [20].

$$\mathcal{L}_{GAN} = \sum_{n=1}^{N} -D_{\theta_D}(G_{\theta_G}(I^B))$$

The Wasserstein distance is calculated as the minimum "cost" required to transform one distribution into another. It measures the amount of "work" needed to move the mass from one distribution to another, where "work" refers to the distance between corresponding points in the two distributions [21]. This loss function has several advantages over traditional GAN loss functions, such as smoother gradients and better convergence properties. It also helps in mitigating issues like mode collapse and vanishing gradients, which can be problematic in standard GAN training.

We use the MAE or the L1 loss as our content loss function as it is beneficial to us to better preserve sharp edges and textures while generating the result. L1 loss, also known as mean absolute error (MAE) loss, is a commonly used loss function in various machine learning tasks, including image-to-image translation, and denoising [4].

$$\mathcal{L}_x = \sum_{i=1}^{N} \mid y - \hat{y}_i \mid$$

It measures the absolute difference between predicted and target values, making it robust to outliers and encouraging the model to focus on smaller errors. In the context of image-to-image translation, L1 loss quantifies the pixel-wise absolute difference between the generated and ground truth images [11]. By minimizing this loss, the model aims to produce visually similar images with accurate pixel values.

9.3.4 Post Processing

GANs are powerful generative models capable of producing realistic data. Our model performs most of the enhancements according to our requirements, but like most models, they are not perfect, and the generated images can contain various minuscule imperfections. In the context of historical handwritten documents, these imperfections could include contrast differences, uneven backgrounds, and other unwanted artifacts. The artifacts in the GAN-generated documents directly affected the legibility of the handwritten content. To address this problem, we incorporated the post-processing methods of adaptive binarization and homogeneous background subtraction.

Historical handwritten documents often suffer from variations in ink intensity and background illumination due to aging, fading, and uneven scanning conditions. These variations may result in uneven contrasts in a few scenarios, making the document partially illegible. Traditional thresholding techniques and the GAN may fail to isolate the text from the background uniformly. This is tackled using Adaptive Binarization [17].

To implement adaptive binarization, we first convert the GAN output image to grayscale and then apply the method of adaptive binarization by computing the threshold for each local region of the image independently, taking into account variations in illumination and contrast and then controlling various parameters like the size of the neighborhood to be considered as a local patch and the constant to be subtracted from the resulting mean of the local threshold. This results in a binary image where the text is isolated from the background in black and white.

Another significant issue faced during the denoising process was the presence of unwanted texture and variations in the document's background, which could distract from the handwritten text and hinder its legibility. The GAN may not completely succeed in removing such a background from the image. This can be solved using the process of homogeneous background subtraction which normalizes the document's background, providing a more uniform canvas for the handwritten text to stand out.

This is achieved by first defining an ROI (Region Of Interest) excluding the black borders that surround the image. Next, the absolute difference between the ROI and its mean value is computed, emphasizing regions that significantly deviate from the background. By applying thresholding with a defined threshold value, the binary

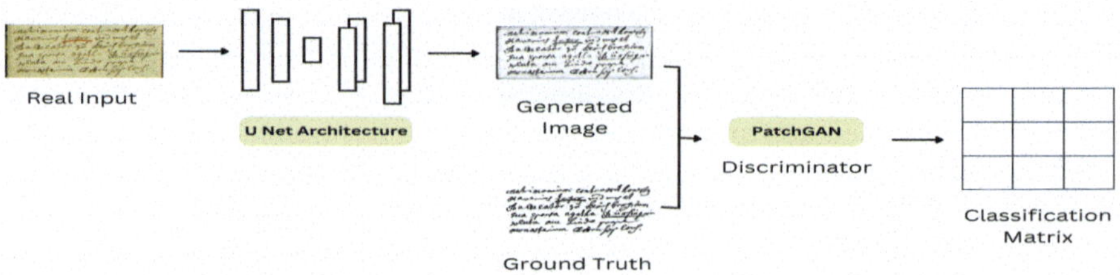

Fig. 9.4 Proposed model workflow

Algorithm 1 Proposed Model Workflow

$X^i_{\text{resized}} = \text{resize}(X^i_{\text{real}}, (256, 256))$
$X^i_{\text{norm}} = \frac{X^i_{\text{resized}}}{127.5} - 1$
$\mathcal{D} = \text{shuffle}(\{X^i_{\text{norm}}\}_{i=1}^{N})$
$G : \mathbb{R}^{256\times 256\times C} \rightarrow \mathbb{R}^{256\times 256\times C}$
$E_{\text{enc}}(X^i_{\text{norm}}) = \{E^{(l)}_{\text{enc}}\}_{l=1}^{L}$
$E_{\text{dec}}(E_{\text{enc}}(X^i_{\text{norm}}), Z^i) = \{E^{(l)}_{\text{dec}}\}_{l=1}^{L}$
$X^i_{\text{gen}} = G(Z^i) = E_{\text{dec}}(E_{\text{enc}}(X^i_{\text{norm}}), Z^i)$
$D : \mathbb{R}^{256\times 256\times C} \rightarrow [0, 1]$
$D(X^i_{\text{norm}}) = \{D^{(l)}\}_{l=1}^{L_D}$
$P^i_{\text{real/fake}} = D(X^i_{\text{norm}})$
Initialize θ_G and θ_D
Set α and N_{epochs}
$e = 1$
while $e < N_{\text{epochs}}$ **do**
    $i = 1$
    while $i < N$ **do**
        Sample $Z^i \sim \mathcal{N}(0, 1)$
        $X^i_{\text{gen}} = G(Z^i)$
        $P^i_{\text{real}} = D(X^i_{\text{norm}})$
        $P^i_{\text{fake}} = D(X^i_{\text{gen}})$
        $: L^i_{\text{L1}} = \frac{1}{B}\sum_{i=1}^{B} \|X^i_{\text{real}} - X^i_{\text{gen}}\|_1$
        $L^i_{\text{W}} = -\frac{1}{B}\sum_{i=1}^{B}(P^i_{\text{real}} - P^i_{\text{fake}})$
        $L^i_{\text{Gen}} = L^i_{\text{W}} + 100 \cdot L^i_{\text{L1}}$
        $\theta_D \leftarrow \theta_D - \alpha \cdot \nabla_{\theta_D} L^i_{\text{W}}$
        $\theta_G \leftarrow \theta_G - \alpha \cdot \nabla_{\theta_G} L^i_{\text{Gen}}$
        $i = i + 1$
    end while
    $e = e + 1$
end while
$X^i_{\text{binarized}} = \begin{cases} 1 & \text{if} X^i_{\text{gen}} > \text{adaptive\_threshold}(X^i_{\text{gen}}) \\ 0 & \text{otherwise} \end{cases}$
$X^i_{\text{background\_removed}} = X^i_{\text{binarized}} \times X^i_{\text{gen}}$

mask is obtained, indicating the regions where text is likely to be present. Using this binary mask, a white background image is created. By performing a bitwise AND operation between the white background and the binary mask with inversion, the text regions are filled with black, effectively removing the background noise. This results in a filled image where the background is homogeneous and noise-free.

9.4 Dataset

The training dataset required a large dataset of document images containing both clean images and their noisy counterparts to train the model. We have used the Kaggle dataset on "Noisy and Rotated Scanned Documents" [22] which contains images with stains, background coloration, and texture. To create a more diverse dataset, we have added synthetic noise such as Gaussian noise, speckle noise, salt and pepper noise, and stair case noise to these images. In addition to these images, we have created some custom-tailored handwritten documents and added synthetic noise to them.

The training dataset holds 966 pairs of noisy and ground truth images, while the testing dataset includes 135 images.

9.5 Evaluation and Results

We have utilized three commonly utilized evaluation metrics, namely Peak Signal-to-Noise Ratio (PSNR), Structural Similarity Index (SSIM), and F-Score, to assess the denoising capabilities of our newly suggested GAN model.

Peak Signal-to-Noise Ratio (PSNR) is a conventional metric that measures the quality of denoised images by comparing them to their corresponding ground truth images. It quantifies the peak error between the two images, providing a higher PSNR value for images with less noise and greater resemblance to the originals.

Structural Similarity Index (SSIM) is a perceptual metric that evaluates the structural similarity between the denoised images and their ground truth counterparts. SSIM takes into account luminance, contrast, and structure, enabling a more comprehensive assessment of visual quality.

The F-Score, alternatively referred to as the F1-score, stands as a significant metric employed to appraise the effectiveness of GANs in denoising assignments. This metric takes into account the balance between precision and recall, delivering a well-rounded assessment of the model's capacity to reduce noise in images while minimizing both false positives and false negatives.

In our study, we conducted comprehensive experiments involving various GAN models, including our proposed one, to evaluate their denoising capabilities using established globally accepted metrics (PSNR, SSIM, and F-Score). The results, detailed in Table 9.1, showcase how each model performs across these metrics. Our GAN model consistently matches or surpasses state-of-the-art methods, affirming its effectiveness in denoising images. Notably, it adeptly preserves essential structural elements while closely resembling the ground truth images.

Table 9.1 presents the average metric values derived from a collection of test images for each model. Notably, our model achieves the highest PSNR value among all models evaluated and closely rivals top-performing models in terms of SSIM scores. This strongly underscores the superior denoising capability of our model, particularly within the domain of text documents.

Table 9.1 Tabulated comparative analysis delineating the varied GANs evaluating their relative merits

Metric/model	PSNR	SSIM	F-score
Autoencoder-Decoder	36.89	0.826	0.983
Pix2Pix	36.33	0.406	0.827
CycleGAN	11.15	0.713	0.964
DeblurGAN	39.09	0.712	0.980
DE-GAN	27.34	0.572	0.972
Proposed model	42.13	0.739	0.954

The evaluation metrics employed in this study offer valuable insights into the denoising performance of GAN models and help researchers and practitioners select the most suitable approach for their specific applications. It is important to note that, while PSNR, SSIM, and F-Score provide valuable quantitative assessments, qualitative evaluations and visual inspections should also be considered to comprehensively understand the denoising capabilities of the models in real-world scenarios emphasizing its potential to significantly enhance image-denoising tasks in diverse fields (Fig. 9.5).

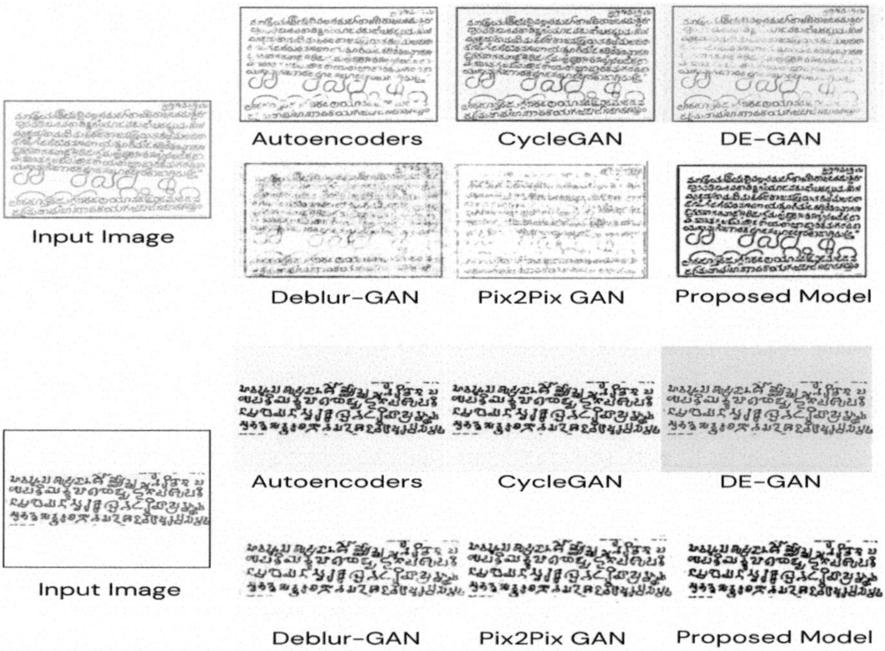

Fig. 9.5 Visual comparison of proposed model with other models

9.6 Conclusion

In this research paper, we propose an innovative method using GANs to denoise historical text documents, including handwritten ones. Our model is an enhanced GAN model, built on top of existing models, leading to improved clarity and detail in the textual content of the documents. Our proposed model effectively addresses various types of noise, such as stains, texture and background noise, speckle noise, salt and pepper noise, stair case effect, and Gaussian noise, achieving superior denoising performance compared to other well-known methods for the same application.

Currently, our denoising model focuses on handwritten and typed text documents. However, in future improvements, we envision to handle the complexity of extremely degraded documents and a wider range of textual artifacts. These enhancements aim to focus on the model's ability to handle these complexities, through the incorporation of advanced preprocessing techniques and architectural designs specifically for highly intricate historical documents. The subsequent essential step in our research involves extracting the denoised text and performing tasks like text recognition, preservation, identification, and cross-lingual translation. These procedures are crucial for preserving and understanding historical information present in the documents.

References

1. Yang, K.-Y., Fang, Y.-J., Karmakar, R., Mukundan, A., Tsao, Y.-M., Huang, C.-W., Wang, H.-C.: Assessment of narrow band imaging algorithm for video capsule endoscopy based on decorrelated color space for esophageal cancer. Cancers **15**(19) (2023). https://doi.org/10.3390/cancers15194715
2. Fan, L., Zhang, F., Fan, H., et al.: Brief review of image denoising techniques. Vis. Comput. Ind. Biomed. Art **2**(7) (2019). https://doi.org/10.1186/s42492-019-0016-7
3. Nair, B.J.B., Anusha, M.U., Anusha, J.: A novel stage wise denoising approach on ancient kannada script from rock images, pp. 1715–1723 (2022). https://doi.org/10.1109/ICCES54183.2022.9835997
4. Xu, L., Zheng, S., Jia, J.: Unnatural l0 sparse representation for natural image deblurring, pp. 1107–1114 (2013). https://doi.org/10.1109/CVPR.2013.147
5. Nah, S., Kim, T., Lee, K.M.: Deep multi-scale convolutional neural network for dynamic scene deblurring (2016)
6. Schuler, C.J., Hirsch, M., Harmeling, S., Scholkopf, B.: Learning to deblur. IEEE Trans. Pattern Anal. Mach. Intell. **38**(7), 1439–1451 (2016). https://doi.org/10.1109/tpami.2015.2481418
7. Gondara, L.: Medical image denoising using convolutional denoising autoencoders, 241–246 (2016). https://doi.org/10.1109/ICDMW.2016.0041
8. Venkataraman, P.: Image denoising using convolutional autoencoder (2022) https://arxiv.org/abs/2207.11771
9. Alshathri, I., Vincent, D.J., Hari, V.S.: Denoising letter images from scanned invoices using stacked autoencoders. Comput. Mater. Contin. **71**(1), 1371–1386 (2022)
10. Zhu, J.-Y., Park, T., Isola, P., Efros, A.A.: Unpaired image-to-image translation using cycle-consistent adversarial networks, pp. 2242–2251 (2017). https://doi.org/10.1109/ICCV.2017.244

11. Isola, P., Zhu, J.-Y., Zhou, T., Efros, A.A.: Image-to-image translation with conditional adversarial networks, pp. 5967–5976 (2017). https://doi.org/10.1109/CVPR.2017.632
12. Kupyn, O., Budzan, V., Mykhailych, M., Mishkin, D., Matas, J.: Deblurgan: Blind motion deblurring using conditional adversarial networks (2018). arXiv:1711.07064 [cs.CV]
13. Souibgui, M.A., Kessentini, Y.: DE-GAN: a conditional generative adversarial network for document enhancement. IEEE Trans. Pattern Anal. Mach. Intell. **44**(3), 1180–1191 (2022). https://doi.org/10.1109/tpami.2020.3022406
14. Neji, H., Halima, M.B., Hamdani, T.M., Nogueras-Iso, J., Alimi, A.M.: Blur2sharp: a gan-based model for document image deblurring. Int. J. Comput. Intell. Syst. (2021). https://doi.org/10.2991/ijcis.d.210407.001
15. Sun, J., Du, Y., Li, C., Wu, T.H., Yang, B., Mok, G.S.P.: Pix2pix generative adversarial network for low dose myocardial perfusion spect denoising. Quant. Imaging Med. Surg. **12**(7), 3539–3555 (2022). https://doi.org/10.21037/qims-21-1042
16. Jadhav, P., Sawal, M., Zagade, A., Kamble, P., Deshpande, P.: Pix2pix generative adversarial network with resnet for document image denoising, pp. 1489–1494 (2022). https://doi.org/10.1109/ICIRCA54612.2022.9985695
17. Saddami, K., Afrah, P., Mutiawani, V., Arnia, F.: A new adaptive thresholding technique for binarizing ancient document, pp. 57–61 (2018). https://doi.org/10.1109/IN'APR.2018.86
18. Gatys, L.A., Ecker, A.S., Bethge, M.: Image style transfer using convolutional neural networks, pp. 2414–2423 (2016). https://doi.org/10.1109/CVPR.2016.265
19. Johnson, J., Alahi, A., Fei-Fei, L.: Perceptual losses for real-time style transfer and super-resolution **9906**, 694–711 (2016). https://doi.org/10.1007/978-3-319-46475-6_43
20. Gulrajani, I., Ahmed, F., Arjovsky, M., Dumoulin, V., Courville, A.: Improved training of wasserstein gans (2017). https://arxiv.org/abs/1704.00028arXiv:1704.00028 [cs.LG]
21. Arjovsky, M., Chintala, S., Bottou, L.: Wasserstein generative adversarial networks **70**, 214–223 (2017)
22. Thabs: Noisy and rotated scanned documents (2020). https://www.kaggle.com/datasets/sthabile/noisy-and-rotated-scanned-documents

Chapter 10
Comprehensive Survey of Audio-to-Text Conversion

Aishwarya Parthasarathi, Almas Banu, and Ashwini Joshi

Abstract Internet has undergone a remarkable evolution, reshaped numerous aspects of our world, and left an indelible mark on society. The Internet has unveiled a new era of rapid and accessible interaction, redefining how we connect and converse. In this paper, our objective is to explore various methodologies employed for the conversion of speech into text. The paper embarks on a journey through the historical evolution of audio-to-text conversion, tracing its roots from rudimentary systems to the current state-of-the-art models. This review dissects what are features of audio-to-text conversion and how machine learning models and neural networks influence this process. We have explored various tools and evaluation metrics which are available for the same process. This review further dissects the real-world applications of audio-to-text conversion, domains such as transcription services, voice assistants, and accessibility technologies. It explores the challenges and open research questions on this topic.

10.1 Introduction

Automatic Speaker Recognition (ASR) plays a pivotal role in security, accessibility, and automation. In a world where seamless communication and security are paramount, ASR stands as a cornerstone technology, enabling both the power of voice and the assurance of identity [1].

Audio-to-text conversion is a technology that transforms spoken language into written text. The process requires complex Machine learning algorithms and neural network architectures to precisely transcribe the verbal content into textual format.

A. Parthasarathi (✉) · A. Banu · A. Joshi
Department of Computer Science and Engineering, PES University, Bengaluru, India
e-mail: pes1ug20cs523@pesu.pes.edu

A. Banu
e-mail: pes1ug20cs535@pesu.pes.edu

A. Joshi
e-mail: ashwinimjoshi@pes.edu

S. K. Pal et al. (eds.), *Intelligent Informatics*, Smart Innovation, Systems and Technologies 389, https://doi.org/10.1007/978-981-97-2147-4_10

The integration of ASR has transformed our interactions with spoken language, unlocking a multitude of practical applications. Although there have been notable strides in speech-to-text conversion technology, achieving absolute accuracy remains a challenge. This challenge arises due to the frequent presence of background noise, intricate vocabulary, and substantial ambient disturbances in audio sources.

10.1.1 Significance of Audio-to-Text Conversion

The significance of audio-to-text transcription and conversion transcends academic realms and is profoundly embedded in numerous real-world applications. This technology plays a pivotal role in the realms of speech recognition, natural language processing, and accessibility. By bridging the gap between spoken language and written text, it enables machines to comprehend human communication effectively. This breakthrough has implications across diverse domains, including healthcare, legal, media, education, and accessibility services, and many more.

10.1.2 Purpose of the Survey Paper

It enables a comprehensive assessment of the current state of the field and emerging trends, shedding light on various methodologies, online services, metrics, and practical applications specific to speech-to-text conversion. By comparing various approaches, we aim to identify common problems, technological gaps, and avenues for future research. Moreover, this survey paper intends to foster collaboration and knowledge sharing among experts, promoting continuous growth and innovation within the audio-to-text conversion domain.

10.2 Historical Perspective

The history of ASR can be traced back to the mid-twentieth century, with the development of Hidden Markov Models (HMMs) and Gaussian Mixture Models (GMMs). These early models laid the foundation for subsequent advancements in speech recognition. Over the years, the field has evolved from rule-based and statistical models to deep learning approaches, such as Recurrent Neural Networks (RNNs) and Convolutional Neural Networks (CNNs). This shift has significantly improved the accuracy and versatility of ASR systems.

10.2.1 Key Milestones and Breakthroughs

The historical trajectory of audio-to-text transcription/conversion has witnessed several key milestones and breakthroughs [2]:

- 1950s–1960s: Early efforts in speech recognition involved rule-based approaches and limited vocabulary tasks. The "Harpy" system developed at Carnegie Mellon University in the 1970s marked a significant step in continuous speech recognition.
- 1980s–1990s: The use of Hidden Markov Models (HMMs) gained prominence in the 1980s, allowing for more robust speech recognition systems. Researchers started to explore statistical approaches for modeling spoken language.
- 1990s–2000s: The introduction of large-scale speech corpora and advances in machine learning paved the way for significant improvements in speech recognition accuracy. Research efforts shifted toward using statistical language models and training on extensive datasets.
- 2000s–2010s: The rise of deep learning, particularly deep neural networks (DNNs) and convolutional neural networks (CNNs), led to remarkable breakthroughs in audio to text transcription. These neural network architectures demonstrated superior performance in both acoustic modeling and language modeling.
- 2010s–2020s: Transformer-based models, such as the "Attention is All You Need" architecture, have become the state of the art in various NLP tasks, including transcription. They have further improved the accuracy and efficiency of audio-to-text conversion.
- 2020s–Present: Artificial intelligence and ML models like deep learning and neural networks are being used in various transcript generation tools. Most of the tools are accurate but still some of them have drawbacks (Fig. 10.1).

10.3 Features for Audio-to-Text Conversion

The key capabilities of a good ASR system are as follows:

Accurate transcription of multiple languages, Support for various audio formats, Reliable performance in noisy environments, Contextual understanding for coherent transcriptions, Capturing speech nuances like pitch and tone, Speaker identification for specific transcriptions, Real-time transcription for live events and interactive applications, Robustness to diverse accents, Compatibility with various platforms and operating systems, Efficient resource usage on different devices, Flexible output formats (plain text, JSON, subtitles), and Scalability for processing large audio data volumes.

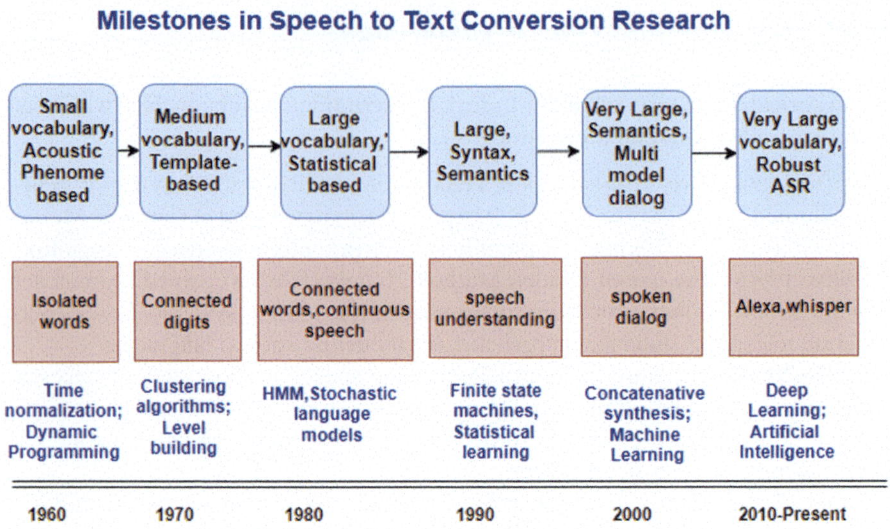

Fig. 10.1 Milestones in speech recognition and understanding technology over the past 40 years

10.4 Comparison Between Various Methods

10.4.1 Various Models for Audio-to-Text Conversion

1. **Hidden Markov Models (HMMs)**

These are subclass of dynamic Bayesian model. HMM has a Markov process having some hidden states. It can be a probabilistic model with static variable S, some observation variable O, and transition between states having probability [3].

Advantages

- Flexible, adaptable, robust, scalable, automatic trainable, efficient, and accurate [4].
- Can be tailored to different levels of granularity, such as phonemes, words, or phrases, and can also incorporate additional information like context, grammar, or pronunciation [4].
- Can handle noise and distortion in speech signals as well as variations in speaker, accent, or environment.
- Can be trained using fast algorithms like the Expectation-Maximization algorithm or the Baum-Welch algorithm and can achieve high accuracy when combined with other techniques like neural networks or deep learning [4].

Disadvantages

- They make strong assumptions about the independence and stationarity of the observations and the states, which may not hold in reality.

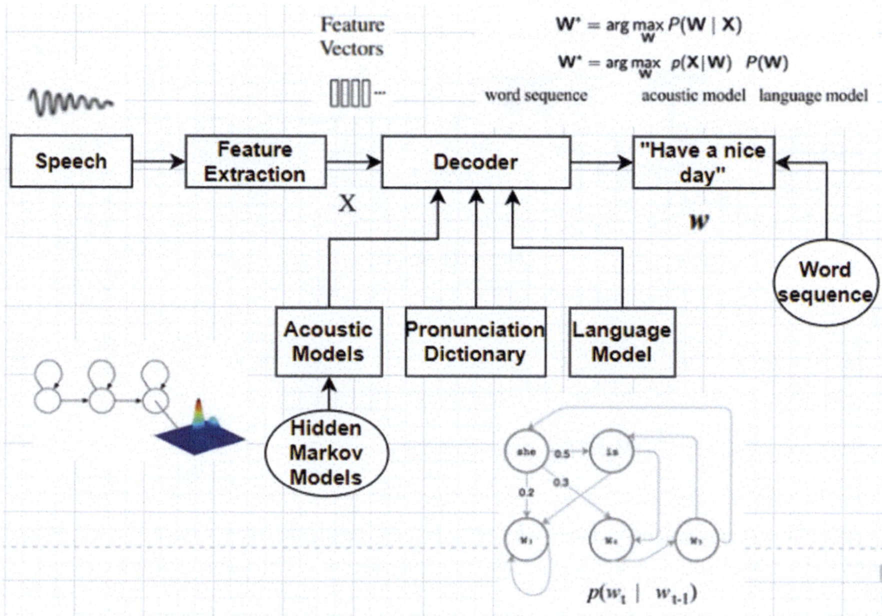

Fig. 10.2 Architecture of a HMM-based speech-to-text recognizer

- Limited capacity to capture complex dependencies in data.
- They are dependent and sensitive, relying heavily on the choice and quality of the features and parameters which may affect their performance.
- Amount of data required to train an HMM is very large [5].
- The number of parameters needed to set up an HMM is huge [5] (Fig. 10.2).

2. **Gaussian Mixture Models (GMMs)**

It is a parametric probability density function represented as weighted sum of Gaussian component densities [6]. Each Gaussian distribution is calculated by mean, variance, and weight of the Gaussian distribution. It is described by mean vectors, covariance matrices, and mixture weights from all component densities [6].

Advantages

- Effective for modeling acoustic features.
- Simplicity and interpretability.

Disadvantages

- May struggle with modeling complex speech patterns.
- Limited capacity for capturing long-range dependencies (Fig. 10.3).

3. **Deep Neural Networks (DNNs)**

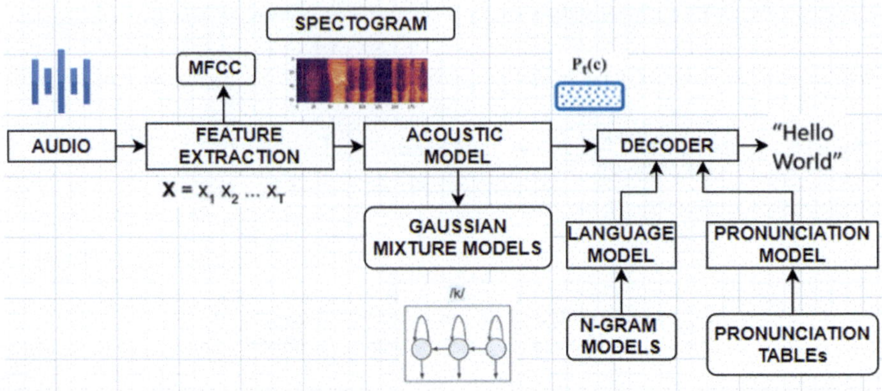

Fig. 10.3 Architecture of a GMM-based speech-to-text recognizer

A type of machine learning networks which have multiple layers between input and output. Each state has probabilistic distribution among all possible output values [7].

Advantages

- High capacity for feature learning.
- Can capture complex, hierarchical patterns in data.
- Reduces the need for manual feature engineering.

Disadvantages

- Requires large amounts of training data.
- Computational resource-intensive.
- May be perceived as a "black box."

4. **Convolutional Neural Networks (CNNs)**

These networks are commonly used for image analysis [8]. It can also be used for audio-to-transcript generation, for training input should be audio-labeled data (like audio spectrogram or MFCCs) which is typically done using gradient-based optimization algorithms [9].

Advantages

- Effective at capturing local acoustic features.
- Scalable to large datasets [8].

Disadvantages

- May struggle with modeling long-range dependencies.
- Limited in capturing context information.

5. **Recurrent Neural Networks (RNNs)**

It is a class of artificial neural network, which has joins between nodes joins as in from more distant node to less distant node. They handle the sequence of words in natural language processing [7].

Advantages

- Suitable for sequential data.
- Can capture long-range dependencies.

Disadvantages

- Vulnerable to vanishing gradient problems.
- Less parallelizable compared to feedforward networks

6. **Transformer-Based Models**

These models use self-attention layers to capture effectively long-range dependencies among the given input string. It also uses positional encoding [10].

Advantages

- Effective at capturing contextual information.
- Capture long-range dependencies in audio sequences effectively.
- Parallelizable, leading to faster training.
- State-of-the-art performance in various NLP tasks.

Disadvantages

- Requires a large amount of data and computational resources.
- Complexity in implementation.

7. **Listen, Attend, and Spell (LAS) Models**

A neural network that learns to transcribe speech to characters this model learns all components of speech together. This particular system consists of two components listener—a pyramidal recurrent network encoder where inputs are given as filter bank spectra and speller—an attention-based recurrent network decoder that gives output as characters without making any independent assumptions between the characters [11].

Advantages

- Dynamically focuses on relevant parts of the input during transcription.
- Effective for end-to-end ASR tasks.

Disadvantages

- Streaming speech recognition is not possible [12]. The attention mechanism of LAS requires the entire audio stream to be processed by the encoder before the decoder can start emitting output labels [11].
- Requires careful hyperparameter tuning.
- Can be computationally expensive (Fig. 10.4).

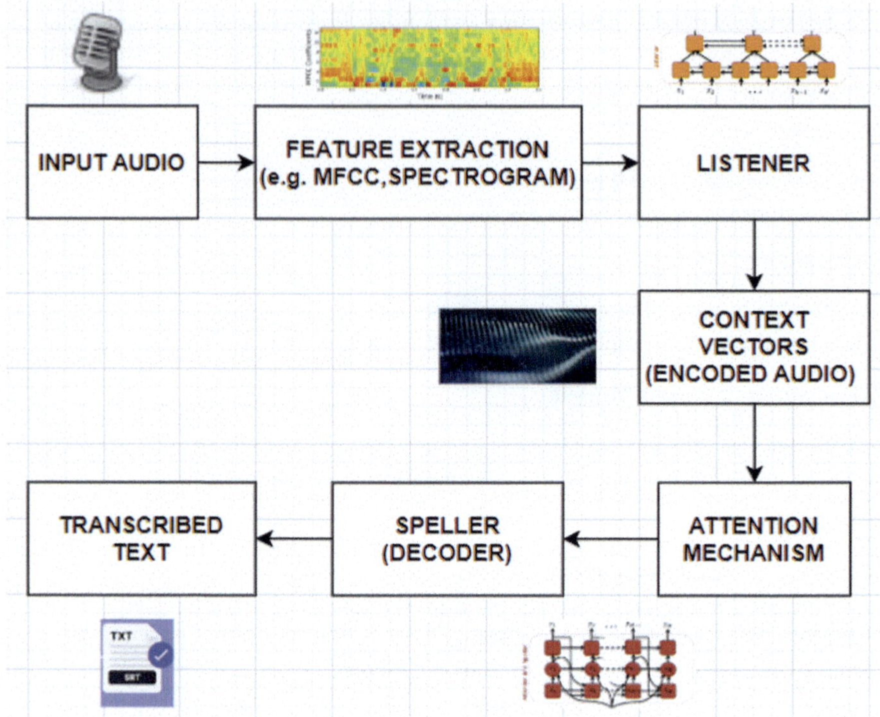

Fig. 10.4 Architecture of a LAS-based speech-to-text recognizer

8. **End-to-End Models**

A system that directly translates a sequence of input acoustic features into a sequence of words or graphemes [13]. It depends on pre-training its encoder and decoder [14].

Advantages

- Simplifies the ASR pipeline by removing intermediate steps.
- Can be tailored to specific applications.

Disadvantages

- May require substantial training data and computational resources [12].
- Challenging to interpret.

9. **Hybrid Models**

The model divides the original model into sound recognition and subsequently converts a sequence of sounds into words. A neural network for phoneme recognition and conversion from phoneme to text requires a graph [15].

Advantages

- Combines the strengths of traditional HMMs and deep learning models [15].
- Improved accuracy compared to standalone approaches.

Disadvantages

- Increased complexity in system design.
- Requires careful model integration.

10. **Connectionist Temporal Classification (CTC)**

CTC is used to synchronize input acoustic features with their corresponding textual transcriptions, allowing for flexible arrangements without the requirement for a strict one-to-one correspondence between them [16].

Advantages

- Allows the model to learn alignment between input and output.
- Suitable for sequence-level labeling tasks.

Disadvantages

- Limited in capturing fine-grained phonetic information.
- Can be sensitive to label imbalance (Fig. 10.5).

11. **Dynamic Time Wrapping**

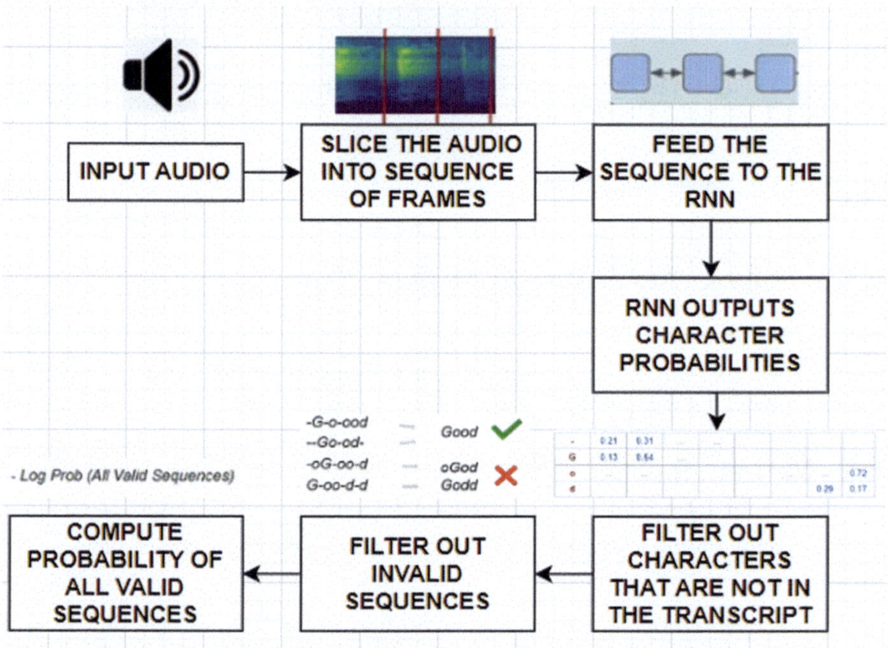

Fig. 10.5 CTC loss algorithm for audio to text

It is a method used to compare two sequences, often of a temporal nature, that do not perfectly synchronize. It enables the calculation of the optimal alignment between these two sequences [3].

Advantages

- The DTW algorithm is used to find the analogy in two-time series events that vary in speed by using dynamic programming.
- Its purpose is to iterate the pair of sequence of feature vectors and finding a feasible match between them.

Disadvantages

- The problem arises in selecting the reference template for comparing the time series events.

12. **CMUSphinx framework**

It is used to train and test the speech processing system. It is dynamic in nature. The main components include acoustic model, language model, and dictionary [17].

Advantages

- Capability for offline speech recognition system.
- It supports multiple languages and systems which have different language requirements [3], 17.

Disadvantages

- Accuracy may not be the same as for other speech recognition systems for certain accents.
- Requires large memory and processing power [3].
- Complex to set up and configure.

10.4.2 Approaches to Various Method

See Table 10.1.

10.5 Various Tools

There are many online tools and APIs for speech recognition. Some of them are as follows:

- Python's pydub module which takes audio files in the form of.wav files to give transcript of audio [18].

Table 10.1 There are three primary approaches to speech-to-text conversion. Each approach has its advantages and disadvantages, which are discussed below

Methods	Advantages	Disadvantages
Rule-based methods: Usepredefined rules to recognize speech patterns and translate them into text [3]	• Fast and efficient • Easy to understand and implement • High accuracy for specific vocabulary	• Limited scalability and poor performance on large datasets • Requires a lot of domain-specific knowledge • Difficult to handle variations in speech
Statistical methods: Use statistical models to analyze speech patterns and identify probabilities of certain words or phrases occurring together [3]	• Can handle variations in speech • Can be trained on existing data	• May not perform well with small datasets • Can be computationally expensive
Deep learning methods: Use neural networks to learn patterns in speech and generate text outputs [3]. These can be further divided into two categories	• Can learn complex patterns and relationships in data • Can handle variations in speech and vocabulary	• Requires large amounts of training data and computing power • Can be difficult to interpret and understand

- Google Speech Recognition system [19] has user-friendly environment for text synthesis.
- AWS Transcribe [20] offered by Amazon which has automatic speech recognition capabilities.
- Pocket Sphinx [17] is a command line process which recognizes speech.
- IBM Watson [19] has advanced speech recognition capabilities and provides good quality and accurate transcripts.
- Whisper [21] A general-purpose speech recognition system. It performs multilingual Speech Recognition, Translation, and Language identification.
- Assembly AI [22] An API which can be used to understand audio files and transcribe them.

10.6 Inferences Made

In this section, we aim to condense the main discoveries and insights derived from our extensive examination of Automatic Speech Recognition (ASR) technology and its practical applications. By methodically delving into ASR's historical development, diverse methodologies, and real-world use cases, we have unveiled significant patterns and connections within this field.

1. The shift toward deep learning models, particularly Transformer-based architectures, has led to significant advancements in ASR accuracy and efficiency.

2. Multilingual support and real-time transcription capabilities are becoming increasingly essential features for ASR systems.
3. Open-source ASR tools and frameworks play a vital role in making this technology accessible to a wider audience.

10.7 Evaluation Metrics

There are various methods to evaluate a generated text from an audio using a speech-to-text synthesizer.

10.7.1 Word Error Rate

The metric which identifies how many words are joined incorrectly after the conversion by the method. WER is based on string edit distance [1].

$$\mathrm{WER} = (\mathrm{S} + \mathrm{D} + \mathrm{I})/\mathrm{N} \tag{10.1}$$

where S stands for substitution (capturing a word which is wrong), D stands for deletions (words that are not included by method), I stands for insertions (words that are not spoken included by method), and N stands for the total number of words spoken.

It penalizes all differences between the method's result and reference transcript. This metric gives accurate results for zero-shot models like the whisper model.

There are a few issues in WER

- It gives 1:1 transcript and it does not take into account the alternatives or word lattices as in ASR systems.
- It gives the same importance to words which are articles compared to verbs and nouns that add more semantic value.

Lower WER values indicate better ASR performance, as they represent fewer errors in the generated output compared to the reference.

A few of WER Variants are described below

1. **Character Error Rate (CER)** [23]
   The rate of characters which are transcribed incorrectly by the text conversion model. CER of 0.1 and less than that are considered effective for automatic transcription [23].
2. **Semantic Word Error Rate (SWER)** [24]: It compares sentences instead of words.

$$\mathrm{SWER}\ (\mathrm{s}_1,\ \mathrm{s}_2) = \left(\mathrm{I} + \mathrm{D} + \sum \mathrm{S}(\mathrm{w}_\mathrm{i},\ \mathrm{w}_\mathrm{j})\right)/(\max\ (|\mathrm{s}_1|,\ |\mathrm{s}_2|)) \tag{10.2}$$

where I is the number of insertions, D is the number of deletions, and S (w_i,w_j) is the similarity of 2 words w_i and w_j.

3. **Match Error Rate (MER)** [25]: It quantifies the likelihood of a particular match being wrong.

$$\begin{aligned} \text{MER} &= (\text{S} + \text{D} + \text{I})/(\text{S} + \text{D} + \text{I} + \text{H}) \\ &= 1 - (\text{H}/(\text{S} + \text{D} + \text{I} + \text{H})) \end{aligned} \tag{10.3}$$

where H is the number of hits. MER is always less than or equal to WER

10.7.2 Levenshtein Distance

A string metric that calculates the difference between two words based on how many characters need to be changed to get as same word that is converting a string A to a string B. It is also known as edit distance [26].

- The distance is calculated by preparing a matrix of (M + 1) × (N + 1) size where M and N are lengths of two words. Iteratively operations are done using loops. It measures the number of insertions and deletions.

10.8 Challenges and Future Work

Name	Description	To address this
Handling diverse accents and languages [27]	ASR systems are typically trained on a specific dialect or language, leading to difficulties in handling diverse accents and languages	– Multilingual training data, Accent adaptation [28], Transfer learning
Multilingual and code-switching ASR	Developing ASR systems capable of handling multilingual conversations and code-switching is a significant challenge	– Code-switching models, Multilingual ASR
Robustness to noise and environmental conditions	ASR systems often struggle with background noise and adverse conditions	– Noise reduction techniques, Robust acoustic models, Environmental adaptation, Real-world data collection
Crosstalk	Speakers can sometimes interrupt each other during the discussion, making it challenging to transcribe accurately	– Speaker diarization, Contextual understanding, multimodal ASR

10.8.1 Audio-to-Text Conversion Potential Applications

The use cases for Audio-to-Text technology, commonly known as Automatic Speech Recognition (ASR), are extensive and influential. ASR has transformed our interaction with spoken language, creating fresh opportunities across numerous sectors and industries. In this survey paper, we explore the multitude of ways ASR technology is shaping our digital landscape, improving efficiency, and enhancing user experiences in countless real-world scenarios. Our goal is to illuminate how ASR has significantly enhanced our digital experiences, making them more efficient and user-friendly.

(1) **Transcription Services**:

- ASR can be used to transcribe audio recordings of meetings, interviews, conferences, and lectures, saving time and effort in manual transcription.
- In healthcare, ASR assists medical professionals by transcribing patient records, clinical notes, and dictations.
- ASR can be integrated into language translation services to transcribe spoken language and translate it into multiple languages.
- ASR simplifies the transcription of legal proceedings, including court hearings, depositions, and legal documentation.

(2) **Voice Assistants**: Popular voice assistants like Siri, Google Assistant, and Alexa rely on ASR to understand and respond to user voice commands [29, 30].

(3) **Podcast Summarization**: ASR can transcribe the spoken content of podcast episodes, converting audio into text. This serves as the foundational step for summarization [20].

(4) **Accessibility Services**:

- ASR improves accessibility by providing live transcription for individuals with hearing impairments during public events, conferences, and lectures [31].
- ASR-powered apps can convert printed text into spoken language, helping visually impaired individuals access written content [32].

(5) **Telematics and In-Car Systems**: In-car voice recognition systems use ASR for hands-free control of navigation, entertainment, and communication [33].

(6) **Voice-Controlled Smart Home Devices**: ASR powers voice-controlled devices like smart speakers, thermostats, and lights, allowing users to interact with their smart homes using voice commands [34].

(7) **Voice Authentication and Biometrics**: ASR technology can be used for voice-based user authentication, enhancing security in applications like banking and access control [35].

(8) **Language Learning**: ASR-assisted language learning applications help learners practice pronunciation and receive feedback on their spoken language skills [36].

(9) **Voice-Enabled Gaming**: ASR enhances the gaming experience by enabling voice commands and interactions within video games and virtual reality environments [37].
(10) **Educational Assessment**: ASR technology is used for automated grading and assessment of spoken responses in language tests and educational assessments [36].

These applications demonstrate the versatility and widespread adoption of ASR technology across various domains, from consumer electronics to healthcare, education, entertainment, and beyond. The continued advancements in ASR algorithms and models are expected to further expand its applications in the future.

10.9 Conclusion

This paper provides an overview of audio-to-text conversion, highlighting its methods, advancements, and ongoing challenges. It emphasizes the growing importance of audio-to-text conversion in various fields. The historical evolution of Automatic Speech Recognition (ASR) is discussed, from rule-based approaches to deep learning and transformer-based models, which have improved accuracy and versatility in Natural Language Processing (NLP).

ASR performance is evaluated using metrics like Word Error Rate (WER), Character Error Rate (CER), and Levenshtein distance to assess transcription accuracy. However, ASR faces challenges, including handling diverse accents and languages, accommodating multilingual and code-switching ASR, and ensuring robustness in noisy environments and crosstalk. To overcome these challenges, future work in ASR should focus on various aspects, including multilingual training data, accent adaptation, transfer learning, audio repair techniques, code-switching models, acoustic model robustness, contextual understanding, speaker diarization improvements, and multimodal ASR.

Despite these challenges, ASR is being increasingly applied in voice assistants, closed captioning, customer support chatbots, and voice-controlled smart home devices, improving efficiency and accessibility. It also aids in content creation, language learning, gaming experiences, and educational assessment. Looking forward, the ongoing development of ASR technology is essential to enhance accessibility, efficiency, and accuracy across different domains in response to evolving communication needs.

References

1. Besacier, L., Barnard, E., Karpov, A., Schultz, T.: Automatic speech recognition for under-resourced languages: a survey. Speech Commun. **56**, 85–100 (2014)

2. Juang, B.H., Rabiner, L.R.: Automatic speech recognition—A brief history of the technology development (2004)
3. Nagdewani, S., Jain, A.: A review on methods for speech-to-text and text-to-speech conversion. Int. Res. J. Eng. Technol. (2020)
4. Elakkiya, A., Jaya Surya, K., Venkatesh, K., Aakash, S.: Implementation of speech to text conversion using hidden markov model. In: 6th International Conference on Electronics, (2022)
5. Issues and limitations of HMM in speech processing: a survey. Int. J. Comput. Appl. (0975 – 8887) **141**(7) (2016)
6. Chauhan, V., Dwivedi, S., Karale, P., Potdar, S.M.: Speech to text converter using Gaussian Mixture Model (GMM). Int. Res. J. Eng. Technol. (IRJET) (2016)
7. Basystiuk, O., Shakhovska, N., Bilynska, V., Syvokon, O., Shamuratov, O., Kuchkovskiy, V.: The Developing of the System for Automatic Audio to Text Conversion. Lviv Polytechnic National University, 12 Bandera str., Lviv, 79013, Ukraine (2021)
8. Abdel-Hamid, O., Mohamed, A.-R., Jiang, H., Penn, G.: Applying convolutional neural networks concepts to hybrid NN-HMM model for speech recognition. In: Proceedings of IEEE International Conference Acoustic, Speech Signal Processing (ICASSP), March (2012)
9. Chan, W., Lane, I.: Deep convolutional neural networks for acoustic modeling in low resource languages. In: Proceedings of IEEE International Conference Acoustic, Speech Signal Processing (ICASSP), April (2015)
10. Latif, S., Zaidi, A., Cuayahuitl, H., Shamshad, F., Shoukat, M., Qadir, J.: Transformers in speech processing: a survey, March (2023)
11. Chan, W., Jaitly, N., Le, Q., Vinyals, O.: Listen, attend and spell: A neural network for large vocabulary conversational speech recognition. In: 2016 IEEE International Conference on Acoustics, Speech and Signal Processing (ICASSP), March, pp. 4960–4964. Shanghai, China (2016)
12. Sainath, T.N., et al.: Two-pass end-to-end speech recognition. In: INTERSPEECH (2019)
13. Sung, T.-W., Liu, J.-Y., Lee, H.-Y., Lee, L.-S.: Towards end-to-end speech-to-text translation with two-pass decoding. In: ICASSP 2019—2019 IEEE International Conference on Acoustics, Speech and Signal Processing (ICASSP), pp. 7175–7179, Brighton, UK (2019)
14. Hannun, A., Case, C., Casper, J., Catanzaro, B., Diamos, G., Elsen, E., Prenger, R., Satheesh, S., Sengupta, S., Coates, A., Ng, A.Y.: Deep speech: scaling up end-to-end speech recognition (2014)
15. Kusumah, R., Hartanto, R., Hidayat, R.: Hybrid automatic speech recognition model for speech-to-text application in smartphones. In: 2019 International Conference on ICT for Smart Society (ICISS), Bandung, Indonesia (2019)
16. Moritz, N., Hori, T., Roux, J.L.: Streaming end-to-end speech recognition with Joint CTC-attention based models. In: 2019 IEEE Automatic Speech Recognition and Understanding Workshop (ASRU), pp. 936–943. Singapore (2019)
17. Shivakumar, K.M., Jain, V.V., Krishna Priya, P.: A study on impact of language model in improving the accuracy of speech to text conversion system. In: International Conference on Communication and Signal Processing, India (2017)
18. Bharti, N., Nadeem Hashmi, S., Manikandan, V.M.: An approach for audio/text summary generation from webinars/online meetings. In: 13th International Conference on Computational Intelligence and Communication Networks, September (2021)
19. Siegert, I., Sinha, Y., Jokisch, O., Wendemuth, A.: Recognition performance of selected speech recognition APIs—A longitudinal study—Speech and computer. In: 22nd International Conference, SPECOM 2020, Proceedings, October 7–9, pp. 520–529. St. Petersburg, Russia (2020)
20. Vartakavi, A., Garg, A., Rafii, Z.: Audio summarization for podcasts. In: 2021 29th European Signal Processing Conference (EUSIPCO), pp. 431–435. Dublin, Ireland (2021)
21. Radford, A., Wook Kim, J., Xu, T., Brockman, G., Mcleavey, C., Sutskever, I.: Proceedings of the 40th International Conference on Machine Learning. PMLR, vol. 202, pp. 28492–28518 (2023)

22. NORDIC APIS. https://nordicapis.com/review-of-assemblyai-speech-to-text-api/. Last Accessed 17 Nov 2020
23. Scott MacKenzie, I., William Soukoreff, R.: A character-level error analysis technique for evaluating text entry methods. In: NordiCHI '02: Proceedings of the Second Nordic Conference on Human-Computer Interaction, NY, United States, October (2002)
24. Spiccia, C., Augello, A., Pilato, G. Vassallo, G.: Semantic word error rate for sentence similarity. In: 2016 IEEE Tenth International Conference on Semantic Computing (ICSC), pp. 266–269. Laguna Hills, CA, USA. https://doi.org/10.1109/ICSC.2016.11 (2016)
25. Cameron Morris, A., Maier, V., Duncan, P.G.: From WER and RIL to MER and WIL: improved evaluation measures for connected speech recognition (2004)
26. Droppo, J., Acero, A.: Context dependent phonetic string edit distance for automatic speech recognition. In: Acoustics Speech and Signal Processing (ICASSP), 2010 IEEE International Conference, pp. 4358–4361 (2010)
27. Babatunde, O.: Automatic speech recognition for Nigerian-accented English. In: 5th Deep Learning Indaba Conference (2023)
28. Behrman, A.: A clear speech approach to accent management. Am. J. Speech-Lang. Pathol. **26**(4), 1178–1192 (2017)
29. Petrock, V.: US Voice Assistant Users 2019—Who, What, Where and Why. eMar-keter (2019). Accessed 15 July 2019
30. Roberts, M.: OK Google, Siri, Alexa, Cortana; can you tell me some stats on voicesearch? The edit blog (2018). Accessed 01 Aug 2018
31. Jeyalakshmi, C., Krishnamurthi, V., Revathi, A.: Speech recognition of deaf and hard of hearing people using hybrid neural network. In: 2010 2nd International Conference on Mechanical and Electronics Engineering, pp. V1-83–V1-87. Kyoto, Japan (2010)
32. Edupuganti, S.A., Durga Koganti, V., Lakshmi, C.S., Naveen Kumar, R., Paruchuri, R.: Text and speech recognition for visually impaired people using Google vision. In: 2nd International Conference on Smart Electronics and Communication (ICOSEC), pp. 1325–1330. Trichy, India (2021)
33. Withanage, P., Liyanage, T., Deeyakaduwe, N., Dias, E., Thelijjagoda, S.: Road navigation system using automatic speech recognition (ASR) and natural language processing (NLP). In: 2018 IEEE Region 10 Humanitarian Technology Conference (R10-HTC), pp. 1–6. Malambe, Sri Lanka (2018)
34. Mittal, Y., Toshniwal, P., Sharma, S., Singhal, D., Gupta, R., Mittal, V.K.: A voice-controlled multi-functional smart home automation system. In: 2015 Annual IEEE India Conference (INDICON), pp. 1–6. New Delhi, India (2015)
35. Zhang, X., Xiong, Q., Dai, Y., Xu, X.: Voice biometric identity authentication system based on android smart phone. In: 2018 IEEE 4th International Conference on Computer and Communications (ICCC), pp. 1440–1444. Chengdu, China (2018)
36. Carrier, M.: Automated Speech Recognition in language learning: potential models, benefits and impact. Train. Lang. Cult. **1**(1), 46–61 (2017)
37. Waqar, D.M., Gunawan, T.S., Kartiwi, M., Ahmad, R.: Real-time voice-controlled game interaction using convolutional neural networks. In: 2021 IEEE 7th International Conference on Smart Instrumentation, Measurement and Applications (ICSIMA), pp. 76–81. Bandung, Indonesia (2021)

Chapter 11
Rainfall Forecasting Using High Spatiotemporal Satellite Imagery and Machine Learning Techniques: A Case Study Using INSAT 3DR Data

V. Deepthi Sasidhar, T. Anuradha, and M. V. Ajay Kumar

Abstract Accurate rainfall forecasting is crucial for various applications such as agriculture, hydrology, and disaster management. Recent advancements in satellite technology have enabled the collection of high spatiotemporal resolution data, which can be used to improve rainfall forecasting models. In this study, we utilize the cloud brightness temperature (CBT) derived from INSAT 3DR satellite images to predict rainfall over the Indian region. The CBT values are calculated using a time series approach, and the random forest algorithm is employed to develop a forecasting model. The performance of the model is evaluated using a set of evaluation metrics, and the results show that the proposed methodology can accurately predict rainfall with a high degree of accuracy. The advantages of using high spatiotemporal satellite imagery and machine learning techniques for rainfall forecasting are discussed, and future research directions in this area are also explored.

11.1 Introduction

Rainfall forecasting stands as a pivotal pillar in weather prediction, wielding significant implications across diverse sectors such as agriculture, hydrology, aviation, and disaster management. The precision of rainfall predictions bears transformative potential, offering farmers the means to optimize crop yields, enabling water resource managers to strategize irrigation and flood control, aiding pilots in navigating safely, and empowering emergency responders with critical insights for disaster preparedness. Despite notable advancements in meteorology and computer science, the intricacies of rainfall forecasting persist as a formidable challenge, particularly when striving for fine spatial and temporal resolutions.

V. Deepthi Sasidhar · T. Anuradha (✉) · M. V. Ajay Kumar
Department of Information Technology, Velagapudi Ramakrishna Siddhartha Engineering College, Vijayawada, Andhra Pradesh, India
e-mail: anuradha_it@vrsiddhartha.ac.in

S. K. Pal et al. (eds.), *Intelligent Informatics*, Smart Innovation, Systems and Technologies 389, https://doi.org/10.1007/978-981-97-2147-4_11

In pursuit of more accurate and reliable rainfall forecasting, the integration of satellite data and machine learning techniques emerges as a promising frontier. Satellites, perched in the celestial expanse, afford an unparalleled vantage point for scrutinizing Earth's atmosphere. From this lofty perch, they capture high-resolution data encompassing cloud properties, atmospheric conditions, and surface characteristics. The marriage of these rich datasets with machine learning algorithms, spanning decision trees, random forests, support vector machines (SVM), and neural networks, holds the potential to forge predictive models of rainfall that transcend existing limitations.

A plethora of studies have underscored the efficacy of amalgamating satellite data and machine learning methodologies for rainfall prediction. For instance, researchers have leveraged satellite-derived cloud brightness temperature and precipitation data to craft probabilistic rainfall models. Other investigations have harnessed machine learning algorithms for classifying rain-bearing clouds, utilizing satellite-borne radar and visible imagery. Additionally, the exploration of deep learning techniques, such as convolutional neural networks (CNNs) and recurrent neural networks (RNNs), has shown promise in elevating rainfall forecasting capabilities [1].

Building upon the foundations laid by these antecedent endeavors, our study embarks on an exploration of the intricate interplay between high spatiotemporal satellite imagery and machine learning techniques for advancing rainfall forecasting accuracy, with a geographical focus on the Indian region. Our methodological approach centers around the utilization of data from the INSAT 3DR satellite, renowned for its provision of high-resolution visible and infrared imagery. By extracting essential atmospheric variables, including cloud brightness temperature, from these satellite-borne observations, we aim to harness the predictive power of various machine learning algorithms. The culmination of our efforts aspires to contribute substantively to the evolution of more precise and dependable rainfall forecasting systems. Ultimately, our research endeavors to furnish society with timely and accurate weather information, catering to the nuanced needs of diverse sectors and fortifying resilience in the face of variable climatic conditions.

About INSAT 3DR satellite: INSAT 3DR, launched by the Indian Space Research Organisation (ISRO) in 2016, stands as a geostationary Earth observation satellite and a vital component of the INSAT satellite series. This satellite plays a pivotal role in providing valuable data and high-resolution imagery through its payloads, including the Visible Infrared Imaging Radiometer Suite (VIIRS). The VIIRS instrument, with its spatial resolution of up to 50 m, contributes significantly to applications like crop monitoring, land cover classification, and cloud property estimation. Despite these advantages, there are potential challenges associated with INSAT 3DR data. These challenges encompass issues such as temporal and spatial resolution limitations, instrument sensitivity concerns, and data latency, all of which can impact the reliability and real-time applicability of the satellite data. Nevertheless, leveraging the high spatiotemporal resolution of INSAT 3DR imagery remains invaluable for studying meteorological phenomena, including rainfall, clouds, and atmospheric circulation patterns. Moreover, the satellite's geostationary orbit ensures continuous monitoring of the same region, facilitating long-term trend detection. In the realm of rainfall forecasting, INSAT 3DR satellite imagery plays a crucial role in estimating cloud brightness temperature—an essential variable for machine learning models predicting rainfall. By acknowledging and addressing potential challenges, researchers can enhance the accuracy and reliability of rainfall forecasting systems through the integration of INSAT 3DR data with other atmospheric variables and surface characteristics.

11.2 Review of Literature

11.2.1 Literature Study

Here are the research works that are done by other researchers. They suggested different methods for predicting the rainfall forecast using different machine learning methods. Researchers have harnessed the power of machine learning to boost rainfall forecasting accuracy. The first study highlights the success of hybrid models, especially GB-Adaboost, in surpassing traditional methods like SVM and neural networks. With a focus on meticulous feature selection using gradient boosting, the GB-Adaboost model stands out for its effectiveness in achieving accurate rainfall predictions [2]. Building on this foundation, the second study introduces a novel stacking ensemble approach, demonstrating its prowess in enhancing daily rainfall prediction accuracy. This innovative technique, evaluated on Indian climate data, achieves a notable accuracy of 81.2%, surpassing individual ML models and offering a valuable contribution to the field of climate forecasting [3]. Simanjuntak et al. pioneer a machine learning approach for high-resolution rainfall forecasting in Indonesia. By combining Multivariate LSTM and random forest, their model accurately predicts rain rates every 10 min, providing valuable updates for early warnings and aviation forecasting [4]. In Australia, the fourth study applies LSTM neural networks for enhanced rainfall prediction. The model, preprocessed with seasonal decomposition, outperforms a baseline ANN model, emphasizing LSTM's potential for accurate time series rainfall prediction and areas for improvement [5]. Ridwan et al.'s studies in Terengganu, Malaysia, present a comprehensive exploration of machine learning models for rainfall forecasting. Leveraging historical and projected rainfall data, the machine learning techniques, including boosted decision tree regression (BDTR), decision forest regression (DFR), Bayesian linear regression (BLR), and neural network regression (NNR), showcase their capacity to yield accurate and quantitative rainfall forecasts across various time horizons. Particularly, BDTR emerges as the most effective model, highlighting the versatility of these techniques [6, 7]. In India, the sixth study proposes a hybrid model combining STARMA, ANN, and SVM for better spatiotemporal rainfall forecasting. This approach outperforms traditional models, highlighting improved precision in modeling and forecasting for spatiotemporal rainfall patterns [8]. Another study takes a practical approach by employing basic machine learning techniques to construct weather forecasting models for predicting rain in major cities based on daily meteorological data. The comparative analysis explores modeling inputs, methodologies, and preprocessing procedures, revealing insights into the performance of different machine learning systems in forecasting rainfall. This study adds a vital perspective, emphasizing the importance of accurate weather predictions for agriculture in India and empowering individuals to make informed decisions based on weather forecasts [9]. These studies collectively showcase the versatility of machine learning [10, 11] in addressing the complexities of rainfall forecasting, offering valuable insights for diverse applications in climate science, agriculture, and societal decision-making.

11.2.2 Comparison with Previous Studies

In the field of weather forecasting, numerical weather models (NWMs) and statistical models have been fundamental components in predictive modeling for many years. NWMs utilize intricate mathematical representations of the atmosphere to simulate and predict weather patterns. In the evolving landscape of forecasting methodologies, the application of time series prediction has emerged as a promising technique for the prediction of temperature and rainfall. This innovative approach utilizes historical data to anticipate forthcoming values of temperature and rainfall, adeptly capturing temporal patterns and adapting to diverse data sources.

Integration of time series prediction with existing forecasting methods, such as NWMs and statistical models, contributes to further refining the accuracy of temperature and rainfall predictions. The implementation of time series prediction in temperature and rainfall forecasting typically involves distinct steps: gathering historical data, preprocessing the data, selecting a suitable time series prediction model, training the model, and employing the trained model for generating forecasts. Time series prediction holds considerable potential for advancing weather forecasting and climate change studies, offering a valuable resource for comprehending and anticipating weather patterns (Table 11.1).

11.3 Proposed Architecture and Methodology

11.3.1 Proposed Architecture

In this comprehensive architecture, data is collected from the INSAT 3DR geostationary meteorological satellite, encompassing diverse atmospheric parameters. The collected data undergoes a series of crucial preprocessing steps, including geometric correction to rectify distortions from satellite orbit and sensor orientation, radiometric correction for standardizing brightness and color, noise reduction to enhance data quality, and image enhancement for improved interpretability. Subsequently, the

Table 11.1 Different methods of the proposed work

Method	Benefits	Limitations
Numerical weather models	Can adapt to diverse elements impacting rainfall	Demanding in terms of computational resources and susceptible to errors in the initial data
Statistical models	Relatively simple and computationally efficient	Constrained by the quality of historical data and responsive to alterations in the surrounding environment
Time series prediction (Our methodology)	Can portray the time-dependent trends in cloud brightness temperatures and rainfall data	Necessitates access to pertinent satellite information and might exhibit decreased accuracy in extended-term predictions

preprocessed data is fed into a sophisticated multivariate Long Short-Term Memory (LSTM) network. This LSTM model is designed to predict the future cloud brightness temperature based on historical data, utilizing a range of input variables, such as cloud cover and temperature (Fig. 11.1).

Following the cloud brightness temperature prediction, the architecture seamlessly integrates the Random Forest algorithm to forecast rainfall. The cloud brightness temperature data, generated by the LSTM, serves as input to the Random Forest model, along with potentially incorporating other relevant meteorological features. The Random Forest algorithm, known for its ability to handle intricate relationships and nonlinear patterns, classifies pixels into rainfall and non-rainfall categories. This dual-step approach enhances the accuracy and interpretability of the rainfall predictions.

In the final phase of the workflow, the architecture produces a comprehensive rainfall forecast. The output from the Random Forest algorithm provides insights into

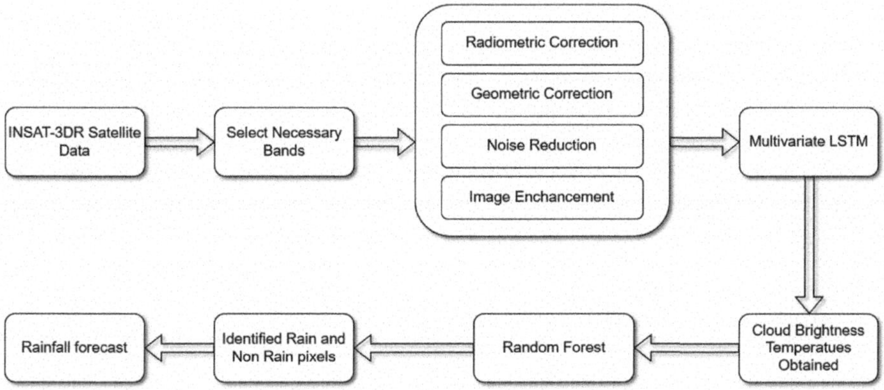

Fig. 11.1 Proposed architecture for rainfall prediction

areas where rainfall is anticipated and regions where it is less likely. This integrated approach not only leverages the strengths of advanced machine learning techniques like LSTM and Random Forest but also ensures a holistic understanding of the atmospheric conditions for accurate and actionable rainfall predictions. Throughout the process, model evaluation and iterative refinement play pivotal roles, allowing for the continuous improvement of the predictive models based on performance metrics and real-world validation.

11.3.2 Proposed Methodology of the Work

11.3.2.1 Algorithm

Input: Collected dataset from ISRO INSAT Satellite

Output: Prediction of rainfall

1. Start
2. Let I be collect a dataset of satellite images of INSAT 3DR from the ISRO website.
3. Load the Dataset I.
4. Perform data preprocessing on D by applying radiometric correction, geometric correction, and noise reduction.
5. Now apply the Multivariate LSTM on the preprocessed images to predict the cloud brightness temperatures by applying time series prediction.
6. After getting brightness temperatures, the random forest algorithm is applied to identify the rainfall and non-rainfall pixels and then predict the rainfall.
7. end

11.3.2.2 Data Preprocessing

For purposes such as geographic analysis, monitoring the environment, and doing spatial analysis, although raw satellite data frequently contains distortions, aberrations, and variations that can make it challenging to interpret the images precisely, satellite photos are an invaluable source of information. Preprocessing is a thorough process that addresses distortions by making a variety of adjustments and repairs. It is an important phase in enhancing the usefulness and quality of satellite data, removing noise, and preparing the data for intelligent analysis.

Radiometric Correction: When it comes to enhancing satellite pictures for accurate interpretation, radiometric correction is essential because air circumstances and sensor features can cause differences in pixel values in raw satellite data. Radiometric correction includes altering these values to guarantee uniformity and provide an accurate portrayal of surface reflectance. In order to successfully support applications like remote sensing, environmental monitoring, and geographical analysis, accurate and consistent satellite images must be produced.

Geometric Correction: In order to precisely represent space in satellite imagery, geometric correction is an essential step in the process. Disturbances in raw satellite data are frequently brought about by topographical features of Earth, sensor orientation, and satellite orbit. In order to precisely align pixels with their corresponding locations on the Earth's surface, geometric correction entails adjusting these distortions. This correction is necessary to ensure spatial accuracy in satellite images and support applications like land cover analysis, mapping, and geographical modeling.

Noise Reduction: Noise reduction is a pivotal preprocessing step in refining satellite imagery and addressing unwanted artifacts and disturbances inherent in raw data. Satellite images often suffer from noise stemming from factors like sensor limitations and atmospheric conditions. Spatial filtering techniques, including median and Gaussian filtering, are employed to selectively smooth images and diminish noise impact. By enhancing the signal-to-noise ratio, this process ensures that relevant features stand out prominently. Noise reduction is especially critical in applications such as remote sensing, environmental monitoring, and geographical analysis, where accurate information extraction is vital. The improved clarity achieved through noise reduction contributes to more reliable and informed decision-making in diverse fields, including agriculture, urban planning, and disaster management.

Image Enhancement: Satellite image improvement is an essential preprocessing technique that aims to improve the raw data's visual quality and interpretability. For a variety of reasons, satellite images frequently have less detailed visibility than ideal. Additionally, spatial filtering techniques like sharpening help to refine the image by emphasizing edges and features. Contrast enhancement techniques, like histogram equalization and contrast stretching, are utilized to accentuate pixel differences, bringing out finer details in the images. A better and more detailed depiction of the landscape is necessary for reliable analysis and decision-making in applications including land cover mapping, environmental monitoring, and disaster response. For these reasons, picture enhancement is critical.

11.3.2.3 Models Used for Training

Multivariate LSTM: Long Short-Term Memory (LSTM) is a specific type of Recurrent Neural Network (RNN), crafted for handling sequential data characterized by multiple variables. It excels in capturing complex patterns and dependencies in time series data by integrating memory cells and gates. LSTM effectively handles long-range dependencies, making it suitable for tasks like financial forecasting where multiple interconnected variables influence the outcome. Multivariate LSTM is adept at modeling intricate relationships within sequential data, offering enhanced predictive capabilities compared to traditional time series models.

Random Forest: Random Forest is an ensemble learning algorithm that constructs a multitude of decision trees during training and outputs the mode of the classes for classification or the average prediction for regression. Each tree is built on a random subset of the dataset, and their collective predictions result in a robust and accurate model. Random Forest excels in handling large datasets with diverse features, avoiding overfitting, and providing feature importance rankings. Its versatility makes it applicable to various domains, from finance to healthcare. This algorithm's strength lies in its ability to harness the collective intelligence of diverse decision trees for robust predictions.

11.4 Results

The graph illustrates the observed and predicted brightness temperatures over time for a specific location in India. The recorded brightness temperature consistently exceeds the predicted brightness temperature by a margin of roughly 2–3 °C. This implies that the model systematically undervalues brightness temperature. Nevertheless, the overall trend of the two curves is similar, suggesting that the model's ability to capture the temporal dynamics of brightness temperature is reasonable. One possible explanation for the underestimation of brightness temperature by the model is that it does not fully account for the effects of cloud cover. Clouds can reflect sunlight, lowering the brightness temperature observed by satellites. Additionally, the model may not accurately capture the effects of other atmospheric phenomena, such as aerosols and haze, which can also influence brightness temperature. Despite these limitations, the model can still provide useful information about brightness temperature. The estimated brightness temperature can serve as a basis for predicting other meteorological parameters, including cloud cover and surface temperature. Furthermore, the model's capability to detect variations in brightness temperature over time enables the monitoring of climate change and other environmental trends (Fig. 11.2).

The performance metrics offer key insights into the accuracy of the rainfall prediction model. A Probability of Detection (POD) at 0.62 indicates that the model successfully predicted around 62% of observed rainfall events; with a False Alarm Ratio (FAR) of 0.33, approximately 33% of forecasted rainfall events were identified as false alarms. The Critical Success Index (CSI) at 0.38 points to opportunities for improvement in accurately predicting both positive and negative events. Together, these metrics offer a thorough assessment of the model, highlighting its strengths and pinpointing areas where refinement could enhance its overall predictive capabilities (Figs. 11.3, 11.4 and 11.5)

The predictive analysis presented here utilizes data extracted from the INSAT 3DR satellite's images captured on the 15th of September 2023. To generate insights into future weather conditions, a focused time span from 12:00 AM to 10:00 AM of the same day is considered. Within this timeframe, a set of 20 images is selected, each serving as a temporal snapshot of atmospheric conditions. The choice of INSAT

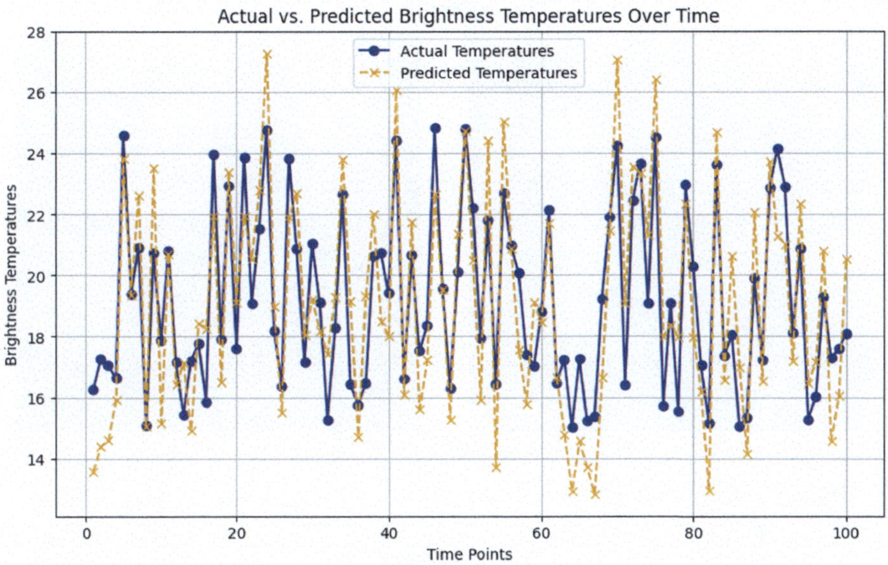

Fig. 11.2 Difference between actual and predicted brightness temperature values

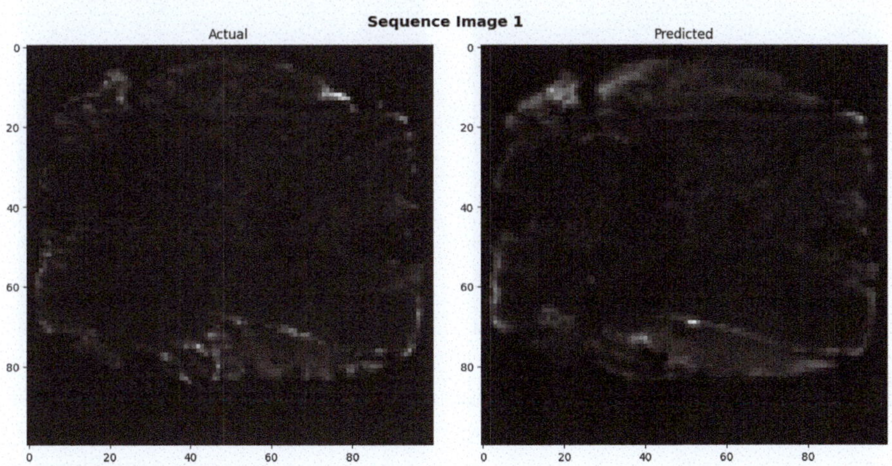

Fig. 11.3 Predicted sequence image 1

3DR is pivotal, given its advanced sensor capabilities, enabling the capture of high-resolution, real-time data crucial for meteorological assessments.

These 20 images collectively form the input dataset for the forecasting model. The model, likely leveraging sophisticated techniques such as Long Short-Term Memory (LSTM) networks, processes this dataset to predict sequential weather outputs. LSTM's proficiency in handling sequential data allows it to capture intricate temporal dependencies within the image series.

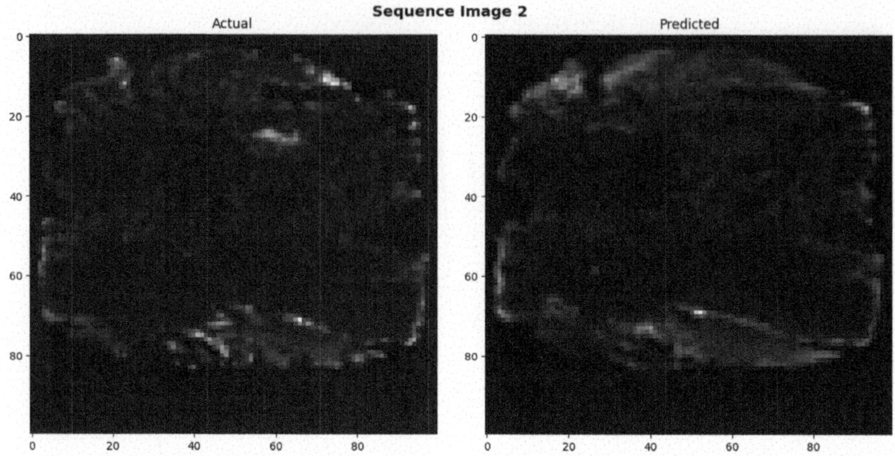

Fig. 11.4 Predicted sequence image 2

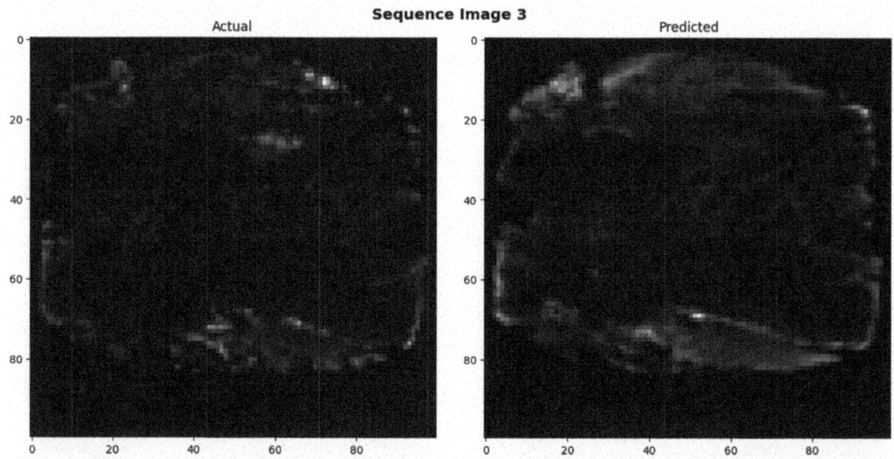

Fig. 11.5 Predicted sequence image 3

The outcome of the model manifests in three consecutive forecasts, each offering a glimpse into the anticipated changes in weather conditions. This tripartite forecast provides a nuanced understanding of how the atmosphere is projected to evolve over time.

This approach aligns with contemporary methodologies in weather prediction, harmonizing satellite imagery, advanced machine learning models, and temporal analysis. The careful selection of the 20-image input sequence ensures a comprehensive representation of atmospheric dynamics during the specified timeframe, facilitating the model's ability to discern patterns crucial for accurate and forward-looking weather forecasts.

11.5 Conclusion and Future Scope

The methodology encompasses a multi-step process, beginning with the calculation of brightness temperature using time series prediction facilitated by Long Short-Term Memory (LSTM) networks. LSTM is a specialized type of recurrent neural network adept at capturing temporal dependencies in sequential data. Leveraging this capability, the model predicts brightness temperature over time, providing a dynamic understanding of the changing atmospheric conditions.

Once the cloud brightness temperature is determined, the next phase involves the identification of rainfall and non-rainfall pixels. This is a crucial step in forecasting, as it enables the discrimination between areas experiencing precipitation and those that are not. The approach likely involves the application of a Random Forest algorithm, known for its effectiveness in handling classification tasks. Random Forest constructs multiple decision trees based on random subsets of the dataset, and their collective predictions contribute to robust classifications. In this context, the algorithm aids in distinguishing between rainfall and non-rainfall conditions, forming a pivotal component in the overall methodology.

Furthermore, the study proposes an extension to the methodology by exploring the calculation of rain rates at individual locations after identifying the rainfall pixels. Rain rate determination adds granularity to the analysis, providing insights into the intensity of precipitation at specific geographic points. This extension suggests a more detailed and localized understanding of rainfall patterns, which can be valuable for applications such as hydrological modeling, water resource management, and disaster preparedness.

In summary, the methodology integrates LSTM for time series prediction of brightness temperature, followed by the application of a Random Forest algorithm to identify rainfall and non-rainfall pixels. The future direction of the study involves advancing the analysis by calculating rain rates at individual locations and enhancing the precision and applicability of the forecasting model.

References

1. Zhang, S., Wang, Z., Yang, Z., Yang, Y., Liu, Z.: Rainfall forecasting using machine learning ensembles with extreme value modeling. J. Hydrol. **624**, 106704 (2022)
2. Singh, G. Kumar, D.: Hybrid prediction models for rainfall forecasting. In: 2019 9th International Conference on Cloud Computing, Data Science & Engineering (Confluence), Noida, India, pp. 392–396 (2019). https://doi.org/10.1109/CONFLUENCE.2019.8776885
3. Jaiswal, P.P.G., et al.: A stacking ensemble learning model for rainfall prediction based on indian climate. In: 2023 6th International Conference on Information Systems and Computer Networks (ISCON), Mathura, India, pp. 1–6 (2023). https://doi.org/10.1109/ISCON57294.2023.10112077
4. Simanjuntak, F., Jamaluddin, I., Lin, T.-H., Siahaan, H.A.W., Chen, Y.-N.: Rainfall forecast using machine learning with high spatiotemporal satellite imagery every 10 minutes. Remote Sens. **14**, 5950 (2022). https://doi.org/10.3390/rs14235950

5. Samad, A., Bhagyanidhi, Gautam, V., Jain, P., Sangeeta Sarkar, K.: An approach for rainfall prediction using long short term memory neural network. In: 2020 IEEE 5th International Conference on Computing Communication and Automation (ICCCA), Greater Noida, India, pp. 190–195 (2020). https://doi.org/10.1109/ICCCA49541.2020.9250809
6. Narejo, S., Jawaid, M.M., Talpur, S., Baloch, R., Pasero, E.G.A.: Multi-step rainfall forecasting using deep learning approach. PeerJ Comput. Sci. **7**, e514 (2021). https://doi.org/10.7717/peerj-cs.514
7. Ridwan, W.M., Sapitang, M., Aziz, A., Faizal Kushiar, K., Ahmed, A.N., El-Shafie, A.: Rainfall forecasting model using machine learning methods: case study Terengganu, Malaysia. Ain Shams Eng. J. **12**(2), 1651–1663 (2021). ISSN 2090-4479. https://doi.org/10.1016/j.asej.2020.09.011
8. Saha, A., Singh, K.N., Ray, M., et al.: A hybrid spatio-temporal modelling: an application to space-time rainfall forecasting. Theor. Appl. Climatol. **142**, 1271–1282 (2020). https://doi.org/10.1007/s00704-020-03374-2
9. Gupta, A., Mall, H.K., Janarthanan, S.: Rainfall prediction using machine learning. In: 2022 First International Conference on Artificial Intelligence Trends and Pattern Recognition (ICAITPR), Hyderabad, India, pp. 1–5 (2022). https://doi.org/10.1109/ICAITPR51569.2022.9844203
10. Grace, R.K., Suganya, B.: Machine learning based rainfall prediction. In: 2020 6th International Conference on Advanced Computing and Communication Systems (ICACCS), Coimbatore, India, pp. 227–229 (2020). https://doi.org/10.1109/ICACCS48705.2020.9074233
11. Sansine, V., Ortega, P., Hissel, D., Ferrucci, F.: Hybrid deep learning model for mean hourly irradiance probabilistic forecasting. Atmosphere **14**, 1192 (2023). https://doi.org/10.3390/atmos14071192

Chapter 12
The Personalization of Justified Recommendations Using the Users Profile Interest and Reviews

Kyelem Yacouba, Tounwendyam Frederic Ouedraogo, and Kiswendsida Kisito Kaboré

Abstract This paper is about the adaptive and personalized justification of the recommenders collaborative filtering system using notices. A method to justify recommendations based on item reviews and the user profile interest is suggested. The reviews with a positive sentiment have been first kept through the sentiment analysis expressed on the reviews and then second, selected potential reviews are candidates for the justification of items. To identify reviews candidates, the frequency calculation of user profile interest terms in the reviews has been done through the TF-IDF weighting method. In order to manage our reviews, an algorithm removing negative sentiment reviews is proposed. However, to test the method, recommendation data already made on a collaborative filtering recommendation system using notices and reviews made on Coursera courses have been used. The data used included 112 recommendations and 11 users. The implementation shows that the inference is constantly evolving and increasingly adapted to the user's profile.

12.1 Introduction

The collaborative filtering is a recommendation method using ratings from users with similar preferences [1, 2]. The system recommends to a user items that have been appreciated by other users who have common interests [2]. The collaborative

K. Yacouba (✉) · K. K. Kaboré
Laboratoire de Mathématiques et d'Informatique (LAMI), Université Joseph Ki-ZERBO, Ouagadougou, Burkina Faso
e-mail: limperialkyelem@gmail.com

K. K. Kaboré
e-mail: kisitokab@gmail.com
URL: https://www.ujkz.bf

T. F. Ouedraogo
Laboratoire de Mathématiques, Informatique et Application (LAMIA), Université Norbert ZONGO, Koudougou, Burkina Faso
e-mail: frederic.ouedraogo@unz.bf
URL: https://unz.bf

S. K. Pal et al. (eds.), *Intelligent Informatics*, Smart Innovation, Systems and Technologies 389, https://doi.org/10.1007/978-981-97-2147-4_12

filtering provides personalized recommendations [3]. There are several types of collaborative filtering algorithms [4, 6, 32, 33]. The evaluation in collaborative filtering recommender systems can be done in different ways: the numerical, the ordinal, the binary notation, etc. [5]. For example, the numerical rating is used in [4] and the recommendation is done through user notices. The notice is the expression of a document recommendation to an user. By giving his/her notice, the user should be able to motivate it in order to convince the receiver of the recommendation about its relevance and accuracy.

The various collaborative filtering recommender system algorithms have only made more specific recommendations of items to users. These algorithms do not promote the understanding of the reasons and the motivations for the recommendation. As time went on, users became interested in the motivation and why they received such a recommendation. This reason and motivation help to improve the satisfaction and the acceptability of the recommendation. Also, the justification is a key factor in guiding the user to consume or not consume a recommendation, and to make a quick and informed decision. Having this in mind, we were interested in justification in recommender systems. The justification provides a rationale that clarifies and positions the user on the recommendation. According to the literature, there are basically two methods of recommendation justification: the recommendation algorithm related to the justification method and the post-hoc method [28].

For the justification method related to the recommendation algorithm, [2] provides a personalized recommendation justification using item features. His method uses user groups, keyword weights, and the neighborhood training. In its approach, we note that there is no question of user's feedbacks but also no personalization and evolution of the recommendation justification. [7] uses the keywords that are written on the items as justification for the recommendation. Also, these keywords are used in the recommendation prediction process. Its strategy is to build an explanation chain and then evaluate it. This technical does not use user reviews as a basis for recommendation justification. In addition, this technic cannot be used directly in another framework, i.e., with another recommendation algorithm.

Several works have addressed the post-hoc method [8–10, 29]. Using the reviews, [8, 10] have generated relevant and distinctive justifications. In [8, 10], the generation of the justification is done in the following way: extractions of relevant terms to be included in the justification, extractions of candidate sentences, and then summaries of the extracted sentences. In [8, 10], the system prepares a justification for each document regardless of the time and the justification remains unchanged for all users of the system. Its method does not allow for personalization or evolution of the justification. The precise and diverse justification was proposed by [9] which constructs the candidate profile of users and items needed for justification prediction. The profile contains the review segments that describe the user and the item. From these review segments, [9] proposes the generation of justifications that contain the aspects that describe the user. Even if this strategy generates personalized justifications, it must be noted that it first requires review left by users on items. Furthermore, the reviews used are not necessarily written by the user, which causes the problem whenever these data really describe the user's profile. Generally, the post-hoc methods we have studied do not allow for personalization and changes in justification as the system evolves over time. As the user's needs change over time, what was relevant today may not be relevant tomorrow. This study is conducted relatively from this perspective.

The collaborative filtering can use notices to recommend documents. In such a context, by which way the user can at the same time express the reasons for his recommendation. This reason can be done more easily through a review and over time users recommend associating new reviews. That is why we address in this work the justification of recommendation using the user reviews. This system is important in online courses and libraries, where students who may or may not know each other can recommend each other. For example, they can exercise to improve their learning. The notices should contain the comments that will be used to generate the personalized, adaptive, and scalable justification. Nevertheless, the existing justification approaches do not allow for the generation of personalized, adaptive, and scalable justification from users' notices. Hence the problem that occurs is: how to provide personalized and scalable justification based on user reviews? Our goal is to generate personalized and relevant justifications to the recommendation. These justifications must also evolve according to the new reviews made by the internauts on the documentary unit for the same user. At this end, we will try to include in the justification the interests of each user. We will use the following:

- The sentiment analysis algorithm to filter out reviews that do not have positive sentiment.
- The interests of the user profile.
- The TF-IDF to compute the occurrences of interests that are in the reviews in order to identify the best review suited to the user's profile.

This paper is structured on the following plan. Firstly, we will review the literature. Secondly, we will propose a method for justifying the recommendations and then implement this method. Finally, we will analyze the collected results.

12.2 Background

In this section, we make explicit the user model, the user interest, the user opinion, the justification using reviews, the TF-IDF weighting, the sentiment analysis, and the offline evaluation metric.

The user model: According to [1] the user model is "a set of knowledge about the user explicitly or implicitly represented, used to improve the user's interaction with the machine". The user model is used to present personalized content to the user. It is a schema of data to be collected to build user profiles. The data collected constitutes the user's interest.

The interest: [1] describes an interest as "any recurring feature in the items with which the user interacts". The interest can be a literary genre, a research topic, or a product category, etc. The user profile is an instance of the user model. Data collection can be explicit or implicit and consists of observing these actions to determine the user's interests. The representation of the interest is the representation by keywords, set of keywords or vectors, the semantic representation, and the multidimensional representation [1, 11, 12]. The Keyword representation is widely used in the information retrieval [11]. The data contained in user profiles is full of user preferences.

The user's notice: The notice is the fact that the user appreciates a documentary unit after consulting it. The notice according to [4] consists of the user group, the documentary unit, and the weight. The weight is a value and it shows the importance of the recommendation. The higher it is, the more relevant the document is for the user. This definition has been implemented on a computer in the work of [14] and later on the mobile [13]. But for the purposes of justifying recommendations, this definition of notice has undergone an evolution [15, 16]. The notice is now constituted in addition to the elements mentioned above of the justification. The justification in recommender systems can be done by using item reviews or features as input data sources.

Justification using reviews: In order to provide justification for the recommendation and present it on an explanation interface, [17] analyzes user reviews to identify relevant features of articles. To justify the recommendation [18] proposed a methodology for generating context-sensitive natural language justification. The justification is the combination of reviews that contain the best "aspect" terms; these terms show the relevance of the context of the item's use. The extraction of terms was done using a natural language pipeline that exploits distributional semantics models. To justify its personalized recommendation, [9] extracts from users' reviews the terms that best express users' intentions to include in the final justification. As for [19], it proposed a method to generate advice by taking into account the user's language style and the features of the items.

TF-IDF is a statistical weighting method used to evaluate the importance of words or a set of words in a document or a corpus. The idea is that if a term or phrase appears frequently in a document or not, then we can say that this term or phrase is used to make the difference between documents and this allows us to make a classification of documents. It is a function for computing the weights of the characteristic words or features of the items. The TF-IDF weighting is composed of two parts, namely the calculation of the term frequency and the inverse frequency of the document. The term frequency calculates the number of occurrences of a word in a document and the inverse document frequency measures the overall importance of a document [20–22]. The calculation of the frequency of terms in a document is given by the following formula:

$$tf_{i,j} = \frac{n_{i,j}}{\sum_k n_{k,j}} \tag{12.1}$$

$n_{i,j}$ is the number of occurrences of the term i in the document j and $\sum_k n_{k,j}$ the sum of the occurrences of all the terms in the document j.

The inverse frequency of the document is given by the following formula:

$$idf_i = log\frac{|D|}{|\{j : t_i \in d_j\}|} \tag{12.2}$$

$|D|$ is the total number of documents and $|\{j : t_i \in d_j\}|$ is the number of documents containing the term t_i.

The TF-IDF formula is given by

$$tfidf_{i,j} = tf_{i,j} * idf_{i,j} \tag{12.3}$$

$tfidf_{i,j}$ is the weight of term i in document j.

Offline evaluation: In offline evaluation, we have metrics that either measure the ability of the system to generate the justification or measure the quality of the justification. We address the ability of the system to generate the justification [26]. The metrics exist for this approach. As an example we cite in the literature MEP (Mean Explainability Precision), MER (Mean Explainability Recall), and model fidelity [26, 27, 30]. The latter is the most recent and is a metric that generalizes the other two. The model fidelity metric is defined by

$$Model\ Fidelity = \frac{|explainable\ items \bigcap recommended\ items|}{|recommended\ items|} \tag{12.4}$$

It is the ratio of the intersection of explainable items and recommended items.

Sentiment Analysis: Sentiment analysis is used to classify the opinions or comments of users on websites or social networks. This classification of opinions is either positive, negative, or neutral [23, 24]. Here is the sentiment classification algorithm with the text blob tool [24].

Algorithm 1: Sentiement Classification of Tweeter comments using Text Blob	
Input	Text File (Tweeter comments which include Nouns, Adjectives, Adverbs)
Output	Values > 0 (Positive), Values<0 (Negative), Values=0 (Neutral)

Begin:	Sentiment Analysis 0 ← File
	For each row in rows
	if Sentiment Polarity Score >0 then
	Sentiment ← Positive
	else
	if Sentiment Polarity Score <0 then
	Sentiment ← Negative
	else
	if Sentiment Polarity Score =0 then
	Sentiment ← Neutral
	end
	end
	end

End:	end

12.3 Positioning

In our work, we assume that users will write their notices on the different documents as they go along. The notice comes from the interaction between the user and the information system. The notice contains the following elements: the user, the document, the weight, and the justification. Our justifications are the users' reviews. So each user will write down his review for the recommended document and the system will personalize it on the profile of the active user. As a user can receive the recommendation of the same document from several users, then the system must be able to dynamically evolve the justification. As the system is dynamic and changes the justification, we tend toward an optimal justification. Our contribution is to suggest a method of dynamic personalization of the document justification by the user. This justification, which is the result of the users' reviews of the different documents, must include the relevant terms from the users' profile. This will generate a relevant justification for the user to facilitate his decision-making.

12.4 Proposed Method of Justifying Recommendations

In this section, we present our method for generating the justification for the recommendations.

12.4.1 Definition of the Problem

We assume that system users are well-organized and have profiles whose interests can be noted $T = \{t_1; t_2; ...; t_q\}$. For each UG (a given user) user we collect all the UD (documents received by UG) documents recommended by the users as well as the reviews $J = \{j_1; j_2; ...; j_p\}$ for each document. All the reviews J for each user and corresponding document are potential candidates for the documents justification. We calculate the sentiment analysis of the reviews and retain those that are positive. Next, we compute the occurrences of the interest T of a given user that are found in the justifications J that are candidates for the justification of a given document.

12.4.2 Collection of Users' Reviews

The user justifies the recommendation after being connected to the information system. We treat this part of the case of a user who wants to recommend a document or a documentary unit. In the case of our study, the user identifies those who receive the recommendation, the document, the weight, and then the review. We note the following actions by the user:

1. User login to the system.
2. The user's notice (document, user, review, weight): as many times as he wants.
3. User validation.
4. User logout.

The documentary units are recommended to a user. If in our information system we have n users then this can be noted by: $UG = \{ug_1; ug_2; ...; ug_n\}$. For m documentary units in our system we can materialize it by: $UD = \{ud_1; ud_2; ...; ud_m\}$. Let us suppose we have n recommendations at a given time t for a given user ud_i, we can have for each documentary unit p justifications noted by: $J = \{j_1; j_2; ...; j_p\}$. The size of the justifications is not necessarily the same for each document per user. Also, at another time t_1, the same documents can be recommended to the same user with different reviews. It may happen that the latter review is more relevant to convince the active user to use the document. The relevance is measured in our framework in terms of the occurrences of the active user's interests.

12.4.3 Management of Reviews by the System

Our strategy divides into two this part. For each user, the system first identifies the justifications to be displayed. Then, it proceeds to the personalization and dynamic adaptation of the justifications. Firstly, the justification comes from the reviews made by the Internet users. The review candidates are those expressing a positive sentiment (sentiment analysis of justifications). Only reviews with a sentiment greater than or equal to zero are maintained. Secondly, the system moves on to the personalization by calculating the frequency of interest terms that are in the reviews of recommendations. Extracting only the reviews that contain the user's profile information as justification allows to personalize. The review with the highest number of interest occurrences is better, i.e., the review with the most interest information. Finally, the interest is obtained from the user's profile. Let us suppose that we have for each user profile the following set of information $T = \{t_1; t_2; ...; t_q\}$ where t_i is a term. For each term t_i of the user profile we calculate its frequency in each review j_i. We suppose that each review constitutes our document. We compute the frequency of interest terms.

$$tf_{t_i,j_i} = \frac{n_{t_i,j_i}}{\sum_k n_{k,j_i}} \tag{12.5}$$

In this formula, n_{t_i,j_i} is the number of occurrences of the term t_i in the justification j_i and $\sum_k n_{k,j_i}$ is the total number of occurrences of the justification j_i.

For each user and corresponding documentary unit, we compute the terms frequency coming from its interest and which are found in the reviews. The global frequency for a given review is nothing else than the sum of the different frequencies of each term of the user's interest ug_i noted tf_{t_i,j_i}. The global frequency is obtained with the following formula:

$$TF(ug_i; ud_i; j_i) = \sum_{\forall t_i \in T} tf_{t_i,j_i} \tag{12.6}$$

This value is the frequency of the ug_i user's interests that are in the j_i review about the document recommendation ud_i. If we suppose that in another time a recommendation of the same document ud_i is made to the same user ug_i with a review j_k then we have

$$TF(ug_i; ud_i; j_k) = \sum_{\forall t_i \in T} tf_{t_i,j_k} \tag{12.7}$$

If $TF(ug_i; ud_i; j_k) > TF(ug_i; ud_i; j_i)$, then the review j_k becomes again the justification of the recommendation of ud_i for ug_i. Below we present an algorithm for identifying candidate reviews for justification. The algorithm removes reviews with negative sentiment.

Algorithm 2: Sentiment_positive_neutral	
Input	All the justifications $J = \{j_1; j_2; ...; j_p\}$
Output	Positive reviews

Begin:	For each justifications j_i
	    Algorithm 1
	    if Sentiment Polarity Score <0 then
	        delete(j_i)
	    end

End:	end

Algorithm removes reviews with negative sentiment.

After the identification of the candidate reviews, it is time to personalize them by calculating the occurrences terms of the interest that are in the reviews.

Algorithm 3: Calculation of the frequency of interests found in the reviews, example of a user ug_i	
Input	Algorithm 2 and interest $T = \{t_1; t_2; ...; t_q\}$ by user
Output	Weighted reviews

Begin:	Algorithm 2
	For each user ug_i and interest T do
	    For each document ud_i do
	        For each justifications j_i
	            $TF(ug_i; ud_i; j_i) = \sum_{\forall t_i \in T} tf_{t_i,j_i}$
	        end
	    end

End:	end

Calculation of the frequency of interests found in the reviews.

Our algorithm uses *TF* weighting to give weights to different user reviews. Now, these reviews have weights, making it easier to extract the justification that best expresses the active user's interest. For a ug_i user who has just received a recommendation from a ud_i document then the system proceeds to calculate the new term frequency. The following processing is necessary:

1. While ug_i receives a recommendation.
2. Algorithm 3.
3. Posting the recommendation with justification.

Now, the new review has an occurrence value and it is compared to the old larger value. If it is found to be greater, then this new review becomes the justification of the corresponding document.

12.4.4 Generation of the Justification

For our study, we select the best of the reviews from each documentary unit as the final justification. A review is considered best if it contains more occurrences of the terms of the user's interest, i.e., the largest TF. This justification will be the best adapted to the user's profile for the reasons of content of the user's interests. Figure 12.1 shows the justification generation process we proposed. From the recommendation, we extract only the reviews for the further processing of the justification generation.

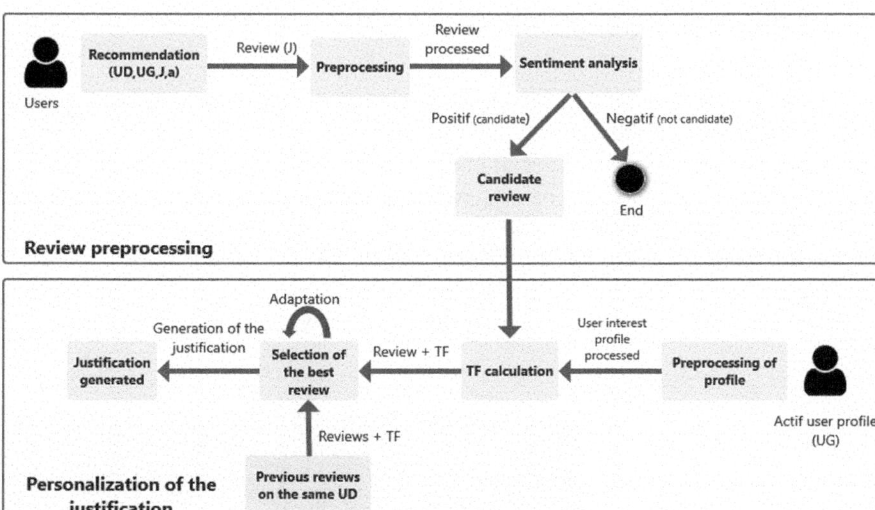

Fig. 12.1 Process of generating personalized and adaptive justification from a recommendation with a review; the example of processing a review for a given user

12.5 Data Collection and Processing

In this section, we present the statistics of our data as well as the preprocessing of the profiles and reviews.

12.5.1 *Data*

We have built our own database and the data came from three sources. We got the recommendation data from the collection in [25]. Also, we downloaded the reviews made on the courses from the site coursera [31].

We did not get any profile data from the Internet because it is a personal data. So we made up our own profiles. To constitute them, we took into account the items on which the reviews of coursera were realized so that the TF is not nil. To do this, the reviews we have uploaded are made on computer science course items so we have proposed profiles that could have computer science students from our university.

The base we built consists of a list of 112 recommendations, recommended to 11 users. We randomly extracted 112 reviews according to the teaching unit that we associated with each line of recommendations. We have taken into account the user's profile to associate the reviews with the appropriate teaching units. This will give us a better chance of finding the user's interests in the reviews.

12.5.2 *Preprocessing of Collected Reviews and Profiles*

Before doing the sentiment analysis and calculating the frequencies of user profile terms, we preprocessed the collected reviews and user profile data. Here, it is about eliminating the stopwords, the special characters, the punctuations, and lemmatization of reviews and profiles. Also, we have transformed all the text into lower case. We used the nltk library of Python to remove stopwords and special characters. For the lemmatization, we used the Spacy library of Python too. The language in which we preprocessed our data is English. We present in Figs. 12.2 and 12.3 the case of the unpreprocessed and preprocessed profile and review data for the user UG1.

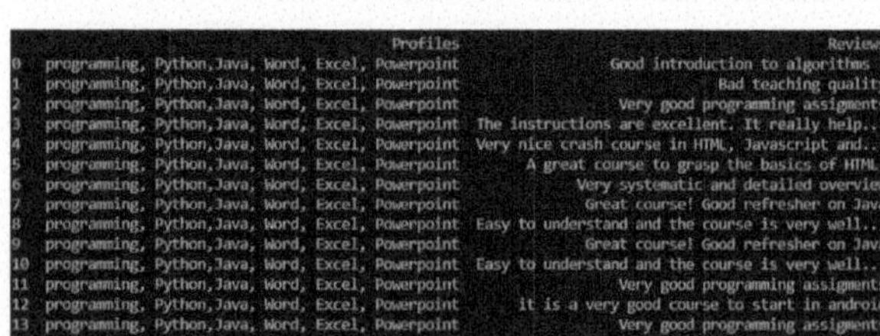

	Profiles	Reviews
0	programming, Python,Java, Word, Excel, Powerpoint	Good introduction to algorithms !
1	programming, Python,Java, Word, Excel, Powerpoint	Bad teaching quality
2	programming, Python,Java, Word, Excel, Powerpoint	Very good programming assigments
3	programming, Python,Java, Word, Excel, Powerpoint	The instructions are excellent. It really help...
4	programming, Python,Java, Word, Excel, Powerpoint	Very nice crash course in HTML, Javascript and...
5	programming, Python,Java, Word, Excel, Powerpoint	A great course to grasp the basics of HTML!
6	programming, Python,Java, Word, Excel, Powerpoint	Very systematic and detailed overview
7	programming, Python,Java, Word, Excel, Powerpoint	Great course! Good refresher on Java
8	programming, Python,Java, Word, Excel, Powerpoint	Easy to understand and the course is very well...
9	programming, Python,Java, Word, Excel, Powerpoint	Great course! Good refresher on Java
10	programming, Python,Java, Word, Excel, Powerpoint	Easy to understand and the course is very well...
11	programming, Python,Java, Word, Excel, Powerpoint	Very good programming assigments
12	programming, Python,Java, Word, Excel, Powerpoint	it is a very good course to start in android
13	programming, Python,Java, Word, Excel, Powerpoint	Very good programming assigments
14	programming, Python,Java, Word, Excel, Powerpoint	Great course! Good refresher on Java

Fig. 12.2 Reviews and users interests unpreprocessed for UG1

	Profiles	Reviews
0	[programming, python, java, word, excel, power...	good introduction algorithm
1	[programming, python, java, word, excel, power...	bad teaching quality
2	[programming, python, java, word, excel, power...	good programming assigments
3	[programming, python, java, word, excel, power...	instruction excellent really help learn analyt...
4	[programming, python, java, word, excel, power...	nice crash course html javascript cs
5	[programming, python, java, word, excel, power...	great course grasp basic html
6	[programming, python, java, word, excel, power...	systematic detailed overview
7	[programming, python, java, word, excel, power...	great course good refresher java
8	[programming, python, java, word, excel, power...	easy understand course well made
9	[programming, python, java, word, excel, power...	great course good refresher java
10	[programming, python, java, word, excel, power...	easy understand course well made
11	[programming, python, java, word, excel, power...	good programming assigments
12	[programming, python, java, word, excel, power...	good course start android
13	[programming, python, java, word, excel, power...	good programming assigments
14	[programming, python, java, word, excel, power...	great course good refresher java

Fig. 12.3 Preprocessed reviews and users interests for UG1

12.6 Implementation of the Proposed Justification Method

After the reviews and profile data were preprocessed, we moved on to the sentiment analysis and the calculation of the terms of the user's interests that are found in the reviews. We used the Python Text blob library to compute the sentiment analysis of the reviews. After the sentiment analysis, we delete the reviews with a negative sentiment. The rest of the reviews are all candidates to justify the item for the user to whom the recommendation is addressed. For the calculation of TF, we have implemented a Python function for this purpose.

12.7 Results

We worked with 112 recommendations, and among these recommendations after deleting those with negative feedback, we obtained 103 recommendations. Eleven users were used in our implementation. We then grouped the data by user. We present in screenshots some of the users' results. Each line of the Figs. 12.4 and 12.5 represents a recommendation. We have the column of the documents followed by the reviews and the penultimate column represents the summation calculation of the frequencies of the user interests, for example, UG1. They are found in each review. We did not show the information of the users interests of each user here. UG1 received the recommendation from the documents UD1, UD2, UD6, and UD16. The first recommendation of UD1 has a review in which term frequency is 0 and after this same document UD1 has been recommended to UG1 but this time the calculation of term frequencies gives a value higher than the first (0,333). Before the second review, the justification that the system was going to use was the first one because there were no other reviews to compare. From this point of view, the review serving as the justification best adapted for the moment to the user's needs is the second review. Hence, the system must proceed to change the justification of item UD1. We deduce that the system makes a change of justification if there is a review that is better adapted to the user's profile. It is the same as the analysis that we have carried out on the rest of the documents for this user. We have UD2 which also went from 0 to 0.125000, so we have a new review that is well suited to the user profile. UD6 received zero reviews and then we have a value of 0.20, then zero TFs, and back to

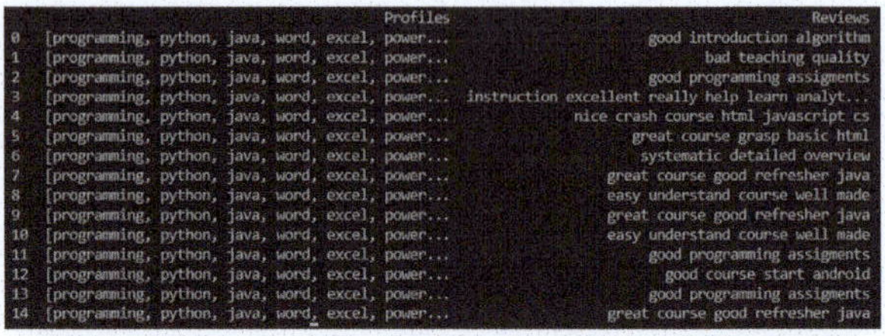

	Profiles	Reviews
0	[programming, python, java, word, excel, power...	good introduction algorithm
1	[programming, python, java, word, excel, power...	bad teaching quality
2	[programming, python, java, word, excel, power...	good programming assigments
3	[programming, python, java, word, excel, power...	instruction excellent really help learn analyt...
4	[programming, python, java, word, excel, power...	nice crash course html javascript cs
5	[programming, python, java, word, excel, power...	great course grasp basic html
6	[programming, python, java, word, excel, power...	systematic detailed overview
7	[programming, python, java, word, excel, power...	great course good refresher java
8	[programming, python, java, word, excel, power...	easy understand course well made
9	[programming, python, java, word, excel, power...	great course good refresher java
10	[programming, python, java, word, excel, power...	easy understand course well made
11	[programming, python, java, word, excel, power...	good programming assigments
12	[programming, python, java, word, excel, power...	good course start android
13	[programming, python, java, word, excel, power...	good programming assigments
14	[programming, python, java, word, excel, power...	great course good refresher java

Fig. 12.4 The results of the recommendations received by the user UG1

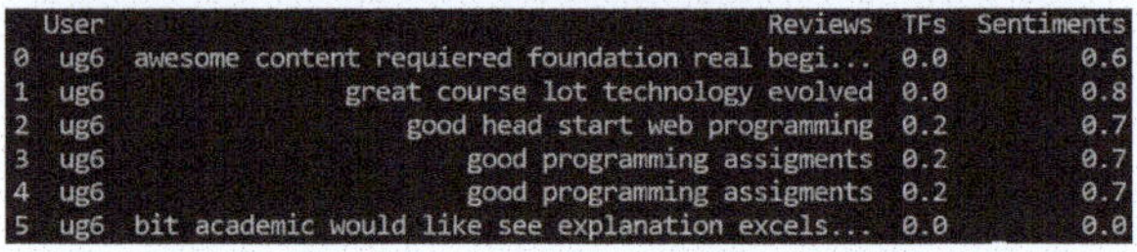

	User	Reviews	TFs	Sentiments
0	ug6	awesome content requiered foundation real begi...	0.0	0.6
1	ug6	great course lot technology evolved	0.0	0.8
2	ug6	good head start web programming	0.2	0.7
3	ug6	good programming assigments	0.2	0.7
4	ug6	good programming assigments	0.2	0.7
5	ug6	bit academic would like see explanation excels...	0.0	0.0

Fig. 12.5 The results of the recommendations received by the user UG6

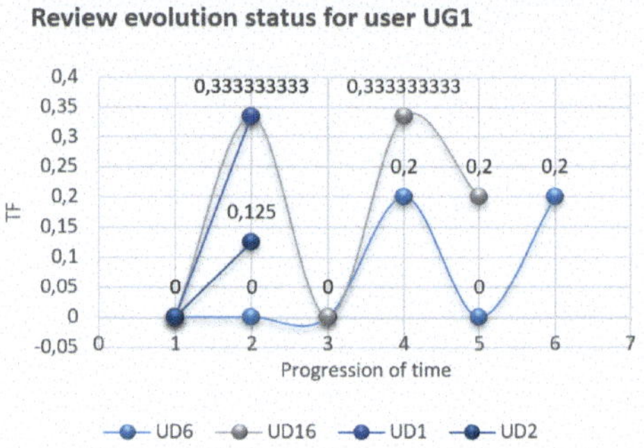

Fig. 12.6 Figure showing the frequency values of the profile terms in the reviews over time, by each document for users UG1

0.20. In this case, according to Algorithm 3, the first review with a TF of 0.20 is the justification. For UD16 it is the first review with TF equal to 0.33 that will be the justification.

The analysis principle remains the same for other users.

Figures. 12.6, 12.7, 12.8, and 12.9 below show the TF values of reviews taken by each given document as a function of time. In the following curves, what center interests us are the points that give the value of personalization of the review to the user profile to the document. In the presentation of the results, we did not mention the users UG2, UG3, UG4, UG8, UG9, UG10, and UG11 because they do not present interesting data in relation to our work. For example, the recommendations received by UG10 are 11 in number but distinct in terms of document. Thus, we could not plot a TF evolution curve of reviews by document. Nevertheless, the case of UG2, UG3, UG4,UG8, UG9, UG10, and UG11 is found in UG5 (Fig. 12.8) which we have analyzed.

The explanation of the Fig. 12.6 is the following. The first reviews all had zero TF, then other reviews came in with TF of 0.33 for UD1 and 0.125 for UD2. So the last reviews will be replaced by the new ones, which are more personalized to the user's profile. The document UD6 gradually received 6 reviews with TFs of 0; 0; 0.2; 0; and 0.2, respectively. As new reviews are received, the system keeps the review with the

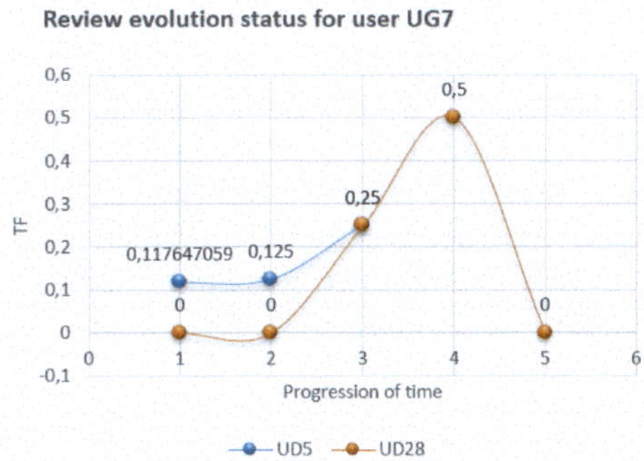

Fig. 12.7 Figure showing the frequency values of the profile terms in the reviews over time, by each document for users UG7

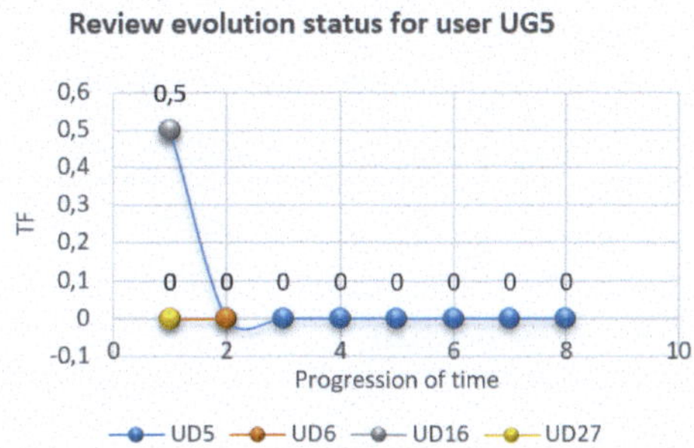

Fig. 12.8 Figure showing the frequency values of the profile terms in the reviews over time, by each document for users UG5

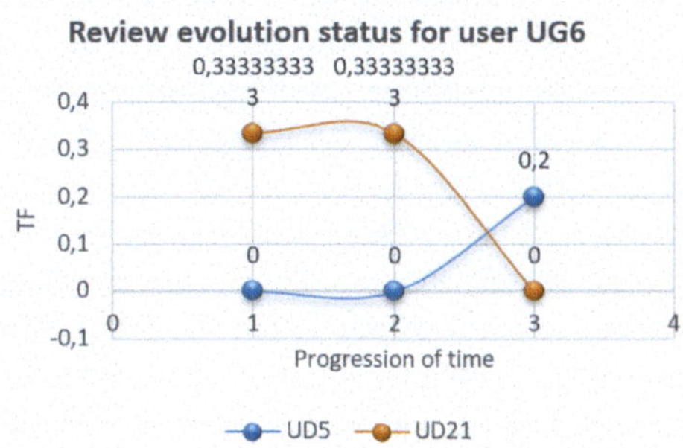

Fig. 12.9 Figure showing the frequency values of the profile terms in the reviews over time, by each document for users UG6

highest TF value, so the first review with a TF of 0.2 is the justification. The document UD16 has received 5 reviews with TF values of 0; 0.33; 0; 0.33; and 0.2, respectively. Like document UD6, the first review with the highest TF is the justification, i.e., 0.33. User UG7 received two documents in recommendations (UD5 and UD28). We can see that for UD5, as time went by, the new reviews came in and were more adapted to the user's profile. So for each new review, the system proceeded to change the justification if it has a larger TF. As for UD28, the system will proceed to the change of the reviews serving as the justification. Only, that between the moment 4 and 5, the new comment has a TF which is lower than 0,5 so there will be no change of justification.

The UG5 user received 4 documents as recommendations (UD5, UD6, UD16, UD27). UD16 and UD27 were the subjects of a single recommendation so the respective reviews of each document are at the same time the justification of UD16 and UD27 for UG5. UD6 received two recommendations with TFs remaining the same so the first review is the justification for UD16 for now pending a new recommendation. UG6 had two items recommended. UD5 had three recommendations as well as UD21. The first two recommendations of UD5 have the same TF and the third recommendation has a TF of 0.2, so the justification will change to take the review with the best TF.

There are two approaches to assessing the explanation of recommendations that are: assessing the percentage of recommendations that can be justified against the justification generation algorithm and assessing the quality of the justification generated [26].

We will evaluate our algorithm through the model fidelity metric [26].

$$Model\ Fidelity = \frac{|explainable\ items \bigcap recommended\ items|}{|recommended\ items|} = \frac{103}{112} = 0.9196$$

12.8 Discussion

The value of 92% for model fidelity is satisfactory even if we had to have 100% because we take our data sources from users' opinions and those users who know their peers are better expected to write reviews that express a positive feeling for the active user. Also, our results are much better compared to the work of [27], which gives a value of 70% for the same metric. This big difference is explained by the fact that in the work of [27] it is the system that generates the justification while in our work, it is the users who suggest justifications for each recommendation. Then, the system only makes calculations and suggests the best review as the justification. The only case in which there is no justification for a given recommendation is when the review written by the user has a negative sentiment. For example, if a document is recommended for the first time and the review that accompanies it has a negative sentiment. In this case, we have a recommendation without any justification since this review is not a candidate for the justification according to Algorithm 2.

12.9 Conclusion

This work deals with the personalized and scalable justification of recommendations in the collaborative filtering recommender systems. Specifically, we have used recommender systems that use notices to make recommendations, i.e., the recommender algorithm uses user's interaction information with the system to recommend. We proposed a scalable justification method for recommendations through Algorithms 2 and 3. So that for the system to personalize the justifications, we use the contents of the user's profile interests in the reviews through the weighting of TF. This TF value for each review measures the relevance of the review to the user due to its content of the profile information of the user. Taking into account that the justification for a given document and user can change, it evolves toward an optimal, unchangeable, unique, and therefore more relevant solution. We did not attempt to build profiles of relevant users and assumed that the profile information was available. This is a limitation of this work. Also, we have not tested our algorithm in a real user situation. These limitations constitute the aspects of our future work.

Acknowledgements This work was carried out under the financial support of the "Pojet d'Appui á l'Enseignement Supérieur(PAES)".

References

1. Lémdani, R.: Système Hybride d'Adaptation dans les Systèmes de Recommandation. Thèse de Doctorat de l'Université Paris-Saclay préparée á CentraleSupelec, pp. 23- 33. 11 juillet 2016. https://www.theses.fr/2016SACLC050.pdf
2. Dudognon, D.: Diversité et système de recommandation: application á une plateforme de blogs á fort trafic thèse de Doctorat de l'Université de Toulouse, pp. 11–12. Accessed 04 April 2014
3. Panagiotis, S., Alexandros, N., Yannis, M.: Providing justifications in recommender systems. IEEE Trans. Syst. Man Cybern.-Part A: Syst. Hum. **38**(6), 1262–1272 (2008). https://ieeexplore.ieee.org/abstract/document/4648950
4. Kabore, K. , Sié, O. , Sèdes, F.: Information access assistant service (IAAS). In: The 8th International Conference for Internet Technology and Secured Transactions (ICITST-2013), IEEE UK/RI Computer Chapter, London, UK, December, pp. 9–12 (2013)
5. Chouaib, Z.: Filtrage des tags dans un environnement collaboratif Mémoire de Fin d'études Master Université de 8 Mai 1945 - Guelma, juillet (2019)
6. Rakotonirina, A.J.: Filtrage Collaboratif Sensible au Contexte: une approche basée sur LDA. Mémoire en vue de l'obtention du diplome de Master 2 Université d'Antananarivo 27 Janvier 2017
7. Rana, A., Bridge, D.: Explanation Chains: recommendation by explanation. In: RecSys '17 Poster Proceedings, Como, Italy, August, pp. 27–31 (2017)
8. Musto, C., Rossiello, G., de Gemmis, M., Lops, P., Semeraro, G.: Natural language justifications for recommender systems exploiting text summarization and sentiment analysis . Copyright c 2019 for this paper by its authors. Use permitted under Creative Commons License Attribution 4.0 International (CC BY 4.0)
9. Ni, J., Li, J., McAuley, J.: Justifying recommendations using distantly-labeled reviews and fine-grained aspects. In: Proceedings of the 2019 Conference on Empirical Methods in Natural Language Processing and the 9th International Joint Conference on Natural Language Processing (EMNLP-IJCNLP), pp. 188–197 (2019)
10. Musto, C., Rossiello, G., de Gemmis, M., Lops, P., Semeraro, G.: Combining text summarization and aspect-based sentiment analysis of users' reviews to justify recommendations. In: RecSys '19: Proceedings of the 13th ACM Conference on Recommender Systems, pp. 383–387 (2019). https://doi.org/10.1145/3298689.3347024
11. Amer, N.O.: Recherche Sociale et Personnalisée d'Information. Thèse de doctorat de l'universite grenoble alpes, pp. 25–39, 16 Décembre 2020
12. Tamine, L., Zemirli, N., Bahsoun, W.: Approche statistique pour la définition du profil d'un utilisateur de système de recherche d'information. Revue I3 - Information Interaction Intelligence, Cépaduès **7**(1), 5–25 (2007). ffhal-00359531ff

13. Kyelem, Y., Kabore, K.K., Bassole, D.: Hybrid approach to cross-platform mobile interface development for IAAS. In: Shakya, S., Bestak, R., Palanisamy, R., Kamel, K.A. (eds.) Mobile Computing and Sustainable Informatics. Lecture Notes on Data Engineering and Communications Technologies, vol. 68. Springer, Singapore (2022). https://doi.org/10.1007/978981161866-6_16
14. Kabore, K. , Peninou, A., Sié, O. , Sèdes, F.: Implementing the information access assistant service (IAAS) for an evaluation. Int. J. Internet Technol. Secur. Trans. **6**(1) (2015)
15. Kyelem, Y., Kabore, K.K., Ouedraogo, T.F., Sèdes, F.: Comparative study of justification methods in recommender systems: example of information access assistance service (IAAS). In 7th International Conference on Computer Science, Engineering and Applications vol. 11, pp. 173–181 (2021). https://doi.org/10.5121/csit.2021.112013
16. Kyelem, Y., Kabore, K.K., Ouedraogo, T.F, Sèdes, F.: Recommendation generation justified for information access assistance service (IAAS): study of architectural approaches. Acad. Ind. Res. Collab. Cent. (AIRCC) Int. J. Comput. Sci. Inf. Technol. **13**(6), 1–17. https://doi.org/10.5121/ijcsit.2021.13601
17. Chen, L., Wang, F.: Explaining recommendations based on feature sentiments in product reviews. In: Proceedings of the 22nd International Conference on Intelligent User Interfaces, pp. 17–28. ACM (2017)
18. Spillo, G., et al.: Exploiting distributional semantics models for natural language context-aware justifications for recommender systems In: Proceedings of the Seventh Italian Conference on Computational Linguistics CLiC-it 2020: Bologna, Italy, pp. 1–3 (2021). https://doi.org/10.4000/ books.aaccademia.8899
19. Liu, H., Yin, Q., Wang, W.Y.: Towards explainable NLP, a generative explanation framework for text classification. In ACL (2019)
20. Liu, C., Sheng, Y., Wei, Z., Yang, Y.: Research of text classification based on improved TF-IDF algorithm. In: IEEE International Conference of Intelligent Robotic and Control Engineering (IRCE), pp. 218–222 (2018). https://doi.org/10.1109/IRCE.2018.8492945
21. Qaiser, S., Ali, R.: Text mining: use of TF-IDF to examine the relevance of words to documents. Int. J. Comput. Appl. **181**(1), 25–29 (2018)
22. Gu, Y., Wang, Y., Huan, J., Sun, Y., Jia, W.: An improved TFIDF algorithm based on dual parallel adaptive computing model. In: 2018 IEEE International Conference on Internet of Things (iThings) and IEEE Green Computing and Communications (GreenCom) and IEEE Cyber, Physical and Social Computing (CPSCom) and IEEE Smart Data (SmartData), pp. 657–663 (2018). https://doi.org/10.1109/Cybermatics_2018.2018.00133
23. El Rahman, S.A., AlOtaibi, F.A., AlShehri, W.A.: Sentiment analysis of twitter data. In: International Conference on Computer and Information Sciences (ICCIS), pp. 1–4 (2019). https://doi.org/10.1109/ICCISci.2019.8716464
24. Biswas, S., Ghosh, S., Roy, S.: A sentiment analysis on tweeter opinion of drug usage increase by TextBlob algorithm among various countries during pandemic. Int. J. Hit. Transc.: Eccn. **6**(2A), 1–9 (2020). www.hithaldia.in/locate/ECCN
25. Kaboré, K.K.: Système d'aide pour l'accès non supervisé aux unités documentaire. Thèse de doctorat du l'Université de Ouaga 1 Pr Joseph KI-ZERBO, Janvier 2018
26. Yongfeng Zhang and Xu Chen: Explainable recommendation: a survey and new perspectives. Found. Trends Inf. Retr. **14**(1), 1–101. (2020). https://doi.org/10.1561/1500000066
27. Peake, G., Wang, J.: Explanation mining: post Hoc interpretability of latent factor models for recommendation systems. In: KDD '18: The 24th ACM SIGKDD International Conference on Knowledge Discovery & Data Mining, August 19-23, 2018, London, United Kingdom. ACM, New York, NY, USA, pp. 2060–2069 (2018). https://doi.org/10.1145/3219819.3220072
28. Du, Y.: Des données aux connaissances: vers des recommandations plus pertinentes, diversifiées et transparentes. 2021. Thèse de doctorat. Imt-Mines Ales-Imt-Mines Alès Ecole Mines-Télécom
29. Balog, K., Radlinski, F., Petrov, A.: Measuring the impact of explanation bias: a study of natural language justifications for recommender systems (2023). arXiv:2303.09498

30. Abdollahi, B., Nasraoui, O.: Using explainability for constrained matrix factorization. In: Proceedings of the 11th ACM Conference on Recommender Systems. ACM, pp. 79–83 (2017)
31. https://www.kaggle.com/datasets/septa97/100k-courseras-course-reviews-dataset
32. Mustafa, M., Sebag, M.: Alors: an algorithm recommender system. Artif. Intell. **244**, 291–314 (2017)
33. Behera, G., Nain, N.: Collaborative filtering with temporal features for movie recommendation system. Procedia Comput. Sci. **218**, 1366–1373 (2023)

Chapter 13
Disease Detection in Tomato Plant Leaf Using Deep Learning Techniques

Piyush Choudhary and A. Vinothini

Abstract The identification of diseases in plants is crucial for sustaining crop yield and securing global food resources. This study delves into the utilization of deep learning methodologies to precisely detect diseases in tomato plant leaves. Focused on the PlantVillage dataset, a comprehensive collection of tomato plant leaf images categorized as healthy or diseased is analyzed. Augmentation techniques, encompassing random transformations like rotations, flips, and zooms, are employed to diversify the training dataset, aiming to improve the models' accuracy and generalization capacities. The deep learning models, including AlexNet, DenseNet, LeNet, and two bespoke models, undergo training with well-suited loss functions and are optimized using appropriate techniques. Assessment of the models' performance involves metrics such as accuracy, loss, recall, precision, and F1 score, calculated on distinct validation sets. The outcomes indicate that the specifically tailored models for disease detection outperform established architectures. Notably, Custom Model 2 achieves an accuracy of 86.10% and an F1 score of 85%, while Custom Model 3 attains an accuracy of 86.84% and an F1 score of 86%. Contrastingly, AlexNet records an accuracy of 79% and an F1 score of 79.29%, DenseNet exhibits an accuracy of 89% and an F1 score of 88%, and LeNet scores an accuracy of 81% and an F1 score of 82%. This study underscores the potential of deep learning in precisely identifying plant diseases in tomato leaves, contributing to the advancement of sophisticated diagnostic techniques. Implementing these models in agricultural systems holds promise for proactive disease identification, safeguarding crops, and ensuring global food stability.

P. Choudhary · A. Vinothini (✉)
School of Computer Science and Engineering, Vellore Institute of Technology, Chennai, Tamil Nadu 600127, India
e-mail: vinothini.a@vit.ac.in

P. Choudhary
e-mail: piyush.choudhary2022@vitstudent.ac.in

S. K. Pal et al. (eds.), *Intelligent Informatics*, Smart Innovation, Systems and Technologies 389, https://doi.org/10.1007/978-981-97-2147-4_13

13.1 Introduction

India's agricultural sector stands as the backbone of the nation's economy, engaging nearly half of its workforce [1]. The imperative to address challenges in tomato plant disease detection arises from the necessity to prevent substantial yield losses [2, 3]. Advancements in techniques like K-Nearest Neighbors (KNN) and machine learning algorithms have significantly bolstered disease recognition, encompassing diverse stages from preprocessing to classification [2, 3]. The integration of machine learning within expansive datasets like PlantVillage has spearheaded a revolution in disease detection across a myriad of plant species. Notably, models like MobileNet have displayed robustness in excelling at disease identification, offering promising avenues for comprehensive agricultural disease management [4, 5].

Emerging technological paradigms, such as computer vision and soft computing techniques, have played a pivotal role in automating disease detection processes utilizing leaf images [6]. These sophisticated methods, spanning from image acquisition to lesion segmentation and feature extraction, contribute significantly to precise disease detection, fostering a nuanced understanding of plant health [6]. Innovative methodologies merging Convolutional Neural Networks (CNNs) with feature extraction mechanisms, such as the Gray Level Co-occurrence Matrix (GLCM), have substantially enhanced disease detection capabilities [3, 7]. These pioneering approaches enable the extraction of distinctive texture features from leaves, facilitating more accurate disease classifications [3, 7].

Integrated techniques encompassing RGB conversion, clustering algorithms, and machine learning models have showcased promising outcomes in discerning between diseased and healthy leaves. SVM, KNN, and CNN, methodologies demonstrate remarkable accuracy rates, indicating their viability for real-world agricultural applications [8]. The infusion of artificial intelligence into agriculture, especially in image acquisition and preprocessing, has significantly augmented disease identification methodologies. Models like DenseNet121 and MobileNetV2 exhibit promise in detecting plant diseases, marking significant strides in leveraging cutting-edge technologies for agricultural sustainability [9–11].

Furthermore, innovative strategies employing transfer learning across diverse CNN architectures exhibit improvements in disease recognition across varied datasets. The utilization of VGG19 networks showcases heightened accuracy in identifying various orange leaf abnormalities, underlining the potential for comprehensive disease diagnosis [12, 13]. Recent advancements also encompass the integration of Faster R-CNN within the Grape Disease Detection Network, showcasing promising outcomes in detecting grape diseases [14]. This convergence of machine learning, deep learning, and image processing technologies marks a transformative era in plant disease detection. These advancements offer scalable and accurate solutions, steering agriculture towards sustainable practices and enhanced crop management strategies.

13.2 Methodology

In this segment, the approach adopted for detecting diseases in plant leaves through deep learning methods is outlined. The research made use of the PlantVillage dataset, which encompasses images of tomato plant leaves classified as healthy or diseased. To expand the training dataset, a strategy involving data augmentation was implemented. Additionally, the study involved employing multiple deep learning models, namely AlexNet, DenseNet, LeNet, and two specifically crafted models.

13.2.1 Dataset

The publicly available "PlantVillage" dataset encompasses ten distinct classes, each representative of various disease categories commonly affecting tomato plants [15] as shown in Fig. 13.1. The "PlantVillage" dataset presents an extensive array of tomato leaf images, covering both healthy leaves and leaves affected by various diseases as shown in Fig. 13.2. This diversity offers a comprehensive and balanced set of visual patterns, which served as the foundation for training and evaluating our deep learning models. The dataset's substantial sample size and the presence of multiple disease categories contribute to its complexity, rendering it a valuable resource for the development of robust and versatile models. This diversity equips the models with the capability to recognize and distinguish a wide range of disease patterns, including variations in severity.

Despite the quality of the dataset, typical challenges in plant disease datasets, such as inconsistent lighting, diverse leaf orientations, and instances of multiple diseases on a single leaf, are encountered. To counter these challenges, thorough preprocessing and augmentation techniques are implemented, aiming to enhance the models' resilience and precision.

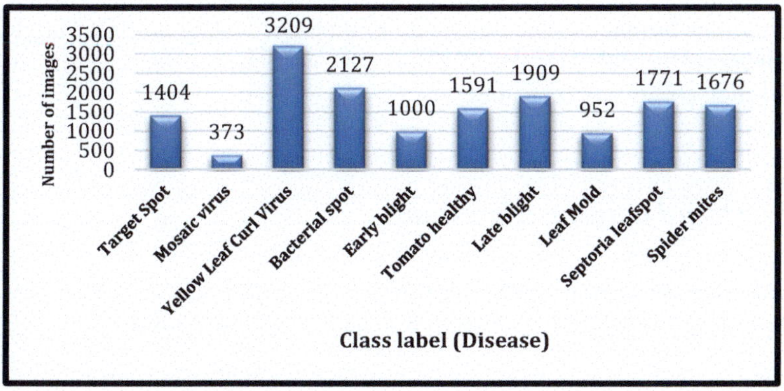

Fig. 13.1 Dataset description

Fig. 13.2 Tomato plant leaf images

Fig. 13.3 Data augmentation

13.2.2 Preprocessing

Before training the deep learning models, we performed preprocessing steps to enhance the dataset's quality and diversity. Data augmentation techniques are applied to generate additional training samples as shown in Fig. 13.3. This involves employing transformations such as rotation, flipping, scaling, and shifting to artificially increase the dataset's size. By doing so, we aimed to improve the model's generalization ability and robustness.

13.2.3 Deep Learning Algorithms

Our disease detection approach involved employing several deep learning algorithms: CNN, AlexNet, DenseNet, LeNet, and two custom-designed models.

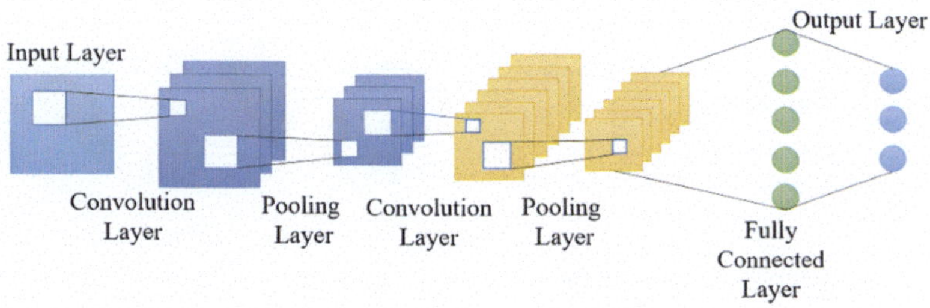

Fig. 13.4 Architecture of a CNN

13.2.3.1 Convolutional Neural Network

In this study, CNN is employed for disease detection in plant leaves. CNN is renowned in image classification tasks for its adeptness in autonomously learning and extracting crucial features from input images [16, 17]. The CNN architecture comprises distinct layers, namely convolutional, pooling, and fully connected layers, as depicted in Fig. 13.4. These layers collectively analyze input images to generate precise predictions.

13.2.3.2 AlexNet

Figure 13.5 illustrates the structure of AlexNet, a prominent deep learning architecture that gained recognition through its success in the ImageNet Large-Scale Visual Recognition Challenge. It encompasses a series of convolutional and fully connected layers, integrating methodologies like local response normalization and dropout to improve overall generalization [18]. In this study, we utilized AlexNet to extract distinguishing features from the plant leaf images and perform classification between healthy and unhealthy specimens.

13.2.3.3 DenseNet

DenseNet, in Fig. 13.6, characterized by dense connections within its convolutional layers, is utilized to exploit feature reuse and mitigate the vanishing gradient issue. The incorporation of skip connections facilitates direct information propagation from preceding to subsequent layers. This architectural approach was applied to capture intricate details and correlations present within the plant leaf images.

13.2.3.4 LeNet

Figure 13.7 illustrates LeNet, a classic CNN architecture that significantly contributed to the inception of deep learning in image classification. Comprising

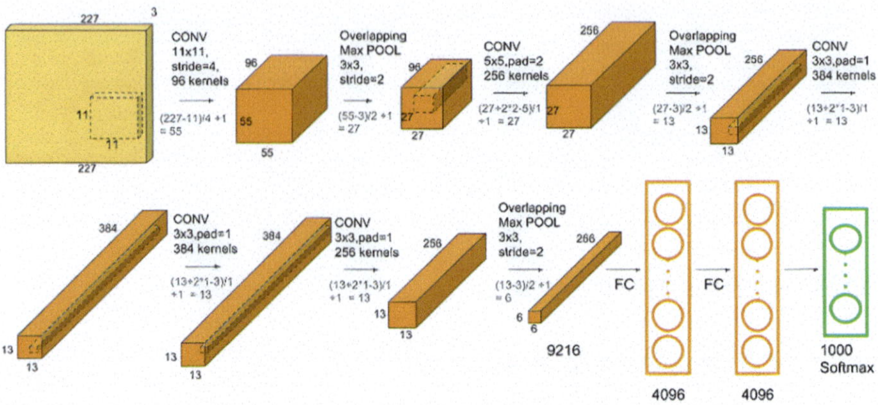

Fig. 13.5 Architecture of AlexNet

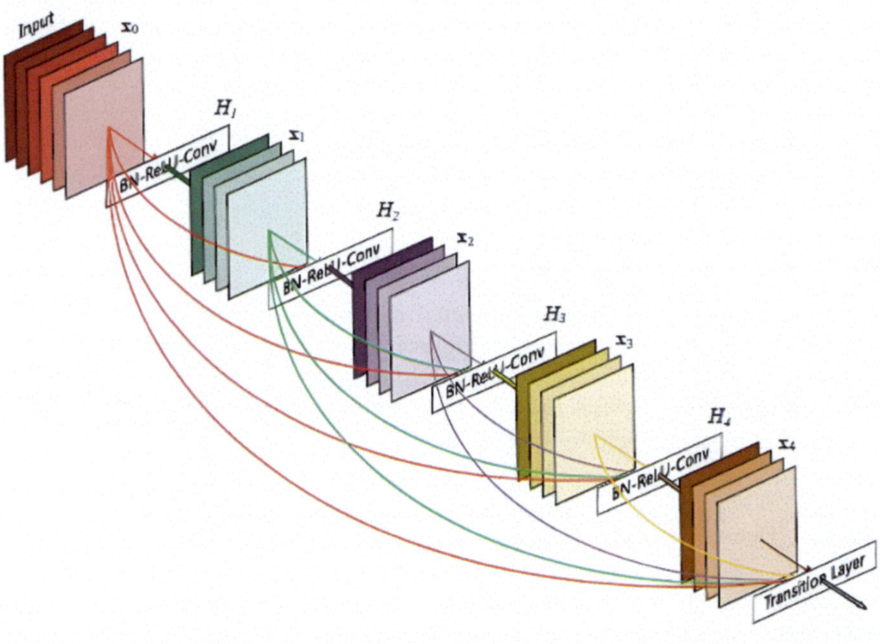

Fig. 13.6 Architecture of DenseNet

convolutional and pooling layers followed by fully connected layers, LeNet has demonstrated notable efficacy across diverse image classification tasks despite its simplicity. Its incorporation in this study facilitated comparative analysis against more sophisticated architectures.

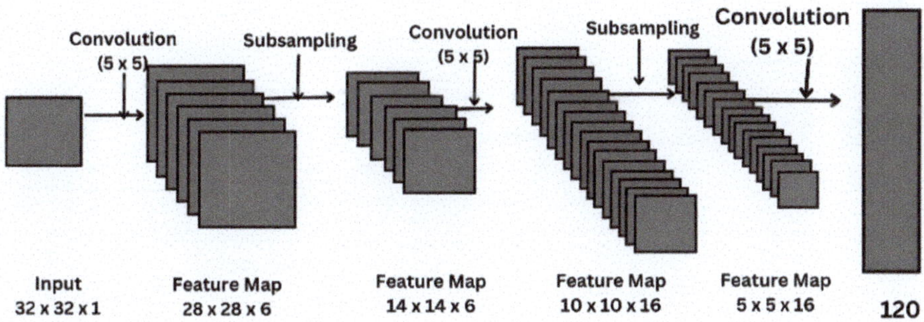

Fig. 13.7 Architecture of LeNet

13.2.3.5 Custom Model 2

In addition to the existing model architectures, we developed two custom models tailored specifically for detecting diseases in plant leaves. Figure 13.8 is the architecture of Custom Model 2, which processes RGB images sized at 224 × 224 pixels. This custom model comprises multiple layers designed to extract features and generate predictions.

The architecture initiates with a convolutional layer utilizing 32 filters, each with a 3 × 3 kernel, employing Rectified Linear Unit (ReLU) activation. A Batch Normalization layer follows, enhancing training stability, succeeded by a max-pooling layer with a size of 2 × 2. This pattern iterates across the subsequent 3 convolutional layers, utilizing 64, 128, and 256 filters with 3 × 3 kernels, each followed by respective max-pooling layers. The flattened data progresses through a sequence of dense layers. The first dense layer integrates 512 units with ReLU activation, succeeded by a dropout layer implementing a dropout rate of 0.5 to mitigate overfitting. Subsequently, the second dense layer adopts 256 units with ReLU activation, followed by another dropout layer employing a dropout rate of 0.5. Finally, the output layer

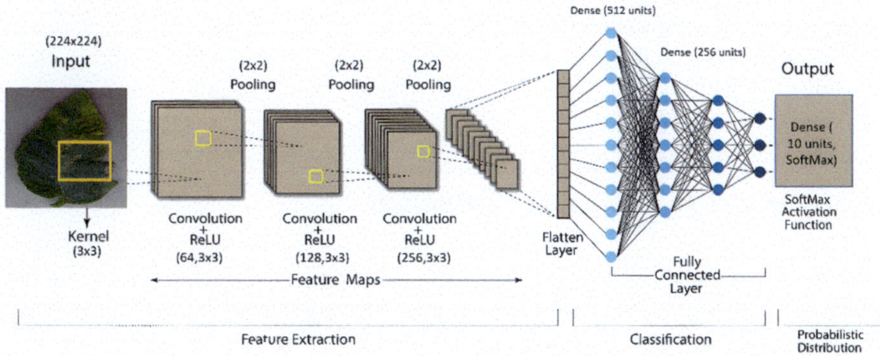

Fig. 13.8 Architecture of custom model 2

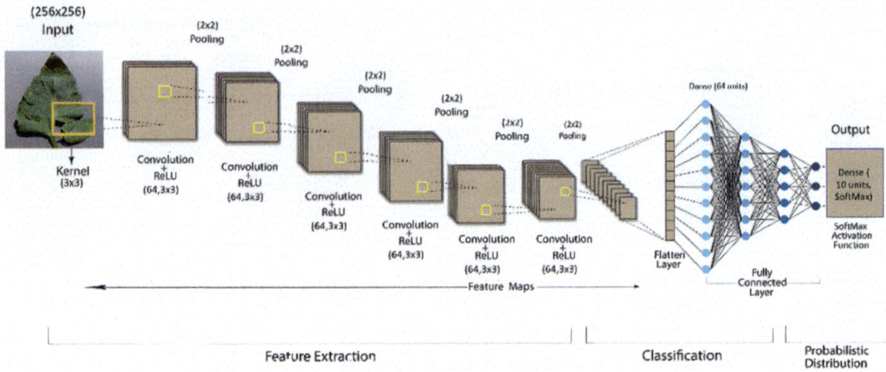

Fig. 13.9 Architecture of custom model 3

encompasses 10 units corresponding to disease classes, utilizing a softmax activation function to present class probabilities.

13.2.3.6 Custom Model 3

Figure 13.9 showcases the architecture of Custom Model 3 designed for our specific task. Our model is structured to handle RGB images sized at 256 × 256 pixels. Initially, a preprocessing layer is integrated to resize and normalize the input images. Subsequently, a sequence of convolutional layers is incorporated to extract salient features. The first convolutional layer consists of 64 filters using a 3 × 3 kernel size and Rectified Linear Unit (ReLU) activation. Following this, max-pooling layers with a 2 × 2 pooling size are introduced to downsample the data. This sequence iterates through 5 additional convolutional layers, each utilizing 64 filters, a 3 × 3 kernel size, ReLU activation, and subsequent max-pooling layers with a 2 × 2 pooling size. The architecture concludes with a flattened layer, preparing the data for fully connected layers. A dense layer employing 64 units and ReLU activation is followed by an output layer tailored to our problem's class count, which stands at 10. The output layer is integrated with a softmax activation function, offering class probabilities.

13.2.4 Model Training and Evaluation

All the deep learning models, comprising AlexNet, DenseNet, LeNet, Custom Model 2, and Custom Model 3, underwent training using the augmented dataset. The training process involved the selection of an appropriate loss function and optimization of the models through a suitable optimizer. Throughout the training phase, continuous monitoring of both loss and validation loss values was conducted to gauge the models' convergence. Furthermore, a separate validation set was employed to evaluate the models' performance, utilizing metrics like accuracy, loss, recall, precision, and F1 score. The acquired results offer significant insights into the efficacy of various deep learning models for disease detection in plant leaves. These findings will be elaborated upon in the Results and Discussion section, facilitating in-depth comparative analysis and the identification of the most proficient model architecture for precise disease detection.

13.3 Results and Discussions

This segment unveils the outcomes derived from our experiments and explores their significance concerning disease detection in plant leaves employing deep learning methodologies.

13.3.1 Experimental Setup

For the investigation into disease detection within tomato plant leaves, our focus centered on employing diverse deep learning models—encompassing AlexNet, Custom Models 2 and 3, DenseNet, and LeNet, alongside a generic CNN model. Each model underwent meticulous configuration, ensuring specific hyperparameters were optimized for robust performance. We meticulously fine-tuned crucial hyperparameters, such as learning rates, batch sizes, and weight decay, tailoring them to achieve optimal results in disease classification. This detailed tuning process played a pivotal role in attaining the highest accuracy possible.

All models underwent training and assessment on the PlantVillage dataset, encompassing both healthy and diseased tomato plant leaf images. To diversify our training data and counter overfitting, a spectrum of data augmentation techniques was implemented, including rotation, flipping, and brightness adjustments. Through the integration of model-specific hyperparameters and strategic data augmentation approaches, our goal was to establish a broad and comprehensive experimental framework, providing valuable insights into the efficacy of disease detection in plant leaves.

13.3.2 AlexNet

AlexNet demonstrated a training loss of 91% and a corresponding validation loss of 89%, as depicted in Fig. 13.10. Its recall rate stood at 78%, accompanied by a precision of 80.52% and an F1 score of 79.29%. Furthermore, the model exhibited a

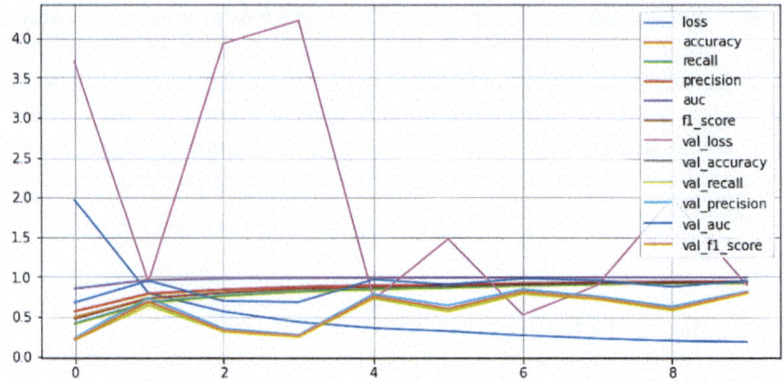

Fig. 13.10 Learning curve of Alexnet

validation accuracy of 80%, while achieving an overall accuracy of 79% on the test set.

13.3.3 Custom 2 Model

The Custom 2 model exhibited a training loss of 43.58% and a corresponding validation loss of 40.14%. According to the findings depicted in Fig. 13.11, this model attained a recall rate of 41%, a precision of 86%, and an F1 score of 85%. Remarkably, the validation accuracy surged to 88.55% while achieving an overall accuracy of 86.10%.

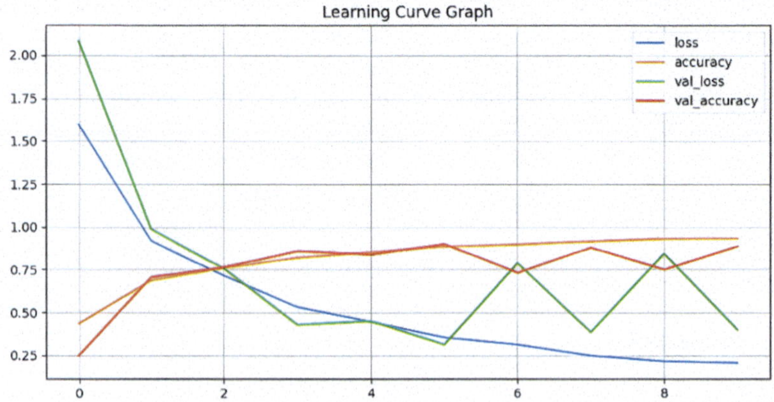

Fig. 13.11 Learning curve of Custom 2

13.3.4 Custom 3 Model

Likewise, Custom 3 recorded a training loss of 46.47% and a corresponding validation loss of 41%. The model showcased a recall rate of 42%, a precision of 85%, and an F1 score of 86%. These outcomes are illustrated in Figs. 13.12 and 13.13. Notably, the validation accuracy reached 87%, while the overall accuracy stood at 86.84%.

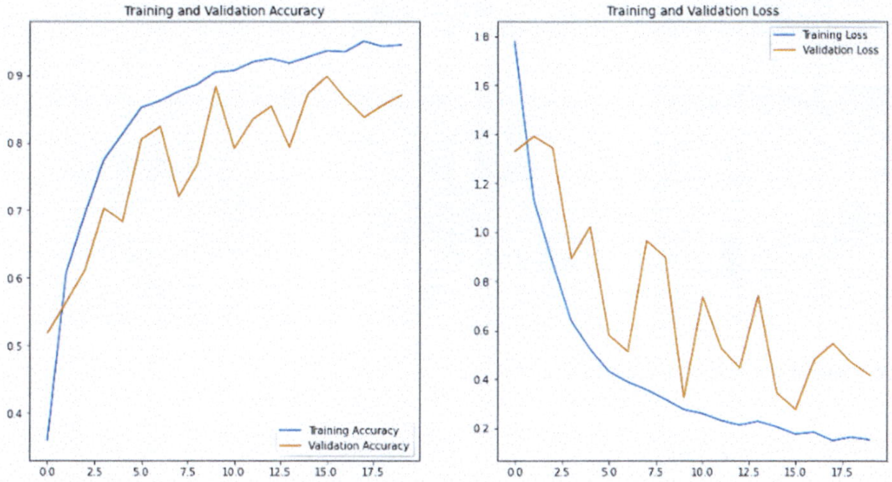

Fig. 13.12 Training and validation curve of Custom 3

Fig. 13.13 Actual and predicted result

```
first image to predict
actual label: Tomato_Late_blight
2/2 [==============================] - 1s 384ms/step
predicted label: Tomato_Late_blight
```

Fig. 13.14 Confidence metrics of DenseNet

13.3.5 DenseNet

The DenseNet model achieved a training loss of 31.77% and a validation loss of 29%. Its performance revealed a recall of 30%, a precision of 88%, and an F1 score of 88%. Validation accuracy stood at 89%, consistent with the overall accuracy of 89%. Visual representation of these metrics is detailed in Fig. 13.14, including confidence metrics.

13.3.6 LeNet

The LeNet model displayed a training loss of 80% and a validation loss of 61%. It presented a recall of 62%, a precision of 82%, and an F1 score of 82%. Figure 13.15 illustrates a validation accuracy of 84%, aligned with the overall accuracy of 81%.

13.3.7 CNN

The generic CNN model displayed a training loss of 7.8% and a validation loss of 10.48%. It demonstrated a recall of 11%, a precision of 94%, and an F1 score of 97%.

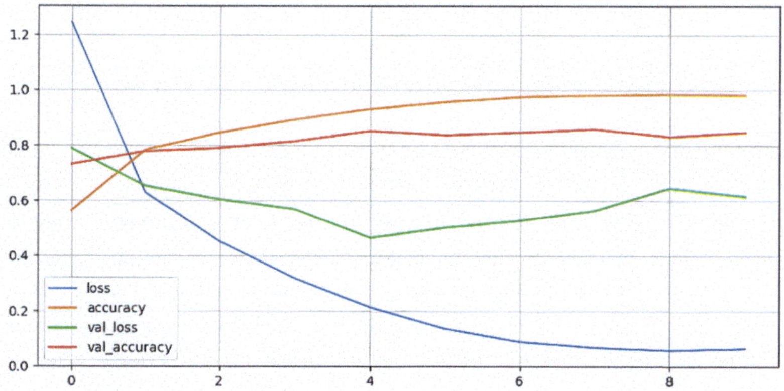

Fig. 13.15 Learning curve of LeNet

Table 13.1 Performance of CNN

Features	Precision (%)	Recall (%)	F1 (%)
Bacterial-spot	100.00	99.08	99.54
Early-blight	95.65	98.88	97.24
Healthy	98.29	100.00	99.14
Late-blight	95.33	98.08	96.68
Leaf-mold	98.26	95.76	97.00
Mosaic virus	100.00	100.00	100.00
Septoria leaf spot	97.59	93.10	95.29
Yellow leaf curl virus	100.00	100.00	100.00

The validation accuracy reached 96%, aligned with an overall accuracy of 94.92%. The performance metrics for the CNN model are detailed in Table 13.1.

13.3.8 Comparative Analysis of Methods

The study conducts a thorough comparative analysis to evaluate the performance of diverse deep learning models employed for disease detection in tomato plant leaves. This comprehensive assessment intends to delineate the strengths and weaknesses of each approach, facilitating a robust evaluation of their disease detection capabilities. The compared methodologies encompass established architectures like AlexNet, DenseNet, LeNet, alongside two customized models denoted as Custom Model 2 and Custom Model 3. Assessment criteria involve pivotal metrics such as accuracy, loss, recall, precision, and F1 score, ensuring a comprehensive grasp of the models' efficacy in discerning between healthy and unhealthy leaves, and elucidating the balance between true positive and false positive predictions.

Table 13.2 A comparative analysis of tomato leaf disease detection methods

Technique/ method	Training loss (%)	Validation loss (%)	Recall (%)	Precision (%)	F1 score (%)	GFlops	Validation accuracy (%)	Accuracy (%)
AlexNet	91.87	89.00	78.15	80.52	79.29	108	80.00	79.00
Custom 2	43.58	40.14	41.00	86.00	85.00	625	88.55	86.10
Custom 3	46.47	41.00	42.00	85.00	86.00	10	87.00	86.84
DenseNet	31.77	29.00	30.00	88.00	88.00	4	89.00	89.00
LeNet	80.00	61.00	62.00	82.00	82.00	176.7	84.00	81.00
CNN	7.81	10.48	11.00	94.00	97.00	636	96.00	97.92

The aim of this comparative analysis is to guide both researchers and practitioners in selecting the most appropriate approach for disease detection in tomato plant leaves. The metrics showcased herein have undergone meticulous evaluation on distinct validation sets, forming the foundation for subsequent discussions and fostering a deep comprehension of the deep learning models' performance. Table 13.2 furnishes a detailed comparative overview of the performance metrics, offering a lucid visual representation that facilitates straightforward interpretation and comparative analysis of the outcomes.

Through this in-depth comparative analysis, our objective is to delineate the strengths and constraints inherent in each methodology, thereby contributing to the advancement of disease detection methodologies in tomato plants. These insights serve as a valuable resource for researchers and practitioners, empowering them to make informed decisions when employing deep learning models for plant disease detection, ultimately fostering improvements in agricultural practices and bolstering crop yield.

13.3.9 Discussion

The outcomes gleaned from our experiments furnish valuable insights into the performance disparities among diverse deep learning models utilized for disease detection in plant leaves. Notably, DenseNet emerged as the frontrunner, boasting an exceptional overall accuracy of 89%. Its equilibrium between precision and recall culminated in an impressive F1 score of 88%, underscoring its proficiency in accurately discerning diseases in plant foliage. Moreover, our custom-crafted models, Custom 2 and Custom 3, surpassed conventional architectures like AlexNet and LeNet, showcasing competitive accuracy, precision, recall, and F1 scores, signifying their potential for disease identification endeavors. Noteworthy is the generic CNN model, which displayed exceptional precision and an outstanding F1 score of 97%. However, its relatively lower recall rate implies the necessity for further enhancements to bolster sensitivity in detecting disease instances.

The insights garnered from our research substantiate the efficacy of employing deep learning methodologies in plant leaf disease detection. By adeptly categorizing healthy and unhealthy leaves, these models facilitate early disease identification, enabling prompt interventions and mitigating potential yield diminutions. The implications derived from our study serve as a guiding beacon for agricultural researchers and practitioners, emphasizing the criticality of tailored deep learning architectures precisely calibrated for the task of detecting diseases in plant leaves.

13.4 Conclusions

This study extensively delved into the effectiveness of deep learning methodologies for detecting diseases in plant leaves, centered on the diverse tomato plant leaf dataset provided by PlantVillage. Our principal aim was to pinpoint the most efficient model architecture for precise disease identification. To fortify the adaptability of the models, we implemented data augmentation techniques, enriching the training dataset with random transformations like rotations, flips, and zooms to enhance image diversity. Employing rigorous training and optimization, incorporating suitable loss functions and optimizers, we developed models like AlexNet, DenseNet, LeNet, and two customized models (Custom Model 2 and Custom Model 3). Continuous monitoring of loss and validation loss values ensured the models' convergence. Assessment using various metrics—accuracy, loss, recall, precision, and F1 score—on separate validation sets offered crucial insights into the models' efficacy.

The experiments yielded promising results: Custom Model 2 attained an accuracy of 86.10% and an F1 score of 85%, while Custom Model 3 showcased an accuracy of 86.84% and an F1 score of 86%, outshining established architectures. However, these custom models, while proficient in capturing disease patterns, may benefit from further fine-tuning for optimal performance. While showcasing the potential of deep learning in disease detection, the study identified challenges. Early implementation of CNN and DNN techniques elevates classification accuracy, yet refinement is necessary to enhance recognition rates during classification. Moreover, ensuring scalability and robustness across varied plant species and environmental conditions is crucial for real-world applicability.

This research significantly contributes to advancing plant disease diagnosis and mitigation, presenting promising prospects for practical integration in agriculture to fortify crop protection and global food security. However, surmounting challenges related to model generalization and adaptability remains imperative for real-world deployment. In essence, this study underscores the transformative potential of deep learning in plant disease detection. While making substantial progress, addressing the outlined challenges will pave the way for more resilient and scalable solutions, fostering sustainable growth in agricultural practices.

References

1. Sawant, C., Shirgaonkar, M., Khule, S., Jadhav, P.: Plant disease detection using image processing techniques. Int. J. Sci. Res. Comput. Sci., Eng. Inf. Technol. (IJSRCSEIT), 295–300 (2020). https://doi.org/10.32628/CSEIT206260
2. Geetha, G., Samundeswari, S., Saranya, G., Meenakshin, K., Nithya, M.: Plant leaf disease classification and detection system using machine learning. J. Phys.: Conf. Series. IOP Publishing (2020)
3. Fulari, U.N., Shastri, R.K., Fulari, A.N.: Leaf disease detection using machine learning. J. Seybold Rep. ISSN NO 1533 (2020)
4. Bhise, N., Kathet, S., Jaiswar, S., Adgaonkar, A.: A Plant disease detection using machine learning. Int. Res. J. Eng. Technol. (IRJET) (2020)
5. Ahmad, M., Abdullah, M., Moon, H., Han, D.: Plant disease detection in imbalanced datasets using efficient convolutional neural networks with stepwise transfer learning. IEEE Access (2021)
6. Vishnoi, V.K., Kumar, K., Kumar, B.: Plant disease detection using computational intelligence and image processing. J. Plant Dis. Protect. (2021)
7. Thangavel. M, Gayathri. P.K., Sabari, K.R., Prathiksha, V.: Plant leaf disease detection using deep learning. Int. J. Eng. Res. Technol. (IJERT) (2022)
8. Harakannanavar, S.S., Rudagi, J.M., Puranikmath, V.I., Siddiqua, A.: Plant leaf disease detection using computer vision and machine learning algorithms. In: Global Transitions Proceedings (2022)
9. Sujawat, G.S.: Application of artificial intelligence in detection of diseases in plants: a survey. Turkish J. Comput. Math. Educ. (TURCOMAT) (2021)
10. Bhattania, Y., Singhal, P., Agarwal, T.: Plant leaf disease detection using deep learning. Int. J. Res. Appl. Sci. Eng. Technol. (IJRASET) (2022)
11. Delnevo, G., Girau, R., Ceccarini, C., Prandi, C.: A deep learning and social IoT approach for plants disease prediction toward a sustainable agriculture. IEEE Internet of Things J. (2021)
12. Barburiceanu, S., Meza, S., Orza, B., Malutan, R., Terebes, R.: Convolutional neural networks for texture feature extraction. Applications to leaf disease classification in precision agriculture. IEEE Access (2021)
13. Gómez-Flores, W., Garza-Saldaña, J.J., Varela-Fuentes, S.E.: A Huanglongbing detection method for orange trees based on deep neural networks and transfer learning. IEEE Access (2022)
14. Dwivedi, R., Dey, S., Chakraborty, C., Tiwari, S.: Grape disease detection network based on multi-task learning and attention features. IEEE Sens. J. (2021)
15. https://www.kaggle.com/datasets/emmarex/plantdisease
16. Jiang, P., Chen, Y., Liu, B., He, D., Liang, C.: Real-time detection of apple leaf diseases using deep learning approach based on improved convolutional neural networks. IEEE Access (2019)
17. Karthik, K., Rajaprakash, S., Ahmed, S.N., Perincheeri, R., Alexander, C. R..: Tomato and potato leaf disease prediction with health benefits using deep learning techniques. In: Fifth International Conference on I-SMAC (IoT in Social, Mobile, Analytics and Cloud) (I-SMAC), pp. 1–3 (2021)
18. Li, L., Zhang, S., Wang, B.: Plant disease detection and classification by deep learning—a review. IEEE Access, vol. 9, pp. 56683–56698 (2021)

Chapter 14
Boosting Precision Agriculture Using Deep Learning Models on Edge Devices

Amarsh Gautam, Mohammad Basil Faruqui, Nadeem Akhtar, and Usama Bin Rashidullah Khan

Abstract In the contemporary landscape, the Internet of Things (IoT) has become a central focus across diverse sectors. Many IoT devices rely on machine learning (ML) models, endowing them with cognitive capabilities and decision-making proficiency. However, the inherent resource constraints of IoT devices often hinder the deployment of intricate ML models, such as deep learning (DL), on these edge devices. This research aims to enhance precision agriculture by employing methodologies that leverage the formidable capabilities of Deep Learning paradigms. The emphasis of the paper is on applying methods to enhance precision agriculture using deep learning on edge devices. In the proposed work, a Deep Neural Network (DNN) is trained and implemented to categorize images as valid or invalid in real time, reducing server load for subsequent tasks in precision agriculture, such as weed detection or crop identification. The DNN for classifying images will be implemented on an edge device, specifically a Microcontroller Unit (MCU). To harness deep learning capabilities on microcontrollers, the project will employ lightweight network modeling, quantizing parameters, and utilizing separable convolution to fit the model within the microcontroller's constraints. The model will be implemented on STM32 series microcontrollers for real-time inferencing. The NUCLEO-G474RE MCU kit and STM32CubeIDE were utilized for interfacing. The results underscore the potential of using deep learning models on edge devices for precision agriculture applications. The approach, involving separable convolution, data augmentation, batch normalization, dropout, and TFLite format conversion, proves effective in overcoming the challenges of limited computational resources on edge devices while achieving high accuracy and efficiency. The model demonstrated promising results on both test and real-world data, and the deployment process was simplified using STM32CubeIDE.

A. Gautam (✉) · M. B. Faruqui · N. Akhtar · U. B. Rashidullah Khan
Aligarh Muslim University, Aligarh, India
e-mail: gj2093@myamu.ac.in

S. K. Pal et al. (eds.), *Intelligent Informatics*, Smart Innovation, Systems and Technologies 389, https://doi.org/10.1007/978-981-97-2147-4_14

14.1 Introduction

Agriculture is vital in India, employing 60–70% of the population. IoT, machine learning, and data analytics on microcontrollers offer opportunities for precise farming to meet food demand [1]. Efficient deep learning on microcontrollers requires lightweight models that fit device constraints [2], particularly on STM32 series microcontrollers for real-time use.

Precision agriculture optimizes production by monitoring environmental variables like temperature, humidity, and soil moisture. AI in precision farming covers fundamentals, engineering, and supporting aspects, reducing costs by inferring variables without direct measurement [3]. Deep learning aids intelligent decision-making in precision agriculture, including disease recognition and quality monitoring. It benefits from data from various sensors, cameras, and smartphones [2, 4], using multi-layer processing to discover complex patterns in high-dimensional data for agricultural tasks.

Modern agriculture aims to enhance crop production and replicate specific environmental conditions, either locally or in controlled environments like greenhouses. It utilizes advanced monitoring and information technologies, including edge device-based machine learning, to mitigate the impact of adverse weather and diseases on crop yield and quality.

Cloud-based machine learning faces three primary challenges: energy conservation, latency, and privacy concerns. TinyML offers solutions to these issues, particularly beneficial for precision agriculture applications [5]. Crop yields are adversely affected by pests, insects, and diseases. Our initiative employs cost-effective advanced technologies to combat these agricultural challenges [1]. TinyML addresses key IoT challenges, such as the need for rapid responses in precision agriculture. By minimizing data transmission delays and reducing server overhead, edge computing using microcontrollers enables real-time machine learning, a highly desirable feature.

This paper will discuss a streamlined model for disease detection in precision agriculture, hosted on a designated server infrastructure [1]. To reduce latency, we'll directly transmit images from an edge microcontroller to the model, alleviating server computational load. This paper is structured as follows: Sect. 14.2 presents the materials and methods employed in this work. Section 14.2.6 describes the experimental results. Section 14.3 discusses the results obtained and Sect. 14.4 concludes this paper.

14.2 Materials and Methods

In this section, we delve into the literature survey, proposed work, model architecture, and solutions for project challenges. We explored problem identification, functionalities, scenarios, and descriptions for both edge devices and servers, along with functional responses in various scenarios.

14.2.1 Literature Survey

Jia et al. [1] describe precision agriculture as a strategic approach to managing agricultural operations. It involves the collection, processing, and analysis of data related to time, space, and individual elements within the agricultural production process. This data is then integrated with other information to enhance decision-making. The goal is to increase resource efficiency, productivity, quality, profitability, and sustainability in agricultural production by considering variability. In the realm of precision agriculture, deep learning techniques are applied to various tasks such as identifying crop pests and diseases, detecting and mapping weeds, monitoring crop growth, estimating crop yields, and classifying crop types. In [1], the authors highlight the utilization of different deep learning algorithms, including YOLO [6] and semantic segmentation [7], for these agricultural tasks. However, it's worth noting that these approaches do not make use of TinyML capabilities; instead, they are implemented using cloud technology.

Numerous researchers across different studies share a common perspective when it comes to designing networks for microcontroller implementation. In their respective works, references [3, 8, 9] all emphasize the importance of developing lightweight network architectures to accommodate the memory constraints of microcontrollers. They provide various strategies for effectively applying these lightweight models. The discussed techniques encompass parameter pruning, lightweight convolutional design, parameter quantization, knowledge distillation, and low-rank factorization. Parameter pruning involves the removal of non-contributing parameters, eliminating redundant and non-informative elements that do not significantly contribute to the model's discriminative performance. Lightweight convolutional design focuses on compactly designing convolutional filters for efficient feature extraction. Parameter quantization [10] reduces the model's storage and transmission volume by reducing the number of bits used for weight representation in the deep model. Knowledge distillation trains a smaller student network from a larger teacher network while preserving its generalization capabilities, thereby achieving a smaller model size and reduced computational requirements. Lastly, low-rank factorization seeks to decompose a large weight matrix into smaller dimension matrices, leading to faster inference and reduced storage needs. When applied to denser fully connected layers, this approach eliminates redundancy and minimizes storage requirements. In their work, Igor et al. [9] introduce Sparse Architecture Search (SpArSe), a technique that combines neural architecture search with network pruning. This approach strikes a balance between generalization performance and stringent memory constraints by enabling the rapid evaluation of numerous sub-networks within a given network. Zhong et al. [11] propose EdgeSegNet, a compact deep convolutional neural network tailored for semantic segmentation tasks on mobile devices and drones. This architecture integrates residual bottleneck macroarchitectures with shortcut connections and non-residual bottleneck macroarchitectures. It effectively reduces channel dimensionality using 1×1 convolutions, thereby lowering computational complexity. EdgeSegNet also employs 8×8 strided convolutions in its non-residual bottleneck reduction module macroarchitecture to aggressively reduce spatial dimensionality in subsequent layers. Despite its success in model size reduction, EdgeSegNet remains impractical for microcontroller implementation, although it offers an innovative approach to address memory constraints in high-end edge devices. Rachel et al. [12] introduce an intriguing concept to make edge computation feasible and real time for non-GPU computers. They emphasize the need for real-time object detection models and discuss the limitations of previous models like YOLO [6] and Regional-based Convolution Neural Networks (R-CNN). While these models achieved high mean average precision (mAP), they fell short in providing practical frames per second (FPS) for real-time applications on non-GPU computers. YOLO-Lite [12] emerges as a solution, showcasing the potential of shallow networks for speedy non-GPU object detection applications. While YOLO-Lite excels in FPS on non-GPU systems, it does come with a trade-off as accuracy gradually decreases. Nevertheless, YOLO-Lite underscores the significance of lightweight real-time object detection in everyday scenarios.

14.2.2 System Architecture

For the system, the device captures the image in real time, it then transfers it to the STM32 MCU (embedded in the device), which acts as a filter for image validation, which is a classification task wherein the proposed TinyML model decides to either validate or discard the current image and forward it to the main server or not. The STM32 kit is used as a platform for generating interface results on edge. The STM32CubeIDE [13] is a special development platform with a peripheral configuration that generates the C code. It is implemented by the model which is a classifier to pass or drop the image. The interfacing is done after the image is classified and action is taken. After the code generation, a TFLite [14] file is generated which is fed to the MCU. The validated images are then sent to the main server model to generate interface results related to precision agriculture. And there rests our main robust model that classifies, predicts, and identifies crops, prediction yields and diseases in the crop, and boosts agriculture in an ML style. Figure 14.1 gives a pictorial view of system architecture.

We employed an innovative strategy merging deep learning models with edge devices, specifically the STM32G474RE Nucleo board [15]. This setup empowers

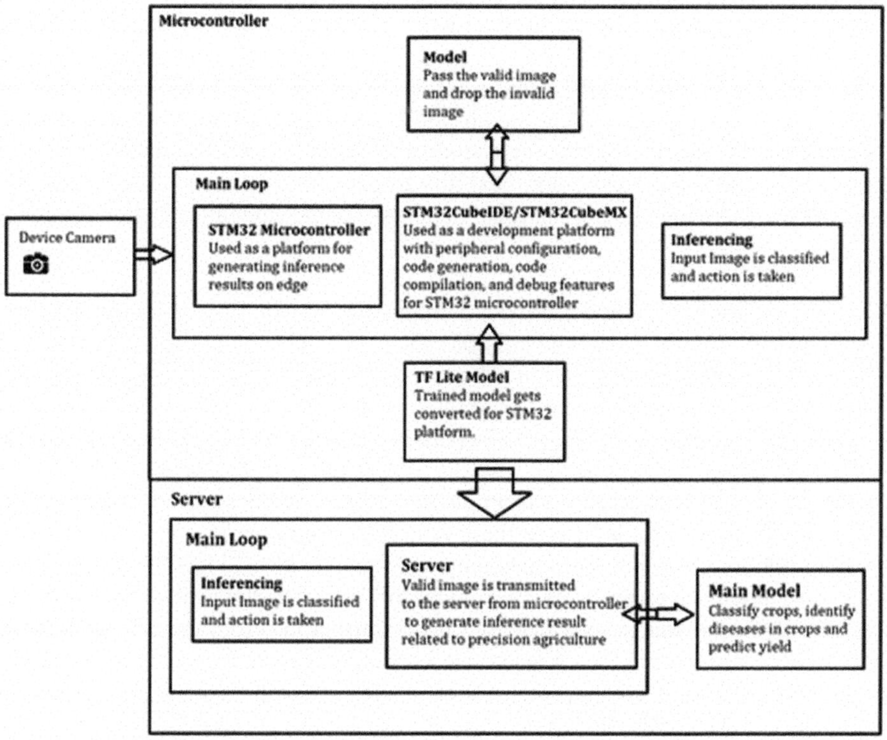

Fig. 14.1 System architecture

us to deliver real-time predictions and optimizations, elevating precision agriculture's efficiency and accuracy via deep learning algorithms and edge computing.

14.2.3 TinyML Model Implementation

The TinyML model utilizes an innovative strategy merging deep learning models with edge devices, specifically the STM32G474RE Nucleo board [15]. This setup empowers us to deliver real-time predictions and optimizations, elevating precision agriculture's efficiency and accuracy via deep learning algorithms and edge computing.

Normal CNN models are designed with deeper and more complex architectures, which eventually require computer-intensive and storage-intensive hardware. As microcontrollers have limited computational and storage power, the computing-intensive CNN models must be dialed down to make the integration process with the microcontroller possible. This can be achieved by compressing the deep neural networks, primarily by employing sophisticated techniques like lightweight network modeling, parameter quantization, parameter pruning, and knowledge distillation [3, 8, 9].

Parameter Quantization [10] minimizes the storage and transmission requirements of the trained model by reducing the bit-width of the model's weights. Knowledge distillation is the process of training a student network using the knowledge from a teacher network while preserving its ability to generalize; this results in a smaller model size with reduced computational demands for the student model.

TinyML Model

A pivotal technique the model utilizes is separable convolution, streamlining data processing by splitting the convolution into two steps: depthwise convolution followed by pointwise convolution. This approach substantially reduced computational demands, enhancing model speed for edge deployment. The model is thoughtfully designed, featuring six layers of separable convolution [16] to strike a balance between accuracy and efficiency. The deep learning model comprises two dense layers. The first extracts features from data processed by the separable convolution layers, while the second leverages these features for predictions. This architecture ensures high accuracy with minimal computational resources, ideal for edge predictions. The model takes an RGB crop image of 28×28 dimensions as input and classifies the input image as a diseased or healthy crop. Programming is done in Python. In summary, the use of separable convolution with six layers followed by two dense layers holds great promise for enhancing precision agriculture on devices like the STM32G474RE Nucleo board [15].

To further optimize the model, various techniques, including data augmentation, batch normalization, and dropout are thoughtfully utilized. Data augmentation generates additional training data through transformations like rotation and scaling to enhance model robustness and prevent overfitting. Batch normalization normalizes inputs to each layer, expediting training and improving accuracy. Dropout randomly deactivates neurons during training to prevent overfitting and enhance model generalization.

Programming the Microcontroller

The model is saved in TFLite [14] format and is integrated with the NUCLEO-G474RE kit [15] through STM32CubeIDE [13]. The model is integer quantized, which is an optimization method that converts 32-bit floating-point numbers (such as activation outputs and weights) to the closest 8-bit fixed-point numbers, resulting in a smaller model and faster inferencing speed [10]. The IDE converts the model into low-level C language functions using a very special X-CUBE-AI [17] library, which must be used to program the microcontroller. The model is to fit in the memory of 512 KB and a RAM of 128 KB, provided by the NUCLEO-G474RE kit [15]. The model will receive an input image from the desktop, and classify the image as a healthy crop and a diseased crop. The diseased crop image will be transmitted to the server model for further classification.

The microcontroller's output is checked on a third-party terminal called TeraTerm [18].

14.2.4 Server Model Implementation

Deep learning models are pivotal in this context due to their capacity to process extensive datasets and offer precise predictions. Notably, the integration of deep learning models into edge devices like microcontrollers is gaining traction in precision agriculture. In this paper, a TinyML model on a microcontroller for validation and a deep learning model on a server for accurate plant disease classification has been integrated. This approach effectively combines edge computing benefits, such as low latency and reduced bandwidth needs, with the computational prowess and storage capacity of cloud computing. The server-based deep learning model comprises multiple layers and employs various advanced techniques, including CNN [19], ViT [20], VGG [21], Swin transformer [22], EfficientNet [23], and Densenet [24], to achieve superior accuracy in plant disease classification. These techniques empower our model to analyze intricate plant images and deliver precise disease classification results, rendering it indispensable for precision agriculture. Additionally, hosting the deep learning model on a server enables it to handle extensive data volumes and execute complex computations, making it well suited for large-scale precision agriculture applications. Ultimately, this approach empowers farmers with accurate data-driven insights, facilitating optimized crop production, waste reduction, and improved efficiency. In conclusion, our project underscores the significance of deep learning models in precision agriculture and highlights the benefits of merging edge and cloud computing for enhanced food security and crop production optimization.

14.2.5 Dataset and Preprocessing

The dataset consisted of images of plants belonging to four different species—Corn, Potato, Rice, and Wheat—that are affected by various diseases. The total number of classes in the dataset is 14, and the total number of images is 13,024. The Corn species has four classes—Common Rust, Gray Leaf Spot, Healthy, and Northern Leaf Blight, with a total of 3,852 images. The Potato species has three classes—Early Blight, Healthy, and Late Blight, with a total of 2,152 images. The Rice species has four classes—Brown Spot, Healthy, Leaf Blast, and Neck Blast, with a total of 4,078 images. Finally, the Wheat species has three classes—Brown Rust, Healthy, and Yellow Rust, with a total of 2,942 images.

Exploratory Data Analysis

Exploratory Data Analysis (EDA) is an important step in understanding the characteristics of a dataset. Here is an EDA of the given dataset: Class distribution: The dataset contains 14 different classes of plants affected by various diseases.

The class distribution is as follows:

- Corn: Common Rust (1192 images), Gray Leaf Spot (513 images), Healthy (1162 images), Northern Leaf Blight (985 images).
- Potato: Early Blight (1000 images), Healthy (152 images), Late Blight (1000 images).
- Rice: Brown Spot (613 images), Healthy (1488 images), Leaf Blast (977 images), Neck Blast (1000 images).
- Wheat: Brown Rust (902 images), Healthy (1116 images), Yellow Rust (924 images).

Image Equalization

Image Equalization served as a preprocessing step designed to improve the contrast of input images [25]. It accomplished this through histogram equalization, a method that redistributes the pixel intensity values within an image to make better use of the entire available range of values. When provided with a list of images, the function independently applies histogram equalization to each color channel (Red, Green, Blue) using the "cv2.equalizeHist" function from the OpenCV library. After this enhancement, the modified channels are recombined into a single image using the "cv2.merge" function. The outcome was a NumPy array containing the preprocessed images.

One Hot Encoding

Image Equalization served as a preprocessing step designed to improve the contrast of input images [25]. It accomplished this through histogram equalization, a method that redistributes the pixel intensity values within an image to make better use of the entire available range of values. When provided with a list of images, the function independently applies histogram equalization to each color channel (Red, Green, Blue) using the "cv2.equalizeHist" function from the OpenCV library. After this enhancement, the modified channels are recombined into a single image using the "cv2.merge" function. The outcome was a NumPy array containing the preprocessed images.

Train Test Split

The dataset splitting took inputs for features and labels. The test size parameter was set to 0.22 (approximately 22%), and the testing data percentage and the random state ensured consistent splitting for reproducibility. The output variables represent the training and testing sets for model training and evaluation. The "get class" function, converts numeric labels to string labels for plant disease classes. It checks the input label and assigns the corresponding string label. This function aids in result interpretation in plant disease classification.

14.2.6 Experiments

In this subsection, we shall discuss our experiments and their results with various deep learning models that we have performed after the pre-processing of the dataset. The models that we have used include CNN, EfficientNet, DenseNet, VGG16, ViT, Swin Transformer, and our final model Inception V3.

CNN

This model comprises a sequence of layers for processing input image data [19]. The initial layer is a Conv2D layer with 32 filters and a (3,3) kernel size. It employs the ReLU activation function and has an input shape defined by the image's height, width, and color channels (3 for RGB). The output from this layer is then directed to a MaxPooling2D layer with a (2,2) pooling size, which reduces the dimensions of the feature maps. Following that, there's a second Conv2D layer with 64 filters and a (3,3) kernel size, followed by another MaxPooling2D layer using a (2,2) pooling size. Subsequently, there's a flattened layer that transforms the 2D feature maps into a 1D feature vector suitable for input into a fully connected layer. The subsequent layer is a Dense layer comprising 64 units with ReLU activation. The final layer is another Dense layer with 4 units, corresponding to the number of classes in the dataset, and it employs the softmax activation function to produce a probability distribution across the classes. The model is configured with the Adam optimizer, categorical cross-entropy loss function, and accuracy metric. When using the summary function, you'll obtain an overview of the model's architecture, including the parameter count for each layer.

Efficient Net

In this approach, a convolutional neural network (CNN) was constructed by leveraging the EfficientNet pre-trained model for image classification [23]. Initially, the "inputs" tensor was established with the dimensions corresponding to the input image. Subsequently, the "EfficientNet" model was invoked, with the "inputs" tensor serving as its input. It's noteworthy that the pre-trained weights were omitted. The model underwent compilation, employing the "Adam" optimizer and adopting the "categorical_crossentropy" loss function. Additionally, the "accuracy" metric was incorporated to assess performance during training. This coding strategy bears resemblance to transfer learning, wherein the pre-trained model is employed as a feature extractor, and the fully connected layers are trained on top of it for image classification. The key advantage of utilizing a pre-trained model lies in its ability to have already acquired valuable features from a substantial dataset, potentially resulting in improved performance when dealing with smaller datasets.

Dense Net

This model was defining and training a deep learning model using the DenseNet121 architecture for image classification [24]. The DenseNet121 function returned the output tensor of the model. The "Model" object is compiled with the Adam optimizer, categorical cross-entropy loss function, and accuracy metric. Finally, the "fit" method was called on the model object to train the model on the training data for 10 epochs. The training history was stored in a variable.

VGG-16

This model consisted of creating a convolutional neural network (CNN) using the VGG16 architecture [21]. First, an input layer is defined with the shape of the input images. The output of the model was obtained by applying the VGG16 architecture to the input layer. The resulting model was compiled using the Adam optimizer and categorical cross-entropy as the loss function, and accuracy as the metric.

ViT

The ViT model represents a neural network structure that excels in delivering leading-edge outcomes across a range of computer vision assignments [20]. In the initial steps, the function delineated the input layer of the model, explicitly designating the input's dimensions and passing it through the function. Subsequently, the function implemented data augmentation techniques on the input images. These enhanced images were then routed through a "Patches" layer, responsible for segmenting the images into smaller patches. To complete the model setup, compilation was accomplished using the Adam optimizer, a cross-entropy loss function, and an accuracy metric.

Swin Transformer

This model was instantiated by specifying its input and output as parameters, denoting the input and output layers of the model, correspondingly [22]. The "compile" method was subsequently employed to configure the loss function, optimizer, and metrics for use during the training process. To train the model, the "fit" method was applied. Upon completion of the training, the "fit" method yielded a "history" object encompassing the loss and accuracy metrics for both the training and validation sets at each epoch. These metrics served as a means to assess the model's performance during training and make informed choices regarding model refinement.

Inception V3

This is a custom deep learning model that was used here for the multi-class classification of plant leaf images [26]. The function takes an optional argument that specifies the size of the input images to the model. Next, the function creates a pre-trained InceptionV3. The model is loaded with pre-trained weights on the ImageNet dataset. The first 291 layers of the model were set to non-trainable. This was to freeze the weights of the first few layers of the pre-trained model which are usually specialized for low-level feature extraction such as edges and corners. The function then defined the input layer of the model. Data augmentation was applied to the input images and the pixel values were preprocessed.

The pre-trained model was then applied to the augmented input data to extract high-level features from the images. A global average pooling layer was added to reduce the number of features and a dropout layer was applied to avoid overfitting. Finally, a dense layer with softmax activation was added as the output layer to classify the input images into one of four possible classes. The function returned the final custom model. Overall, this function created a custom deep learning model for plant leaf classification using transfer learning from a pre-trained InceptionV3 model. The pre-trained model was fine-tuned by freezing some layers and adding custom layers on top to learn a specific task of classifying plant leaves into different categories. After training, the model was evaluated on the validation dataset. The loss was 0.0609 and the accuracy was 0.9792, indicating that the model was performing well on the validation dataset.

14.3 Results and Analysis

This section discusses the results achieved in both the components of the project. Our main objective was to develop a deep learning model that could accurately classify crop images as healthy or diseased using the STM32G474RE Nucleo board. To achieve this, we used a combination of techniques such as separable convolution, data augmentation, batch normalization, dropout, and TFLite [14] format conversion. Plant disease detection is an important application of computer vision and deep learning. In this context, several deep learning models have been proposed in the literature, including CNN, VGG16, Densenet, EfficientNet, ViT, Swin transformer, and Inception V3.

14.3.1 TinyML Model

The performance of the model was evaluated using a dataset of crop images containing both healthy and diseased samples. The model achieved an accuracy of 83%, which is a promising result, considering the limited computational resources of the STM32G474RE Nucleo board [15]. Figure 14.2 depicts the accuracy of the model.

One interesting finding of the project was the effectiveness of separable convolution in reducing the computational cost and improving the efficiency of our model. The use of separable convolution allowed us to achieve high accuracy while using a smaller number of parameters and computations compared to traditional convolution. This result highlights the potential of separable convolution as a key technique for deploying deep learning models on edge devices.

The results proved that data augmentation, batch normalization, and dropout were effective techniques for improving the robustness of our model and reducing overfitting. Data augmentation helped to generate additional training samples, while batch normalization and dropout helped to improve the generalization ability of our model. These findings demonstrate the importance of using multiple techniques to improve the performance of deep learning models on edge devices.

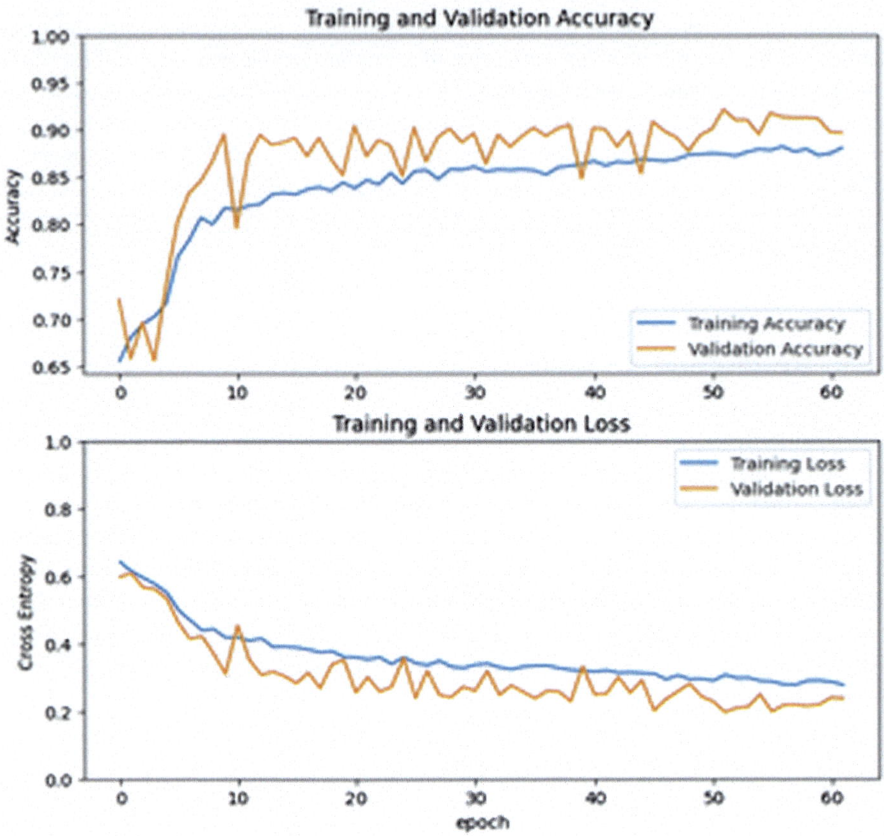

Fig. 14.2 Model accuracy and model loss plot

The model's performance was further tested on real-world data collected from farms, and the results were consistent with the performance on the test dataset. This indicates that our model can effectively generalize to new data and has potential for deployment in real-world precision agriculture applications.

However, one limitation of the microcontroller model is its ability to classify only two categories of crop images, healthy and diseased, and then transmit the diseased crop image to the server for further heavy classification tasks. In the future, this can be extended to the model classifying multiple categories of crop diseases and pests, which would make it more applicable to a wider range of precision agriculture applications.

In terms of deployment, the model was saved in TFLite [14] format and imported into STM32CubeIDE. The model was deployed onto the STM32G474RE Nucleo board and ran real-time inference on the available crop images dataset. This demonstrates the feasibility of deploying deep learning models on edge devices and highlights the potential of our approach for precision agriculture applications.

14.3.2 Server Model

In the context of plant disease detection, these deep learning models have displayed encouraging outcomes. Nevertheless, their effectiveness hinges on several factors dataset size and quality, available computational resources, and the optimization methods employed. In this context, Inception V3 has exhibited remarkable performance in accurately identifying plant diseases from images while maintaining a low count of trainable parameters, rendering it efficient in terms of computational training demands.

Following assessments of various deep learning models, including CNN, VGG16, Densenet, Efficient Net, ViT, and Swin Transformer, the central server model achieved a relatively high level of accuracy. However, it was observed that the Swin Transformer model attained the highest accuracy, albeit still falling short of the desired threshold. This discrepancy might be attributed to the Swin Transformer's need for an extensive training dataset, which the dataset used in this experiment may not have provided in adequate quantity for optimal performance.

Consequently, a decision was made to experiment with the Inception V3 model instead. This choice proved highly effective, yielding significantly elevated accuracy rates. Inception V3, a deep convolutional neural network, had undergone training for large-scale image classification tasks and exhibited notable success across various image recognition assignments. Notably, it featured a relatively limited parameter count in comparison to other deep learning models, resulting in expedited training.

The attainment of such high accuracy with the Inception V3 model underscores its efficacy in plant leaf classification. Moreover, its modest parameter count facilitates deployment on less powerful hardware, positioning it as an advantageous choice for edge devices or mobile applications.

This experiment's findings underscore the pivotal importance of selecting an appropriate deep learning model tailored to the specific task at hand. While certain models may possess greater complexity and potency, they may not universally suit all datasets or applications. In this particular scenario, the Inception V3 model emerged as the optimal choice for plant leaf classification, achieving exceptional accuracy and delivering relatively swift training times.

The performance of various deep learning models in plant disease detection is summarized as follows: CNN achieved 71% accuracy, EfficientNet and Densenet both achieved 84%, VGG16 reached 69%, and ViT and Swin Transformer scored 84% and 87% accuracy, respectively. The Swin Transformer stands out as the top performer, while the VGG16 performs the least effectively. A comparison of accuracies is depicted in Fig. 14.3. Model architecture and complexity play a crucial role in these results. For instance, the Swin Transformer and ViT models employ self-attention mechanisms, whereas EfficientNet uses scaling techniques for optimization. In contrast, CNN and VGG16 are simpler, older architectures, possibly explaining their lower accuracies. Dataset size and quality are also influential factors. Larger and more diverse datasets tend to enhance model accuracy, as do high-resolution, clear images. Deeper and more complex models generally outperform simpler ones

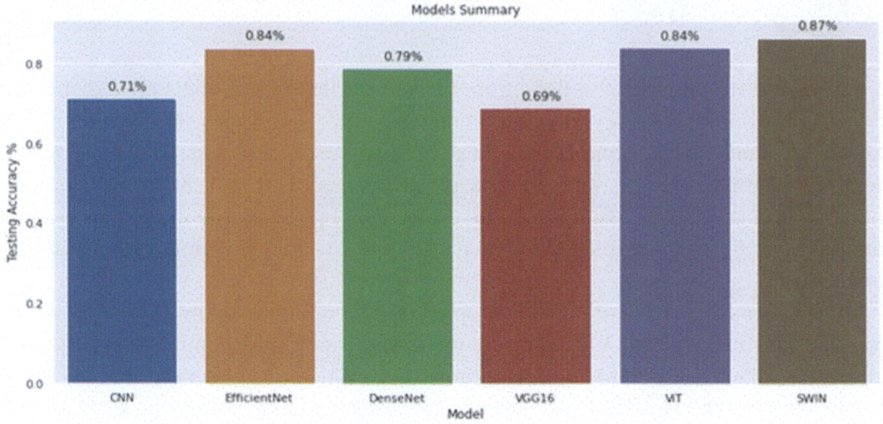

Fig. 14.3 Server model results compilation

in plant disease detection. However, the difference in accuracy between models is not substantial, suggesting that factors like computational efficiency and ease of training should be considered. Notably, Inception V3 achieved an outstanding accuracy of 97.9%, likely due to its unique architecture, pre-training on a large dataset, and various advanced techniques. Nevertheless, its computational demands may limit practicality in certain applications.

The Inception module, a fundamental component of Inception V3, enables simultaneous learning of multiple feature maps through filters of varying sizes, optimizing resource utilization and enhancing pattern recognition. The model's high parameter count and training on a large, diverse dataset contribute to its robustness and accuracy. Advanced techniques like batch normalization and dropout regularization prevent overfitting and enhance generalization.

In conclusion, the choice of a deep learning model significantly impacts plant disease detection accuracy, with factors such as complexity, dataset characteristics, and computational resources playing crucial roles in model selection.

14.4 Conclusion

The study underscores the promise of employing deep learning models to enhance the precision and efficiency of agricultural tasks. By deploying these models on edge devices like drones and sensors, farmers can access real-time data and insights to inform decision-making and enhance crop yields. This observation holds substantial merit and practicality, given that the application of deep learning models, particularly convolutional neural networks (CNNs) and recurrent neural networks (RNNs), has yielded marked enhancements in crop detection accuracy, disease identification, and yield estimation. These models can undergo training on extensive agricultural image and sensor data sets, enabling them to discern patterns and make highly accurate predictions. Overall, this project has brought attention to the potential of deep learning and edge computing to revolutionize precision agriculture, empowering farmers to make more informed decisions. With ongoing research and development efforts, these technologies can be further fine-tuned to address the unique requirements and challenges of the agricultural sector.

14.4.1 Future Scope

In the realm of advancing precision agriculture (PA) with intelligent edge devices, there have been several promising developments. Firstly, there is a strong emphasis on enhancing model efficiency and size metrics. This involves taking into account data and algorithm characteristics and employing advanced model compression techniques such as knowledge distillation and combinations of compression methods. These methods offer significant potential for tasks like crop classification, plant disease detection, and yield prediction when applied to edge devices equipped with real-time environmental sensors. Secondly, future endeavors aim to boost the performance of PA models on edge devices through techniques like transfer learning and incremental learning. This strategy involves harnessing pre-trained models trained on extensive datasets, reducing the reliance on abundant labeled data, and enabling continuous adaptation to changing crop and soil conditions. Thirdly, the incorporation of federated learning into the PA system is poised to simplify model training across multiple edge devices while safeguarding data privacy, which is especially crucial in the context of precision agriculture. Fourthly, the utilization of Unmanned Aerial Vehicles (UAVs) outfitted with sensors significantly enhances the PA system by rapidly gathering high-resolution data. This aids in real-time predictions and crop analysis while simultaneously reducing labor and costs. As future work, experiments could be conducted to measure and compare the precision and recall values obtained from both systems when categorizing images as valid or invalid in real time.

This comparative analysis would aim to provide insights into the adaptability and effectiveness of the system across different computing environments. By quantifying the performance disparities between portable and non-portable systems, a more comprehensive understanding of the proposed methodology's robustness and potential limitations can be achieved. This comparative evaluation will be crucial for validating the generalizability and effectiveness of the deep learning model across diverse deployment scenarios in precision agriculture. Lastly, the integration of edge computing and cloud computing addresses resource constraints in edge devices. Exploring an alternative approach to enhance our system's capabilities by incorporating real-time cloud usage for immediate data analysis and classification, this potential extension presents distinct advantages, including leveraging substantial computational power, scalability for managing extensive datasets, and centralized processing capabilities offered by cloud resources. By transmitting data to the cloud, the system could execute complex analyses and deploy advanced machine learning models swiftly, addressing the inherent limitations of IoT devices. While our current research predominantly focuses on edge computing, emphasizing the benefits of this alternative approach would add valuable insights to our analysis.

Future development could further investigate the integration of real-time cloud usage to enhance system adaptability and responsiveness, especially in scenarios necessitating immediate and extensive data processing, such as precision agriculture applications where rapid image validation is crucial for subsequent tasks. This exploration of a hybrid architecture, combining real-time edge computing with on-demand cloud utilization, represents a promising avenue for future research to achieve an optimal balance between computational efficiency and cloud-based processing advantages when needed. These strategies offer promising prospects for the future of precision agriculture, with the potential for increased efficiency, data privacy, and enhanced decision-making capabilities for farmers.

References

1. Liu, J., Xiang, J., Jin, Y., Liu, R., Yan, J., Wang, L.: Boost precision agriculture with unmanned aerial vehicle remote sensing and edge intelligence: a survey
2. Patrício, D.I., Rieder, R.: Computer vision and artificial intelligence in precision agriculture for grain crops: a systematic review. Comput. Electron. Agric. **153**, 69–81 (2018)
3. Lin, J., Chen, W.-M., Lin, Y., Cohn, J., Gan, C., Han, S.: MCUNet: tiny deep learning on IoT devices. In: Advances in Neural Information Processing System (2020)
4. Allende, A., Monaghan, J.: Irrigation water quality for leafy crops: a perspective of risks and potential solutions. Int. J. Environ. Res. Public Health **12**, 7457–7477 (2015)
5. TinyML Foundation. https://www.tinyml.org/about/
6. Redmon, J., Divvala, S., Girshick, R., Farhadi, A.: You only look once: unified, real-time object detection. In: Proceedings of the IEEE Conference on Computer Vision and Pattern Recognition (2016)
7. Girshick, R., Donahue, J., Darrell, T., Malik, J.: Rich feature hierarchies for accurate object detection and semantic segmentation. In: Proceedings of the IEEE Conference on Computer Vision and Pattern Recognition (2016)
8. Soro, S.: TinyML for ubiquitous edge AI. CoRR, vol. abs/2102.01255 (2021)
9. Federov, I., Adams, R.P., Mattina, M., Whatmough, P.N.: SpArSe: sparse architecture search for CNNs on resource-constrained microcontrollers. In: Advances in Neural Information Processing Systems (2019)
10. Post-Training Integer Quantization, Tensorflow. https://www.tensorflow.org/lite/performance/post_training_integer_quant
11. Qiu Lin, Z., Chwyl, B., Wong, A.: EdgeSegNet: a compact network for semantic segmentation
12. Huang, R., Pedoeem, J., Chen, C.: YOLO-LITE: a real-time object detection algorithm optimized for non-GPU computers. In: Proceedings of the IEEE Conference on Big Data (2018)
13. Integrated Development Environment for STM32, STMicroelectronics. https://www.st.com/en/development-tools/stm32cubeide.html
14. TensorFlow Lite. https://www.tensorflow.org/lite
15. Nucleo-G474E,STMicroelectronics. https://www.st.com/en/evaluation-tools/nucleo-g474re.html
16. Chollet, F.: Xception: deep learning with depthwise separable convolutions. In: IEEE Conference on Computer Vision and Pattern Recognition (2017)

17. X-CUBE-AI Package, STMicroelectronics. https://www.st.com/en/embedded-software/x-cube-ai.html
18. Tera Term, Wikipedia. https://en.wikipedia.org/wiki/Tera_Term
19. O'Shea, K., Nash, R.: An introduction to convolutional neural networks (2015)
20. Dosovitskiy, A., Beyer, L., Kolesnikov, A., Weissenborn, D., Zhai, X., Unterthiner, T., Dehghani, M., Minderer, M., Heigold, G., Gelly, S., Uszkoreit, J., Houlsby, N.: An image is worth 16x16 words: transformers for image recognition at scale (2020)
21. Simonyan, K., Zisserman, A.: Very deep convolutional networks for large-scale image recognition. In: 3rd IAPR Asian Conference on Pattern Recognition (2015)
22. Liu, Z., Lin, Y., Cao, Y., Hu, H., Wei, Y., Zhang, Z., Lin, S., Guo, B.: Swin transformer: hierarchical vision transformer using shifted windows (2021)
23. Tan, M., Le, Q.V.: EfficientNet: rethinking model scaling for convolutional neural networks (2019)
24. Huang, G., Liu, Z., van der Maaten, L., Weinberger, K.Q.: Densely connected convolutional networks. In: IEEE Conference on Computer Vision and Pattern Recognition (2017)
25. Mustafa, W.A., Abdul Kader, M.M.M.: A review of histogram equalization techniques in image enhancement application. J. Phys.: Conf. Series Vol. 1019, International Conference on Green and Sustainable Computing (2017)
26. Szegedy, C., Vanhoucke, V., Ioffe, S., Shlens, J., Wojna, Z.: Rethinking the inception architecture for computer vision (2015)

Chapter 15
Comprehensive Review of Capsule Networks with a Case Study on Potato Leaf Disease Detection Using CapsNet and Attention Mechanism

Rajalakshmi Shenbaga Moorthy, K. S. Arikumar, Sahaya Beni Prathiba, and P. Pabitha

Abstract Accurate solutions are needed for a variety of computer vision applications, including medical imaging, object detection, and recognition. Such complicated challenges are beyond the capabilities of artificial intelligence and machine learning, which mainly rely on data and algorithms to learn. The path to Deep Learning (DL) is thus paved. Even though deep learning is incredibly effective at solving complex problems, it is invariant, meaning that it ignores the spatial relationship between the features and instead searches for features only. This can cause the model's performance to suffer. Additionally, because the model is invariant, gathering a large volume of training data is a difficult undertaking. To address this, the Capsule Network (CapsNet), whose performance outperforms Deep Learning techniques, was introduced. Despite its success, applications of CapsNet and working remain a mystery. Thus, the performance of CapsNet across many applications and obstacles is examined in this study together with the predecessors of CapsNet. Additionally, this paper explores the use of Capsule Networks with an attention mechanism (CapsNet-ATM) for predicting potato leaf diseases. The performance of CapsNet-ATM in this crucial agricultural application is thoroughly assessed and contrasted with the well-known models VGG-16 and VGG-19, providing insight into the system's efficacy and promise as a reliable tool for the detection and prediction of potato leaf disease.

R. Shenbaga Moorthy (✉)
Sri Ramachandra Faculty of Engineering and Technology, Sri Ramachandra Institute of Higher Education and Research, Chennai 600 116, India
e-mail: srajiresearch@gmail.com

K. S. Arikumar
Department of Data Science and Business Systems, SRM Institute of Science and Technology, Kattankulathur, 603 203 Chennai, India

S. B. Prathiba
Centre for Cyber Physical Systems, School of Computer Science and Engineering, Vellore Institute of Technology, Chennai 600127, India

P. Pabitha
Department of Computer Technology, Anna University, Madras Institute of Technology Campus, Chennai 600044, India

S. K. Pal et al. (eds.), *Intelligent Informatics*, Smart Innovation, Systems and Technologies 389, https://doi.org/10.1007/978-981-97-2147-4_15

15.1 Introduction

Object identification, face recognition, action, and activity recognition are just a few of the many computer vision applications that call for real-time solutions. Additionally, the amount of data created by these apps is substantial enough for human analysis. Deep learning [1] and neural networks [2] can be used to analyze such a large amount of data. Convolutional neural network (CNN) [3] has dominated the field of computer vision for a number of years, and it offers exceptional insights. The depth (number of levels) and width (number of neurons) of each level of CNN can be increased to improve performance. Additionally, a significant amount of training data is a crucial prerequisite for CNN. Due to the automatic feature extraction, CNN is invariant, meaning it will still provide the intended results even if the inputs are changed.

In conventional CNNs, neurons are arranged in layers, and each neuron stands for a feature or scalar value. However, in CapsNet, neurons are arranged into "capsules," which are assemblages of neurons that stand in for vectors. These vectors represent various attributes of an entity, such as its position, orientation, or appearance of an object. In [4], the MNIST dataset was employed using CapsNet, which classifies the digits more precisely than CNN. Two convolutional layers and one fully connected layer make up the shallow network of the CapsNet that was created on [4]. An algorithm known as routing by agreement was employed in [4]. The essential elements of Capsule Network are shown in Fig. 15.1. The capsule essential elements of capsule network are as follows:

- Capsule: The likelihood and parameters of a feature are captured by a set of neurons known as a capsule.
- Dynamic routing: Dynamic routing, also known as a parallel attention mechanism, is a technique for connecting active capsules in one level to capsules in another level. This characteristic enables the capsule network to recognize several objects even when they overlap in the image.
- Capsule activation: Each capsule generates a vector whose magnitude indicates the likelihood that an instance belongs to a specific class. Additionally, the magnitude exposes the representations of higher order.
- Reconstruction: The reconstruction component of the capsule network uses higher order capsule representation to rebuild the input data.

The order of the features and their relationships to one another are preserved in Capsule networks because dynamic routing is used instead of pooling. The CNN layers at a higher level search for the retrieved features. The model is classified in accordance with the presence of all extracted features, without regard to their appearance order. CNN does not preserve the spatial relationship between the features where the model is invariant and not equivariance. The paper aims to bring the limitations of CNN and the excellent performance of CapsNet in various domains as represented in the state-of-the-art methods.

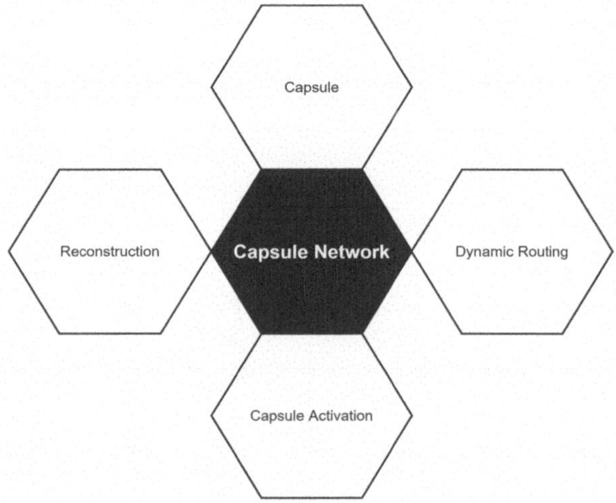

Fig. 15.1 Elements of capsule network

The goal of this study is to examine the shortcomings of CNNs and the exceptional capabilities of Capsule Networks in several fields, as demonstrated by cutting-edge techniques. We will highlight the use of capsule networks, specifically using the proposed CapsNet with an attention mechanism (CapsNet-ATM), in addressing the prediction of potato leaf disease as part of this larger investigation, bridging technological advancements in computer vision with the critical issues in agriculture.

The rest of the paper is organized as follows: Sect. 15.2 describes the concept of artificial neural network. Section 15.3 describes the working of deep learning algorithms. Section 15.4 outlines the limitations of deep learning. Section 15.5 highlights the capsule network and various state-of-the-art methods. Section 15.6 highlights the application of capsule network with attention mechanism in identifying the plant leaf disease prediction. Section 15.7 highlights the significance of the proposed CapsNet-ATM with other existing methodologies. Section 15.8 highlights research areas for possible applications of capsule networks. Finally, Sect. 15.9 concludes the research work.

15.2 Artificial Neural Network

Neuron is the central component of a neural network where information processing takes place [5]. For example, considering the handwritten digit recognition where the size of the image is 28*28, the number of neurons in input layer is 784 as shown in Fig. 15.2. The first layer of the neural network feeds each of the 784 pixels to a neuron. Through connecting channels, which are sometimes referred to as weighted channels because each of them has a value assigned to them, information is moved from one layer to another. There can be one or more hidden layers with varying

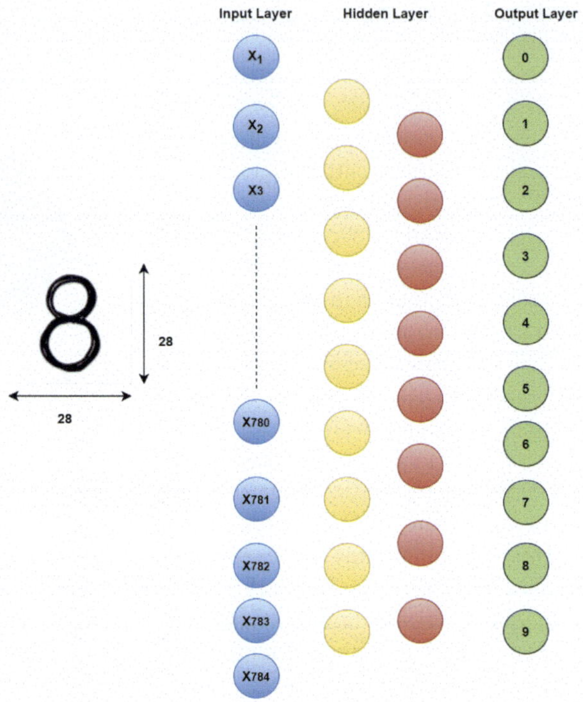

Fig. 15.2 Representation of neurons for digit recognition

numbers of neurons. The output layer followed by the hidden layer contains the number of neurons equivalent to the number of classes in the classification problem. The network is fully connected, i.e., each neuron in a layer is connected with each neuron in the successive layer. Various applications of ANN include medical science [6], energy systems [7], financial economics [8], and chemical industries [9]. The major limitations of ANN are that it consumes more time for training and also the possibility to stuck in the local optimal solution [10].

15.3 Deep Learning

An artificial neural network called a convolutional neural network (CNN) is very good at processing and analyzing organized grid-like input, such as videos and images [11]. Different computer vision tasks, such as image classification [12], object recognition [13], and image segmentation [14], have been revolutionized by CNNs. Convolutional layers, pooling layers, and fully connected layers are used in CNN to automatically learn hierarchical representations of data.

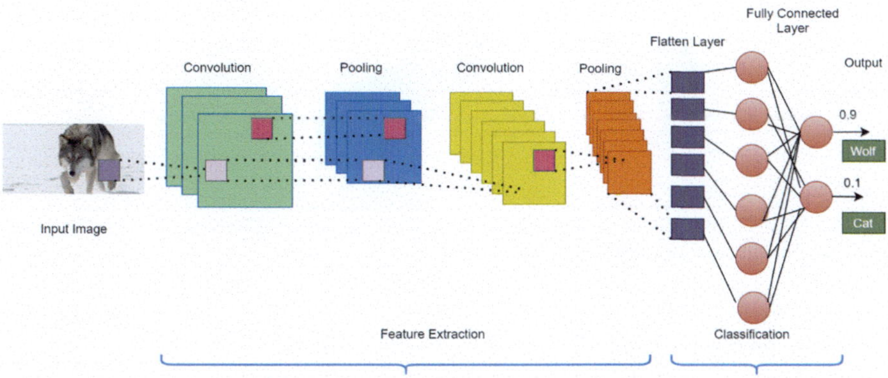

Fig. 15.3 Architecture of CNN

A convolutional layer's function is to apply convolutional kernels of various sizes in order to extract features from the image. A mathematical process called convolution is used to take the features of the image. With the help of kernel/filter, the action is accomplished. Despite the fact that the filter's size can vary, 3×3 is the most common. The dot product of the input pixel value and the kernel pixel value, together with their product, is the convolution matrix. Since the kernel is a learning parameter, its values are first set at random.

In order to shrink the size of the feature map by removing the most important features, the pooling layer, which sits between convolutional layers, is used. The model is prevented from over fitting by the pooling, which keeps only the essential features. Utilizing the characteristics acquired using earlier convolutional and pooling layers, one or more Fully Connected layers can be present to do the actual classification of images. Figure 15.3 represents the basic architecture of CNN to predict whether the given image is cat or wolf. The features are extracted by the convolution layer and pooling layer. The Fully connected layer does the process of classification. The neurons in the output layer give the probability of likelihood for a particular class.

15.4 Limitations

The prime limitation of CNN is that it does not maintain the spatial information about the image instance. As a result, the model is highly vulnerable to predict the objects when it is rotated. Even when the items in the object are dislocated CNN is prone to classify the instance which is actually a crucial issue when solving real-world problems. In other words, CNN is vulnerable to rotation and spatial affinities. And thus, requires a huge dataset for training, i.e., in order to recognize the cat which is rotated left by 20 °C, CNN requires a greater number of rotated images of cat [15]. Thus, the major limitation of Deep Learning is that

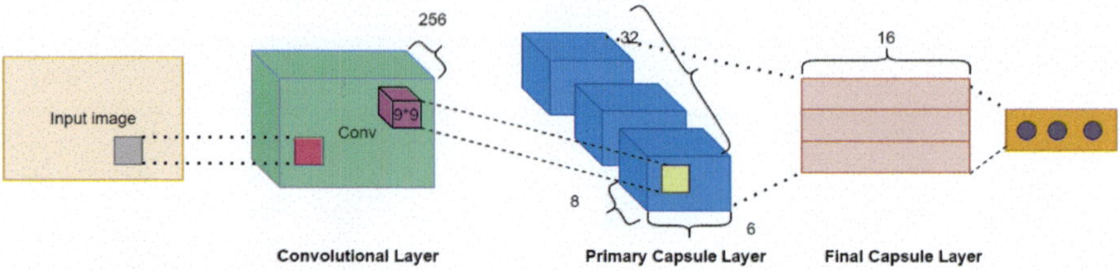

Fig. 15.4 Architecture of CapsNet

- Failed to preserve spatial relationship between features.
- Lack of equivariance to rotation or other transformations.

15.5 CapsNet

CapsNet is an upgrade to CNN that aims to address the pooling-related issue. Information is lost as a result of the pooling operation failing to maintain the spatial link between the features. In other words, the pooling layer avoids structural orientation when extracting the lower-level characteristics. As a result, higher-level layers of CNN simply search for the presence of features without taking into account their position. However, each capsule in the CapsNet learns the characteristics' sequence during training. As a result, the CapsNet examines both the features and their appearance order. Additionally, the CapsNet output is shown as a vector with magnitude that aims to recognize deformed images as well.

The capsule network was created to maintain spatial information and rotational invariance using the concepts of dynamic routing and reconstruction. The basic structure of CapsNet is depicted in Fig. 15.4. The convolution layer receives the input image as input. In general, a convolutional layer may have one or more levels. The objective of the convolutional layer is to extract low level information from the image and express them as a feature map. The feature maps that the convolution layer provides are processed by the main capsules. Every capsule aims to identify a specific feature. In addition to dealing with the existence or absence of the feature, it also reflects the instantiation characteristics like posture, orientation, and scale. In between Primary Capsule layer and final capsule layer, there can be one or more convolutional capsule layers to learn feature maps using the concept of dynamic routing. The final capsule layer contains the capsules which are equivalent to number of classes in the problem of classification. Each capsule outputs the vector which includes the probability, together with instantiation parameters. Table 15.1 elaborates the various implementations of CapsNet for real-time problems.

Table 15.1 Summary of capsule network applications

Reference	Problem	CapsNet implementation	Approach	Dataset	Accuracy
[17]	Emotion recognition	Transformer capsule network	Patch partition module segments EEG signals into small patches	DEAP (32 subjects) and DREAMER (23 subjects) 98.59% on DREAMER	98.76% on DEAP
[18]	Data hiding in the image	CapsNet	Two convolution layers: first layer: 256 kernels of size 9×9. Second layer: 32 capsules, each with 8 kernels of size 9×9	ImageNet	Embedding capacity is 100 bits for image sized 128×128
[19]	Analyzing chest X-rays for COVID-19 diagnosis	VGG 16 + CapsNet	Primary Capsule layer: 10 capsules. COVID Caps—contains 3 capsules	Italian Society of Medical and Interventional Radiology (SIRM) Dataset	96.58% accuracy
[20]	Military object recognition	Multi-level CapsNet framework	Class capsule layer utilizes the concept of dynamic routing	Self-built dataset	93.37% accuracy
[21]	Security of CapNet against adversarial attacks	DeepCaps	Convolutional layer and two capsule layers with dynamic routing	GTSR Benchmark and CIFAR10 datasets 91.3% on CIFAR10	97.6% accuracy on GTSR
[22]	Grape leaves disease classification	Convolutional capsule network	Convolutional layer prior to primary caps layer	Self-built dataset + Plant Village Dataset	99.12% accuracy
[23]	Prediction of antibacterial peptides	ABPCaps	Convolutional neural network + Long Short-Term Memory + CapsNet	GRAMPA	93.33% accuracy

(continued)

Table 15.1 (continued)

Reference	Problem	CapsNet implementation	Approach	Dataset	Accuracy
[24]	Diagnosis of COVID-19 using X-ray images	DPDH CapsNet	Depthwise convolution (D), pointwise convolution (P), Dilated convolution (D), Homogeneous classification layer (H)	Two chest X-ray images with 994 and 3886 instances, respectively	97.99% accuracy
[25]	Fault location in power grid	Convolutional neural network based on capsule network	Convolutional layer, primary capsule layer, Digit Caps	Datasets are simulated in MATLAB	97.87% accuracy
[26]	Diagnosing skin cancer	FixCaps	Large receptive field at the convolutional layer where the kernel size is 31×31	HAM10000	96.49% accuracy
[27]	Plant disease prediction	CapPlant	Last convolutional layer is replaced by capsule layer	PlantVillage	93.01% accuracy
[28]	Alzheimer's disease detection	Faster R-CNN using capsule network	Dynamic routing	Alzheimer's disease dataset from Kaggle	93.55% accuracy
[29]	Detection of COVID using chest X-ray images	COVID—CAPS	Four convolutional layers, 3 capsule layers	Datasets are generated from the existing dataset	95.7% accuracy
[15]	Detection of Pneumonia	CHX CapsNet (CHX—Chest X-ray)	Deep capsule network with transfer learning	Benchmark pneumonia dataset [30]	94.84% accuracy
[31]	International classification of diseases	Hybrid capsule network model	Bi-directional long short-term memory	MIMIC—III	67.5% micro F1-score

(continued)

Table 15.1 (continued)

Reference	Problem	CapsNet implementation	Approach	Dataset	Accuracy
[32]	Potato leaf disease detection	CapsNet	Primary caps followed by convolutional layer	PlantVillage dataset	91.83% accuracy
[33]	Mango leaf disease detection	Multi-level CapsNet	Multiple layers of Capsule layer with dynamic routing	Self-built dataset	98.5% accuracy
[34]	Performance analysis on complex Data	CapsNet	Dynamic routing and reconstruction regularization	MNIST	68.93% validation accuracy
[35]	Performance analysis on complex data	Multi-scale capsule network	Multi-scale convolution and multi-dimension capsule	Fashion MNIST and CIFAR-10	92.2% accuracy on fashion MNIST and 75.1% accuracy on CIFAR-10

15.6 Proposed CapsNet-ATM for Potato Leaf Disease Prediction

The equilibrium between equivariance and invariance is addressed by attention procedures in capsule networks. Better object recognition and pose estimation are made possible by them because they enable capsules to be invariant to specific transformations while maintaining spatial links. The working of the proposed CapsNet-ATM on potato leaf disease prediction is given as follows:

Primary Capsules: These capsules extract salient characteristics from images of potato leaves. The output y_i for each primary capsule i is calculated by first applying an activation function (AF) and then a weight matrix W_i to the input features x_i which is shown in Eq. 15.1.

$$v_i = \sigma(W_i x_i) \tag{15.1}$$

Dynamic Routing: To come to a consensus regarding the existence of disease features, capsules converse with each other. For primary capsules i and illness capsules j, the routing weights c_{ij} indicate the strength of the connection as shown in Eq. 15.2.

$$c_{ij} = \frac{e^{b_{ij}}}{\sum_k e^{b_{ik}}} \tag{15.2}$$

Disease Capsules: Predictions for several potato leaf diseases are produced by these capsules. An agreement score is calculated by comparing the predictions with the actual features. The dot product between the output v_i from the primary capsule and the prediction u_j yields the agreement score a_{ij} which is represented in Eq. 15.3.

$$a_{ij} = v_i \cdot u_j \tag{15.3}$$

Attention Mechanism: The attention method is used to calculate scores for choosing the class label for the instances. A softmax function is used to compute attention weight s_{ij} and it is represented in Eq. 15.4.

$$s_{ij} = \frac{e^{a_{ij}}}{\sum_k e^{a_{ik}}} \tag{15.4}$$

Weighted Sum (Disease Probability Prediction): Using attention weights, a weighted sum of disease predictions is used to determine the final forecast for potato leaf disease as shown in Eq. 15.5.

$$s_j = \sum_i s_{ij} u_j \tag{15.5}$$

The outcome helps with accurate disease prediction by providing a weighted aggregate that represents the likelihood of particular potato leaf diseases.

15.7 Result Analysis

The dataset used in this research work is taken from PlantVillage dataset [16] , which includes three classes of potato leaf diseases viz. Early blight, late blight, and healthy. The original dataset consists of 1500 image of which 500 images belong to Early blight, 500 images belong to late blight, and 500 images belong to healthy. All the images are of resolution 256*256. For experimental analysis, the research work had been carried out to identify whether the leaf is healthy or unhealthy, i.e., the images in the late blight and early blight are combined to form unhealthy.

The proposed CapsNet is integrated with the Attention mechanism and it is compared with other deep learning models like VGG-16 and VGG-19. Metrics taken into account for comparison include accuracy, F1-Score, Precision, Recall, False Positive Rate, and Sensitivity (True Negative Rate). The number of epochs is set as 20 and the learning rate is set as 0.001. Figure 15.5 shows the confusion matrix obtained for the proposed CapsNet with Attention mechanism. It is evident from the confusion matrix that the proposed method classifies the healthy and unhealthy instances thereby maximizing the accuracy than VGG-16 and VGG-19, respectively.

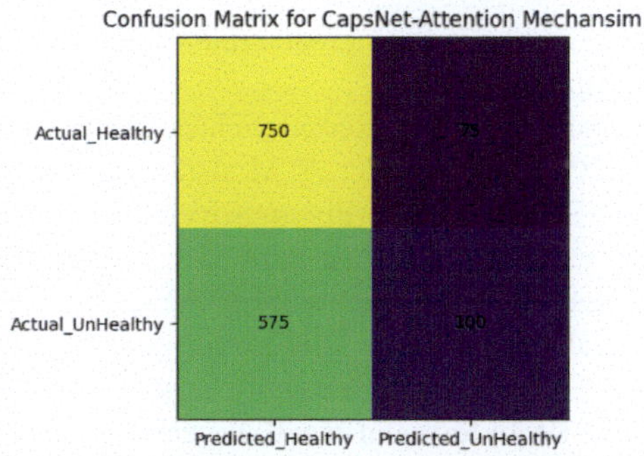

Fig. 15.5 Confusion matrix of proposed CapsNet-ATM on potato leaf disease prediction

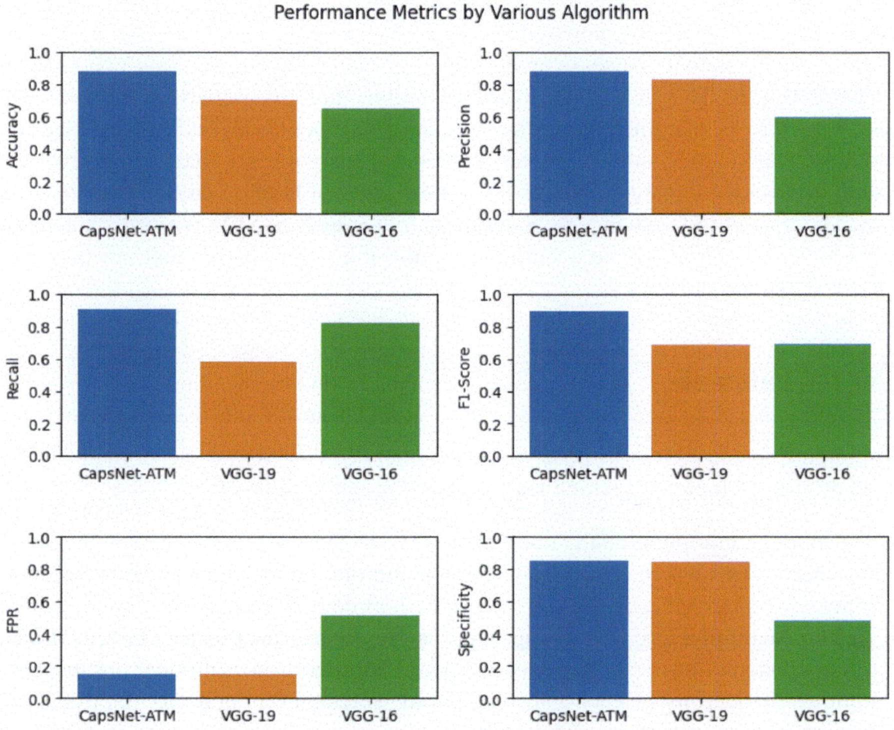

Fig. 15.6 Performance comparison of various algorithms

From Fig. 15.6, it is inferred that between the three approaches, CapsNet-ATM produces a substantially higher accuracy, measuring around 26.19% greater than that of VGG-19. This significant improvement shows that CapsNet-ATM predicts outcomes far more accurately than VGG-19. It outperforms VGG-19 by a wide margin, which can be crucial in applications where accurate classification is vital.

With an accuracy that is roughly 35.90% greater than VGG-16, CapsNet-ATM shows an even more notable improvement.

The precision of CapsNet-ATM is exceptional, exhibiting a noteworthy 47.06% gain in comparison to VGG-16. CapsNet-ATM also obtains a 5.88% greater precision compared to VGG-19. This noteworthy improvement emphasizes its capacity to correctly and efficiently categorize positive examples, which makes it a better option for tasks requiring precision. When compared to VGG-19, CapsNet-ATM significantly improves its recall by 54.55%. This tremendous improvement shows that CapsNet-ATM excels at capturing a substantially larger percentage of positive cases when compared to VGG-19. In terms of recall, it performs impressively better than VGG-19 (54.55%). Also, CapsNet-ATM improves recall of VGG-16 by 9.84%. The proposed CapsNet-ATM achieves 29.85% greater F-Score than VGG-19 and 28.73% than VGG-16. The False Positive rate of the proposed CapsNet-ATM is reduced to 71.30% and 3.70% when compared to VGG-16.

In terms of potato leaf disease prediction, CapsNet-ATM outperforms VGG-19 by a meager 0.67%. This minor enhancement implies that, in comparison to VGG-19, CapsNet-ATM is marginally accurate in classifying potato leaves that are healthy and those that are unhealthy. CapsNet-ATM shows a significant 76.049% increase in specificity when compared to VGG-16 in the context of Potato Leaf Disease Prediction. The noteworthy enhancement suggests that CapsNet-ATM performs substantially better than VGG-16 in accurately identifying potato leaves that are infected and those that are healthy.

15.8 Challenges

Although Capsule Networks pose various applications across various domains [17], there are some challenges which lead to active research in this field.

- The model is computationally complex than CNN since it involves a dynamic routing mechanism which involves iterative manipulations which increases training time
- Design of architecture and tuning hyperparameter is a challenging task as it has direct impact on the performance of the model. The decision on number of capsules, number of neurons in each capsule, initial weight, and iterations for dynamic routing need to be chosen effectively.

15.9 Conclusion

The objective of computer vision is to enable computers to perceive and interpret visual information from digital images or video. The processing of unstructured data for several computer vision applications, such as object recognition, augmented reality, and robotics systems, has been revolutionized by deep learning techniques.

However, the issue with deep learning is that it is not equivariant, suggesting that it does not consider the spatial information of the features, thereby lowering the performance. To get around this, a new network called CapsNet was developed to address the issues with deep learning methods. The goal of this research project is to analyze various state-of-the-art techniques and the effective use of CapsNet in literature in order to comprehend the potential and significance of the CapsNet architecture across a variety of sectors. Despite the fact that the methodology is new and effective, it uses the iterative dynamic routing concept, which makes the model more complex. In order to attain success in the field of computer vision, future research may concentrate on incorporating new approaches as an alternative to iterative algorithms. Additionally, this research investigates the use of Capsule Networks with an attention mechanism, CapsNet-ATM, in the crucial field of potato leaf disease prediction—an extremely significant agricultural concern. Together with a comparison analysis with other well-known models, such as VGG-16 and VGG-19, the performance of CapsNet-ATM is convincing and provides a trustworthy instrument for the identification and forecasting of potato leaf disease.

References

1. LeCun, Y., Bengio, Y., Hinton, G.: Deep learning. Nature **521**(7553), 436–44 (2015)
2. Abiod un, O.I., Jantan, A., Omolara, A.E., Dada, K.V., Mohamed, N.A., Arshad, H.: State-of-the-art in artificial neural network applications: a survey. Heliyon **4**(11) (2018)
3. Li, Z., Liu, F., Yang, W., Peng, S., Zhou, J.: A survey of convolutional neural networks: analysis, applications, and prospects. IEEE Trans. Neural Netw. Learn. Syst. Accessed 10 June 2021
4. Sabour, S., Frosst, N., Hinton, G.E.: Dynamic routing between capsules. Adv. Neural Inf. Process. Syst. 30
5. Krogh, A.: What are artificial neural networks? Nat. Biotechnol. **26**(2), 195–7 (2008)
6. Patel, J.L., Goyal, R.K.: Applications of artificial neural networks in medical science. Curr. Clin. Pharmacol. **2**(3), 217–26 (2007)
7. Kalogirou, S.A.: Applications of artificial neural-networks for energy systems. Appl. Energy **67**(1–2), 17–35 (2000)
8. Li, Y., Ma, W.: Applications of artificial neural networks in financial economics: a survey. In: 2010 International Symposium on Computational Intelligence and Design 2010 Oct 29, vol. 1, pp. 211–214. IEEE
9. Maltarollo, V.G., Honório, K.M., da Silva, A.B.: Applications of artificial neural networks in chemical problems. Artif. Neural Netw.-Arch. Appl. **16**, 203–23 (2013)
10. Kumar, P., Lai, S.H., Wong, J.K., Mohd, N.S., Kamal, M.R., Afan, H.A., Ahmed, A.N., Sherif, M., Sefelnasr, A., El-Shafie, A.: Review of nitrogen compounds prediction in water bodies using artificial neural networks and other models. Sustainability **12**(11), 4359 (2020)
11. Li, Z., Liu, F., Yang, W., Peng, S., Zhou, J.: A survey of convolutional neural networks: analysis, applications, and prospects. IEEE Trans. Neural Netw. Learn. Syst. Accessed 10 June 2021
12. Guo, T., Dong, J., Li, H., Gao, Y.: Simple convolutional neural network on image classification. In: 2017 IEEE 2nd International Conference on Big Data Analysis (ICBDA), pp. 721–724. IEEE. Accessed 10 March 2017
13. Liang, M., Hu, X.: Recurrent convolutional neural network for object recognition. In: Proceedings of the IEEE Conference on Computer Vision and Pattern Recognition, pp. 3367–3375 (2015)

14. Milletari, F., Navab, N., Ahmadi, S.A.: V-net: fully convolutional neural networks for volumetric medical image segmentation. In: 2016 Fourth International Conference on 3D Vision (3DV), pp. 565–571. IEEE. Accessed 25 Oct 2016
15. Bodapati, J.D., Rohith, V.N.: ChxCapsNet: deep capsule network with transfer learning for evaluating pneumonia in paediatric chest radiographs. Measurement **1**(188), 110491 (2022)
16. https://www.kaggle.com/datasets/abdallahalidev/plantvillage-dataset
17. Liu, S., Wang, Z., An, Y., Zhao, J., Zhao, Y., Zhang, Y.D.: EEG emotion recognition based on the attention mechanism and pre-trained convolution capsule network. Knowl.-Based Syst. **8**(265), 110372 (2023)
18. Wang, Z., Feng, G., Wu, H., Zhang, X.: Data hiding during image processing using capsule networks. Neurocomputing **7**(537), 49–60 (2023)
19. AbouEl-Magd, L.M., Darwish, A., Snasel, V., Hassanien, A.E.: A pre-trained convolutional neural network with optimized capsule networks for chest X-rays COVID-19 diagnosis. Cluster Comput. **26**(2), 1389–403 (2023)
20. Janakiramaiah, B., Kalyani, G., Karuna, A., Prasad, L.N., Krishna, M.: Military object detection in defense using multi-level capsule networks. Soft Comput. **27**(2), 1045–59 (2023)
21. Marchisio, A., Nanfa, G., Khalid, F., Hanif, M.A., Martina, M., Shafique, M.: SeVuc: a study on the Security Vulnerabilities of Capsule Networks against adversarial attacks. Microprocess. Microsyst. **1**(96), 104738 (2023)
22. Diana Andrushia, A., Mary Neebha, T., Trephena Patricia, A., Umadevi, S., Anand, N., Varshney, A.: Image-based disease classification in grape leaves using convolutional capsule network. Soft Comput. **27**(3), 1457–70 (2023)
23. Yao, L., Pang, Y., Wan, J., Chung, C.R., Yu, J., Guan, J., Leung, C., Chiang, Y.C., Lee, T.Y.: ABPCaps: a novel capsule network-based method for the prediction of antibacterial peptides. Appl. Sci. **13**(12), 6965 (2023)
24. Yuan, J., Wu, F., Li, Y., Li, J., Huang, G., Huang, Q.: DPDH-CapNet: a novel lightweight capsule network with non-routing for COVID-19 diagnosis Using X-ray images. J. Digit. Imaging **22**, 1–3 (2023)
25. Mirshekali, H., Keshavarz, A., Dashti, R., Hafezi, S., Shaker, H.R.: Deep learning-based fault location framework in power distribution grids employing convolutional neural network based on capsule network. Electr. Power Syst. Res. **1**(223), 109529 (2023)
26. Lan, Z., Cai, S., He, X., Wen, X.: Fixcaps: an improved capsules network for diagnosis of skin cancer. IEEE Access **8**(10), 76261–76267 (2022)
27. Samin, O.B., Omar, M., Mansoor, M.: CapPlant: a capsule network based framework for plant disease classification. PeerJ Comput. Sci. **5**(7), e752 (2021)
28. Vasukidevi, G., Ushasukhanya, S., Mahalakshmi, P.: Efficient image classification for alzheimer's disease prediction using capsule network. Ann. Rom. Soc. Cell Biol. **2**, 806–15 (2021)
29. Afshar, P., Heidarian, S., Naderkhani, F., Oikonomou, A., Plataniotis, K.N., Mohammadi, A.: Covid-caps: a capsule network-based framework for identification of covid-19 cases from x-ray images. Pattern Recogn. Lett. **1**(138), 638–43 (2020)
30. Kermany, D.S., Goldbaum, M., Cai, W., Valentim, C.C.S., Liang, H., Baxter, S.L., McKeown, A., Yang, G., Wu, X., Yan, F., Dong, J., Prasadha, M.K., Pei, J., Ting, M.Y.L., Zhu, J., Li, C., Hewett, S., Dong, J., Ziyar, I., Shi, A., Zhang, R., Zheng, L., Hou, R., Shi, W., Fu, X., Duan, Y., Huu, V.A.N., Wen, C., Zhang, E.D., Zhang, C.L., Li, O., Wang, X., Singer, M.A., Sun, X., Xu, J., Tafreshi, A., Lewis, M.A., Xia, H., Zhang, K.: Identifying medical diagnoses and treatable diseases by image-based deep learning. Cell **172**(5), 1122-1131.e9 (2018). https://doi.org/10.1016/j.cell.2018.02.010. PMID: 29474911
31. Bao, W., Lin, H., Zhang, Y., Wang, J., Zhang, S.: Medical code prediction via capsule networks and ICD knowledge. BMC Med. Inform. Decis. Mak. **21**(2), 1-2.F (2021)
32. Verma, S., Chug, A., Singh, A.P.: Exploring capsule networks for disease classification in plants. J. Stat. Manag. Syst. **23**(2), 307–15 (2020)
33. Janakiramaiah, B., Kalyani, G., Prasad, L.V., Karuna, A., Krishna, M.: Intelligent system for leaf disease detection using capsule networks for horticulture. J. Intell. Fuzzy Syst. **41**(6), 6697–713 (2021)

34. Xi, E., Bing, S., Jin, Y.: Capsule network performance on complex data. Accessed 10 Dec 2017. arXiv:1712.03480
35. Xiang, C., Zhang, L., Tang, Y., Zou, W., Xu, C.: MS-CapsNet: a novel multi-scale capsule network. IEEE Signal Process. Lett. **25**(12), 1850–4 (2018)

Part III
Healthcare and Medical Diagnostics

Chapter 16
DenseFed-PSO: Particle Swarm Optimization-Based DenseNet Federated Model in Alzheimer's Detection

Ananya Ghosh and S. Gayathri

Abstract The research presents DenseFed-PSO, a novel approach for Alzheimer's disease prediction, harnessing the power of DenseNet, Particle Swarm Optimization (PSO), and Federated Learning. Alzheimer's disease is a pressing global health concern, and accurate early detection is crucial for effective intervention. In this innovative model, DenseNet serves as the foundation, capitalizing on its proficiency in image-based tasks. PSO is applied at the client level, enabling local parameter optimization tailored to each dataset's unique characteristics. This individualized fine-tuning enhances the model's precision and ensures adaptability across diverse data sources. Federated Learning orchestrates the collaboration between multiple clients, preserving data privacy and decentralizing the learning process. Clients' devices remain the custodians of their sensitive medical data, mitigating privacy risks associated with centralized systems. This decentralized approach enhances scalability and fault tolerance. The synergy of these components results in a robust and accurate Alzheimer's prediction model. Local PSO optimizations yield refined parameters, aggregated at a central server to enhance the global model iteratively. This process not only ensures continuous model improvement but also minimizes communication overhead. Furthermore, this approach extends beyond Alzheimer's prediction, offering versatility for other medical image analysis tasks. It encourages community collaboration among healthcare institutions, fostering a collective effort to combat Alzheimer's disease. While promising, the model's efficacy relies on data quality, system design, and PSO's optimization capabilities. Rigorous validation and evaluation are essential to gauge its real-world impact. 'DenseFed-PSO' signifies a significant step toward early Alzheimer's detection, privacy-preserving AI in healthcare, and collaborative medical research.

A. Ghosh · S. Gayathri (✉)
Vellore Institute of Technology, Vellore, Tamil Nadu 632014, India
e-mail: gayathri.s@vit.ac.in

A. Ghosh
e-mail: ananya.ghosh2020@vitstudent.ac.in

S. K. Pal et al. (eds.), *Intelligent Informatics*, Smart Innovation, Systems and Technologies 389, https://doi.org/10.1007/978-981-97-2147-4_16

16.1 Introduction

Alzheimer's disease, a neurodegenerative condition affecting millions worldwide, poses a formidable healthcare challenge. Early and accurate diagnosis is critical for effective intervention and treatment. The advent of deep learning and federated learning has ushered in new opportunities for developing powerful diagnostic tools. In this context, our research presents a groundbreaking approach—DenseFed-PSO, a novel framework for Alzheimer's prediction that integrates DenseNet-based transfer learning with Federated Learning (Fed) and Particle Swarm Optimization (PSO). The motivation behind this research stems from the pressing need for more accurate and privacy-preserving Alzheimer's prediction models. Existing systems often face challenges in ensuring data security and scalability. Our approach addresses these issues by harnessing the collaborative power of multiple data sources while preserving individual privacy, thus motivating the development of DenseFed-PSO.

DenseFed-PSO distinguishes itself through its unique combination of DenseNet-based transfer learning, Federated Learning, and Particle Swarm Optimization. This novel amalgamation harnesses the strengths of each technique, resulting in a robust, accurate, and privacy-conscious predictive model. By effectively amalgamating DenseNet, Fed, and PSO, our system achieves higher accuracy rates while preserving the confidentiality of sensitive patient data. Main Contributions of this Research Paper are as follows:

- DenseFed-PSO Framework: Introduction of a novel framework that leverages DenseNet, Federated Learning, and Particle Swarm Optimization for Alzheimer's prediction as in Fig. 16.1.
- Privacy Preservation: Robust data privacy mechanisms through federated learning, ensuring secure data sharing among multiple healthcare institutions.
- Enhanced Accuracy: Achieving a remarkable accuracy rate of 94.20% on the ADNI MRI Alzheimer's Prediction dataset, surpassing existing models.
- Future Potential: The identification of opportunities for further research in multimodal data integration, model interpretability, and real-world deployment.

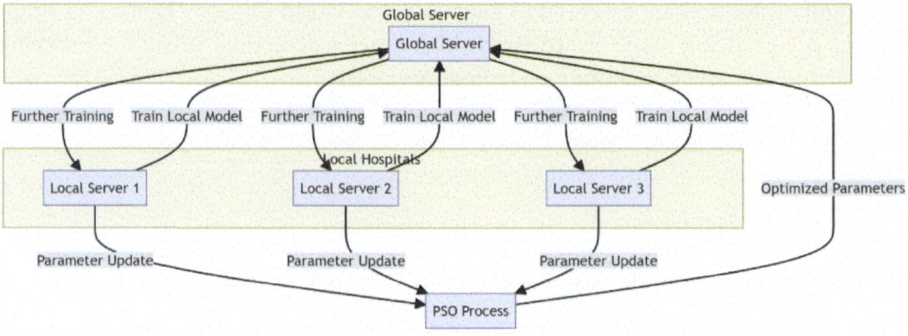

Fig. 16.1 DenseFed-PSO overall architecture

16.2 Related Work

A federated learning-based technique has been presented in an effort to address Alzheimer's disease (AD) diagnosis while protecting data privacy [1]. Results from experiments using the ADReSS dataset have been encouraging in terms of accuracy and privacy protection. Additionally, fairness in client aggregation mechanisms in federated contexts has been taken into account. Another work used Federated Learning via Conditional Mutual Learning for Alzheimer's Disease Classification on T1w MRI to address the drawbacks of small, distributed medical datasets [2]. In comparison to other frameworks, this method, known as FedCM, performed better, providing high recognition rates and insights into prospective brain regions of interest. An evolutionary deep convolutional neural network scheme (EDCNNS) was proposed to solve computational issues in Alzheimer's disease detection as it becomes a more urgent health concern [3]. EDCNNS significantly outperforms earlier methods by optimizing computing time, security, and prediction accuracy. A multimodal sensor and federated learning system were created to detect Alzheimer's biomarkers in actual situations [4]. This creative method overcame difficulties such constrained data labels and computational limitations, showcasing its potential to advance AD monitoring.

Model performance in multi-site Alzheimer's disease studies may be impacted by data discrepancies between locations. While lowering data heterogeneity, FedDAvT, a multi-site federated domain adaption system employing Transformers [5], provides data privacy. Achieving promising accuracy rates in AD classification tasks. Traditional deep learning algorithms are constrained by data scarcity and privacy issues in the field of brain MRI-based AD detection. A 3D-CNN was trained using data from each of the three phases of the ADNI using a federated learning approach [6]. This federated CNN demonstrated the promise of federated learning in neuroimaging applications with high AD classification performance. Through chest X-rays (CXR), Federated Learning has also been vital in the diagnosis of COVID-19 infections [7]. In terms of correctness and consistency, the privacy-preserving protocol SecureFed consistently outperformed competing FL frameworks. Convolutional neural networks (CNN) and federated learning were used to classify skin diseases and preserve data privacy [8]. This method demonstrated its potential for reliable skin disease classification by achieving high accuracy.

In centralized learning for breast cancer diagnosis, privacy problems persisted. The classification performance was improved using a federated learning strategy that concentrated on feature extraction from diverse environments [9]. It made use of FeAvg-CNN + MobileNet, SMOTE data processing, and transfer learning. A federated framework including explainable AI and deep CNNs was presented in the field of arrhythmia classification utilizing ECG data [10]. It successfully categorized arrhythmias while improving the clinical practitioners' ability to understand the results. Through PFTL-DDD, a federated transfer learning method with a privacy-preserving protocol, driver sleepiness detection was handled [11]. To improve accuracy, efficiency, and privacy protection, this strategy integrated transfer learning with a privacy protocol based on CKKS. Federated Learning [12] made it possible to exchange private medical information among many sources. Electronic Health Records (EHRs) were examined using models like graph networks, advancing cooperative data analysis in healthcare.

Federated Learning provided a solution to the problems with data security and privacy for heart failure detection [13]. Patient information remained dispersed across regional databases, models were trained on client hardware, and only model changes were centrally transmitted. Federated learning utilized data from several sources in the context of cancer detection while protecting data privacy [14]. In comparison to centralized learning, local VGG16 models were trained, and then parameters were combined for a global model. Federated Learning and Particle Swarm Optimization (PSO) were used in a machine learning framework for the early prediction of brain stroke [15]. The confidentiality of patient data was maintained while this method improved prediction accuracy. The study also introduced PAASO, a framework that combines Particle Swarm Optimization (PSO) and Federated Learning and has different privacy requirements for agents [16]. Different privacy models were investigated, including differential privacy. These researches show how Federated Learning has enormous potential for addressing data privacy issues and achieving high accuracy in a range of healthcare applications, from Alzheimer's disease diagnosis to cancer detection and beyond. This strategy has the potential to revolutionize group data analysis in the healthcare industry.

A previous study evaluates CNN models (VGG16, Inception, ResNetV2) for MRI-based Alzheimer's disease identification. Cross-validation is used, and the most accurate model is the fine-tuned VGG16, which achieves 98.810% [17]. AlzheimerNet [18], a CNN that has been fine-tuned, beats conventional methods on brain MRI and achieves 98.67% accuracy in classifying five stages of Alzheimer's disease and Normal Control. By employing 2D MRI images, the 'Biceph-net' [19] framework for diagnosing Alzheimer's disease (AD) was able to gather intra- and inter-slice information and exceed 2D CNNs in terms of accuracy and computational efficiency, attaining an accuracy rate of over 97%. An effective patch-based deep learning network for Alzheimer's disease (AD) diagnosis using structural magnetic resonance imaging (MRI) was first shown in a previous study [20]. It demonstrated increased accuracy and efficiency by utilizing a unique patch-based network and an explainable patch selection mechanism. Deep Transfer Ensemble (DTE), a DL-architecture agnostic model for Alzheimer's disease categorization, was also introduced in earlier works [21]. DTE proved effective in many datasets, exceeding previous models and achieving 99.05% accuracy for Non-Cognitively Impaired versus Alzheimer's Diseased. A prior study [22] presented a technique for early Alzheimer's detection with U-Net and EfficientNet, producing accurate MRI analysis results. With a deep learning framework, the early-stage prediction approach enabled prompt preventative measures.

An innovative early feature fusion approach for the prediction of Alzheimer's disease using PET and MRI data was previously presented [23]. On the ADNI database, the improved Resnet18 produced a classification accuracy of 73.90%. Furthermore, to address heterogeneity issues in multimodal data, an Explainable

Artificial Intelligence (XAI) model was used for result interpretation. A unique CNN architecture called ADD-Net [24] is used to classify Alzheimer's disease (AD) based on MRI data. ADD-Net outperformed previous models with a focus on efficiency and early AD stage classification, achieving 98.63% accuracy, 99.76% AUC, and 0.0549% loss. Using deep structured architectures and the Internet of Things (IoT), a novel Alzheimer's prediction model was presented in a previous paper [25]. It used the parameter-improved horse herd optimization (PI-HHO) algorithm for feature selection in deep convolutional networks (DCN) and deep residual networks (DRN), and an upgraded deep residual network–long short-term memory (DRN-LSTM) for patient detection. The suggested system included audio, data, and video-based sensor data for a thorough assessment and prompt hospital alarms, and it was able to monitor Alzheimer's patients with 98% accuracy and 97% precision. Using synthetic MRI images created by cascading DCGANs to simulate different disease stages, a model based on Convolutional Neural Networks (CNN), Deep Convolutional Generative Adversarial Networks (DCGANs), and Super-Resolution Generative Adversarial Networks (SRGANs) [26] for Alzheimer's disease detection achieved 99.7% classification and prediction accuracy while addressing data scarcity and improving data resolution. With the use of SHapley Additive exPlanations, a hybrid deep learning framework for diagnosing Alzheimer's disease that included Principal Component Analysis and 3D convolutional neural networks [27] obtained 91% accuracy and 0.97 AUC, emphasizing the important importance of apathy among behavioral symptoms.

16.3 Methodology

16.3.1 DenseFed-PSO Model for Alzheimer's Detection

In the comprehensive federated learning model, each hospital operates its own server alongside a global server. The individual hospital servers undertake the training of Alzheimer's prediction models from MRI patient images utilizing DenseNet Federated Learning. Figure 16.2 shows the architecture. Only the updated model parameters are transmitted to the global server. At the global server, these parameters undergo further training and refinement using DenseNet. Subsequently, these enhanced parameters are sent back to the local hospital servers for additional training iterations. Importantly, during the parameter transmission from local to global servers, they traverse through a Particle Swarm Optimization (PSO) process. This PSO step enables hyperparameter tuning and parameter optimization, ensuring that the finest parameters are conveyed from local models to the global model. This meticulous approach culminates in the most accurate Alzheimer's disease predictions from MRI scans. Figure 16.4 describes the model.

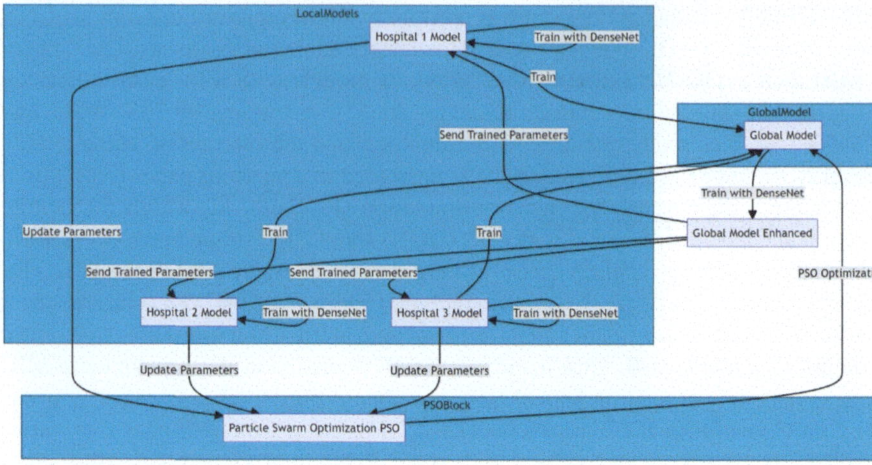

Fig. 16.2 DenseFed-PSO model

16.3.2 Dataset and Data Preprocessing

In our research for Alzheimer's detection, we leveraged the Alzheimer's Disease Neuroimaging Initiative (ADNI) dataset, which provides a rich repository of MRI images. These MRI images were instrumental in training and evaluating our federated learning model, enhancing its accuracy and robustness for Alzheimer's prediction. It involved skull stripping using the 'Deepbrain' library, padding for batch encoding, bias correction with SimpleITK's N4BiasFieldCorrection(), segmentation into cerebrospinal fluid (CSF), white matter (WM), and gray matter (GM), and 2D image extraction from 3D volumes. Data augmentation (horizontal and vertical flip) was applied using Keras' ImageDataGenerator to get 6000 images in each class for enhanced model training.

16.3.3 Data Collection and Distribution

In the proposed methodology, data collection and distribution are pivotal components of the Alzheimer's prediction system, ensuring the availability of diverse and comprehensive datasets while maintaining data privacy. This phase involves the collaboration of multiple hospitals, each operating its own data server, and a central global server.

- Hospital Data Servers: Each participating hospital contributes to the dataset by hosting its own data server. These servers store MRI images of patients, which serve as the primary source of data for Alzheimer's prediction. These images are collected from a wide range of patients, encompassing various demographics, disease stages, and conditions.

- Global Server: To facilitate collaborative learning and model enhancement, a central global server is established. This server acts as a coordination hub where data from all participating hospitals is aggregated. However, it's important to note that the global server does not store raw patient data. Instead, it orchestrates the Federated Learning (FL) process, transmitting and receiving model parameter updates to and from local hospital servers.

The data collection and distribution phase forms the foundation for collaborative research among hospitals, ensuring a rich and diverse dataset for accurate Alzheimer's prediction while upholding strict data privacy standards.

16.3.4 DenseFed—DenseNet with Federated Learning

DenseNet201 is a convolutional neural network (CNN) architecture designed for image classification tasks. In the context of TensorFlow's Keras API, the configuration for DenseNet201 involves specific layers and parameters to achieve optimal performance for your image classification problem with four classes. Shape 224, 224, 3 accommodates RGB images with a resolution of 224 × 224 pixels. A convolutional layer followed by batch normalization and ReLU activation. Each of the 4 dense blocks contains multiple densely connected layers. Growth Rate determines the number of filters added to each layer's output. Bottleneck Structure utilizes 1 × 1 convolutions to reduce the number of input channels and conserve computational resources. The 3 Transition Layers consist of batch normalization, ReLU activation, 1 × 1 convolution for compression, and 2 × 2 average pooling for downsampling. Global Average Pooling aggregates spatial information by taking the average over the spatial dimensions. In the output dense layer a fully connected layer with a softmax activation function is there. Four neurons in that correspond to the four classes in your image classification task. DenseNet (Fig. 16.3) Federated Learning is a key component of the proposed methodology for Alzheimer's prediction. It leverages the collaborative power of multiple hospitals while preserving data privacy.

- Hospital-Specific Local Models: Each participating hospital hosts its own local model based on DenseNet architecture. These local models are initialized with a common base architecture but have unique model parameters that adapt to the specific patient data available at each hospital.
- Data Partitioning: The Alzheimer's patient data within each hospital is partitioned into training and validation subsets. These subsets remain within the respective hospital's data server, ensuring data privacy and security.
- Local Model Training: Each hospital independently trains its local DenseNet model using its partitioned data. Training involves multiple epochs of forward and backward passes, updating the model parameters to minimize the loss function. This process enables each hospital to refine its model based on its specific dataset.

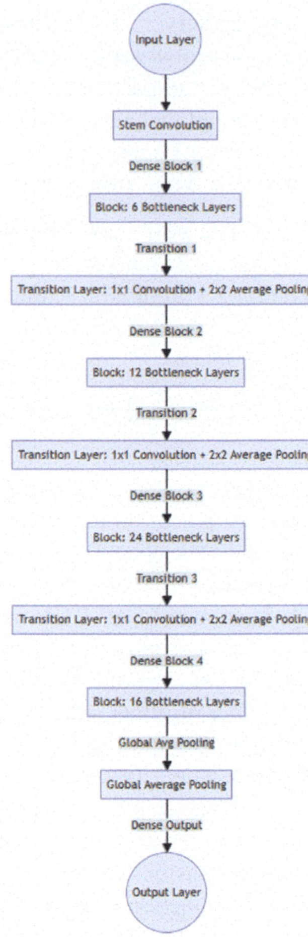

Fig. 16.3 DenseNet architecture

- Model Parameter Update: After local training, the model parameters (weights and biases) are not shared directly. Instead, only the parameter updates, which represent the differences between the initial model parameters and the optimized ones, are transmitted from each hospital to the central global server. This step ensures that sensitive patient data remains on the local server.
- Global Model Aggregation: The central global server aggregates the parameter updates received from all hospitals using Federated Averaging. This aggregation process creates a refined global model that captures collective learning from all hospitals while preserving data privacy. The global model is the culmination of knowledge from diverse datasets and reflects the collaborative effort of all hospitals.
- Model Distribution: The improved global model is then distributed back to each hospital's local server. It serves as the starting point for the next round of local

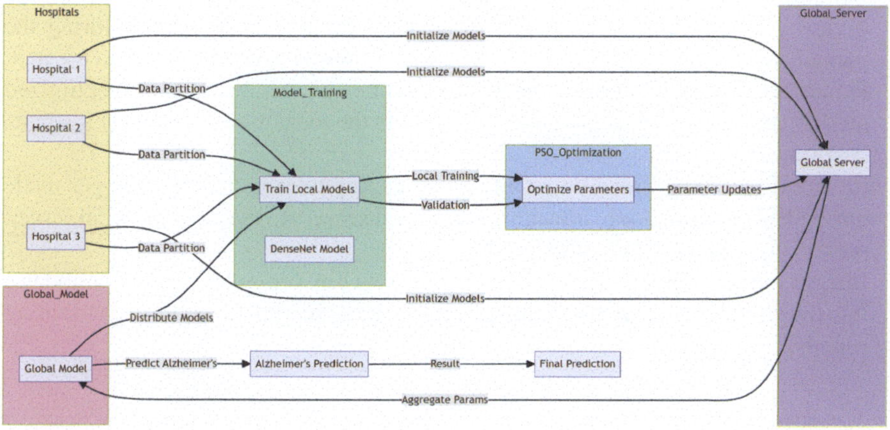

Fig. 16.4 DenseFed-PSO: Particle Swarm Optimization-based DenseNet federated model in Alzheimer's detection

model training, ensuring that the global model reflects the collective knowledge acquired from all participating hospitals.

16.3.5 Particle Swarm Optimization (PSO) with the DenseNet Federated Model

Particle Swarm Optimization (PSO) plays a pivotal role in fine-tuning the model parameters in our federated learning framework for Alzheimer's prediction at each hospital's local server. Following the training of local models using DenseNet, these models often possess suboptimal parameters. PSO steps in to enhance their performance:

- Local Model Optimization: After each local training round, hospitals have sets of non-optimal model parameters. PSO's objective is to refine these parameters for improved local model performance.
- Particle Initialization: PSO represents each model parameter as a particle in a multi-dimensional space. These particles begin with random initializations, with each particle's position denoting a potential solution (i.e., parameter values).
- Fitness Evaluation: Particle fitness is determined by evaluating their ability to minimize the loss function on local validation data. Lower loss values signify superior particles.
- Hyperparameter tuning and Velocity Update: Particles adapt their positions (model parameters) based on their own best-known positions (local best) and information from neighboring particles (global best). This adaptation involves cognitive and social learning components.

- Iterative Refinement: PSO iteratively updates particle positions, exploring the parameter space to minimize the loss function until convergence criteria are met.
- Parameter Update: The particle with the best-known position represents optimized model parameters. These parameters fine-tune the local model, ensuring it's in its best state.
- Privacy-Preserving Transition: Only parameter updates, reflecting changes made during PSO optimization, transition from local servers to the central global server, preserving patient data confidentiality.
- Integration with Global Model: The central server incorporates parameter updates into the global model, disseminating it back to local servers for subsequent training rounds.

16.4 Results and Analysis

In the comparative analysis of various transfer learning methods within the Federated Learning (Fed) and Particle Swarm Optimization (PSO) framework for Alzheimer's prediction, four pre-trained models, namely ResNet101, DenseNet, InceptionV3, and VGG19, were evaluated. Figure 16.5 and Table 16.1 show the comparison between different Transfer learning models. Each model was fine-tuned using the ADNI MRI Alzheimer's Prediction dataset. Among these methods, the DenseNet-based approach, combining both Fed and PSO techniques, exhibited the highest accuracy, achieving an impressive 94.20% accuracy on the test dataset. Additionally, the model showcased strong performance across various evaluation metrics, with an F1 score of 0.92, recall of 0.93, sensitivity of 0.94, precision of 0.91, and an AUC-ROC score of 0.98 as shown in Table 16.3. The loss and accuracy are depicted in Fig. 16.6. The confusion matrix further highlighted the model's capability, showing minimal misclassifications. These results underscore the effectiveness of the proposed DenseFed-PSO model, not only in achieving high accuracy but also in demonstrating robustness in distinguishing Alzheimer's cases from MRI scans, making it a promising tool for early Alzheimer's disease diagnosis and prediction. The hyperparameters for the PSO algorithm is shown in Table 16.2. The parameters for the DenseNet structure is shown in Table 16.3.

16.5 Conclusion and Future Scope

This study introduces a new approach, DenseFed-PSO, combining Federated Learning (Fed) and Particle Swarm Optimization (PSO) for Alzheimer's disease prediction. Achieving 94.20% accuracy on the ADNI MRI dataset, the model excels in early diagnosis. DenseNet, Fed, and PSO integration outperform transfer learning methods. Robust performance across metrics highlights its efficacy. Future directions include multimodal data integration, Explainable AI (XAI) for interpretability,

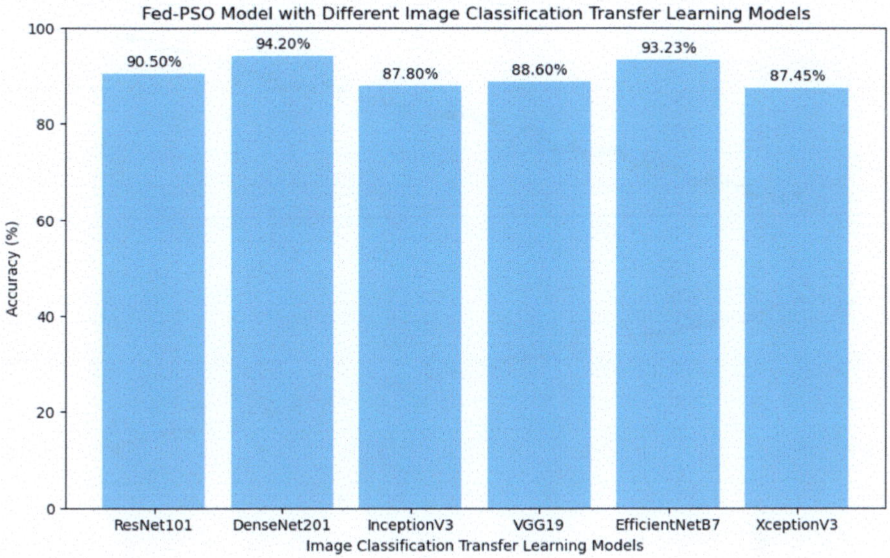

Fig. 16.5 Comparison of different transfer learning models

Table 16.1 Fed-PSO model with different image classification transfer learning models

Model	Accuracy with Fed-PSO (%)
ResNet101	90.50
DenseNet201	94.20
InceptionV3	87.80
VGG19	88.60
EfficientNetB7	93.23
XceptionV3	87.45

real-world deployment collaboration with healthcare institutions, and cross-domain applications. Ensuring privacy through advanced techniques like differential privacy, addressing biases, and seamless integration into clinical workflows will enhance its impact on Alzheimer's prediction and broader healthcare applications.

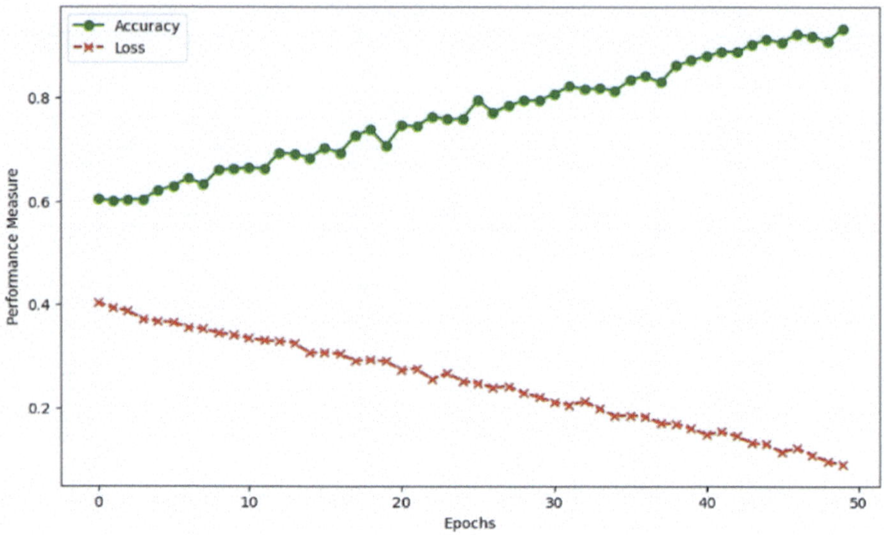

Fig. 16.6 Loss versus epochs, accuracy versus epochs graphs

Table 16.2 Hyperparameters of the PSO

Hyperparameter	Value
Total number of epochs	50
Total number of client's iterations	5
Batch size	10
Learning rate	0.001

Table 16.3 DenseNet parameters

Parameter	Value
Growth rate	32
Bottleneck configuration	(1 × 1, 0.5)
Number of dense blocks	4
Compression factor	0.5

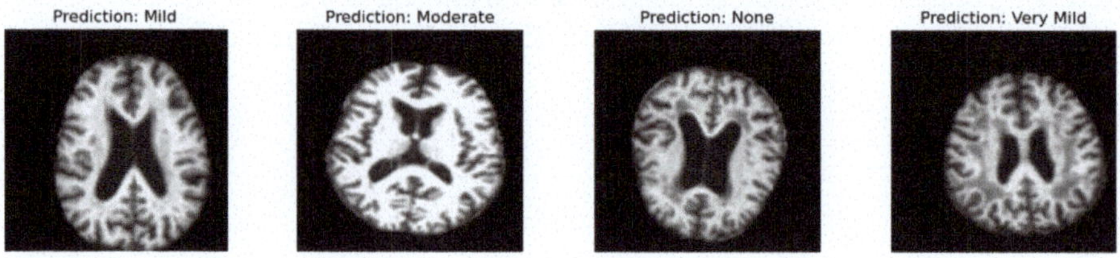

Fig. 16.7 Predicted MRI images

Table 16.4 DenseFed-PSO model performance in prediction of Alzheimer's disease from MRI

Metrics	Score (%)
Accuracy	94.20
F1 score	92
Recall	93
Sensitivity	94
Precision	91
AUC-ROC score	98

Table 16.5 Comparative analysis

Model	Citation	Accuracy	Advantages and disadvantages	Application
DenseFed-PSO	Proposed	94.20%	High accuracy, potential computational complexity	Alzheimer's disease prediction
FedCM	[2]	High	Improved recognition rates, insights into brain regions	Alzheimer's disease classification on T1w MRI
FedDAvT	[5]	Promising	Data privacy, promising accuracy in AD classification	Multi-site federated domain adaptation
Federated 3D-CNN	[6]	High	Demonstrates promise in neuroimaging, high AD classification	AD detection via brain MRI
SecureFed	[7]	Consistent	Privacy-preserving protocol, outperforms competing FL frameworks	Diagnosis of COVID-19 infections
Federated breast cancer diagnosis	[9]	Improved	Improved classification performance, enhanced data privacy	Breast cancer diagnosis
Federated Arrhythmia classification	[10]	Effective	Successful arrhythmia classification, improved interpretability	Arrhythmia classification using ECG data
Federated data exchange in healthcare	[12]	Secure	Secure exchange of private medical information	Collaborative data analysis in healthcare
Federated heart failure detection	[13]	Secure	Patient information remains distributed, models trained locally	Heart failure detection
Federated cancer detection	[14]	High	Local models trained, parameters aggregated for a global model	Cancer detection

(continued)

Table 16.5 (continued)

Model	Citation	Accuracy	Advantages and disadvantages	Application
Fed-PSO for early brain stroke prediction	[15]	Improved	Maintains patient data confidentiality, enhances prediction	Early prediction of brain stroke
PAASO	[16]	Different	Combines PSO and federated learning, explores differential privacy	Addressing varying privacy requirements for agents

References

1. Meerza, S.I.A., Li, Z., Liu, L., Zhang, J., Liu, J.: Fair and privacy-preserving Alzheimer's disease diagnosis based on spontaneous speech analysis via federated learning. In: 2022 44th Annual International Conference of the IEEE Engineering in Medicine & Biology Society (EMBC), July, pp. 1362–1365. IEEE (2022)
2. Huang, Y.L., Yang, H.C., Lee, C.C.: Federated learning via conditional mutual learning for Alzheimer's disease classification on T1w MRI. In: 2021 43rd Annual International Conference of the IEEE Engineering in Medicine & Biology Society (EMBC), November, pp. 2427–2432. IEEE (2021)
3. Lakhan, A., Grønli, T.M., Muhammad, G., Tiwari, P.: EDCNNS: federated learning enabled evolutionary deep convolutional neural network for Alzheimer disease detection. Appl. Soft Comput. 110804 (2023)
4. Ouyang, X.: Design and deployment of multi-modal federated learning systems for Alzheimer's disease monitoring. In: Proceedings of the 21st Annual International Conference on Mobile Systems, Applications and Services, June, pp. 612–614 (2023)
5. Lei, B., Zhu, Y., Liang, E., Yang, P., Chen, S., Hu, H., ... Han, H.: Federated domain adaptation via transformer for multi-site Alzheimer's disease diagnosis. IEEE Trans. Medical Imag. (2023)
6. Stripelis, D., Dhinagar, N.J., Romero, R.V.S., Thomopoulos, S.I., Thompson, P.M., Ambite, J.L.: Federated deep learning for detecting Alzheimer's disease in multi-cohort brain MRI. Alzheimers Dement. **19**, e065998 (2023)
7. Makkar, A., Santosh, K.C.: SecureFed: federated learning empowered medical imaging technique to analyze lung abnormalities in chest X-rays. Int. J. Mach. Learn. Cybern. 1–12 (2023)
8. Hossen, M.N., Panneerselvam, V., Koundal, D., Ahmed, K., Bui, F.M., Ibrahim, S.M.: Federated machine learning for detection of skin diseases and enhancement of internet of medical things (IoMT) security. IEEE J. Biomed. Health Inform. **27**(2), 835–841 (2022)
9. Tan, Y.N., Tinh, V.P., Lam, P.D., Nam, N.H., Khoa, T.A.: A transfer learning approach to breast cancer classification in a federated learning framework. IEEE Access **11**, 27462–27476 (2023)
10. Raza, A., Tran, K.P., Koehl, L., Li, S.: Designing ECG monitoring healthcare system with federated transfer learning and explainable AI. Knowl.-Based Syst. **236**, 107763 (2022)
11. Zhang, L., Saito, H., Yang, L., Wu, J.: Privacy-preserving federated transfer learning for driver drowsiness detection. IEEE Access **10**, 80565–80574 (2022)
12. Halim, S.M., Khan, L., Hamlen, K.W., Thuraisingham, B., Hossain, M.D.: A federated approach for learning from electronic health records. In: 2022 IEEE 8th Intl Conference on Big Data Security on Cloud (BigDataSecurity), IEEE International Conference on High Performance and Smart Computing (HPSC) and IEEE Intl Conference on Intelligent Data and Security (IDS), May, pp. 218–223. IEEE (2022)
13. Dhavamani, M., Niranjan, K.: A federated learning based approach for heart disease prediction. In: 2022 6th International Conference on Computing Methodologies and Communication (ICCMC), March, pp. 1117–1121. IEEE (2022)

14. Agbley, B.L.Y., Li, J., Haq, A.U., Bankas, E.K., Adjorlolo, G., Agyemang, I.O., ... Khan, J.: Federated approach for lung and colon cancer classification. In: 2022 19th International Computer Conference on Wavelet Active Media Technology and Information Processing (ICCWAMTIP), December, pp. 1–8. IEEE (2022)
15. Victor, N., Bhattacharya, S., Maddikunta, P.K.R., Alotaibi, F.M., Gadekallu, T.R., Jhaveri, R.H.: FL-PSO: a federated learning approach with Particle Swarm Optimization for brain stroke prediction. In: 2023 IEEE/ACM 23rd International Symposium on Cluster, Cloud and Internet Computing Workshops (CCGridW), May, pp. 33–38. IEEE (2023)
16. Torra, V., Galván, E., Navarro-Arribas, G.: PSO+ FL= PAASO: Particle Swarm Optimization+ federated learning= privacy-aware agent swarm optimization. Int. J. Inf. Secur. **21**(6), 1349–1359 (2022)
17. Ayman, M., Darwish, F., Mohammed, T. Mohammed, A.: Deep learning-based Alzheimer's disease classification: an experimental study. In: 2023 Intelligent Methods, Systems, and Applications (IMSA), Giza, Egypt, pp. 352-357 (2023). https://doi.org/10.1109/IMSA58542.2023.10217418
18. Shamrat, F.J.M., Akter, S., Azam, S., Karim, A., Ghosh, P., Tasnim, Z., ... Ahmed, K.: AlzheimerNet: an effective deep learning based proposition for alzheimer's disease stages classification from functional brain changes in magnetic resonance images. IEEE Access **11**, 16376–16395 (2023)
19. Rashid, A.H., Gupta, A., Gupta, J., Tanveer, M.: Biceph-Net: a robust and lightweight framework for the diagnosis of Alzheimer's disease using 2D-MRI scans and deep similarity learning. IEEE J. Biomed. Health Inform. **27**(3), 1205–1213 (2022)
20. Zhang, X., Han, L., Han, L., Chen, H., Dancey, D., Zhang, D.: sMRI-PatchNet: a novel efficient explainable patch-based deep learning network for Alzheimer's disease diagnosis with structural MRI. IEEE Access (2023)
21. Tanveer, M., Rashid, A.H., Ganaie, M.A., Reza, M., Razzak, I., Hua, K.L.: Classification of Alzheimer's disease using ensemble of deep neural networks trained through transfer learning. IEEE J. Biomed. Health Inform. **26**(4), 1453–1463 (2021)
22. Sekhar, B.V.D.S., Jagadev, A.K.: Efficient Alzheimer's disease detection using deep learning technique. Soft Comput. 1–8 (2023)
23. Odusami, M., Maskeliūnas, R., Damaševičius, R., Misra, S.: Explainable deep-learning-based diagnosis of Alzheimer's disease using multimodal input fusion of PET and MRI images. J. Med. Biol. Eng. 1–12 (2023)
24. Fareed, M.M.S., Zikria, S., Ahmed, G., Mahmood, S., Aslam, M., Jillani, S.F., ... Asad, M.: ADD-Net: an effective deep learning model for early detection of Alzheimer disease in MRI scans. IEEE Access **10**, 96930–96951 (2022)
25. Arunachalam, R., Sunitha, G., Shukla, S.K., pandey, S.N., Urooj, S., Rawat, S.: A smart Alzheimer's patient monitoring system with IoT-assisted technology through enhanced deep learning approach. Knowl. Inf. Syst. **65**(12), 5561–5599 (2023)
26. SinhaRoy, R., Sen, A.: A hybrid deep learning framework to predict Alzheimer's Disease progression using generative adversarial networks and deep convolutional neural networks. Arab. J. Sci. Eng. 1–18 (2023)
27. Liu, S., Zheng, Y., Li, H., Pan, M., Fang, Z., Liu, M., ... Ge, X.: Improving Alzheimer diagnoses with an interpretable deep learning framework: including neuropsychiatric symptoms. Neuroscience **531**, 86–98 (2023)

Chapter 17
A Machine Learning-Based Marine Vessel/Ship Classification Using Passive Sonar Signals—A Multi-class Problem

Sai Kiran Malkapurapu, Venkat Guntupalli, Bhanu Nivas Manapaka, and Venkata Sainath Gupta Thadikemalla

Abstract Passive sonar signal detection and classification play pivotal roles in underwater surveillance, providing invaluable insights into the aquatic environment and aiding in the identification of submerged objects. This comprehensive paper explores the utilization of machine learning techniques in the context of passive sonar signal classification. We focus our efforts on the *ShipsEar* dataset, which contains a total of 90 instances representing 11 different types of sonar signals from ships, classified into four distinct classes. To train the models, features such as spectral contrast and chroma were extracted from these instances. Additionally, to enhance the dataset's utility, we have partitioned the original samples into multiple segments, each spanning 10 s. Our work aims to assess the effectiveness of various machine learning algorithms in classifying these ship signals. In this pursuit, we utilize a range of classifiers, including k-nearest neighbor's (KNN), decision tree (DT), random forest (RT), and logistic regression, with the goal of identifying the most proficient algorithm for ship classification. The results of our analysis reveal that the Random Forest Classifier emerges as the most accurate, achieving an impressive test accuracy of 82%.

17.1 Introduction

Effective communication plays an indispensable role in our daily lives, acting as the cornerstone of interaction for individuals across the globe. This fundamental need for communication extends even beneath the ocean's surface, encompassing those who operate submarines and ships. In the depths of the ocean, where sound travels differently, Santantos-Dominguez [1] proposed about specialized techniques like sonar signals come into play to facilitate underwater communication. Sonar signals, a technology primarily harnessed for underwater communication, serve as

S. K. Malkapurapu (✉) · V. Guntupalli · B. N. Manapaka · V. S. G. Thadikemalla
Department of Electronics and Communication Engineering, Velagapudi Ramakrishna Siddhartha Engineering College, Vijayawada, Andhra Pradesh, India

S. K. Pal et al. (eds.), *Intelligent Informatics*, Smart Innovation, Systems and Technologies 389, https://doi.org/10.1007/978-981-97-2147-4_17

a pivotal tool for maritime operations. These signals are essential for submariners and seafarers who rely on them to navigate, avoid obstacles, and communicate with others in the submerged realm.

The two primary categories of sonar signals are "active" and "passive". Active sonar signals, a creation of human ingenuity, involve the intentional transmission of sound waves into the water. These waves travel through the medium, interact with objects in their path, and return as echoes. Active sonar is primarily used to detect the presence of objects and determine their distance from the transmitter. However, it's important to note that active sonar signals can generate noise that is more harmful to marine life. On the other hand, passive sonar signals encompass ambient sounds in the underwater environment, including those emitted by marine creatures, vessel engines, propellers, and echoes of external sources. Passive sonar is not just used to detect the noise generated by sonar signals but also to detect objects by analyzing the acoustic information present in the environment. It provides invaluable insights into the marine vessel activities without actively disturbing the marine life.

Intriguingly, passive sonar signals can take diverse forms, including reflected signals originating from various sources and naturally generated signals from marine organisms. In summary, sonar signals play a crucial role in underwater operations, with active sonar being used to detect objects and their distance from the transmitter, while passive sonar is employed to detect both the noise generated by sonar signals and objects in the underwater environment. Passive sonar allows for non-invasive observation and monitoring, while active sonar, though useful, can potentially generate harmful noise. Accurate identification and classification of passive sonar signal types hold paramount significance within the realm of underwater communication. To achieve this, the integration of advanced deep learning and machine learning techniques emerges as the cornerstone of efficient detection.

In this pursuit, the utilization of real-time datasets becomes a pivotal asset. These datasets encompass a rich tapestry of distinct classes of sonar signals, each harboring unique characteristics. Previously, numerous deep learning techniques were employed for the classification of passive sonar signals. Two distinct datasets were utilized for this purpose, Wang [2] implemented this work with each instance being segmented into durations of 1 s and 200 ms. These datasets were then subjected to various deep learning models, including Darknet 53, Densenet-121, RepVGG-A0, and AmNet-S, in order to classify the data. Remarkably, AmNet-S outperformed the other models, demonstrating the highest accuracy in classifying the sonar signals. The exceptional performance of AmNet-S highlights its proficiency in handling the complex task of passive sonar signal classification. Its robust ability to discern intricate patterns within the data sets it apart as a promising choice for the task at hand.

In addition to the direct application of deep learning techniques, Li [3] proposed various feature extraction methods such as spectral contrast, chroma, tonnetz, zero crossing rate, and MFCC (Mel frequency cepstral coefficients) which have been incorporated. Here, the dataset comprises five distinct classes, with each instance segmented into 1-s durations, resulting in a substantial dataset. These datasets were then subjected to several deep learning models, including CNN, DNN, CRNN, and

AResnet, with performance metrics computed as averages of precision, recall, and F1-score. Impressively, AResnet emerged as the most effective technique for classification based on these metrics. However, it's important to note that deep learning approaches demand a significant amount of data for effective model training, and this requirement constitutes one of their primary drawbacks. This limitation can pose significant challenges when attempting to apply such techniques in real-time applications. Consequently, we have turned to machine learning techniques to classify the data within the same dataset. This transition to machine learning methods offers a potential solution to the data-intensive nature of deep learning, making it more feasible to apply these classification techniques in real-time scenarios. This organized dataset forms the foundation for trying out different algorithms that can learn from the data, like logistic regression, k-nearest neighbor's (KNN), decision trees, and the powerful random forest.

The remaining paper presents some more related works in the field of sonar signal classification, followed by Sect. 17.3, discussing methods employed, encompassing a comprehensive examination of the datasets, a detailed exploration of implementation particulars, and a thorough investigation of various machine learning algorithms. In Sect. 17.4, we unveil the outcomes and initiate a comprehensive discussion. Finally, the concluding section outlines our overall findings and prospects for future research.

17.2 Related Works

Santos-Dominguez et al. [1] proposed a database named *Ships ear*, consisting of 90 recordings covering noises from 11 vessel types that are divided into four groups. The four classes had a 75% overall categorization rate. Wang et al. [2] worked on different deep learning models like DenseNet-121, DarkNet53, RepVGG-A0, and AMNet-S and got higher accuracy for AMNet-S with 99.4%. In 2022, Li et al. [3] compared deep learning models like AResnet, CNN, DNN, and CRNN. Using the deep ship dataset, it was observed that AResnet is the best model, with 99% accuracy. In 2022, Wang et al. [4] presented the GRU-CKF method to solve the problem of low accuracy in target estimation. They compared the GRU-CKF algorithm to the CKF and SCKF methods. In 2022, de Castro Vargas Fernandes et al. [5] presented about ship identification via passive sonar using ship noise. Traditional data acquisition for such tasks is costly and time-consuming. The study makes use of deep learning methods to train neural networks based on convolution (CNNs), especially by generating fake data using generative adversarial networks (GANs). The approach achieves an impressive accuracy of 99.0. In 2021, Lui et al. [6] proposed a MCNN-DAN model using different features like MLCC and LM. The classification accuracy of MLCC is 95.6% which is higher than LM. The top accuracy of 94.3% was obtained. In 2021, Hong et al. [7] introduced a novel method for underwater target classification, evaluated using a computer system featuring four Nvidia GeForce RTX 2080Ti GPUs and a Core i7-6900K CPU. They employed a training strategy with adaptive learning rates, early stopping, and batch sizes of 128 and 200 epochs. Their

ResNet18 model, using three-dimensional features, achieved impressive results, with an average recall, F1-score, and precision of 0.941. In 2021, Li [8] proposes the concept of passive multiple underwater target tracking, which utilizes cost-effective covert sensor measurements (bearing and Doppler data) to estimate both target quantity and state. The approach combines CPHD for data association uncertainty and EKF to address measurement nonlinearity, resulting in precise tracking, especially in cluttered conditions, as indicated by the minimal tracking OSPA. In 2020, Yang et al. [9] presented that underwater acoustics play a vital role in SONAR systems, marine research, and environmental mapping. Remote sensing through acoustic data is a key objective, with machine learning increasingly applied for target detection, identification, and localization. In 2002, Jiang et al. [10] compared different features: MFCC, MFCC + energy, and spectral contrast that are useful for signal detection. The classification accuracy with spectral contrast is 82.3%.

After thoroughly comprehending and analyzing the aforementioned research works, a prevalent trend emerges: the majority of these papers focus on various deep learning techniques and neural networks. Surprisingly, no one has proposed the utilization of machine learning techniques in this context, despite the fact that deep learning typically demands a substantial amount of dataset samples. Therefore, our approach seeks to leverage machine learning techniques to efficiently mitigate the complexity of this problem. By doing so, we aim to provide a more streamlined and resource-efficient solution, reducing the burden of data requirements often associated with deep learning approaches.

17.3 Methods

This section presents the details of datasets, ML models, and their implementation. We leverage a real-time dataset known as "*ShipsEar*" [1] for the utilization of machine learning models to effectively classify the data. The dataset includes 90 audio files with lengths ranging from 15 s to 10 min, all saved in WAV format. Each audio file has matching data, which together make up unique entries in the *ShipsEar* dataset.

Initially, we undertake data preprocessing techniques to clean and prepare the dataset. A variety of popular machine learning methods including logistic regression (LR), K-nearest neighbors (KNN), decision trees (DT), and random forest (RF) are used to present the performance analysis. Once trained, we assess their performance using test data. Further, to enhance model accuracy, we employ hyperparameter tuning via grid search. The schematic diagram below provides an overview of our proposed methodology (Table 17.1).

Table 17.1 Related works on passive sonar signal classification using ML/DL

S. No	Year	Authors	Methodology/ models	Key findings	Disadvantages
1.	2023	Wang et al. [2]	DenseNet-121, DarkNet53, RepVGG-A0, AMNet-S	AMNet-S achieved the highest accuracy among models; dataset information not provided	Lack of information on computational requirements and model complexity
2.	2022	Li et al. [3]	AResnet, CNN, DNN, ResNet18	AResnet outperformed other models on Deepship; AResnet achieved 98% on Shipsear compared to ResNet18	Inadequate exploration of AResnet interpretability; unaddressed dataset biases
3.	2022	Wang et al. [4]	GRU-CKF, CKF, SCKF	Introduced GRU-CKF to improve target estimation accuracy; compared with CKF and SCKF methods	Limited exploration of hyperparameter sensitivity; no comparative computational analysis
4.	2022	de Castro Vargas Fernandes et al. [5]	CNNs, GANs	Used deep learning, specifically CNNs and GANs, for ship identification via passive sonar; achieved 99.0% accuracy	Reliance on generated fake data, potential bias introduced through GANs
5.	2021	Lui et al. [6]	MCNN-DAN, MLCC, LM	Proposed MCNN-DAN model with MLCC features achieving 95.6% accuracy; emphasized data augmentation	Limited exploration of model generalization to diverse underwater environments

(continued)

Table 17.1 (continued)

S. No	Year	Authors	Methodology/ models	Key findings	Disadvantages
6.	2021	Hong et al. [7]	ResNet18, Adaptive learning rates	Introduced a novel underwater target classification method using ResNet18 with adaptive learning rates	Computational intensity; limited exploration of model robustness across different datasets
7.	2021	Li [8]	CPHD, EKF	Introduced passive multiple underwater target tracking using CPHD and EKF; achieved precise tracking	Sensitivity to variations in target movement patterns; limited scalability
8.	2020	Yang et al. [9]	Acoustic data, Machine learning	Highlighted the role of underwater acoustics in SONAR systems; emphasized the application of machine learning	Minimal discussion on potential biases in the training data; challenges in scalability
9.	2016	Santos-Dominguez et al. [1]	Shipsear database	Proposed Shipsear database with 11 vessel types; achieved a 75% overall categorization rate	Limited information on specific vessel types and potential biases in the dataset
10.	2002	Jiang et al. [10]	MFCC, MFCC + Energy, Spectral contrast	Compared different features for signal detection; spectral contrast achieved 82.3% classification accuracy	Incomplete analysis of robustness across different signal types; susceptibility to noise

17.3.1 Dataset

The *Telecommunications Faculty of the Universidad de Vigo*, Santos-Domínguez [1] has provided this dataset, which is useful for classifying and identifying vessels based on size using passive sonar signal data. It includes a wide variety of acoustic signals produced by many kinds of watercraft and is organized under eleven different vessel categories, four experimental classes, and a separate category for background noise.

You can find the following vessel groups in this dataset:

Group A: This group comprises dredgers, tugboats, mussel boats, trawlers, and fishing boats.
Group B: consisting of sailboats, motorboats, and pilot boats.
Group C: Ferries that carry passengers.
Group D: Ro-ro boats and ocean liners.
Group E: Reserved for recordings of ambient noise.

In order to provide a thorough resource for vessel identification and categorization using passive sonar sound data, this database provides detailed information fields. In our present work group E was not considered. We only considered the A, B, C, and D groups only.

17.3.2 Implementation

Initially, we obtained a dataset comprising 90 instances of passive sonar signals from various vessels. To ensure data consistency, we extracted two distinctive and important features: *spectral contrast and Chroma*, from each sample. These features were subsequently merged into a comprehensive CSV file. The pivotal task of manually labeling each sample based on its original sample name followed, effectively transforming the dataset into a supervised. Our dataset now comprises 90 labeled instances.

With the labeled data in hand, we proceeded with meticulous training and testing of models. Four distinct classifier models: Decision Tree, Random Forest, K-Nearest Neighbors (KNN), and Logistic Regression were employed. During model training, we utilized 90% of the dataset, which equates to 81 samples, while reserving the remaining 10%, consisting of 9 samples, for testing. To evaluate model performance, we rigorously applied a range of metrics.

In our quest to optimize the classifiers' performance, we undertook the crucial step of hyperparameter tuning. This phase involved leveraging the grid search method to pinpoint the most effective hyperparameters for each model. Additionally, we utilized grid search in conjunction with cross-validation to ensure robust metric results. This meticulous approach enabled us to enhance both the accuracy and overall effectiveness of our classification models as shown in Fig. 17.1.

Subsequently, to further bolster the training of our models, we augmented the dataset by segmenting each sample into 10-s durations. Once segmented, the above discussed process, to create train and test datasets and hyperparameter tuning, was also applied to these segmented instances. Consequently, our training data now consists of 991 samples, while the test data comprises 111 samples. The same set of

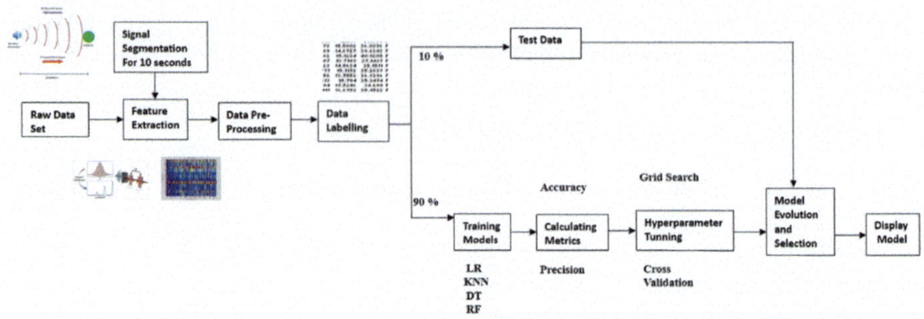

Fig. 17.1 Process flow

metrics was employed to evaluate model performance in this extended dataset. Next, we will be presenting the basic working details of models applied in the present work as shown in Fig. 17.2.

Fig. 17.2 System diagram

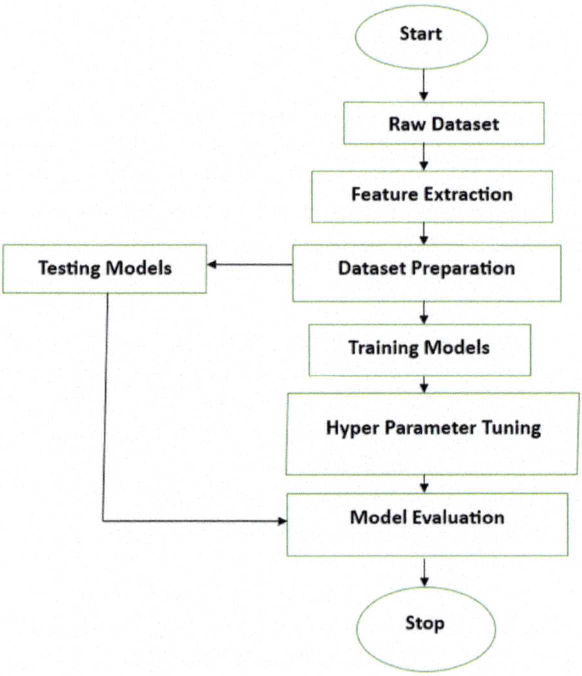

17.3.3 KNN Algorithm

The K-Nearest Neighbors (KNN) algorithm is a simple and efficient supervised machine learning method. Instead of immediate learning from the training data, it saves all of the previously collected data and categorizes incoming data points using a similarity score. Because it stores the training data and uses it during the classification phase rather than immediately learning, KNN is sometimes referred to as a "lazy learner."

The following actions are taken when the KNN algorithm is in practice:

- Start by classifying the data into two separate groups or categories.
- Choose a value for 'K', which establishes the number of nearest neighbors to take into account depending on their proximity to the relevant data point.
- There are two methods for calculating distance to find out nearest neighbors.
  - Manhattan Distance
  - Euclidean Distance
- The K-nearest neighbors are determined for each test data point that is provided.
- Next, take the test data point in the group with the most neighbors. The point's neighbors are determined.

17.3.4 Logistic Regression

Logistic regression is a kind of supervised machine learning method. It is employed to address categorization issues using probability.

The Logistic Regression approach consistently provides a probabilistic outcome ranging from 0 to 1.

Taking the Logistic Regression Algorithm into Practice:

- In logistic regression, an S-shaped logistic function, also known as a sigmoid function, is fitted to the data rather than a line.
- The range of the curve is 0 to 1.
- If a test data point has a probability of more than 50%, it falls into one group, and if it has a probability of less than 50%, it falls into the other category.

17.3.5 Decision Tree

Decision Tree approach may be used for both classification and regression problems, Classification tasks are the most common scenarios where it is often employed. Decision nodes and leaf nodes are two different sorts of nodes that may be found in a decision tree. While leaf nodes indicate the results of decision nodes and do not branch further, decision nodes collect dataset properties and may lead to many branches.

Implementation of the Decision Tree Algorithm:

- To choose the best feature, start by computing the Information Gain or Gini Index.
- Construct a node for the decision tree that includes the most important feature.
- The decision tree node is now divided into subgroups, each of which has the potential to develop into a leaf node or another decision node.
- Continue this approach until additional node splitting is no longer viable. The last node, called as the leaf node, represents the algorithm's result.

17.3.6 Random Forest

A machine learning (ML) technique known as ensemble learning aims to improve predictive performance by mixing predictions from many models. The Bagging approach, which makes use of several base learners, is used by the Random Forest in the context of ensemble techniques. Decision trees serve as these basic learners in the Random Forest Algorithm, which uses them to predict outcomes.

- When implementing the Random Forest Algorithm, the following steps are typically followed:
- A decision tree is given a random sample of data with replacement, and the tree is trained on these samples.
- Using samples from the dataset that are randomly selected, all decision trees are trained in a similar way.
- Repeat this procedure until the full dataset has been examined.
- Every decision tree produces an output in response to test data. A majority vote, as shown in Fig. 17.3, decides the outcome in the end.

17.3.7 Evaluation Criteria

True Positives (TP)—Accurate Ship Classification: TP relates to the successful detection and classification of ships as authentic based on their acoustic signatures, confirming their genuineness.

True Negatives (TN)—Correct Absence of Ships: TN represents precise recognition of ship absence via passive sonar, ensuring no false alarms were triggered.

False Positives (FP)—False Alarms: FPs occur when passive sonar wrongly identifies real vessels due to noise or amplitude variations, leading to unnecessary alerts.

False Negatives (FN)—Missed Ship Classification: FN is when passive sonar can't classify real ships due to minimal auditory signature differences or technological limitations, resulting in missed identifications.

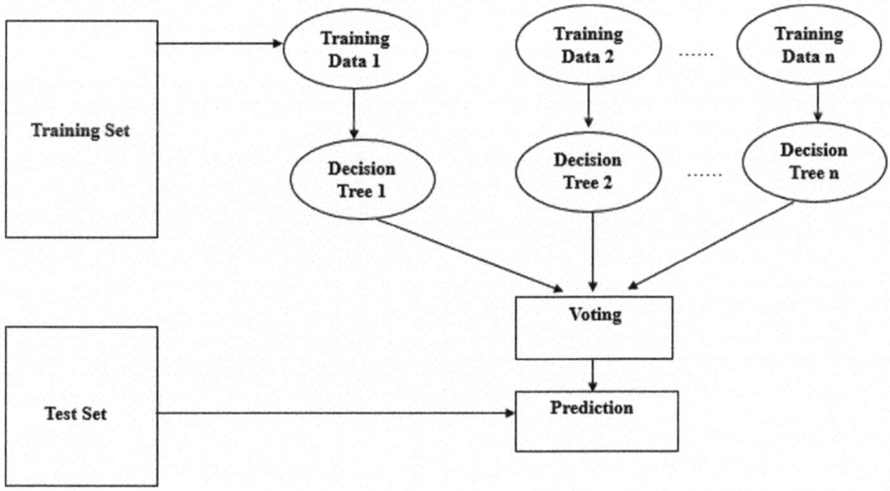

Fig. 17.3 Random Forest algorithm

$$\text{Accuracy} = \text{TP+TN/(TP+FP+TN+FN)} \tag{17.1}$$

$$\text{Precision} = \text{TP/(TP+FP)} \tag{17.2}$$

$$\text{Recall} = \text{TP/(TP+FN)} \tag{17.3}$$

$$\text{F1-score} = 2*(\text{precision}*\text{recall} /\text{precision}+\text{recall}) \tag{17.4}$$

17.4 Results

In order to determine the best method for vessel/ship classification using Sonar signals, we performed an in-depth evaluation of four classifier models, including metrics such as accuracy, precision, recall/sensitivity, and F1-score, as shown in Eqs. (17.1–17.4) and the confusion matrix (see Fig. 17.4). Furthermore, we fine-tuned the hyperparameters for all the classifiers.

Tables 17.2, 17.3, and 17.4 provide a full analysis of the metrics for each of these classifiers, including accuracy, precision, recall/sensitivity, F1-score, and the confusion matrix. To examine performance further, we separated the original samples into 10-s segments and assessed the results for sampled signals, as shown in Tables 17.5, 17.6, and 17.7. Again, in this case, the Random Forest Algorithm showed greater

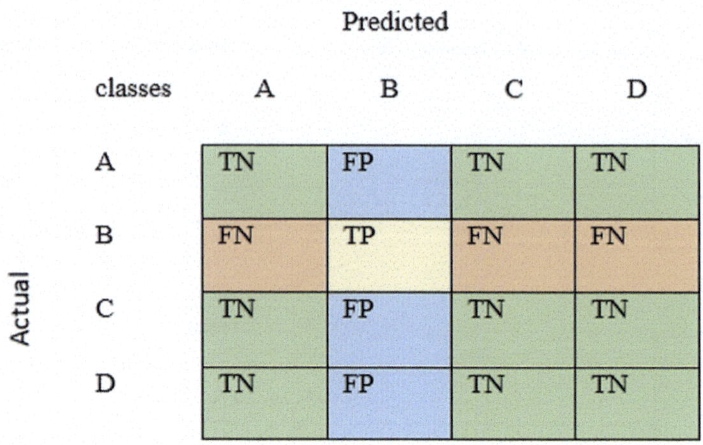

Fig. 17.4 Confusion matrix

performance. Table 17.8 lists the Random Forest model's best parameters and displays the parameters that produced the best results.

Random Forest proves its supremacy in sonar-based vessel classification, showcasing excellence in accuracy, precision, recall, F1-score, and confusion matrix metrics on the *ShipsEar* dataset. Its robust performance in handling unbalanced data, coupled with fine-tuned hyperparameters, distinguishes it from other classifiers. The sustained superiority extends to 10-s sample segments, reinforcing Random Forest's

Table 17.2 Training results of original dataset (90 samples)

Metrics	Classifiers			
	KNN	LR	DT	RF
Accuracy	1.0	0.99	1.0	***1.0***
Precision	1.0	0.69	1.0	***1.0***
Recall	1.0	0.70	1.0	***1.0***
F1-score	1.0	0.70	1.0	***1.0***

Table 17.3 Test results of original dataset (90 samples)

Metrics	Classifiers			
	KNN	LR	DT	RF
Accuracy	0.66	0.66	0.55	**0.78**
Precision	0.64	0.59	0.48	**0.68**
Recall	0.66	0.62	0.55	**0.78**
F1-score	0.61	0.66	0.51	**0.72**

Table 17.4 Confusion matrices for training and test data (90 samples)

Classifiers	Confusion Matrix on Training Data	Confusion Matrix on Test Data
k-Nearest Neighbors	[12 19 30 20] [69 62 51 61] [0 0 0 0] [0 0 0 0]	[1 0 3 2] [8 7 5 4] [0 0 0 3] [0 2 1 0]
Logistic Regression	[12 15 25 17] [69 59 44 59] [0 3 7 2] [0 4 5 3]	[0 1 3 2] [8 6 4 6] [0 1 1 1] [1 1 1 0]
Decision Tree	[12 19 30 20] [69 62 51 61] [0 0 0 0] [0 0 0 0]	[0 2 2 1] [8 5 4 6] [0 2 1 1] [1 0 2 1]
Random Forest	**[12 19 30 20] [69 62 51 61] [0 0 0 0] [0 0 0 0]**	**[0 2 4 1] [8 7 4 6] [0 0 1 1] [1 0 0 1]**

Table 17.5 Training results of original dataset (1102 samples)

Metrics	Classifiers			
	KNN	LR	DT	RF
Accuracy	1.0	0.59	1.0	**0.99**
Precision	1.0	0.60	1.0	**0.99**
Recall	1.0	0.59	1.0	**0.99**
F1-score	1.0	0.59	1.0	**0.99**

Table 17.6 Test results of original dataset (1102 samples)

Metrics	Classifiers			
	KNN	LR	DT	RF
Accuracy	0.70	0.62	0.48	**0.82**
Precision	0.70	0.62	0.48	**0.83**
Recall	0.70	0.62	0.48	**0.82**
F1-score	0.70	0.61	0.48	**0.82**

Table 17.7 Confusion matrix on training and test data for 1102 samples

Classifiers	Confusion Matrix on Training Data	Confusion Matrix on Test Data
k-Nearest Neighbors	[231 280 197 283] [760 711 794 708] [0 0 0 0] [0 0 0 0]	[22 18 16 20] [78 73 83 66] [7 7 6 13] [4 11 6 12]
Logistic Regression	[143 168 107 176] [667 586 735 588] [93 125 59 120] [88 112 90 107]	[18 22 07 21] [79 63 84 65] [6 17 5 14] [7 9 15 11]
Decision Tree	[231 280 197 283] [760 711 794 708] [0 0 0 0] [0 0 0 0]	[9 17 14 14] [73 65 73 65] [12 15 16 14] [17 14 08 18]
Random Forest	**[230 279 197 283]** **[759 710 794 708]** **[1 1 0 0]** **[1 1 0 0]**	**[23 29 15 25]** **[81 72 85 76]** **[4 8 4 3]** **[3 2 7 7]**

Table 17.8 Random Forest grid search parameters

Parameters	
Max_depth	10
Max_features	auto
Min_samples_split	2
n_estimators	2000

effectiveness in ship classification. Table 17.8 revelation of the best parameters solidifies Random Forest as the ideal choice for achieving the classification goals specific to the *ShipsEar* dataset.

17.5 Conclusion and Future Scope

In this paper, performance of various machine learning-based vessel/ship identification models using passive sonar signals was presented. Our investigation reveals that the training accuracy and precision of K-Nearest Neighbor (KNN), Decision Tree, and Random Forest are quite similar. However, there is a significant difference in their test accuracy and precision. Additionally, all models initially exhibited low metric scores prior to the grid search. By adjusting some of the default parameters (hyperparameter tuning) of the models during the grid search, we were able to increase the metric scores. Notably, the test accuracy of the Random Forest model surpassed that of the other models, making it the optimal choice for detecting and classifying passive sonar signals, particularly vessel classes. In the future, we plan to employ advanced deep learning techniques to enhance feature extraction and data sampling. We intend to implement autoencoder techniques to improve the overall classification/detection accuracy. Furthermore, we will utilize other available datasets for evaluation purposes.

Further, the future roadmap involves integrating advanced deep learning, employing autoencoder techniques for enhanced feature extraction and robustness. Additionally, the plan includes hardware implementation on Raspberry Pi to deploy the model in real-world scenarios, ensuring practical applicability beyond evaluation on diverse datasets.

References

1. Santos-Dominguez, D., Torres-Guijarro, S., Cardenal-López, A., Pena-Gimenez, A.: ShipsEar: an underwater vessel noise database. Appl. Acoust.Acoust. **113**, 64–69 (2016)
2. Wang, B., et al.: An underwater acoustic target recognition method based on AMNet. IEEE Geosci. Remote Sens. Lett.Geosci. Remote Sens. Lett. **20**, 1–5 (2023)
3. Li, J., et al.: Underwater acoustic target recognition based on attention residual network. Entropy **24**(11), 1657 (2022)
4. Wang, Y., et al.: Passive sonar target tracking based on deep learning. J. Marine Sci. Eng. **10**(2), 181 (2022)
5. de Castro Vargas Fernandes, J., de Moura Junior, N.N., de Seixas, J.M.: Deep learning models for passive sonar signal classification of military data. Remote Sens. **14**(11), 2648 (2022)
6. Liu, C., et al.: Underwater acoustic target recognition based on dual attention networks and multiresolution convolutional neural networks. In: OCEANS 2021: San Diego–Porto. IEEE (2021)
7. Hong, F., et al.: Underwater acoustic target recognition with resnet18 on shipsear dataset. In: 2021 IEEE 4th International Conference on Electronics Technology (ICET). IEEE (2021)
8. Li, X., et al.: Passive tracking of multiple underwater targets in incomplete detection and clutter environment. Entropy **23**(8), 1082 (2021)
9. Yang, et al.: Underwater acoustic research trends with machine learning: Ocean parameter inversion applications. J. Ocean Eng. Technol. **34**(5), 371–376 (2020)
10. Jiang, D.-N., et al.: Music type classification by spectral contrast feature. In: Proceedings. IEEE International Conference on Multimedia and Expo. Vol. 1. IEEE (2002)

Chapter 18
A Computer-Aided Diagnosis System for the Detection of Parkinson's Disease

K. P. Abhijith, R. Sarath, Partha Santhosh, Jesna Mohan, and Bejoy Abraham

Abstract Parkinson's disease (PD) is a neurological condition that worsens over time and causes accidental or uncontrolled movements, stiffness, and problems with balance and coordination. Usually, symptoms are minor, to begin with, and worsen with time. As the condition develops, individuals may have trouble speaking and moving about. They could also have mental and psychological issues, such as fatigue, sadness, sleeplessness, and cognitive impairment. The development of useful, technologically supported methods for monitoring the development of PD symptoms in everyday life has the potential to change disease assessment and hasten diagnosis. The creation of simple, technology-based techniques for tracking PD symptoms over time in daily life has the potential to revolutionize disease evaluation and speed up diagnosis. The proposed work aims to detect Parkinson's disease from a patient's vocal features at a very early stage. This can be achieved by training machine learning models to detect the vocal characteristics distinct in patients with PD from the data of patients who are already diagnosed with Parkinson's Disease.

18.1 Introduction

Parkinson's disease is a neurological condition that affects mobility. It is caused by the death of cells in certain parts of the brain that generate a neurotransmitter called dopamine. Without enough dopamine, the brain is unable to properly control movement and coordination. Tremors, stiffness, slow movement, and difficulties with balance and walking are all symptoms of Parkinson's disease. The disease is typically diagnosed in people over the age of 50, but it can occur in younger people as well. These symptoms are produced by the loss of cells in certain parts of the brain

K. P. Abhijith (✉) · R. Sarath · P. Santhosh · J. Mohan
Department of Computer Science and Engineering, Mar Baselios College of Engineering and Technology, Trivandrum, Kerala, India
e-mail: abhijithkp773@gmail.com

B. Abraham
Department of Computer Science and Engineering, College of Engineering Muttathara, Trivandrum, Kerala, India

S. K. Pal et al. (eds.), *Intelligent Informatics*, Smart Innovation, Systems and Technologies 389, https://doi.org/10.1007/978-981-97-2147-4_18

that make dopamine, a neurotransmitter. Dopamine is in charge of sending messages in the brain that control movement. When dopamine levels are low, the brain has difficulty controlling movement, leading to the symptoms of Parkinson's disease. In addition to the main symptoms, people with Parkinson's disease may also experience other problems, such as difficulty with balance and walking, difficulty speaking and swallowing, constipation, and changes in mood and behavior. The severity of the symptoms might vary greatly between people, and they can also fluctuate over time. Although there is no cure for Parkinson's disease, medicine and other therapies can help control symptoms and improve quality of life. The most common medications used to treat Parkinson's disease are levodopa and dopamine agonists, which help to increase dopamine levels in the brain. Physical therapy, speech therapy, and occupational therapy are some other therapies that may be beneficial. Surgery may be a possibility in some circumstances to help control specific symptoms of Parkinson's disease.

With the advancement of technology, the methods based on machine learning techniques have been employed to automate Parkinson's disease detection. Artificial intelligence (AI)-based technologies have shown great promise in recent years for improving the diagnosis and prognosis of Parkinson's disease (PD) and have had a significant impact on automated seizure detection, atrial fibrillation, and computer-aided diagnostics. Additionally, PD biomarkers, namely posture analysis during the gait cycle, can be used by medical devices with machine learning or deep learning based on automated detection. Additionally, AI-based gait assessment has shown promise in anticipating and averting impending falls-related injuries and anxiety, as well as enhancing PD patients' independence. The proposed work aims to address the existing issues in automated PD. The method investigates a deep learning model that uses vocal characteristics to identify Parkinson's disease early on as well as a deep learning algorithm that uses MRI pictures to identify Parkinson's disease early on. The results clearly show that CNN architecture is able to attain high accuracy compared to traditional machine learning models using MRI images.

18.1.1 Literature Survey

A number of works exist for the computer-aided diagnosis of Parkinson's disease. After Alzheimer's disease, Parkinson's is the second-worst neurological condition in the world. In the US alone, its annual occurrence rate in 2019 ranged from 40.37 to 53.89 per 100,000 people. Early PD diagnosis is crucial to reducing consequences from the disease. To definitively diagnose it in the early stages, there is no medical test available. In a conventional clinical setting, the doctor might urge the patient to move around and perform some mental and physical exercises [1] or undergo a brain scan using magnetic resonance imaging (MRI) or positron emission tomography-computed tomography (PET/CT) [2]. The ability to discern and accurately identify PD depends on the radiologist's experience because it might be difficult to separate it from other neurological illnesses. As a result, a computer-aided diagnostic (CAD) system aids in the interpretation of MRI scans by the radiologist. The authors of [3] developed a CAD system in 2003 to track body acceleration and identify PD patients' gait freezing.

It was suggested in [4] that voice recording with metrics for central tendency and dispersion is helpful. Given that PD affects patients' handwriting motor skills, handwriting samples from PD patients are gathered and examined. Machine learning is used to predict PD, and it has been discovered that sustained vowels include PD-discriminative information [5].

Senturk et al. in [6] present a review of the use of machine learning algorithms for early diagnosis of Parkinson's disease. The author discusses various machine learning techniques that have been applied to the diagnosis of Parkinson's disease, including decision tree algorithms, support vector machines, and neural networks. The author also discusses the various types of data that have been used to train these algorithms, including clinical data, imaging data, and genetic data. It is evident from the results that machine learning algorithms have the potential to improve the accuracy and efficiency of Parkinson's disease diagnosis, particularly when used in combination with other diagnostic methods. Gunduz et al. in [7] present a deep learning-based method for classifying Parkinson's disease based on vocal feature sets. The method is based on a convolutional neural network (CNN) trained on a dataset of vocal features extracted from individuals with and without Parkinson's disease. The results of the study showed that the proposed CNN model was able to achieve high accuracy in classifying Parkinson's disease from the vocal features, with an accuracy of 95.4 and an F1 score of 0.94. The model also outperformed other machine learning algorithms tested on the same dataset. The study suggests that deep learning-based approaches have the potential to improve the accuracy and efficiency of Parkinson's disease diagnosis based on vocal features. Xiaodong et al. in [8] additionally investigate the validity and viability of incorporating 3D-dopaminergic binding parameters into the clinical scoring system for Parkinson's disease (PD) in order to examine the relationship between dopamine transporter (DAT) PET/CT and the clinical traits and scales of Parkinson's disease (PD) patients. The study reveals that in PD patients, 3D parameters in the neostriatum had a better correlation with activities of daily living, the severity and duration of disease, and cognition than plane parameters. The quantitative parameters based on plane and 3D images of 11C-CFT PET/CT had good consistency.

Clinical research can make reference to a video that was taken while the patient was engaged in physical activity, such as a PD bed test. As mentioned in [9–11] a neural network was able to recognize PD symptoms in a patient video sample. Future clinical studies may examine any video taken of other patients while they were hospitalized, such as those taken during therapy sessions, and forecast whether or not this patient will be suspected of having Parkinson's disease.

Despite significant progress, challenges persist. Limited labeled datasets, data variability among individuals, and interpretability of complex ML models remain areas of concern. Future research should focus on large-scale, diverse datasets, collaborative efforts to create standardized datasets, and the development of explainable AI techniques for clinical acceptance.

The proposed work has the following contributions to overcome the above challenges.

- A deep learning model to detect Parkinson's disease at an early stage from vocal features.
- A deep learning model to detect Parkinson's disease at an early stage from MRI images.

18.2 Materials and Methods

18.3 Voice Data

To develop a machine learning-based algorithm for Parkinson's disease (PD) detection using vocal features, the first step is to record vocal input from subjects as they pronounce a particular alphabet. This dataset should include recordings from both healthy individuals and those diagnosed with PD, ensuring consent and adhering to ethical guidelines. Once the recordings are obtained, necessary features must be selected and extracted from the speech tests. Common vocal features used for PD detection include Mel-Frequency Cepstral Coefficients (MFCC), Fundamental Frequency (F0), and Formants, which capture spectral characteristics, average pitch, and resonant frequencies of the vocal tract, respectively. After feature extraction, the next step is to develop a machine learning algorithm using the extracted vocal features. This algorithm can be trained on the dataset, with appropriate labels indicating PD or healthy status. Various machine learning techniques such as Support Vector Machines (SVM), Random Forests, or Deep Learning models can be employed. Finally, the trained model can be used to classify new subjects as either PD or healthy based on their vocal features, providing a potential tool for early detection and monitoring of Parkinson's disease.

18.3.1 Dataset

The study has been performed utilizing two datasets.

18.3.1.1 Dataset1

The dataset [5] has been taken from the UCI Machine Learning Repository. It consists of 1040 instances having 26 different attributes. The training data was gathered from 20 Parkinson's Disease (PD) patients and 20 healthy people. For each PWP and healthy subjects, 26 recordings were taken. The attributes are shown in Fig. 18.1.

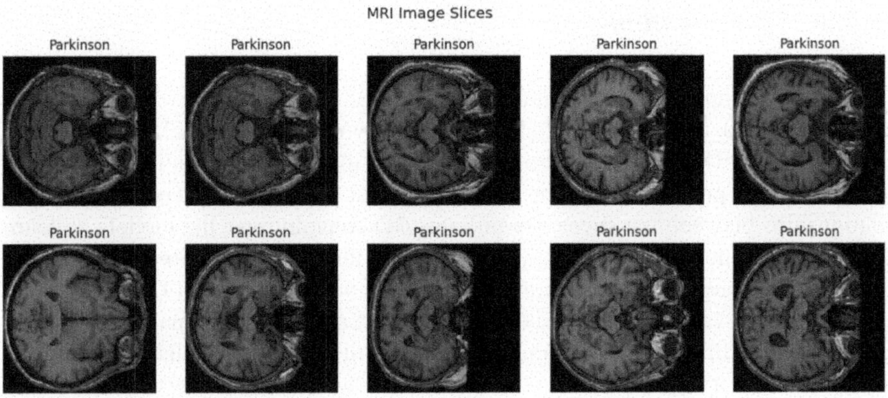

Fig. 18.1 MRI dataset

18.3.1.2 MRI Neurocon Dataset

The dataset comprises of 27 Parkinson's disease patients. The data is collected from multiple sites across the United States and Europe.

- 16 age-matched healthy controls.
- Contained T1 and resting-state scans.

18.3.2 Pre-processing

The MRI data, initially stored in NIfTI format, enables the storage of volumetric data capturing three-dimensional information. The MRI data was transformed into a 2D image sequence for training the model using the NiBabel library. This conversion allows for the extraction of individual slices from the volumetric data. In this case, the middle 5 slices were specifically selected and extracted for further processing. By focusing on these middle slices, the model can potentially capture relevant anatomical information while reducing computational complexity compared to using the entire volume. To facilitate correlation and spatial relationships between the slices, the extracted slices were then converted into a 3D tensor representation. This conversion maintains the sequential order of the slices while enabling the model to recognize patterns and dependencies across the different slices. The resulting 3D tensor serves as input data for training the model, allowing it to learn and make predictions based on the correlated information contained within the middle slices of the MRI data.

18.3.3 Classification

To effectively capture the correlations between the extracted slices from the volumetric MRI data, a 3D Convolutional Neural Network (CNN) model was employed as shown in Fig. 18.1. Unlike traditional 2D CNNs that operate on individual images, a 3D CNN can process the entire volume as input, taking into account the spatial relationships between the slices. By utilizing 3D convolutions, the model can learn features and patterns across multiple slices simultaneously, enabling it to capture complex structures and information present in the volumetric data (Table 18.1).

K-fold cross-validation was used to evaluate the model's performance and generalization capabilities. This technique involves partitioning the available data into K subsets or folds. The model is then trained K times, with K-1 folds serving as training data and the remaining fold serving as validation data. By systematically rotating the folds, the model is evaluated on different subsets of the data, providing a more robust estimate of its performance. The evaluation metrics, such as accuracy or area under the receiver operating characteristic curve (AUC-ROC), are computed across the K iterations, providing a thorough assessment of the model's performance.

For the output layer of the model, a sigmoid function was used. This choice of activation function allows the output to be a single value representing the predicted class probability. The sigmoid function converts the model's final activation to a number between 0 and 1, indicating the likelihood that the input belongs to the positive (e.g., Parkinson's disease) or negative (e.g., healthy) class. This output probability can be further thresholded to make binary predictions, such as classifying subjects as PD or healthy based on a predefined threshold.

By utilizing a 3D CNN architecture, employing K-fold cross-validation, and using a sigmoid activation function in the output layer, the model can effectively learn and

Table 18.1 Model Summary

Layer (Type)	Output Shape	Param #
Conv3d_9(Conv 3d)	(None,9,255,255,32)	288
Max_pooling3d_9 (Max pooling 3D)	(None,9,127,85,32)	0
Batch_normalization_9	(None,9,127,85,32)	128
Conv3d_10(Conv 3d)	(None,8,126,84,64)	16448
Max_pooling3d_10(Max pooling 3D)	(None,8,63,42,64)	0
Batch_normalization_10	(None,8,63,42,64)	256
Conv3d_11(Conv 3d)	(None,7,62,41,128)	65664
Max_pooling3d_11 (Max pooling 3D)	(None,7,31,42,64)	0
Batch_normalization_11	(None,7,31,20,128)	12
flatten_3	(None,555520)	0
Dense_9	(None,128)	71106688
Dense_10	(None,64)	8256
Dense_11	(None,1)	65

```
1    520
0    520
Name: class, dtype: int64
```

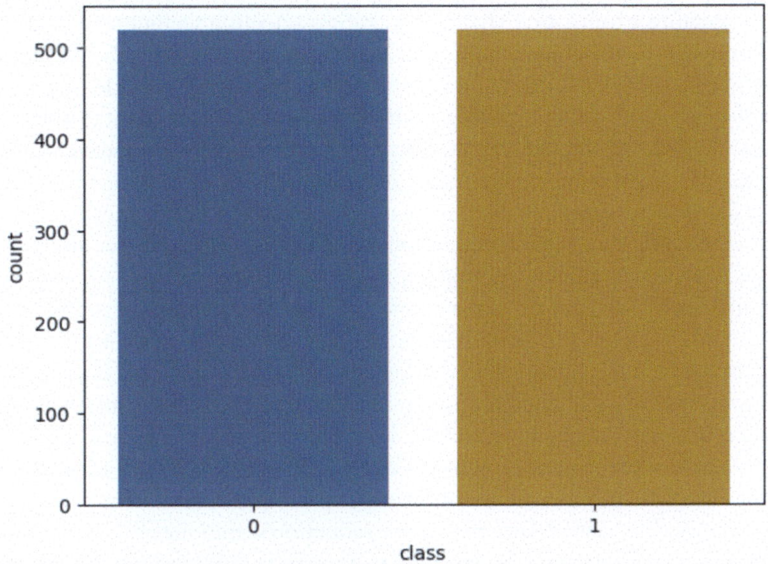

Fig. 18.2 Data instances of healthy subjects(0) and PD patients(1)

predict the presence of Parkinson's disease based on the correlated information from the volumetric MRI data.

The dataset was properly balanced dataset, so concerns about class imbalance causing your model to be biased can be avoided as seen in Fig. 18.2.

18.3.4 Classification Model Training

In this project, the ensemble learning techniques of SVM RBF and XGBoost were employed to identify Parkinson's disease using vocal data. The dataset is divided into two parts: training and testing. First step involved fitting an SVM model to the training data using the RBF kernel. The model is then trained to learn the decision boundaries that best separate the classes. Then fitting is done on an XGBoost model to the training data. The model will learn the decision boundaries by sequentially adding trees and adjusting the tree predictions based on the error of the current model. To implement ensemble learning we have used a voting classifier. Here each model makes a prediction and the final prediction is the one that receives the most votes.

18.3.5 Feature Reduction

Recursive Feature Elimination (RFE) is a feature selection technique used to determine the most crucial features in a dataset. It operates by recursively eliminating the least important features based on the importance ranking provided by a chosen machine learning algorithm. Starting with all features, the algorithm evaluates their importance and eliminates the least significant one. The process is repeated iteratively until a predetermined number of features or a desired level of importance is reached. RFE assists in identifying a subset of features that are most relevant for the target variable, enhancing model performance and decreasing computing complexity.

PCA is a dimensionality reduction approach that is often used to transform high-dimensional data into a lower-dimensional space while keeping the significant patterns contained in the original data. This is accomplished by determining the principal components, which are linear combinations of the original features that capture the most variance in the data. These components are orthogonal to one another, with the first few components accounting for the vast majority of the variance. PCA allows for a more compact representation of the data while minimizing information loss by projecting the data onto a smaller collection of principal components.

After feature selection or dimensionality reduction techniques like RFE and PCA, it is often necessary to normalize the data to ensure that all features are on a similar scale. Normalization transforms the features to have zero mean and unit variance, which prevents certain features from dominating the learning process due to their larger scales. Standardization (subtracting the mean and dividing by the standard deviation) and min-max scaling (scaling the data to a preset range, frequently between 0 and 1) are two common normalization approaches. Normalizing the data before applying machine learning algorithms helps to ensure fair comparisons between different features and improves the stability and performance of the models.

Overall, the combination of feature selection techniques like RFE, dimensionality reduction methods like PCA, and data normalization can enhance the effectiveness of machine learning algorithms by focusing on relevant features, reducing dimensionality, and ensuring consistency in feature scales.

18.4 Results and Analysis

18.4.1 Voice Data

Table 18.2 presents the performance metrics of three machine learning methods, namely Support Vector Machine (SVM), XG Boost, and Ensemble Learning, on the RFE Data. These metrics include test accuracy, precision, F1 score, and recall. Among the three methods, SVM demonstrated the highest test accuracy and performed the best in this particular scenario.

From Table 18.3, after applying Principal Component Analysis (PCA) for dimensionality reduction, the performance of all three classifiers experienced an overall

Table 18.2 Recursive feature elimination

Recursive feature elimination

	Test accuracy	Precision	F1 score	Recall
SVM	72.022	100	83.73	72.02
XGBoost	71.42	100	83.33	71.42
Ensemble	70.83	100	82.92	70.83

Table 18.3 Principal component analysis on RFE data

Principal component analysis on RFE data

	Test accuracy	Precision	F1 score	Recall
SVM	68.45	100	81.27	68.45
XGBoost	63.69	100	77.81	63.69
Ensemble	64.28	100	78.26	64.28

drop. The accuracy and F1 score of the Support Vector Machine (SVM) and Ensemble classifiers both decreased slightly, indicating a slight decline in their predictive power. On the other hand, the XGBoost classifier demonstrated a more significant decrease in accuracy and F1 score, suggesting a larger impact of dimensionality reduction on its performance. However, precision remained at 100%, indicating that the classifiers maintained a high level of accuracy in correctly identifying positive cases. Nevertheless, the drop in recall suggests that more false negative predictions occurred, indicating a higher likelihood of misclassifying individuals with Parkinson's disease as healthy. These findings emphasize the trade-off between dimensionality reduction using PCA and the predictive performance of the classifiers, highlighting the need for careful consideration and experimentation when applying dimensionality reduction techniques to machine learning models.

Table 18.4 provides an overview of the performance metrics, including test accuracy, precision, F1 score, and recall, for three machine learning methods: Support Vector Machine (SVM), XG Boost, and Ensemble Learning. These metrics were evaluated using Principal Component Analysis (PCA) on the original dataset. Interestingly, a similar trend was observed in Table 18.4, wherein all three classifiers experienced an overall decrease in performance compared to Table 18.2.

Table 18.4 PCA on original dataset

PCA on original dataset

	Test accuracy	Precision	F1 Score	Recall
SVM	63.09	100	77.37	63.09
XGBoost	67.26	100	80.42	67.26
Ensemble	62.5	100	76.92	62.5

18.4.2 MRI

Deep learning libraries like tensorflow and keras in colab environment were used to develop the model. The classification model was built using a 3D convolutional neural network (CNN) model. K-fold cross-validation was used to evaluate the model. The results were visualized via boxplot and violin plot through libraries such as Matplotlib and seaborn. NiBabel library was used to manipulate MRI scans. Tkinter library was used to develop the desktop software and Flask web framework was used to develop webapp.

In our study, we evaluated the performance of our classification model using cross-validation, and the resulting accuracies across different folds are as follows: [55.56%, 33.33%, 66.67%, 55.56%, 33.33%, 66.67%, 62.5%, 62.5%]. Figure 4.1 represents the performance measure of the model, and to gain insights into the distribution and variability of these accuracies, we employed two commonly used statistical visualization techniques: box plots and violin plots.

Figure 4.2 represents the box plot, which provides a concise summary of the accuracy distribution. The line inside the box represents the median accuracy, which in our case is approximately 54.51%. The box itself represents the interquartile range (IQR), indicating that 50% of the accuracy values fall within this range. The whiskers extending from the box depict the minimum and maximum values, excluding outliers. From the box plot, we can observe that the accuracy is relatively spread out, with some variability in performance across different folds (Figs. 18.3 and 18.4).

To further understand the accuracy distribution, we utilized violin plots as shown in Fig. 4.3. The width of the violin plot represents the density of accuracy values, with wider sections indicating higher density. The height of the plot indicates the probability density estimate. We can observe that the violin plots for our accuracy show a relatively symmetrical distribution, with the peak density occurring around the median accuracy. Additionally, the plots provide a visual representation of the variability in accuracy, as wider sections indicate higher dispersion of values (Fig. 18.5).

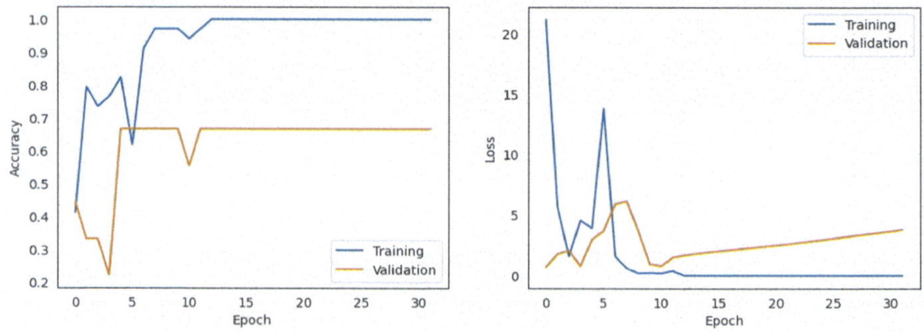

Fig. 18.3 Performance measure

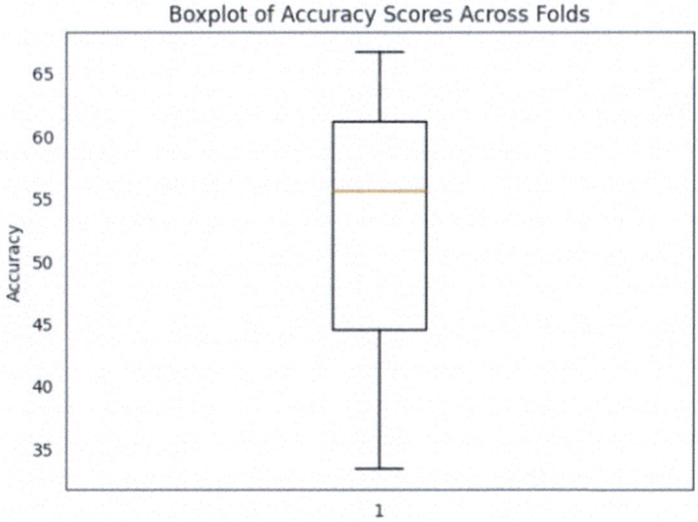

Fig. 18.4 Boxplot of accuracy scores across folds

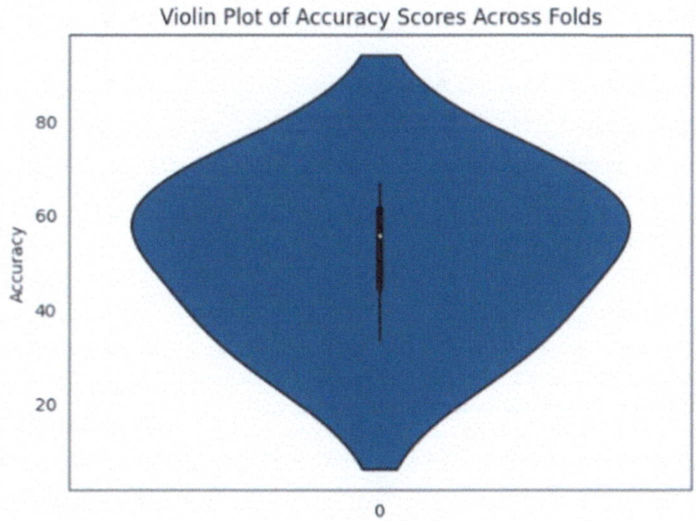

Fig. 18.5 Violin plot of accuracy scores across folds

Overall, the box plot and violin plot visualizations help us gain a comprehensive understanding of the distribution and variability of accuracy across different folds, providing valuable insights into the performance of our classification model.

18.5 Conclusion

For the voice data, after the experimental analysis, the model which was created in SVM after the feature reduction with recursive feature elimination was found to be the most accurate with an accuracy of 72.0238 A web and mobile-based application was developed utilizing a machine learning model. The model's performance was evaluated using k-fold cross-validation. The mean accuracy across the folds was found to be 54.5%, with a standard deviation of 12.8%. This indicates that, on average, the model achieved an accuracy of 54.5% across different subsets of the data. However, the presence of a standard deviation of 12.8emphasizes the significance of using cross-validation. It demonstrates the variability in accuracy scores observed across the different folds, highlighting the necessity of evaluating the model's performance on multiple subsets of the data. Cross-validation can provide a more robust and trustworthy assessment of the model's performance on fresh and previously unknown data. It also suggests that there is potential for further improving the model's accuracy by obtaining a better dataset or exploring other strategies to enhance its performance.

References

1. Mahlknecht, P., Hotter, A., Hussl, A., Esterhammer, R., Schocke, M., Seppi, K.: Significance of MRI in diagnosis and differential diagnosis of Parkinson's disease. Neurodegener. Dis. **7**(5), 300–318 (2010)
2. Mirzadeh, Z., Chapple, K., Lambert, M., Evidente, V.G., Mahant, P., Ospina, M.C., Samanta, J., Moguel-Cobos, G., Salins, N., Lieberman, A., et al.: Parkinson's disease outcomes after intraoperative ct-guided "asleep" deep brain stimulation in the globus pallidus internus. J. Neurosurg. **124**(4), 902–907 (2016)
3. Alzubaidi, M.S., Shah, U., Dhia Zubaydi, H., Dolaat, K., Abd-Alrazaq, A.A., Ahmed, A., Househ, M.: The role of neural network for the detection of Parkinson's disease: a scoping review. In: Healthcare, vol. 9, p. 740 (2021). MDPI
4. Isenkul, M., Sakar, B., Kursun, O., et al.: Improved spiral test using digitized graphics tablet for monitoring Parkinson's disease. In: The 2nd International Conference on E-health and Telemedicine (ICEHTM-2014), vol. 5, pp. 171–175 (2014)
5. Sakar, B.E., Isenkul, M.E., Sakar, C.O., Sertbas, A., Gurgen, F., Delil, S., Apaydin, H., Kursun, O.: Collection and analysis of a Parkinson speech dataset with multiple types of sound recordings. IEEE J. Biomed. Health Inform. **17**(4), 828–834 (2013)
6. Senturk, Z.K.: Early diagnosis of Parkinson's disease using machine learning algorithms. Med. Hypotheses **138**, 109603 (2020)
7. Gunduz, H.: Deep learning-based Parkinson's disease classification using vocal feature sets. IEEE Access **7**, 115540–115551 (2019)
8. Wu, X., Wei, F., Gan, J., Li, Z., Wu, P., Liang, S., Ma, Y., Ding, L., Zuo, C., Liu, Z., et al.: Correlation analysis between 3d and plane DAT binding parameters of 11C-CFT pet/CT and the clinical characteristics of patients with Parkinson's disease (2023)
9. Rahman, W., Lee, S., Islam, M.S., Antony, V.N., Ratnu, H., Ali, M.R., Mamun, A.A., Wagner, E., Jensen-Roberts, S., Waddell, E., et al.: Detecting Parkinson disease using a web-based speech task: observational study. J. Med. Internet Res. **23**(10), 26305 (2021)

10. Filtjens, B., Nieuwboer, A., D'cruz, N., Spildooren, J., Slaets, P., Vanrumste, B.: A data-driven approach for detecting gait events during turning in people with Parkinson's disease and freezing of gait. Gait Posture **80**, 130–136 (2020)
11. Reyes, J.F., Montealegre, J.S., Castano, Y.J., Urcuqui, C., Navarro, A.: LSTM and convolution networks exploration for Parkinson's diagnosis. In: 2019 IEEE Colombian Conference on Communications and Computing (COLCOM), pp. 1–4 (2019). IEEE

Chapter 19
Impact of the Use of Social Media on the Addiction and Social Isolation Levels of Adolescents After the COVID-19 Pandemic

V. S. Kochukrishna Kurup, P. Rangasami, Bhagya V. Pillai, and V. C. Geetha

Abstract The COVID-19 pandemic has had far-reaching consequences on almost every human way of life by interrupting the usual flow of existence. The first case of Coronavirus infection was identified in Wuhan, China, and it spread like wildfire throughout the globe, forcing life to come to a halt. Because of the spread of these viruses, schools have been shuttered, and students are expected to stay at home and learn online. This leads to a significant increase in internet usage among children and teenagers. This study investigates the impact of social media and the internet on addiction and social isolation levels among adolescent children during and after the pandemic. A convenience sample of 90 school children was used in this investigation. The information was gathered using online questionnaires that comprised a demographic question, a social media addiction scale, and a social isolation scale. There was a substantial medium negative and significant relationship between social media use and social isolation, indicating that social isolation decreases as social media addiction rises. There was no significant difference in offender scores discovered. Both boys and girls are moderately addicted to social media, which impacts the reduction of social isolation.

V. S. Kochukrishna Kurup (✉)
School of Social and Behavioural Sciences, Amrita Vishwa Vidyapeetham, Amritapuri, India
e-mail: kochukrishnan@am.amrita.edu

P. Rangasami
School of Social and Behavioural Sciences, Amrita Vishwa Vidyapeetham, Coimbatore, India

B. V. Pillai
Jal Jeevan Mission Project, Kerala, India

V. C. Geetha
Higher Secondary Education, Kerala, India

S. K. Pal et al. (eds.), *Intelligent Informatics*, Smart Innovation, Systems and Technologies 389, https://doi.org/10.1007/978-981-97-2147-4_19

19.1 Introduction

Adolescence is an important time in an individual's life, as it shows the special uniqueness of behaviour development. Adolescents' interactions with social signals will significantly affect their behaviour and how they engage with social settings [1]. Social contact is a vital component of a young person's proper social behaviour. COVID-19 has transformed the current social normalcy in an incomparable way, which has led to people's separation from physical social interactions. Separation and social distance to protect against the threat of illnesses have been major methods of preventing COVID-19 infection worldwide. As a result, several nations have been launching local and public legislation or lockdowns since January 2020.

19.1.1 Educational Closures and Social Isolation

The implementation of lockdown measures, including the closure of educational institutions and the suspension of other activities, has been identified as a significant strategy for mitigating the transmission of COVID-19. Nevertheless, the impact of these measures on the broader populace, namely the education of Children, has been significant. The extent to which young people are socially separate, the inspiration behind social distance, and how inspirations relate to measurements of social distance are seldom considered [2].

The prolonged cessation of educational institutions and the enforcement of measures to maintain physical distance have had a substantial influence on the interpersonal interactions and social experiences of young individuals, namely, children and teenagers. Educational institutions have been closed for an extended period of time, resulting in prolonged social isolation of children from both the wider population and their classmates. The implications of such social isolation on the psychological well-being and mental health of teenagers are significant issues that require comprehensive investigation.

Prevailing attempts to control the spread of COVID-19 have needed strange and widely ordered physical distance, eliminating various social linkage sources from life, and such measures will likely have a significant impact, not only on the society and the economy but also on the mental well-being of individuals through such things as reduced contact with others [3].

In the twenty-first century, there was a huge growth in the use of social networking sites worldwide. A 2018 worldwide blog-digital research states that there are 3,196 billion social media users, up 13% from last year. Social networking services are becoming more and more popular among all people, especially youngsters and adolescents [4].

19.1.2 Internet and Social Networking

The COVID-19 scenario has been a stimulus for individuals throughout the world to use the internet and social networking. Communication technology has changed social networking from an action-based platform to one centred on identification. Today, social networking is characterized by our identities, not our actions. Authentically expressing and connecting with people based on who we are may lead to a more comprehensive view of interpersonal connections. Since the late 1990s, children have grown up in a society that depends on technology as an important part of their existence, which makes it hard to imagine living without connections. This was called "always-on" living, and "on" has become the status norm [5].

Mobile communication and accessibility to smartphones facilitate the use of social media by everybody. Almost all internet services may be utilized as smartphones for daily use since mobile technology and communication are progressing and are not seen as addictions. Social networking sites are extremely popular on smartphones, with around 80% of social media use via mobile technology [6]. Excessive social media use may lead to addiction as it is seen as the single most important activity in which a person engages on an everyday basis. It controls behaviour patterns [7]. However, it also taps the very fundamental human need by offering social support and self-expression [8]. Some other studies early studies show that internet/social media use for social purposes reduces loneliness and depression among users [9].

Because of the spread of these viruses, schools have been shuttered, and students are expected to stay at home and learn online. This leads to a significant increase in internet usage among children and teenagers [10].

19.1.3 Online Education During COVID-19

The COVID-19 standards mandate the confinement of students inside their residences, requiring them to depend on internet access to engage in remote education via mobile devices and laptops. It is worth mentioning that a considerable number of educational institutions have adopted prolonged online teaching methods, which has resulted in teenage students being exposed to the widespread impact of social media platforms like Facebook, Instagram, and WhatsApp.

Unfortunately, a concerning truth arises as several prevalent social networking platforms demonstrate their inadequacy in catering to the vulnerable population of children and adolescents. Therefore, it is essential for parents to get an exhaustive understanding of the intricate attributes of these platforms and actively encourage prudent and beneficial patterns of involvement. Significantly, it is noteworthy that some social media platforms provide beneficial channels for children and young persons, enabling them to engage in offline educational experiences and get access to enriching online activities [11].

However, the post-COVID-19 environment reveals a complicated interaction between beneficial and detrimental effects from the younger generation's widespread use of social media and the Internet. Many impacts include the facilitation of knowledge retrieval, the facilitation of socializing and communication, and the maintenance of social involvement throughout the epidemic. The complex scenario sheds light on the intersection of incentives pertaining to state and local lockdowns, parental controls, and social duty with the overarching idea of social removal. Furthermore, reasons pertaining to the lack of alternatives exhibit unique characteristics when compared to those related to teenage stress, empirical findings, onerous manifestations, oppressive encounters, and the inherent feeling of belonging [2].

Given the complex context, the current research study aims to examine the complex impact of social media on the degrees of social isolation experienced by teens during a certain period. The aim is to determine if online platforms have alleviated or intensified the prevailing feeling of social isolation. This academic investigation holds importance in understanding the extensive impacts of the pandemic on the younger population. It provides valuable guidance to policymakers responsible for developing effective strategies for future crises. Additionally, it offers insightful perspectives on the potential addictive and social consequences that arise from increased dependence on digital platforms during periods of confinement.

19.2 Method

A convenience sampling approach for male and female adolescent social media users has been used in this study (n = 90). The participants were from both genders and age ranges 12–19 years old (Boys n = 43/48%; girls n = 47/52%), and they were divided into two age groups. 93% of participants fall in the age group of 16–19. The sampling approach was employed for the participants from five different class groups of two government-aided schools.

The respondents are distributed in three class divisions viz. 10th (n−10/11%), plus 1 (n = 33/36%), and plus 2 (n = 47/52%). Most of the participants use social media, that is, 84 respondents (93%). Informed consent is requested from the school's authorities to conduct the study. The sample was selected from the rural area of the southern part of Kerala. All selected adolescents are enrolled in different government-aided schools and are natives of Kerala.

The respondents were supplied with structured questionnaires online. In the study, the UCLA loneliness scale is employed. A 20-point scale was developed to evaluate subjective experiences of solitude and social isolation. Each item is classified as O (I often feel that way), S (I occasionally feel like that), R (I seldom feel that way), and N (I never feel this way). The questionnaire corresponded with the scale, the native language of the respondents and the verified translation phase was translated to the local language (Malayalam).

Arslan and Kırık's 'Social Media Addiction Scale (SMAS)' (2013) was used in the study to assess social media addiction. The measure consists of 25 items on a Likert scale of five points. Scale items contain comments like "I love to spend time on social networking sites." Five choices were provided to participants, ranging from strongly disagreements. For assessing the level of social isolation, the UCLA loneliness scale is used in the study. The questionnaire corresponding to the scale was translated to Malayalam the native language of the respondents, and phase validated to back translation. For the gathering of data, an online Google form survey was utilized. The questionnaire consisted of a fact sheet and agreement from which the subject consented to give his consent.

19.3 Results and Discussion

This study investigated the notion that social media addiction and social isolation are negatively correlated. Questionnaires also collected data on the extent of social isolation.

19.3.1 Social Media Addiction

It is important to note that a very large majority of respondents (n = 84, 93%) use social media, particularly in the case of COVID-19; children spend 4–5 h in class and spend time talking and watching on social media between and at other times (Fig. 19.1).

Findings show three levels of social media addiction among school students. A majority of respondents (58%) are mildly hooked, and 22% are addicted. The

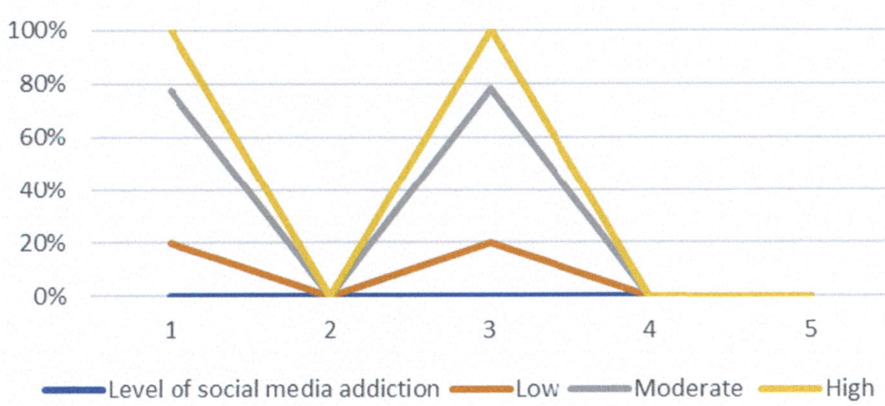

Fig. 19.1 Shows the social media addiction among adolescents

Table 19.1 Level of social media addiction by gender

Gender		Level of social media addiction			Total
		Low	Moderate	High	
Male	Count	12	15	16	43
	% within gender	27.9%	34.9%	37.2%	100.0%
Female	Count	18	15	14	47
	% within gender	38.3%	31.9%	29.8%	100.0%
Total	Count	30	30	30	90
	% within gender	33.3%	33.3%	33.3%	100.0%

engagement on mobile phones to listen to online lessons for hours constantly leads to the development of social communication on several platforms such as Facebook, WhatsApp, Instagram, YouTube, etc.

19.3.2 Level of Social Media Addiction by Sex of the Subject

Table 19.1 shows the level of social media addiction by sex of the participants. One can see that addiction is higher among boys than girls.

19.3.3 Class and Level of Social Media Addiction

Table 19.2 analyses the social media addiction of adolescents according to their educational level, which shows the significant result that the addiction level is increasing with the increasing level of classes.

Table 19.2 Level of social media addiction by class

			Level of social media addiction			Total
			Low	Moderate	High	
Class	10th	Count	1	6	3	10
		%	10.0%	60.0%	30.0%	100.0%
	Plus 1	Count	15	8	10	33
		%	45.5%	24.2%	30.3%	100.0%
	Plus 2	Count	14	16	17	47
		%	29.8%	34.0%	36.2%	100.0%
Total		Count	30	30	30	90
		%	33.3%	33.3%	33.3%	100.0%

Table 19.3 Level of social isolation level by gender

			Level of social isolation			Total
			Low	Medium	High	
Gender	Male	Count	31	12	0	43
		% within gender	72.1%	27.9%	0.0%	100.0%
	Female	Count	24	19	4	47
		% within gender	51.1%	40.4%	8.5%	100.0%
Total		Count	55	31	4	90
		% within gender	61.1%	34.4%	4.4%	100.0%

$P = <0.05$

19.3.4 Social Isolation Level by Gender

The social isolation of the participants is assessed by using the loneliness scale, and it is a common factor for all children due to the COVID-19 restrictions and regulations. Hypnotically, social media addiction leads to social isolation due to a lack of engagement with social activities or social relations (Table 19.3).

Table 19.2 shows a clear picture of the negative correlation level of social isolation among boys and girls with social media addiction. The majority (72.1%) of boys show a low level of social isolation, and none of them have a high level.

Chi-Square tests

	Value	df	Asymp. Sig. (two-sided)
Pearson Chi-Square	6.306[a]	2	0.043
Likelihood ratio	7.855	2	0.020
No. of valid cases	90		

[a]Two cells (33.3%) have an expected count of less than 5. The minimum expected count is 1.91

However, in the case of females, 51.1% have a low level of social isolation, 40.4% have moderate feelings of social isolation, and 8.5% have a high level. The result showed that social isolation and gender have a significant correlation. There was a statistically significant medium negative correlation between the participant's level of addiction and social isolation ($p < 0.043$).

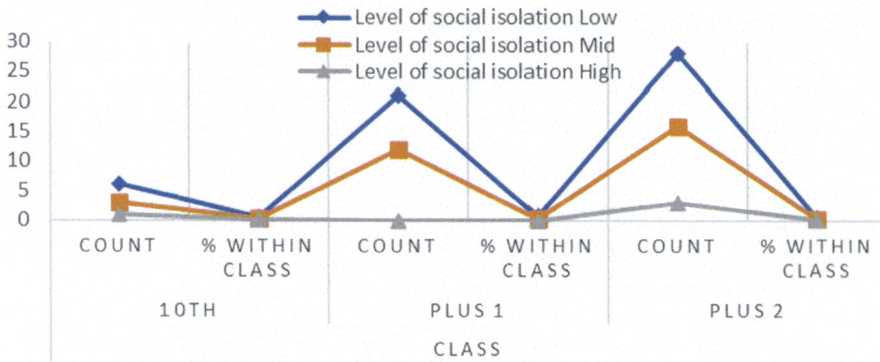

Fig. 19.2 Showing the level of social isolation by class

19.3.5 Social Isolation Level by Class

The social isolation level is comparably more among the children in the lower class i.e.; 10th class (Fig. 19.2).

19.4 Discussion

The main purpose of this study was to evaluate the amount of dependence on social media during the COVID-19 phase and its impact on social isolation. The first investigation was to determine if boys and girls are social media addicts owing to the continuing use of internet class cell phones. The second objective was to find out whether the impact of addiction was different based on the sex, age and class of the user they are studying.

The Social Media Addiction Scale (SMAS) created by Arslan and Kirik (2013) has been used to quantify participant social media addiction. There was significant variation in the scores of social media dependence by gender following the analysis results. Male children are showing more dependency on social media than females. A study investigating this showed that women rely on social media to keep friends while men use it to create new connections [12].

The degree of social media addiction and social isolation of respondents indicate a strong trend to increase the class level and to reduce the amount of social isolation. The outcome of the study may show that girls are more likely to be socially isolated than boys.

19.5 Limitations

There was an age preference with fewer children under the age of 15 in relation to the sample of this study. This might have a detrimental effect on the generalization and comparison of age differences. Another weakness of this study was the absence of control over other factors, for example, whether individuals had previous experience in the use of social networking sites or personal problems at that time.

19.6 Conclusion

This study's findings reflect the perspective of society that social media dependence causes social isolation among both genders, which has changed after the COVID-19 situation. This study indicates that both genders were equally affected by social media. Earlier research focused on the usage of social media leading to addiction and, hence, social isolation. However, social media is a necessary source for preserving and growing social contacts and minimizing social isolation under the circumstances after COVID-19. The gender variation in addiction and social involvement is substantial. This study sheds light on the gap between online and offline social involvement of people and on the detrimental effects of social media dependence.

References

1. Jones, R.M., et al.: Adolescent-Specific Patterns of Behavior and Neural Activity During Social Reinforcement Learning. Harvard University's DASH Repository (2019). http://nrs.harvard.edu/urn-3:HUL.InstRepos:dash.current.terms-ofuse#OAP
2. Oosterhoff, B., Palmer, C.A., Wilson, J., Shook, N.: Adolescents motivation to engage in social distancing during Covid 19 pandemic associations with mental and social health. J. Adolesc. Health (2020). https://doi.org/10.1016/j.jadohealth.2020.05.004
3. Orben, A., Tomova, L., Blakemore, S.-J.: The Effect of Social Deprivation on Adolescent Development and Mental Health (2020)https://doi.org/10.1016/S2352-4642(20)30186-3
4. Gilbert, A.: The impact of social media on the self-esteem levels of its users, Norma Smurfit Library National College of Ireland's Institutional Repository TRAP (2019). http://trap.ncirl.ie/
5. Kuss, D.J., Griffiths, M.D.: Social networking sites and addiction: ten lessons learned. Int. J. Environ. Res. Public Health **14**(3), 311 (2017)
6. Markeingt Land.: Nearly 80 per cent of social media time is now spent on mobile devices (2017). https://martech.org/facebook-usage-accounts-1-5-minutes-spent-mobile. Accessed 09 July 2021
7. Griffiths, M.D., Kuss, D.: Adolescent social media addiction (revisited). Educ. Health **35**(3), 49–52 (2017)
8. Riva, G., Widerhold, B.K, Cipresso, P.: Psychology of Social Media: From Technology to Identity. De Gruyter Open, Warsaw, Poland (2016)
9. Kraut, R., Patterson, M., Lundmark, V., Kiesler, S., Mukophadhyay, T., Scherlis, W.: Internet paradox: a social technology that reduces social involvement and psychological well-being? Am. Psychol. **53**(9), 1017–1031 (1998)

10. Dong, H., Yang, F., Lu, X., Hao, W.: Internet Addiction and Related Psychological Factors among Children and Adolescents in China During the Coronavirus Disease 2019(COVID-19) Epidemic, Published: 02 September 2020 (2020). https://doi.org/10.3389/fpsyt.2020.00751
11. Gwenn Schurgin O'Keeffe, Kathleen Clarke-Pearson and Council on Communications and Media, Clinical Report- The Impact of Social Media on Children, Adolescent And Families. https://doi.org/10.1542/peds.2011-0054 originally published online March 28, 2011
12. Mazman, S.G., Usluel, Y.K.: Gender differences in using social networks. Turkish Online J. Educ. Technol.-TOJET **10**(2), 133–139 (2011)

Part IV
Biotechnology and Environmental Applications

Chapter 20
Preliminary Testing of a Color-Based Test Kit Detector for Bioplastics

Farrah Wong, Noor Fazilah Binti Rahmansyah, Sariah Abang, Seng Kheau Chung, Aroland Kiring, Jamal Ahmad Dargham, and Rosalam Sarbatly

Abstract Plastic was invented in 1907 by Leo Baekeland who is a Belgian Chemist. Since then, his creation has paved the way for the future of polymers. Nowadays, plastics come in different types with multitude of uses ranging from household storage purpose to medical packaging means. However, it also adversely affects humanity, particularly on the environment and bioplastics is noticeably would be the way forward to achieve a sustainable environment. Bioplastics is synthesized from biomass or other natural material as the new alternative to plastic as it degrades much faster. Eventually, a bioplastic testing kit will be necessary, especially in the market where the use of plastic will be regulated. A prototype test kit based on Arduino and a color sensor was developed to distinguish different plastic types based on their distinct color reactions to specific chemical reagents. The fundamental question was how to create a feasible way to distinguish between cellulose-based, starch-based, biodegradable, and conventional plastics and deal with the accompanying challenges. The reagents applied to the samples included iodine, iodine-$CaCl_2$, and Schultze reagents. Notably, the cellulose-based and starch-based straw samples exhibited a dark purple color change with iodine and dark blue with iodine-$CaCl_2$ and Schultze reagent. In contrast, starch-based singlet bags exhibited brown transforming into purple with iodine and Schultze reagents. Besides, biodegradable, and conventional plastics displayed no color changes with any reagents. The test kit has shown a promising way to assist consumers to make a more informed decision through a simple test.

F. Wong (✉) · N. F. B. Rahmansyah · S. K. Chung · A. Kiring · J. A. Dargham · R. Sarbatly
Faculty of Engineering, Universiti Malaysia Sabah, 88400 Kota Kinabalu, Sabah, Malaysia
e-mail: farrah@ums.edu.my

S. Abang
Faculty of Engineering, Universiti Malaysia Sarawak, 94300 Kota Samarahan, Sarawak, Malaysia

S. K. Pal et al. (eds.), *Intelligent Informatics*, Smart Innovation, Systems and Technologies 389, https://doi.org/10.1007/978-981-97-2147-4_20

20.1 Introduction

20.1.1 Bioplastics

Over the last decades, the market has developed various materials within various manufacturing fields; overtaking wood, glass, ceramic and metal, plastics have become an essential material in the manufacturing environment [1]. There is also an increasing impetus for bioplastics manufacturing, growing demand, and the introduction of more complicated applications and products. The worldwide bioplastics manufacturing capacity is predicted to increase dramatically from 2.2 million tons in 2022 to over 6.3 million tons in 2027 [2]. Bioplastics refers to bio-based plastics, which comprise biogenic materials, such as crop-based feedstock [3] or organic waste [4, 5].

20.1.2 Bioplastics as Alternative to Plastics

Verma and Fortunati [6] stated that bioplastics have emerged as a viable candidate with the potential as an alternative to petroleum-based plastics due to their biodegradability. Alternative materials regarded bioplastics as promising because their feedstocks are renewable and, in theory, can be composed and recycled. Besides, their manufacturing method might be more energy efficient, abundant, and inexpensive than petroleum-based ones [7–9].

According to the World Economic Forum [10], in 2014 alone, the sector produced 311 million metric tons and is predicted to triple by 2050, yet less than 15% of it gets recycled. These are the primary reasons behind the recent transition to bio-based plastics and significant growth in polymer production.

20.1.3 Needs for Testing Option

However, there is currently a lack of standardized and reliable methods for evaluating the biodegradability of bioplastics, which is crucial for developing and implementing more sustainable and environmentally friendly products. As a result, there is a gap in the market for an accurate and accessible biodegradability testing option.

20.2 Relevant Research

20.2.1 Biodegradability Testing

Koh and Khor [11] provided an overview of the current state and potential future advancements in the sensors used for assessing the biodegradability of polymers and sensors made from biodegradable polymers. The research concentrated on standard tests, analytical processes for biodegradability testing, and sensors for assessing biodegradable organic matter. They examined various methods and techniques for evaluating biodegradation, including gravimetric analysis, microbial tests, spectroscopy, and electrochemical sensors.

Chowdhury et al. [12] studied the development and characterization of tamarind seeds, berry seeds, and licorice roots to produce bioplastic materials. The researchers synthesized the bioplastic material and conducted various characteristics to evaluate its properties. In this paper, laboratory-based techniques were applied to test the biodegradability of bioplastic materials. Techniques such as scanning electron microscopy (SEM), Fourier-transform infrared spectroscopy (FTIR), and mechanical testing were employed to assess the bioplastics' surface morphology, thermal stability, and tensile strength. The results indicated that the bioplastics derived from these natural sources exhibited favorable properties. Interestingly, the bioplastics produced without licorice root showed superior biodegradability and had improved mechanical, morphological, thermal, and antibacterial characteristics.

The biodegradability of bioplastics is often evaluated using standard test methods. According to Ashter [13], ASTM D6866 is a standardized test technique created by the American Society for Testing and Materials (ASTM) to assess bio-based content. The test technique is based on determining the quantity of radiocarbon (carbon-14) in the sample, as bio-based materials include carbon generated from modern renewable sources and will contain a higher percentage of carbon-14 than fossil-based materials. The ASTM D6866 test applies to many goods, including plastics, coatings, adhesives, lubricants, fuels, and other materials. It provides a reliable method of assessing bio-based content, allowing consumers and the industry to make educated decisions regarding the environmental impact and sustainability of the goods they use.

20.2.2 Research Activities on Testing

Film-Based Testing Analysis. Pucci et al. [14] applied this to monitor the deformation of polymer produced by film manufacturing or any structural alteration of plastic material using a luminescence mechanochromic sensor. The melt processing technique was used to create films made of poly(propylene) (PP) with varied concentrations of bis(benzoxazolyl)stilbene (BBS). During the observation, the films noticed a change, with the emission shifting from blue to green. Subsequently, the tensile deformation detection in PP films confirmed and demonstrated the existence of the polymer's properties.

Another approach by Pucci et al. [15] focused on using the BBS compounds as luminescent sensors to measure temperature and deformation in biodegradable polybutylene succinate (PBS) films. The BBS molecules dispersed in a PBS matrix to achieve changes in emission properties. The concentration of BBS in films influences the emission color, with lower concentrations exhibiting blue emission from isolated BBS molecules and higher concentrations showing green emission from supramolecular aggregates called excimers. In the case of the tested PLA, the monomer band (representing blue light) exhibited a significant presence during tensile stress. In contrast, the emission band (indicating green light) showed notable prominence under thermal stress.

Capacitance Testing Analysis. Schusser et al. [16] introduce the idea of real-time and monitoring of the degrading process of biopolymers using capacitive field-effect sensors comprising electrolyte-insulator-semiconductor (EIS). Their research findings demonstrated the practicality of capacitive field-effect sensors for evaluating polymer biodegradation under the influence of pH and enzymes. However, the research was deemed unsuitable for examining the degradation of bulk polymers due to the enzymes' size, which hindered their ability to penetrate the polymer. Hence, for future research, they suggest the formulation of a more detailed model of the sensor for a precise description of the degradation process.

Inductance Testing Analysis. Salpavaara et al. [17] proposed a technique for detecting biodegradable polymer changes during hydrolysis using an inductively coupled resonance sensor. The sensor, equipped using capacitive sensing elements digital, was utilized to track modifications in two biodegradable poly(lactide-co-glycolide)s (PLGs) over an 8-week testing period. However, this approach solely offers a qualitative evaluation of the biodegradability of the tested polymers, lacking quantitative data like the biodegradation rate, which cannot be deduced. Hence, for future research, it was suggested in this paper that the sensor encapsulation techniques should be improved.

Mechanical Testing Analysis. Karthiani et al. [18] signified the development and characterization of biodegradable algae-based bioplastics based on alginate derived from brown seaweeds of *Sargassum* sp. One of the crucial variables during the research is the mechanical testing of synthesized bioplastics to determine tensile strength and elongation at break. The results demonstrated that bioplastics mixed with inverted sugar improved the properties of the bioplastics by making them more flexible, as opposed to the control bioplastics, which were fragile. Then, with good mechanical properties, a soil burial test was performed, resulting in bioplastics with 6% alginate and 5% inverted sugar having a rapid degradation within 4 days.

20.2.3 Color-Sensing Analysis

Due to the lack of research regarding color-based testing analysis for biodegradability, this section explains the color-sensing approach, which will detect the color of the prospective object. Seelye et al. [19] developed an automated technique for determining the color of plant leaves using low-cost color sensors. The research focused on calibrating and validating the TCS3200 sensor to provide accurate plant leaf red, green, and blue (RGB) readings. Integrating the sensor with a proximity sensor and a robotic arm enabled autonomous and fixed-height color measurements. The study, however, was limited to calibrating the sensor for a particular range of green-yellow colors, focusing just on plant leaves. Future research opportunities include broadening the sensor's applicability, combining it with other sensors, and developing image-processing approaches to optimize plant development based on color information.

Mogi et al. [20] presented a study focused on developing a method to detect insertion errors in electronic boards by utilizing the hue, saturation, and value (HSV) color format. The notion of using HSV format instead of RGB format is due to its benefits in identifying color, simplicity of conversion, and resemblance to human color perception. This paper described extracting regions based on hue values and detecting regions using borderline tracing. Experimental results demonstrated the effectiveness of the HSV-based recognition method compared to RGB. However, the limitations showed the computational complexity and sensitivity to lighting conditions. These outcomes acknowledge improvement areas, such as utilizing machine learning and Bayesian detection theory for more intelligent and efficient recognition.

Feng et al. [21] proposed a novel detection for identifying rare-colored capsules (RCC) mixed with regular-colored capsules in pharmaceutical manufacturing. The method utilizes RGB and HSV color spaces to detect accurately and efficiently. The histogram-based approach to extracting RGB values speeds up detection while reducing data processing. Based on the results, the proposed method outperforms the conventional HSV algorithms in terms of time and accuracy. The method's effectiveness may be affected by lighting and camera setting variations, necessitating further research against image noise and variances.

Malkurthi et al. [22] developed an automated color detection system to analyze liquid reagents qualitatively. The system was designed to be simple and cost-effective, utilizing a light-emitting diode (LED) and a light-dependent resistor (LDR), which was compared to a camera-based system (ESP32-CAM, OV2640 camera). As a result, the LED-LDR system outperformed the camera-based module, with a maximum error that was less than 8%. However, the LDR is subject to influence from outside light, necessitating an opaque enclosure. They suggested fabricating a single substrate, LED-LDR pair, for future studies to produce a small real-time color detection on a monolithic system-on-chip.

A trichromatic-color-sensing (TCS) metasurface featuring reprogrammable electromagnetic functions was proposed by Chen et al. [23]. The design incorporated a photodiode sensor (TCS3200) to recognize the color of light in the environment and a one-bit programmable device to control the reflected phase response using the field-programmable gate array (FPGA) hardware system. This system allowed the detection and conversion of RGB colors into desired metasurface scattering patterns. However, the TCS metasurface may require considerable light intensity to activate the matching patterns, which might cause unwanted interference. Future research on integrating photodiode sensors with metasurfaces can be extended to sensor arrays for material characterization or strain sensing.

Panuganti et al. [24] proposed an embedded color segregation system using Arduino with a multi-rate sensor and color identification for various applications, such as waste management and fruit and vegetable packing. The sensor measurements were collected from the TCS34725 RGB sensor, which will detect and organize color according to its category. The significant benefits from the approach showed that the accuracy is above 95% with a response time of fewer than 3 ms. Despite this, the research did not specify the system's effectiveness under difficult lighting conditions or the categories of items that can be effectively categorized based on color. Future studies might improve the color classification system to handle a greater variety of colors.

Therefore, the applications of color detection for determining bioplastics and non-bioplastics in the prototype test kit will be described in the next section.

20.3 Test Kit Development

20.3.1 System Architecture

The color detection will be implemented using three main components: (i) microcontroller as the program control; (ii) color sensor; and (iii) LCD display to fulfill the bioplastic test kit requirements. Figure 20.1 shows the block diagram of the proposed project. The color sensor will detect the color of the solution formed from mixing the reagent with the test substrate (bioplastic or non-bioplastic material). Next, the microcontroller algorithm will process the sensed color to detect the present color. An external power supply will power the microcontroller. After processing, the information in bioplastic or non-bioplastic will be displayed on the LCD screen.

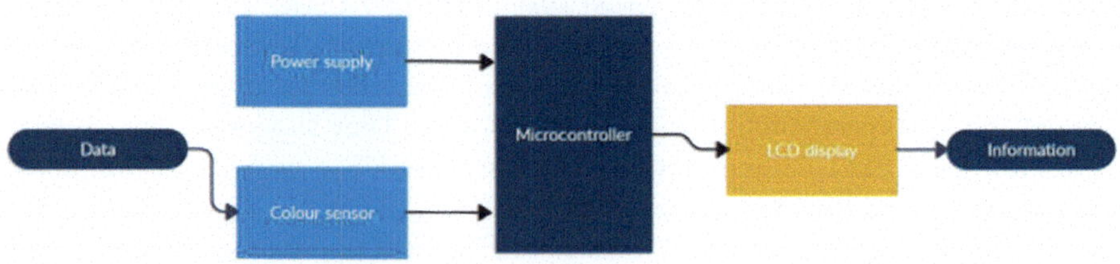

Fig. 20.1 Block diagram of the system

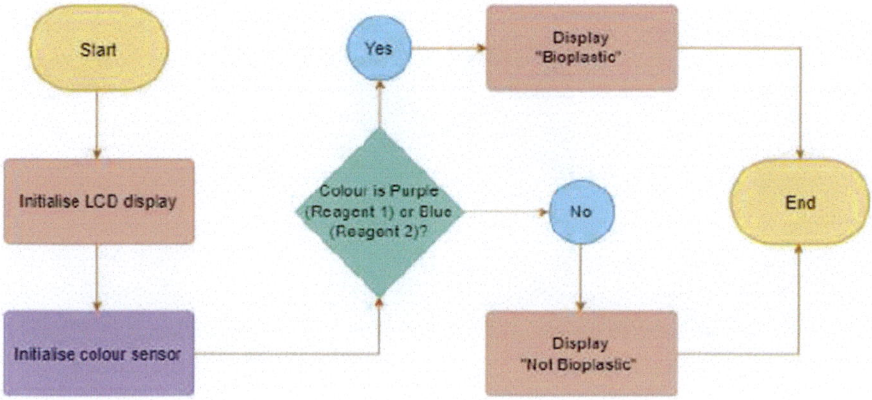

Fig. 20.2 Flowchart of the system

The flowchart in Fig. 20.2 represents the more detailed processing of the block diagram. The LCD and the color sensor will be initially initialized. Once the reagent has been poured onto the test material, color change in the form of purple (Reagent 1—iodine) or blue color (Reagent 2—iodine-$CaCl_2$ and Schultze solution) will be detected and if these colors are present, it will indicate as "bioplastics" else, it is "not bioplastics" on the LCD. Each of the components will be explained in detail in the following subsections.

20.3.2 Test Samples

Cellulose-Based Bioplastic. A laboratory-made bioplastics shown in Fig. 20.3 was produced using tapioca starch and contained 20 and 30 g of cellulose extracted from palm empty fruit, respectively, were used as test samples.

Starch-Based Bioplastic. Two types of starch-based plastics were used, namely, the bio singlet bags from Dunkin' Donuts and straws called RiceStraws manufactured by NLYTech Biotech Penang. The bio singlet bag (Fig. 20.4a) is made from starch mixes such as cornstarch, tapioca, and sweet potato. Meanwhile, the straw (Fig. 20.4b) is made using rice flour (63%) and tapioca starch.

Biodegradable Plastic. Figure 20.5a shows the biodegradable garbage bag found in any grocery store. The ingredients to make it are unknown.

Conventional Plastic. The conventional plastic used in this experiment is regular plastic found everywhere and used daily, as shown in Fig. 20.5b.

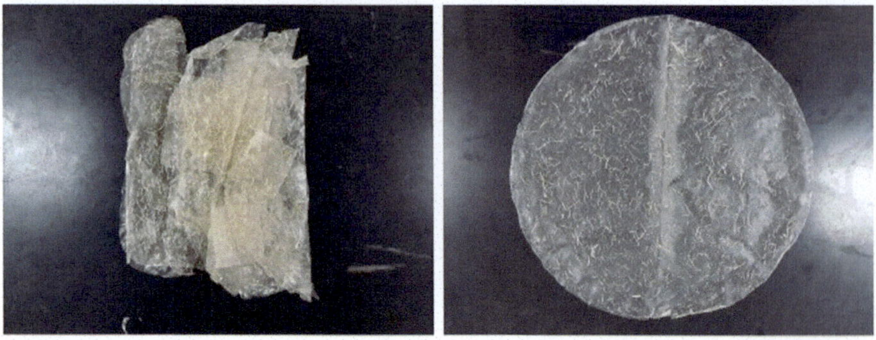

(a) 20g Cellulose Bioplastics (b) 30g Cellulose Bioplastics

Fig. 20.3 Cellulose-based bioplastics

(a) Bio singlet Bag (b) RiceStraws

Fig. 20.4 Starch-based bioplastics

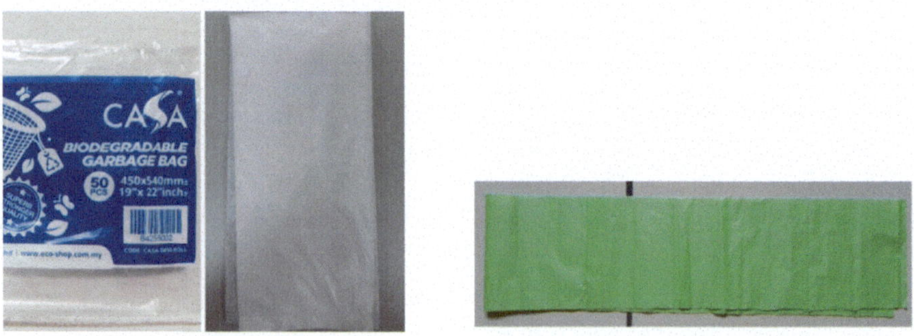

(a) Biodegradable Plastic (b) Conventional Plastic

Fig. 20.5 Biodegradable and conventional plastic samples

Table 20.1 Color reaction after exposure to reagents

Samples		Reagents		
		Iodine	Iodine-$CaCl_2$	Schultze
Cellulose-based	10 g	Dark purple	Dark blue	Dark blue
	20 g	Dark purple	Dark blue	Dark blue
Starch-based	Singlet bag (corn)	Brown → purple	Slightly brown	Brown → purple
	Straws (rice)	Dark purple	Dark blue	Dark blue
Biodegradable		No change	No change	No change
Conventional		No change	No change	No change

20.4 Experimental Testing

The experiments were conducted through the testing of four types of samples, namely, cellulose-based, starch-based, biodegradable, and conventional plastic. These samples have been outlined in the previous section. The reagents applied to the samples included iodine, iodine-$CaCl_2$, and Schultze reagents. Table 20.1 shows the results of the experimental testing of the test kit.

Table 20.1 shows the color reaction of each sample after being exposed to the reagents. The cellulose-based samples (both 10 and 20 g) and starch-based samples (straws) exhibit a "dark purple" color reaction after being exposed to iodine solution. Meanwhile, the starch-based sample in a singlet bag shows a "brown" color and then changes to "purple" after a brief time.

As for the iodine-$CaCl_2$ and Schultze reagents, the reactions show a "dark blue" color reaction for the cellulose-based (both 10 and 20 g) and starch-based (straws). However, the starch-based (singlet bag) reaction differed for both reagents. The iodine-$CaCl_2$ shows a "slightly brown" while the Schultze shows the change from "brown to purple" color reactions.

The reaction for the biodegradable and conventional plastics shows "no change" in color for all the reagents. The reaction of the biodegradable garbage bag supports the statement that not all biodegradable bioplastics are derived from bio-based sources.

20.5 Conclusions

A prototype test kit using Arduino and color sensors to differentiate which plastic is biodegradable, bio-based, or both, based on color reactions. The experiments demonstrated that chemical reagent-induced color changes provided a method for distinguishing between cellulose-based, starch-based, biodegradable, and conventional plastics.

However, difficulties in color sensor calibration were encountered, resulting in RGB value inaccuracies. Another limitation of the detector is that it still indicates other blue or purple conventional plastics as bioplastics if they fulfill the RGB range values. The prototype has the potential to identify plastic types, but further refinement and research are required to improve sensor calibration and accuracy. This study demonstrates the potential of using technology to address plastic identification issues, paving the way for potential applications in waste management and environmental sustainability.

For improvement, several key areas can be considered to enhance the effectiveness and reliability of the developed test kit. Firstly, in terms of accuracy, it is needed to tackle the calibration challenges faced during the project. It is to ensure the consistency and accuracy of RGB readings. Exploring alternative calibration methods and sensor configurations could be done to mitigate inaccuracies. Secondly, consider incorporating a camera module and utilizing the HSV color space for color detection. By integrating a camera, the system can capture images of plastic samples and perform color analysis based on the HSV color model. This method has advantages such as higher color resolution, lower lighting sensitivity, and better color discrimination.

Acknowledgements This work was financially supported by the Ministry of Higher Education Malaysia through the Fundamental Research Grant Scheme (Number FRGS/1/2019/TK10/UMS/02/3) and the Universiti Malaysia Sabah through the Niche Fund Scheme (Number SDN0071-2019).

References

1. Karana, E.: Characterization of 'natural' and 'high-quality' materials to improve perception of bio-plastics. J. Clean. Prod. **37**, 316–325 (2012)
2. European Bioplastics: Bioplastics market development update (2022). https://docs.european-bioplastics.org/publications/market_data/2022/Report_Bioplastics_Market_Data_2022_short_version.pdf. Last accessed 03 July 2023
3. Karan, H., Funk, C., Grabert, M., Oey, M., Hankamer, B.: Green bioplastics as part of a circular bioeconomy. Trends Plant Sci. **24**(3), 237–249 (2019)
4. Burgos, N., Valdés, A., Jiménez, A.: Valorization of agricultural wastes for the production of protein-based biopolymers. J. Renew. Mater. **4**(3), 165–177 (2016)
5. Abidin, N.D.Z., Azhar, N.S., Sarip, M.N., Hamid, H.A., Nasir, N.A.H.A.: Production of bioplastic from cassava peel with different concentrations of glycerol and $CaCO_3$ as filler. In: AIP Conference Proceedings, vol. 2332(1). AIP Publishing (2021)
6. Verma, D., Fortunati, E.: Biobased and biodegradable plastics. In: Handbook of Ecomaterials, vo. 4, pp. 2955–2976 (2019)
7. Álvarez-Chávez, C.R., Edwards, S., Moure-Eraso, R., Geiser, K.: Sustainability of bio-based plastics: general comparative analysis and recommendations for improvement. J. Clean. Prod. **23**(1), 47–56 (2012)
8. Liu, W., Misra, M., Askeland, P., Drzal, L.T., Mohanty, A.K.: 'Green' composites from soy based plastic and pineapple leaf fiber: fabrication and properties evaluation. Polymer **46**(8), 2710–2721 (2005)
9. Ren, X.: Biodegradable plastics: a solution or a challenge? J. Clean. Prod. **11**(1), 27–40 (2003)
10. World Economic Forum: Plastic is a global problem. It's also a global opportunity (2019, January 25). https://www.weforum.org/agenda/2019/01/plastic-might-just-be-the-solution-to-its-own-problem/. Last accessed 03 July 2023

11. Koh, L.M., Khor, S.M.: Current state and future prospects of sensors for evaluating polymer biodegradability and sensors made from biodegradable polymers: a review. Anal. Chim. Acta **1217**, 339989 (2022)
12. Chowdhury, M.A., Badrudduza, M.D., Hossain, N., Rana, M.M.: Development and characterization of natural sourced bioplastic synthesized from tamarind seeds, berry seeds and licorice root. Appl. Surf. Sci. Adv. **11**, 100313 (2022)
13. Ashter, S.A.: Introduction to Bioplastics Engineering. William Andrew (2016)
14. Pucci, A., Bertoldo, M., Bronco, S.: Luminescent bis(benzoxazolyl)stilbene as a molecular probe for poly(propylene) film deformation. Macromol. Rapid Commun. **26**(13), 1043–1048 (2005)
15. Pucci, A., Di Cuia, F., Signori, F., Ruggeri, G.: Bis(benzoxazolyl)stilbene excimers as temperature and deformation sensors for biodegradable poly(1,4-butylene succinate) films. J. Mater. Chem. **17**(8), 783–790 (2007)
16. Schusser, S., Poghossian, A., Bäcker, M., Leinhos, M., Wagner, P., Schöning, M.J.: Characterization of biodegradable polymers with capacitive field-effect sensors. Sens. Actuators, B Chem. **187**, 2–7 (2013)
17. Salpavaara, T., Hänninen, A., Antniemi, A., Lekkala, J., Kellomäki, M.: Non-destructive and wireless monitoring of biodegradable polymers. Sens. Actuators, B Chem. **251**, 1018–1025 (2017)
18. Kanagesan, K., Abdulla, R., Derman, E., Sabullah, M.K., Govindan, N., Gansau, J.A.: A sustainable approach to green algal bioplastics production from brown seaweeds of Sabah, Malaysia. J. King Saud Univ.-Sci. **34**(7), 102268 (2022)
19. Seelye, M., Gupta, G.S., Bailey, D., Seelye, J.: Low cost colour sensors for monitoring plant growth in a laboratory. In: 2011 IEEE International Instrumentation and Measurement Technology Conference Proceedings, Hangzhou, China, pp. 1–6 (2011)
20. Mogi, T., Namekawa, M., Ueda, Y.: Detection of insertion error parts in electronic boards using HSV color format. In: 12th International Conference on Intelligent Systems Design and Applications (ISDA) Proceedings, pp. 357–362. IEEE (2012)
21. Feng, L., Xiaoyu, L., Yi, C.: An efficient detection method for rare colored capsule based on RGB and HSV color space. In: IEEE International Conference on Granular Computing (GrC) Proceedings, pp. 175–178. IEEE (2014)
22. Malkurthi, S., Yellakonda, K.V.R., Tiwari, A., Hussain, A.M.: Low-cost color sensor for automating analytical chemistry processes. In: IEEE Sensors Proceedings, pp. 1–4. IEEE (2021)
23. Chen, L., Ye, F.J., Ruan, Y., Cuo, M., Luo, S.S., Cui, H.Y.: Trichromatic-color-sensing metasurface with reprogrammable electromagnetic functions. Opt. Mater. **123**, 111892 (2022)
24. Panuganti, S.L., Gajula, N.H., Rathnala, P., Patnaik, M.P.K., Sura, S.R.: Embedded color segregation using Arduino. In: Sixth International Conference on I-SMAC (IoT in Social, Mobile, Analytics and Cloud) (I-SMAC) proceedings, pp. 832–838. IEEE (2022)

Chapter 21
Applications of Artificial Intelligence in Biosensors

Behnaz Shirgir, Kamil Dimililer, and Suleyman Asir

Abstract Healthcare, constituting approximately 10% of the global GDP, faces transformative opportunities through technological innovations, particularly data-driven methods and computational power. Biosensors, capable of detecting biological and chemical analytes, have gained attention in academic circles. However, conventional biosensors have limitations, which can be overcome by integrating machine learning (ML) algorithms. Electrochemical biosensors, in particular, have evolved from traditional to wearable forms, offering applications in healthcare, disease diagnosis, and environmental monitoring. AI-enhanced biosensors, known as AI-biosensors, enable early disease diagnosis, bacteria detection, and hazardous compound measurement. Machine learning optimizes biosensor performance and enhances parameters like electrode design and analyte selection. The use of machine learning in biosensors has witnessed significant growth, facilitating efficient data analysis. This fusion of artificial intelligence with biosensors bridges the gap between data acquisition and analysis, resulting in enhanced diagnostic precision. This evaluation reviews the use of ML in conjunction with non-invasive biosensors, highlighting their potential to revolutionize healthcare and medicine. Future research directions and opportunities are also discussed.

Keywords Biosensor · Artificial intelligence · Nano-biosensor · AI-biosensor · Machine learning

B. Shirgir (✉) · S. Asir
Biomedical Engineering, Research Center for Science, Technology and Engineering (BILTEM), Near East University, Via Mersin 10, Nicosia, N. Cyprus, Turkey
e-mail: 20233395@std.neu.edu.tr

S. Asir
e-mail: suleyman.asir@neu.edu.tr

K. Dimililer
Electrical and Electronic Engineering, Research Center for Science, Technology and Engineering (BILTEM), Applied Artificial Intelligence Research Centre (AAIRC), Near East University, Via Mersin 10, Nicosia, N. Cyprus, Turkey
e-mail: kamil.dimililer@neu.edu.tr

S. K. Pal et al. (eds.), *Intelligent Informatics*, Smart Innovation, Systems and Technologies 389, https://doi.org/10.1007/978-981-97-2147-4_21

21.1 Introduction

Approximately 10% of the worldwide gross domestic product (GDP), which amounts to 10 trillion USD, is allocated to the domain of healthcare per year [1]. The latest developments in technological innovations, particularly methods driven by data and the computational capacity for processing, have the potential to bring immense advantages to both patients and the medical field, while also diminishing the substantial costs [1]. Lately, there has been a growing academic interest in biosensors. Biosensors are sophisticated analytical mechanisms capable of detecting and measuring the existence of various biological and chemical analytes [2]. Biological diagnostics are frequently employed in clinical settings as an essential component of disease diagnosis. However, conventional analytical techniques possess certain drawbacks such as their time-consuming nature, expensive nature, and their high demands on laboratory researchers. These drawbacks underscore the significance of developing a novel, expeditious, cost-effective, and readily available test in the clinical realm [3]. A biosensor can be described as a concise analytical equipment that encompasses a biological or biologically derived sensing element, which is either integrated within or closely affiliated with a physicochemical transducer [4]. The initial biosensor was devised by the American biochemist L.L Clark in the year 1950, while the designation "biosensor" was initially proposed by Cammann in 1977 [2]. Biosensors possess multifarious utilities encompassing clinical diagnosis, environmental surveillance, food safety, and the exploration of pharmaceutical agents. The simple design of biosensor instruments generates significant fascination in numerous applications owing to their effortless functionality, portability, swift identification, commendable specificity, perceptivity, replicability, cost-effectiveness, and so forth [5]. It is widely acknowledged that biosensors are categorized into two primary groups, which are based on the transducer and nature of bio receptors. To illustrate, if a biosensor device incorporates an optical transducer, it is classified as an optical sensor. Similarly, when electrical transducers are utilized, the biosensor is referred to as an electrochemical biosensor. As one of the most crucial classifications among them, biosensors have witnessed a long growth from typical electrochemical biosensors to wearable and implantable biosensors, and have been universally appealed in food security, healthcare, disease diagnosis, environmental monitoring, and biosafety [6]. Biosensors present a promising and potentially transformative technology in the field of medicine, like glucose biosensors have traditionally been focused on the regulation of diabetes [2]. The significance of biosensors persists as they enable individuals to manage their blood sugar levels effectively and ascertain ecological responses to the disease. In recent times, computers have become widespread in all areas of science and technology. The combination of AI and biosensors has produced a groundbreaking development in biosensors referred to as AI-biosensors. Machine learning (ML), an essential component of AI, has emerged as a useless tool for data analysis and processing in the realm of materials science [7]. Machine learning holds great significance in the field of computer programming and has been extensively utilized for monitoring and refining diverse data outputs. As complex structures and entities are sensed, large data are obtained, and it becomes difficult to manually interpret all the data. Machine learning aids in the interpretation of large sensing data and has found applications in various domains. Furthermore, the collected data often suffer from noise and disturbances, making them unclear. Machine learning can effectively analyze such data and derive the desired outcomes [8]. The purpose of this evaluation is to bridge a necessary gap in the literature regarding the utilization of Machine Learning in combination with biosensors in the medical field and to provide a comprehensive review of recently published articles. These types of evaluations are imperative for the advancement of the field. In this Review Article, we will focus on the application of biosensors and the role that Artificial intelligence has played in the diagnosis of diseases. Particular attention is given to the enhancement of biosensors which has a vital role in the realm of digital healthcare. Lastly, we propose challenges and prospects to elucidate the gaps in existing research and necessary improvements.

21.2 Biosensor

A biosensor serves as a refined instrument utilized for the precise determination of analytes in a living sample. This remarkable device possesses a delicate nature and boasts a unique measurement configuration. As it is specifically employed for the detection of living samples, it comprises living components such as micro-organisms, organelles, cell receptors, enzymes, or nucleic acids. Due to its interaction with the sample, a signal is generated, which could potentially appear as electric, optical, or thermal in nature. These signals are subsequently subjected to transformation via a transducer, ultimately yielding a measurable parameter [9]. Biosensing in the context of patient monitoring is an innovative field that integrates the disciplines of biology, sensing technologies, and data analysis to consistently or periodically evaluate the health parameters of patients [10]. It does not involve the utilization of biosensors, frequently in the form of wearable or implantable devices, to collect real-time physiological data, enabling healthcare professionals to monitor patient health, identify irregularities, and make informed clinical judgments [11, 12]. The implementation of biosensing in patient monitoring represents a significant advancement over conventional approaches. By implementing biosensors, healthcare providers can gather precise and dynamic measurements of vital signs, biomarkers, and other physiological parameters [13, 14]. This comprehensive comprehension of a patient's health condition permits the early detection of any alterations or deviations from standard values [2, 15]. Continuous surveillance is particularly useful in the management of chronic illnesses as it provides valuable insights into disease progression, response to treatments, and overall patient well-being [2, 13, 16]. Biosensors engaged in the monitoring of patients have the ability to evaluate a variety of critical physiological indicators, including heart rate, blood pressure, body temperature, respiratory rate, oxygen saturation, and various others [17]. In addition, they possess the capability to detect specific biomarkers, such as glucose, electrolytes, hormones, and enzymes, which are essential for managing conditions like diabetes, cardiac disorders, kidney diseases, and infectious diseases [18, 19]. The information gathered by the biosensors is of immense value to healthcare experts. By employing basic algorithms, data analysis, and traditional statistical methodologies, it is impossible to unveil patterns, trends, and irregularities in the data, which in turn can provide insights into a patient's state of health [20, 21]. By incorporating data collected via biosensors with electronic health records and other clinical data, it is possible to establish a comprehensive profile of the patient, allowing for the delivery of personalized medicine and customized treatment plans [22, 23]. Ultimately, this could result in improved patient outcomes and a general advancement in healthcare. The advantages of employing biosensing for patient monitoring are manifold. It permits the identification of deteriorating health conditions at an early stage, facilitates timely interventions, and enables personalized care. Moreover, it boosts patient security, cuts down healthcare expenses, and contributes to enhanced patient results [24, 25]. Biosensing technology is at the forefront of proactive and preventive healthcare by enabling continuous monitoring, which can recognize subtle changes that may indicate the onset of a medical issue before symptoms become apparent [26, 27]. As biosensing technologies advance, patient monitoring will become more sophisticated, convenient, and seamlessly integrated into daily routines [28]. With smaller dimensions, improved precision, and wireless connectivity, biosensors are positioned to play a pivotal role in the future of healthcare. They will empower patients to take control of their own health and enable healthcare providers to deliver tailored, data-driven care [14, 27].

21.3 Types of Biosensors

Biosensors have the ability to be categorized based on either their biological component or their transduction component. The biological components encompass enzymes, antibodies, micro-organisms, biological tissue, and organelles. The manner in which transduction occurs is contingent upon the specific type of physicochemical alteration that arises from the sensing event. Remarkably, biosensors predominantly employ transducer elements that are mass-based (such as piezoelectric mechanisms), electrochemical in nature (potentiometric or amperometric), or have an optical design (such as fiber optics) [29]. The usage of biosensors offers various benefits, including their economical nature, compact size, convenient and effortless operation, and their enhanced sensitivity and selectivity compared to current instruments. Biosensors find manifold applications in clinical analysis and the monitoring of general health care. The most widely recognized instance is the deployment of glucose oxidase-based sensors by individuals afflicted with diabetes to assess glucose levels in their blood. Biosensors have discovered prospective implementations in the realm of industrial processing and monitoring, environmental pollution control, as well as the agricultural and food industries [29]. In recent times, researchers in the medical and biotechnology fields have directed their attention toward diagnostics and the understanding of diseases, with a specific focus on the identification of biomarkers at low concentrations. This accentuation has resulted in noteworthy progress in both basic and applied science, where biomarkers play a vital role in the diagnosis, prognosis, and development of diseases. The ability to identify biomarkers with high sensitivity, precision, quantification, and multiplexing relies heavily on advancements in diagnostic technologies. Within medical experiments, biosensors are utilized to swiftly detect diseases, including cancer. Electrochemical biosensors, in particular, enable faster and more accurate diagnosis. These biosensors employ metal-specific electrodes to detect the presence of harmful metal concentrations in water, as well as to identify dangerous pathogens and bio-recognition components such as enzymes, antibodies, or biomolecules. Optical transducer biosensors have the ability to record changes in light during biological interactions, making them indispensable in the detection of molecules of interest. The applications of biosensors span across various areas of healthcare, including the management of diabetes and the detection of cardiac troponin for diagnosing heart muscle injuries. These sensors are renowned for their rapid processing capabilities and are commonly employed for point-of-care monitoring in portable devices, encompassing glucose monitoring, addiction tracking, and pregnancy tests. Implantable biosensors hold promise in revolutionizing patient care and the production of personalized drugs, although they do encounter technical challenges in the validation of biomarkers. Advancements in biotechnology have given rise to compact and cost-effective medical diagnostic instruments that allow for at-home health monitoring, which has proven to be particularly advantageous for patients with diabetes. This trend has fueled an increased interest in patient-centered healthcare and the development of diagnostic tools based on biosensors. Energy harvesting technologies are pivotal in the next generation of biosensors, as they offer sensors that are both cost-effective and easily producible, thereby reducing the risks associated with contamination and sterilization costs in medical devices. Flexible sensors, due to their compatibility with the natural curvature of the body, are highly favored as they enhance patient comfort [30] (Fig. 21.1).

21.3.1 Electrochemical Biosensor

The electrochemical biosensors (EC) are widely utilized in medical diagnosis [31, 32], making them one of the most common biosensors. Electrochemical sensing and biosensing systems offer extraordinary capabilities for detecting multiple analytes in complex samples, such as serum and clinical specimens. These systems demonstrate high selectivity and sensitivity. In this sensing platform, bio/electrochemical events occur at the interface of the electrochemical transducer surface, specifically the working electrode surface. Moreover, the electrochemical signals generated in this process allow for the screening and determination of the selective binding affinity and catalytic activity between an analyte and a fixed or immobilized bio-receptor. The primary electrochemical approaches employed for constructing, adjusting, and optimizing electrochemical sensors and biosensors include potentiometric, amperometric, conductometric, impedimetric, and voltammetric techniques [33]. However, the utilization of novel machine learning (ML) procedures in current EC biosensors

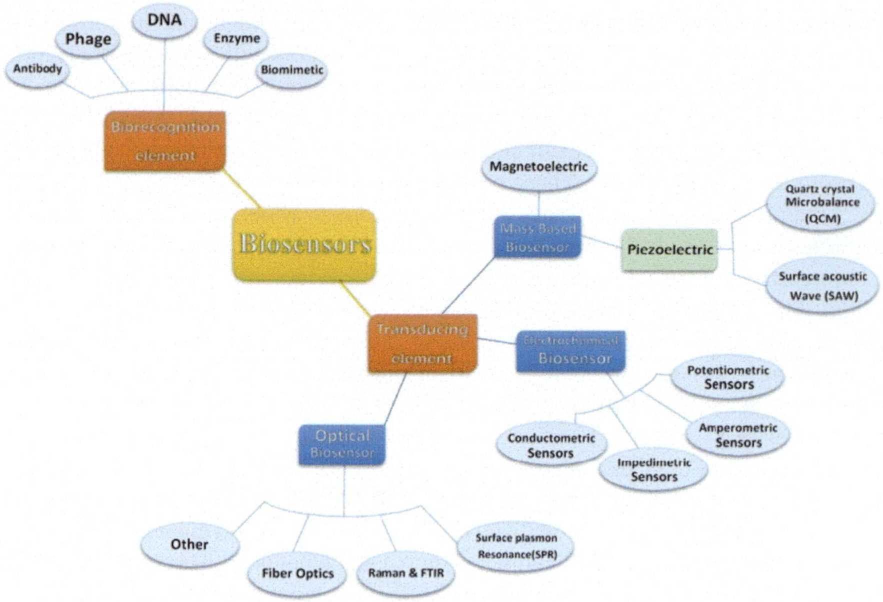

Fig. 21.1 The categorization of biosensors can be determined by the component used for transduction as well as the element responsible for bio-recognition

is still in its early stages. Another factor to consider is that the electrode used in EC biosensors deteriorates over time. As a result, one-dimensional data analysis is insufficient for acquiring sensitive signals that are closely related to the type and quantity of analytes. This is where the chance of merging ML with electrochemical biosensors comes into play. By doing so, it becomes possible to explore how ML can enhance sensor accuracy and reliability in real sample estimations [34, 35]. Notably, Massah and colleagues employed an SVM regression model to improve the functionality of a portable EC biosensor based on cyclic voltammetry [36]. The use of the SVM regression model extended the useful life of the EC biosensor, allowing it to continue functioning for 10 d after the enzyme has been immobilized. Combining single-molecule (SM) identification with ML can significantly enhance accuracy and precision. Furthermore, this combination allows for the optimization of design constraints of electrical biosensors and the quantitative assessment of molecule recognition capabilities [37]. SVM electrical detecting techniques can be divided into two categories: nanopore and nanogap. These techniques are commonly employed for viral detection, peptide sequencing, DNA analysis, carbohydrate analysis, and RNA analysis [38–40]. However, when analytes have similar molecule volumes and border orbital energies, the maximum current (Ip) and current duration (td) signals tend to be similar. This overlap of current signals makes it difficult to detect and identify multiple analytes. To overcome this challenge, ML techniques like SVM,

RF, and CNN are applied to analyze the current-time waveform [41]. Kawai and colleagues demonstrated the effectiveness of the Rotation Forest model in distinguishing between electrical signals and similar microbiological forms [42]. Additionally, this model allows for the recovery of subtle features for distinguishing bacterial species, such as E. coli and Bacillus subtilis, based on the width td and height Ip of the ongoing waveform.

21.3.2 Nano Biosensor

Nano-biosensors are exquisite devices utilized to observe, measure, and analyze the intricate happenings of the biological world. These sensors are meticulously crafted using the artistry of nanotechnology, often employing the brilliance of quantum dots, nanoparticles, nanowires, and nanofilms. Considering that a multitude of biological processes occur at the minute nanoscale, nano-biosensors confer upon us a vast array of benefits in reference to the comparative performance metrics, the novel biosensing approach demonstrates enhanced sensitivity and specificity, alongside a reduction in the duration required to elicit responses, relative to conventional biosensing methodologies. They find themselves woven into the fabric of enzymatic glucose detection, captivating fluorescence-based binding detection through the utilization of quantum dots, and the exquisite realm of specific biomolecular detection with bio-conjugated nanomaterials [43]. At the core of these nano-biosensors lies the essential components, these being nanomaterials of the utmost brilliance such as quantum dots, nanoparticles, and nanotubes. These materials dance harmoniously, amplifying the electrochemical signals that emanate from the biocatalytic events transpiring at the electrode/electrolyte interface. Various methods of synthesis are employed, ranging from the finesse of physical techniques like ball mixing and physical vapor deposition to the enchantment of chemical methods like the sol-gel process. Nanostructures are brought to life through the synergy of both bottom-up and top-down approaches, an artful dance of creation. Furthermore, this passage beckons us to acknowledge the paramount importance of addressing safety concerns that enshroud the world of nanomaterials prior to their widespread adoption in various applications [43]. The realm of nano-biosensors encompasses a vast spectrum of research arenas, including the ethereal pursuit of air monitoring, the noble quest for point-of-care diagnosis, and the pioneering expedition toward the development of advanced personal protective equipment (PPE) to bravely combat the perils of infectious diseases such as the formidable COVID-19. In essence, this passage weaves together a tapestry that grants us a comprehensive overview of the principles, materials, and applications that reside within the realm of nano-biosensors. It gracefully accentuates their profound significance in various fields of research, while also reminding us of the pressing need to address the intricate safety issues that intertwine with the mesmerizing world of nanomaterials [43].

21.3.3 Wearable Biosensor

The present passage examines the growing interest in wearable biosensors, motivated by increasing consumer demand for health and fitness awareness and advancements in material technology. Wearable biosensors are garnering attention due to their provision of minimally/non-invasive monitoring, continuous tracking, and ease of portability and use. They have witnessed a remarkable surge in popularity, with the U.S. market expected to surpass tens of billions of dollars by 2022. While wearable technology has been accessible for monitoring heartbeats and biophysical activity, the identification of individual biophysical and biochemical markers necessitates further exploration. Hence, efforts have been focused on developing wearable biosensors that can detect biomarkers in easily accessible biofluids like sweat, saliva, tears, interstitial fluid (ISF), and wound fluids without any invasive procedures. The passage notes that the catalog of biomarkers found in biofluids is extensive and can be found in comprehensive literature reviews. As the Internet of Things (IoT) and big data concepts transition from theoretical frameworks to practical implementation, wearable biosensors find themselves presented with novel opportunities for advancement, fueled by their distinct technical capabilities [3]. Various categories of on-body biosensors and their applications for the monitoring of diverse biomarkers and health-related parameters can be identified. Tattoo Sensors: The primary focus of utilizing tattoo sensors is to detect biomarkers present in sweat and interstitial fluid (ISF). These sensors are not expected to detect substances such as glucose, lactate, and electrolyte ions. Furthermore, they are able to stimulate the secretion of sweat and extract ISF for simultaneous monitoring. Moreover, they are employed for the measurement of hydration levels. Contact Lenses: The presence of contact lenses provides a non-invasive platform for tear sensors. These lenses have been employed in the continuous monitoring of glucose levels, wherein optical sensors are integrated into commercially available contact lenses. The introduction of laser-inscribed contact lenses has facilitated the sensing of tear pH, glucose, protein, and nitrite ions. Another notable application of contact lenses involves the monitoring of tear glucose utilizing various mechanisms. Skin Patch Biosensors: The utilization of skin patch biosensors permits the real-time tracking of biomarkers present in sweat. These biosensors have been specifically designed for the analysis of glucose in sweat and may also incorporate pH and temperature correction mechanisms. Clothing and Textile-Based Biosensors: These biosensors are seamlessly integrated into wearable textiles and serve the purpose of monitoring wound fluid, sweat, and ambient air. One noteworthy example is the employment of smart bandages, which possess the capability to monitor wound status and prevent infection. Furthermore, omniphobic paper-based smart bandages have been developed for the detection of wound status and pressure ulcers. Fiber-Based Biosensors: Fiber-based biosensors are skillfully woven into textiles, thereby offering a comfortable wearing experience. These biosensors have been successfully demonstrated for sweat glucose monitoring, utilizing gold fibers for electrochemical biosensing. Accessories: Accessories represent wearable biosensors that are loosely attached to clothing or the body. An exemplary instance of this is eyeglasses-based biosensing, in which metabolite and electrolyte sensors are integrated into nose-bridge sticker pads. These glasses are capable of monitoring sweat electrolytes and metabolites in real time and allow for multianalyte sensing through the alteration of the sticker pads. These on-body biosensors cater to a wide range of applications, altering from health monitoring to the assessment of biomarkers in various bodily fluids. Besides, they offer non-invasive and ongoing tracking capabilities [7]. The convergence of integrated wearable sensors, AI, big data, and IoT has led to a notable opportunity in the form of AI biosensors. This emergent trend is expected to generate a heightened interest in the creation of wearable AI-biosensor networks (WAIBN), enabling swift and cost-effective access to comprehensive data on human subjects. In China, there exists a focus on the development of smart health electronic products and mobile application services for healthcare, where the gradual enhancement of the AI technology system, coupled with the integration of AI and IoT, will catalyze the rapid advancement of intelligent biosensors [6].

21.4 Artificial Intelligence

The incorporation of artificial intelligence (AI) into the realms of healthcare and biomedical research is progressively transforming these fields. This alteration is exemplified by the collaborative endeavors between ophthalmologists and computer scientists at the Aravind Eye Care System in India. At this institution, an automated system for classifying images is being employed to screen millions of retinal photographs of diabetic patients. Diabetic retinopathy, a condition that affects over 90 million individuals worldwide and is a major cause of blindness in adults, can be effectively monitored and diagnosed through the utilization of fundus photography. However, the scarcity of ophthalmologists in many regions impedes comprehensive screening efforts. It is worth mentioning that AI systems created by teams at Google Inc. and the University of Iowa have demonstrated the capacity to achieve diagnostic accuracy for diabetic retinopathy that is on par with that of physicians. Moreover, these systems have unveiled new connections between retinal images and cardiovascular risk factors. Consequently, they have been seamlessly integrated into clinical practice in India and have been granted approval by the United States Food and Drug Administration (FDA). The recent resurgence of AI has been marked by significant advancements in image classification, which have suppressed innovation and applications in various domains. AI, which has its foundation in computer science, aims to comprehend and create intelligent entities, often in the form of software programs. Its historical development can be followed back to the Dartmouth conference in 1956, where the term was initially presented. However, the definition of "real AI" has evolved over time, making it challenging to track progress. Once considered advanced AI, technologies like automated route planners in the 1970s have now become ordinary, showcasing the ever-changing character of this field. Consequently, past achievements in AI, such as the automated interpretation of electrocardiograms, although valuable, no longer meet the standards of true AI in contemporary times. More recently, AI has expanded its reach into medical image diagnostics and has extended its influence into diverse medical domains, including clinical practice, translational medical research, and basic biomedical research. This expansion has transcended the limitations imposed by human expertise and continues to shape the future of AI applications in healthcare and biomedical sciences [44]. The utilization of artificial intelligence (AI) in the realm of medicine encompasses two main branches: the virtual aspect and the physical aspect. The virtual component involves the application of Machine Learning, particularly Deep Learning, which employs mathematical algorithms to enhance the process of acquiring knowledge through experience. Machine learning encompasses three distinct categories: unsupervised (deducing patterns), supervised (making classifications and predictions based on previous data), and reinforcement learning (making strategic decisions based on rewards and penalties). AI has made remarkable contributions to the fields of genetics and molecular medicine, as evidenced by the success of unsupervised protein-protein interaction algorithms in identifying potential therapeutic targets. Furthermore, AI assists in the identification of DNA variations for the purpose of predicting diseases, utilizing innovative evolutionary algorithms that mitigate the problem of overfitting [45].

21.4.1 AI in Biosensors

During the course of the previous decade, there has been a noticeable focus on the integration of intelligence in biosensors, particularly within the domain of AI biosensors. This interdisciplinary field comprises three fundamental components: information collection, signal conversion, and AI-data processing. Information collection involves the continuous monitoring of diverse data types using biosensors. Signal conversion translates this information into electrical output with well-defined sensitivity. AI-data processing encompasses various tasks, including interface design, data classification, data modeling, analysis, and decision-making [6]. AI, a discipline within computer science that amalgamates intelligence, has witnessed substantial growth, with projections indicating a fourfold increase in its trillion-dollar industry within three years. The upcoming progressions of AI are mainly concentrated on enhancing healthcare and quality of life. China's "Healthy China 2030" initiative, for example, underscores the significance of wearable devices, smart health electronic products, and mobile healthcare applications. As AI technology progresses, its integration with IoT is poised to drive the rapid advancement of intelligent biosensors [6].

21.4.2 Machine Learning

AI-biosensors possess value solely when their data is usable and understandable. The process of AI-data processing encompasses the acquisition of knowledge from data in order to appropriately analyze it, draw precise inferences, and identify misinterpretations. Machine learning plays a pivotal role in this procedure, with two primary classifications: supervised learning, which employs labeled data for the identification of patterns, and unsupervised learning, which addresses unlabeled data. AI-biosensors employ a variety of machine learning algorithms, such as Support Vector Machines (SVMs), Principal Component Analysis (PCA), Hierarchical Cluster Analysis (HCA), Artificial Neural Networks (ANNs), and Decision Trees (DTs), to facilitate effective and accurate decision-making. SVMs are utilized for pattern recognition, PCA reduces dimensionality, HCA clusters data, ANNs simulate intricate relationships, and DTs handle complex behaviors. Ensemble algorithms have the capability to enhance classification accuracy, although this may come at the cost of increased storage and computation requirements. The selection of the most appropriate machine learning algorithm is contingent upon the specific problem encountered. For instance, in the realm of diabetes management, companies are employing machine learning to continuously monitor blood glucose levels and provide healthcare recommendations. DreaMed Diabetes, for instance, utilizes event-driven, machine learning, and fuzzy logic technologies for data processing, while Bigfoot Biomedical focuses on automated insulin delivery systems utilizing Model-Based Design. Medilync employs machine learning to continuously monitor blood sugar levels. In conclusion, the application of machine learning in AI-data processing is crucial for endowing AI-biosensors with value and practicality in a multitude of healthcare and biosensing applications [6].

21.4.3 Machine Learning for Biosensors

The field of biosensors has observed significant and diverse advantages through the integration of machine learning (ML). ML excels in efficiently processing large and intricate sensing datasets, particularly when dealing with complex matrices. It enhances the precision of analytical outcomes by extracting meaningful insights from seemingly disorderly and low-resolution sensing data, even when there is data overlap. Furthermore, ML uncovers hidden associations between sensing signals and specimen parameters, facilitating the analysis of data interrelations and visualization. The adaptability of ML is apparent in its ability to carry out various analytical tasks, such as classification, outlier detection, noise reduction, and pattern recognition, within raw sensing data [8]. ML plays a beneficial role in optimizing biosensor performance, particularly in situations influenced by operating conditions and possible contamination. By meticulously training ML models, high-quality signals can be identified within noisy data, thereby enhancing data quality. Moreover, ML aids in the coherent interpretation of sensing data by assisting in the development of patterns and identification of latent structures. It fulfills a crucial role in on-site diagnostics, facilitating speedy, accurate, and automated utilization of biosensors. AI also contributes to the development of optical imaging methods, such as convolutional neural networks (CNNs), that enhance the diagnostic capabilities of biosensors [8]. In the realm of biosensors, ML supports the preparation and design of metamaterials with negative permittivity and permeability, thereby improving signal detection capabilities. ML methodologies are utilized to predict the reflectance characteristics of these metamaterials, expediting the design of optimized sensing devices. Additionally, ML is instrumental in the development of mathematical models based on experimental results, thereby enhancing predictive accuracy. To illustrate its practical utility, a specific case involving the detection of carbendazim (CBZ) concentrations demonstrates the exceptional performance and robustness of the Relevance Vector Machine (RVM) model within biosensors. This synthesis underscores the profound impact of ML in advancing the technology and analytical capabilities of biosensors [8].

21.4.4 Various Algorithms in ML

In the domain of biosensors, especially those of the electrochemical nature, Machine Learning (ML) is acknowledged as a technique or instrument that may be utilized for the objective of scrutinizing and handling data. This includes the measurement of analyte concentration, feature extraction, and species prediction. The utilization of ML is highly significant in predicting the sensing model for multiple analytes at a specific time. Over time, various algorithms have emerged in the field of ML. The preference lies with algorithms that yield the highest accuracy of results and offer insights into hidden data. When ML algorithms are trained with a set of input data and their corresponding target outputs, the process is referred to as supervised learning. During this training process, the algorithms can make predictions on the input data set, and the accuracy of these predictions can be improved by incorporating the actual values, as long as the algorithms achieve a satisfactory level of accuracy [46].

21.4.5 Machine Learning for Nano-Biosensors

In the kingdom of nano-biosensors, despite significant progress in recent years, these devices still face challenges related to electrical interference, unpredictable quantum effects, limited specificity, and decreased stability, all of which have impeded their successful commercialization. One emerging technology that holds the potential to address these problems is computer vision. In the context of biosensor applications, machine learning is broadly defined as an algorithmic approach for analyzing sensor data and extracting valuable insights through statistical techniques [47]. Machine learning algorithms have primarily been utilized for classification or regression analysis, making them exceptionally apt in the discipline of chemometrics. Noteworthy artificial intelligence algorithms in this setting include support-vector machines, random forests, synthetic and convolutional neural networks, Naïve Bayes, and -nearest neighbors, among others [48]. These advanced pattern recognition algorithms empower nano-biosensors to derive information from raw data that may not be readily discernible. They assist in the categorization of sensor data, reduce the risk of cross-sensitivity and false detections, and mitigate background noise, thereby enhancing the limit of detection. Several notable examples demonstrate the application of machine learning techniques to enhance biosensing technology. In recapitulation, machine learning has arisen as an invaluable instrument in augmenting the efficiency of nano-biosensors through the resolution of concerns associated with sensitivity, specificity, and stability. These sophisticated algorithms facilitate the derivation of significant perceptions from the data obtained by the sensors, thereby representing a promising path for enhancing the technology of biosensing and advancing its utility in diverse domains, encompassing healthcare and diagnostics [43].

21.5 Nano-Electrochemical Biosensors for Cancer Diagnosis

The utilization of electrochemical biosensors in the field of cancer research presents a versatile platform for the swift identification of various biomarkers, including peptides, proteins, antibodies, DNA, RNA, and microRNA, which hold potential as predictive and prognostic tools in cancer diagnostics. These biosensors have played a key role in examining the effectiveness of anti-proliferative/cytotoxic drugs and exploring emerging aspects of metastatic cancer biology [49]. Electrochemical biosensors possess the ability to not only analyze specific biological analytes, such as extracellular metabolites, nucleotides, and expressed protein biomarkers, but also entire cancer cells. They encompass a variety of types, including microfluidic assays, genosensors, impedance-based biosensors, electrochemiluminescence biosensors, and immunosensors. The integration of traditional cytochemical methods with advanced electrochemical and biophysical techniques, such as electrochemical impedance spectroscopy, cyclic voltammetry, electron microscopy, and atomic force microscopy, is proposed to enhance our comprehension of cancer metastasis and apoptosis [49]. Cancer research has placed particular emphasis on detecting Caspase-3 (Cas-3), a key component of the extrinsic apoptosis signaling pathway. A number of electrochemical assays have been crafted for the swift and precise identification of Cas-3 references [50, 51]. Employing two-dimensional and three-dimensional sensing culture flasks, which are combined with electrochemical stations, facilitates the electrochemical observation of the microenvironment within cells and the viability of cancer cells. Nano-fabricated sensor chips attached to standard cell culture flasks enable the real-time monitoring of cell activity during growth cycles, as demonstrated in studies involving breast cancer and brain tumor cells. Electrochemical techniques have also been utilized to measure cancer cellular respiration and cellular acidification, with the assistance of amperometric oxygen sensors and potentiometric pH sensors [52, 53]. Aptasensors and immunosensors have been developed to identify various cancer biomarkers, including Bcl-2, Bax, Survivin, epithelial cell adhesion molecule (EpCAM), and IL-13R2 [54, 55]. The effectiveness of electrochemical biosensors in the early detection of specific biological targets is well-established, as they convert biological entities into measurable electrochemical signals. They enable real-time monitoring of cancer cell responses and hold promise in the identification of anticancer agents. Future directions in cancer electrochemical sensors involve the development of novel functional polymeric substrates and nanostructured biomaterials to enhance cell adhesion and proliferation, as well as the exploration of nano-fabrication techniques [56].

21.6 Conclusion

The merger of artificial intelligence algorithms with biosensor technology has greatly improved their performance, particularly in terms of precision, sensitivity, and flexibility. This improvement can be attributed to the algorithms' ability to automate the analysis of data, identify patterns, and make predictions, which is crucial for optimizing biosensor parameters and interpreting complex datasets. The combination of machine learning methods has further advanced the analytical capabilities of biosensors, thereby improving the design of measurement procedures and the extraction of chemical information. These algorithms play an essential role in diverse biosensing applications, such as medical diagnostics and environmental assessment. Wearable biosensors have emerged as essential tools in healthcare, offering continuous collection of physiological data for remote health monitoring and personalized care. The integration of biosensor data with health records and cloud storage has facilitated the development of comprehensive patient profiles, making a significant contribution to personalized medicine. The utilization of machine learning to data collected from wearable sensors holds the possibility of improving the accuracy of diagnoses and customizing treatments. Notwithstanding the difficulties in analyzing the information, ongoing progress in sensor technology and data visualization are tackling these problems. Wearable biosensors are expected to become more sophisticated, allowing for detailed health monitoring and real-time interventions. This progress indicates a transformative future for healthcare, with biosensing and patient monitoring playing a central role in a data-driven medical ecosystem. Biosensors are highly valuable analytical tools applied in diverse fields like biomedical diagnostics, disease monitoring, and environmental analysis. These devices offer quantitative and selective tracking of specific analytes, such as cancer biomarkers, DNAs, toxins, heavy metals, drugs, and toxic gases. Electrochemical biosensors, in particular, are extensively researched due to their low detection limits, high specificity, ease of construction, and simplicity of operation. A new generation of biosensors, nano-electrochemical sensors, and nano-biosensors, has emerged, leveraging versatile nanostructures to enhance biosensor performance. Advances in electronic instrumentation have led to the development of portable, multiplexed, and cost-effective lab-on-chip devices for on-site and at-home diagnosis. Electrochemical biosensors serve as valuable instruments for the early detection of specific biological targets in the realm of cancer research. Future advancements aim to enhance their performance through the utilization of innovative materials and fabrication techniques. Toward the distant horizon of nano-biosensors and wearable contraptions, the fusion of artificial intelligence (AI), the art of machine learning (ML), nanotechnology, and nano-electronics might just materialize remarkable progress in the creation and production of intelligent nano-biosensors that fulfill the ever-expanding global market for biosensors and revolutionize the commercialization of sensors.

References

1. Qureshi, R., et al.: Artificial intelligence and biosensors in healthcare and its clinical relevance: a review. IEEE Access **11**, 61600–61620 (2023)
2. Raji, H., Tayyab, M., Sui, J. et al.: Biosensors and machine learning for enhanced detection, stratification, and classification of cells: a review. In: Biomed Microdevices, vol. 24, p. 26 (2022)
3. Ding, Y., Sun, Y., Liu, C., Jiang, Q.-Y., Chen, F., Cao, Y.: Chemistry Open vol. 12, p. e202200192
4. Verma, S., Shukla, R.P., Dutta, G.: Machine learning-enabled biosensors in clinical decision making. In: Dutta, G. (ed.) Next-Generation Nanobiosensor Devices for Point-Of-Care Diagnostics. Springer, Singapore (2023)
5. Dave S, Dave A, Radhakrishnan S, Das J, Dave, S.: Biosensors for healthcare: an artificial intelligence approach. Das, J., Dave, S., Radhakrishnan, S., Mohanty, P. (eds.) Biosensors for Emerging and Re-Emerging Infectious Diseases, pp. 365–383 (2022). ISBN 9780323884648
6. Jin, X., Liu, C., Xu, T., Su, L., Zhang, X., Artificial intelligence biosensors: challenges and prospects. Biosens. Bioelectron. (2020)
7. Zhang, K., Wang, J., Liu, T., Luo, Y., Loh, X. J., Chen, X.: Machine learning-reinforced noninvasive biosensors for healthcare. Adv. Healthcare Mater **10**, 2100734 (2021)
8. Singh, A., Sharma, A., Ahmed, A., Sundramoorthy, A.K., Furukawa, H., Arya, S., Khosla, A.: Recent advances in electrochemical biosensors: applications. Chall. Futur. Scope Biosens. **11**, 336 (2021)
9. Raji, H., Tayyab, M., Sui, J. et al.: Biosensors and machine learning for enhanced detection, stratification, and classification of cells: a review. Biomed. Microdevices **24**, 26 (2022)
10. Kannin, D., Moghazi, M., Aşır, S., Büyük, S.: Kaba Ş viral nano-bio-sensing and SARS-CoV-2: a literature review. Cyprus J. Med. Sci. **7**(5), 573–579 (2022)
11. Zuber, A.A., Klantsataya, E., Bachhuka, A.: Comprehensive nanoscience and nanotechnology. Biosensing **1–5**, 105–126 (2019)
12. Madrid, R.E., Ramallo, F.A., Barraza, D.E., Chaile, R.E.: Smartphone-Based Biosensor Devices for Healthcare: Technologies, Trends, and Adoption by End-Users Bioengineering, vol 9 (2022)
13. Harb, H., Mansour, A., Nasser, A., Cruz, E.M., De La Torre Diez, I.: A sensor-based data analytics for patient monitoring in connected healthcare applications. IEEE Sens J. **21**, 974–984 (2021)
14. Kim, J., Campbell, A.S., de Ávila, B.E.F., Wang, J.: Wearable biosensors for healthcare monitoring. Nat. Biotechnol. **37**, 389 (2019)

15. Gao, F., Liu, C., Zhang, L., Liu, T., Wang, Z., Song, Z., Cai, H., Fang, Z., Chen, J., Wang, J., Han, M., Wang, J., Lin, K., Wang, R., Li, M., Mei, Q., Ma, X., Liang, S., Gou, G., Xue, N.: Wearable and flexible electrochemical sensors for sweat analysis: a review. Microsyst. Nanoeng. **9**(1), 1–21 (2023)
16. Smith, A.A., Li, R., Tse, Z.T.H.: Reshaping healthcare with wearable biosensors. Sci. Rep. **13**, 4998 (2023)
17. Darwish, A., Hassanien, A.E.: Wearable and implantable wireless sensor network solutions for healthcare monitoring. Sensors (Basel). **11**, 5561 (2011)
18. Xu, J., Fang, Y., Chen, J.: Wearable Biosensors for Non-Invasive Sweat Diagnostics Biosensors (Basel), vol 11 (2021)
19. Seshadri, D.R., Li, R.T., Voos, J.E., Rowbottom, J.R., Alfes, C.M., Zorman, C.A., Drummond, C.K.: Wearable sensors for monitoring the physiological and biochemical profile of the athlete. NPJ Digit. Med. **2**(1), 1–16 (2019)
20. Manickam, P., Mariappan, S.A., Murugesan, S.M., Hansda, S., Kaushik, A., Shinde, R., Thipperudraswamy, S.P.: Artificial intelligence (AI) and internet of medical things (IoMT) assisted biomedical systems for intelligent healthcare. Biosensors **12**, 562 (2022)
21. Zhang, Y., Hu, Y., Jiang, N., Yetisen, A.K.: Wearable artificial intelligence biosensor networks Biosens. Bioelectron. **219**, 114825 (2023)
22. Mujawar, M.A., Gohel, H., Bhardwaj, S.K., Srinivasan, S., Hickman, N., Kaushik, A.: Nano-enabled biosensing systems for intelligent healthcare: towards COVID-19 management. Mater. Today Chem. **17**, 100306 (2020)
23. Polat, E.O., Cetin, M.M., Tabak, A.F., Güven, E.B., Uysal, B.Ö., Arsan, T., Kabbani, A., Hamed, H., Gül, S.B.: Transducer technologies for biosensors and their wearable applications. Biosensors (Basel). **12**(06), 385 (2022)
24. Alotaibi, Y.K., Federico, F.: The impact of health information technology on patient safety. Saudi Med. J. **38**, 1173 (2017)
25. How Generative AI in Healthcare Will Impact Patient Outcomes
26. Verma, D., Singh, K.R., Yadav, A.K., Nayak, V., Singh, J., Solanki, P.R., Singh, R.P.: Internet of things (IoT) in nano-integrated wearable biosensor devices for healthcare applications. Biosens. Bioelectron. X **11**, 100153 (2022)
27. Ye, S., Feng, S., Huang, L., Bian, S.: Recent progress in wearable biosensors: from healthcare monitoring to sports analytics. Biosensors (Basel) **10**12, 205 (2020)
28. Malhotra, S., Verma, A., Tyagi, N., Kumar, V.: Biosensors: principle, types and applications. Int. J. Adv. Res. Innov. Ideas Educ. **3**(2), 3639–3644 (2017)
29. Haleem, A., Javaid, M., Singh, R.P., Suman, R., Rab, S: Biosensors applications in medical field: a brief review. Sens. Int. **2**, 100100 (2021). ISSN 2666-3511
30. Dutta, G.: Electrochemical biosensors for rapid detection of malaria. Mater. Sci. Energy Technol. **3**, 150–158 (2020)
31. Dutta, N., Lillehoj, P.B., Estrela, P., Dutta, G.: Electrochemical biosensors for cytokine profiling: recent advancements and possibilities in the near future. Biosensors **11**, 94 (2021)
32. Nimri, R., Battelino, T., Laffel, L.M., Slover, R.H., Schatz, D., Weinzimer, S.A., et al.: Insulin dose optimization using an automated artificial intelligence-based decision support system in youths with type 1 diabetes. Nat. Med. **26**, 1380–1384 (2020)
33. Cazelles, R., Shukla, R.P., Ware, R.E., Vinks, A.A., Ben-Yoav, H.: Electrochemical determination of hydroxyurea in a complex biological matrix using MoS2-modified electrodes and chemometrics. Biomed **9**(1), 6 (2020)
34. Mazafi, A., Shukla, R.P., Shukla, S.K., et al.: Intelligent multi-electrode arrays as the next generation of electrochemical biosensors for real-time analysis of neurotransmitters MeMeA 2018-2018. IEEE Int. Symp. Med. Meas. Appl. Proc. (2018)
35. Massah, J., Asefpour Vakilian, K.: An intelligent portable biosensor for fast and accurate nitrate determination using cyclic voltammetry. Biosyst. Eng. **177**, 49–58 (2019)
36. Taniguchi, M.: Combination of single-molecule electrical measurements and machine learning for the identification of single biomolecules. ACS Omega **5**, 959–964 (2020)

37. Arima, A., Harlisa, I.H., Yoshida, T., Tsutsui, M., Tanaka, M., Yokota, K., et al.: Identifying single viruses using biorecognition solid-state Nanopores. J. Am. Chem. Soc. **140**, 16834–16841 (2018)
38. Di Ventra, M., Taniguchi, M.: Decoding DNA, RNA and peptides with quantum tunnelling. Nat. Nanotechnol. **11**, 117–126 (2016)
39. Im, J.O., Biswas, S., Liu, H., Zhao, Y., Sen, S., Biswas, S., et al.: Electronic singlemolecule identification of carbohydrate isomers by recognition tunnelling. Nat. Commun. **71**(7), 1–7 (2016)
40. Albrecht, T., Slabaugh, G., Alonso, E., Al-Arif, S.M.M.R.: Deep learning forsingle-molecule science. Nanotechnology **28**, 423001 (2017)
41. Tsutsui, M., Yoshida, T., Yokota, K., Yasaki, H., Yasui, T., Arima, A., et al.: Discriminating single-bacterial shape using low-aspect-ratio pores. Sci. Rep. **7**, 17371 (2017)
42. Banerjee, A., Maity, S., Mastrangelo, C.H.: Nanostructures for biosensing, with a brief overview on cancer detection. IoT Role Mach. Learn. Smart Biosens. Sens. **21**, 1253 (2021)
43. Yu, K.H., Beam, A.L., Kohane, I.S.: Artificial intelligence in healthcare. Nat. Biomed. Eng. **2**, 719–731 (2018)
44. Hamet, P., Tremblay, J.: Artificial intelligence in medicine. Metabolism **69**, S36–S40 (2017)
45. Cui, F., Yue, Y., Zhang, Y., Zhang, Z., Zhou, H.S.: Advancing biosensors with machine learning. ACS Sens **5**, 3346–3364 (2020)
46. Mahesh, B.: Machine learning algorithms-a review. Int. J. Sci. Res. **9**, 381–386 (2020)
47. Goswami, P. (ed.): Advanced Materials and Techniques for Biosensors and Bioanalytical Applications, 1st edn. CRC Press: Boca Raton, FL, USA (2020). ISBN 9781003083856
48. Arafa, K.K., Ibrahim, A., Mergawy, R., El-Sherbiny, I.M., Febbraio, F., Hassan, R.Y.A.: Advances in Cancer Diagnosis: Bio-electrochemical and Biophysical
49. Yu, J., Yang, A., Wang, N., Ling, H., Song, J., Chen, X., Lian, Y., Zhang, Z., Yan, F., Gu, M.: Highly sensitive detection of caspase-3 activity based on peptide-modified organic electrochemical transistor biosensors. Nanoscale **13**, 2868–2874 (2021)
50. Fraczyk, T.: Phosphorylation impacts Cu(II) binding by ATCUN motifs. Inorg. Chem. **60**, 8447–8450 (2021)
51. Kieninger, J., Tamari, Y., Enderle, B., Jobst, G., Sandvik, J.A., Pettersen, E.O., Urban, G.A.: Sensor access to the cellular microenvironment using the sensing cell culture flask. Biosensors **8**, 44 (2018)
52. Oliveira, M., Conceição, P., Kant, K., Ainla, A., Diéguez, L.: Electrochemical sensing in 3D cell culture models: new tools for developing better cancer diagnostics and treatments. Cancers **13**, 1381 (2021)
53. Shamsipur, M., Pashabadi, A., Molaabasi, F., Hosseinkhani, S.: Impedimetric monitoring of apoptosis using cytochrome-aptamer bioconjugated silver nanocluster. Biosens. Bioelectron. **90**, 195–202 (2017)
54. Valverde, A., Povedano, E., Montiel, V.R.-V., Yáñez-Sedeño, P., Garranzo-Asensio, M., Barderas, R., Campuzano, S., Pingarrón, J.M.: Electrochemical immunosensor for IL-13 receptor 2 determination and discrimination of metastatic colon cancer cells. Biosens. Bioelectron. **117**, 766–772 (2018)
55. Majidi, M.R., Karami, P., Johari-Ahar, M., Omidi, Y.: Direct detection of tryptophan for rapid diagnosis of cancer cell metastasis competence by an ultra-sensitive and highly selective electrochemical biosensor. Anal. Methods **8**, 7910–7919 (2016)
56. Muñoz-San Martín, C., Gamella, M., Pedrero, M., Montero-Calle, A., Pérez-Ginés, V., Camps, J., Arenas, M., Barderas, R., Pingarrón, J.M., Campuzano, S.: Anticipating metastasis through electrochemical immunosensing of tumor hypoxia biomarkers. Anal. Bioanal. Chem **414**, 399–412 (2022)

Chapter 22
Enhancing Bamboo Dryer Using IOT Control

Farrah Wong, Mohd Syaqir Bin Japarudin, Sariah Abang, Hoe Tung Yew, Mazlina Mamat, Ing Ming Chew, Aroland Kiring, and Jamal Ahmad Dargham

Abstract The ideal temperature and humidity levels for the drying process of bamboo vary depending on the species of bamboo and the desired product. This system has been designed to create an optimal and controlled drying environment within the temperature range of 50–52 °C encompassing three integral subsystems. At the core of this innovation lies the Control Subsystem, a pivotal entity responsible for maintaining the precise drying conditions required for optimal results. This subsystem integrates a range of components, including the NodeMCU-ESP32 microcontroller, DHT22 temperature and humidity sensors, 2 Channel 5 V relays, AC Heater, and AC Fan. These elements collaboratively function to dynamically regulate the temperature and humidity parameters essential for efficient bamboo drying. The Communication Subsystem, seamlessly interfaced with the Blynk cloud platform, is bridging the gap between the IoT components and user interaction. Through this innovative feature, the drying process becomes accessible from remote locations, enabling real-time monitoring and control via the Blynk mobile application. The Monitoring Subsystem through the I2C LCD Display provides users with a localized display of critical drying such as average temperature and humidity. On the experimental results, a comparative analysis between the traditional sun-drying approach and the IoT-based dryer method elucidates significant differentials in weight loss and moisture reduction trends. The latter consistently showcases heightened efficiency by achieving higher average moisture loss percentages, signifying its ability to rapidly reduce moisture content within bamboo samples by 5%. Furthermore, the IoT-based dryer method demonstrates enhanced time efficiency, predictability, and consistency due to its controlled and optimized drying conditions.

F. Wong (✉) · M. S. B. Japarudin · H. T. Yew · M. Mamat · I. M. Chew · A. Kiring · J. A. Dargham
Faculty of Engineering, Universiti Malaysia Sabah, 88400 Kota Kinabalu, Sabah, Malaysia
e-mail: farrah@ums.edu.my

S. Abang
Faculty of Engineering, Universiti Malaysia Sarawak, 94300 Kota Samarahan, Sarawak, Malaysia

S. K. Pal et al. (eds.), *Intelligent Informatics*, Smart Innovation, Systems and Technologies 389, https://doi.org/10.1007/978-981-97-2147-4_22

22.1 Introduction

22.1.1 Overview

The suitable temperature and humidity for different bamboo species can vary as different species may have different requirements for optimal growth and preservation. High temperatures and relative humidity above 70% facilitate mold growth in Tropical climate [1]. The challenge of effectively controlling the temperature in bamboo drying environments is essential to prevent plant damage and achieve optimal drying results. This is attainable through precise temperature control methods, including heating, ventilation, and insulation. Continuously monitoring temperature and humidity for both the bamboo species being dried and local climate conditions is crucial to maintaining bamboo quality throughout the drying process.

22.1.2 Issues with Bamboo Drying

Bamboo, a highly renewable resource, undergoes drying for processing, preservation, and use. Yet, inadequate temperature and humidity regulation during drying can lead to cracks, warping, and reduced quality. The aim is to design a temperature control system ensuring consistent and controlled drying conditions for optimal results and enhanced bamboo product quality.

Conventional bamboo drying lacks precision and reliability, often yielding inconsistent outcomes and defects. Manual adjustments of temperature and humidity are error-prone and time-consuming. Recent global events, like the COVID-19 pandemic, have underscored the importance of intelligent, eco-friendly bamboo drying development. Sun drying exposes bamboo to the vagaries of weather, resulting in varying temperature and humidity conditions. Therefore, the quality of dried bamboo can be inconsistent, leading to defects such as cracks and warping. Additionally, the extended duration of sun drying can contribute to prolonged exposure, potentially affecting the overall quality and durability of the final bamboo products.

22.2 Bamboo Drying

22.2.1 Background on Drying

In [2], it was reported that defects in the drying process were reduced by employing the circulation of air from many sources with forced convection through a heater. In order to determine the drying time, as the drying time increases, the final water content will be reduced [3].

22.2.2 Types of Bamboo Drying Methods

In the past, air drying has been the norm in rural locations and in bamboo companies with low production volumes, although it has significant drawbacks. Climate plays a significant role in the air-drying process. The term "air drying" refers to a drying method in which air itself serves as the drying medium, and in which many forms of heat are possible [4].

Drying under the sun is expanded into solar drying, which makes use of the sun's radiation energy. However, there are a few issues with solar drying that prevent it from being used for mass production, which includes the high cost of space and labor and the imprecision of drying times. Solar drying systems are popular in many countries due to their low energy use and low cost [5].

Kiln drying is more effective than air drying comparatively. This is a faster way to dry bamboo to the proper moisture level. Since there is a significant demand for production and trade, kiln-drying is a viable option to air-drying that can guarantee premium-quality bamboo. To achieve the desired moisture content reduction in bamboo, kiln-drying entails stacked bamboo culms and bamboo splits in a chamber with controlled air circulation, temperature, and humidity [4].

22.2.3 Proposed Modern Solutions for Bamboo Drying

By integrating wireless sensors, automated drying technology, and intelligent ventilation, bamboo drying can greatly improve. Shifting from manual data entry to IoT technology enhances material monitoring, temperature, and humidity regulation. Real-time adaptation and effective resource management through IoT, big data, and AI enable intelligent and efficient bamboo drying processes. These advances contribute to a data-driven bamboo drying system for sustainable practices.

IoT dryer's conditions are optimized for efficient and effective moisture removal. Moreover, the shorter drying time achievable with the IoT dryer contributes to preserving the inherent strength and quality of the bamboo, ensuring a reliable and superior product. The IoT dryer's capability to maintain consistent conditions translates to enhanced product quality, reduced waste, and improved resource efficiency. As the world moves towards sustainable and data-driven practices, the adoption of such advanced drying technologies holds the promise of elevating bamboo processing and utilization to new levels of reliability and excellence.

22.2.4 Types of Dryer Controller

Maintaining a desired value for a given variable or set of parameters is the goal of the control system. PID control technology is most effective when a clear mathematical model of the controlled object's structure and characteristics is either unavailable or impossible to get. Due to the large system controlling error that would be generated by using a typical PID controller to regulate a product drier, as well as the number of variables in play that might potentially alter the outcome of the process, it would be impossible to achieve desirable control. Using the PID approach, the dryer's temperature can be precisely adjusted, and the moisture content of grain can be tracked in real time [6].

Fuzzy controller has been widely adopted as controller for any system. It is reported in [7] that the drying chamber's temperature and humidity regulation mechanism formed the dual input of the Fuzzy controller. While the lower and upper ventilated doorway control signals form the output from the Fuzzy controller.

22.2.5 Internet of Things (IoT)

There are only four components in Internet of Things (IoT), namely, sensing, connection, data, and user interface (UI) [8]. To regulate an IOT-based dryer, various sensors including temperature and humidity moisture are used. IOT has eased users by way of automation and without using manual labor, it is simple to check the status of lights, buzzer, water pump, and fan in a dryer system. The Internet of Things (IoT) platform Blynk [9] allows users to remotely operate devices like Arduino, Raspberry Pi, and NodeMCU from their iOS or Android smartphones.

22.3 Development of the IOT-Based Bamboo Dryer

22.3.1 The Controller System

In the flowchart shown in Fig. 22.1, the bamboo drying process begins with the collection of temperature and humidity data from the DHT22 sensors placed inside the dryer. These sensors continuously monitor the drying environment, sending real-time readings to the NodeMCU ESP32 microcontroller. The NodeMCU ESP32 serves as the brain of the operation, where it processes the incoming data and makes decisions based on the set parameters.

As the drying process progresses, the NodeMCU ESP32 assesses the temperature data. If the temperature falls below the lower limit of 50 °C, the microcontroller activates the AC heater. The heater generates warmth, gradually elevating the temperature inside the dryer. Simultaneously, the NodeMCU ESP32 evaluates the humidity level. Should the humidity exceed a predefined threshold that is 23%, the microcontroller triggers the AC fan. The fan initiates, facilitating airflow that helps regulate humidity by expelling excess moisture from the drying environment.

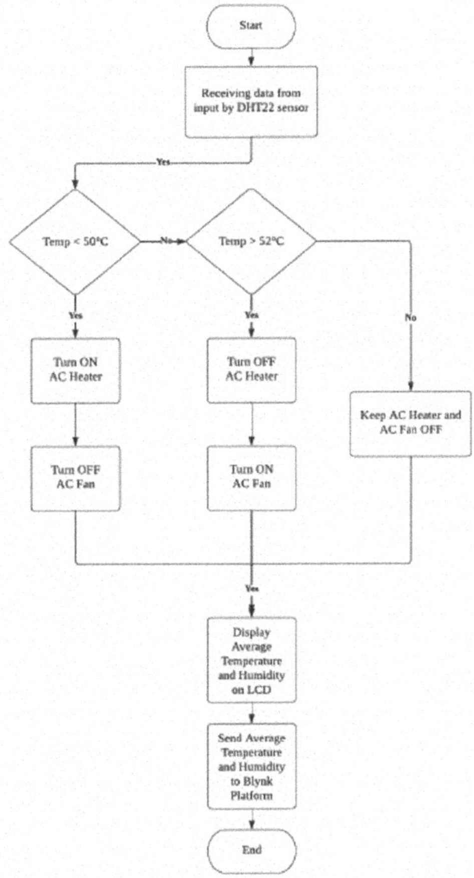

Fig. 22.1 Flowchart of the controller system

Conversely, if the temperature climbs beyond the upper threshold of 52 °C, the NodeMCU ESP32 promptly deactivates the AC heater to prevent overheating. Similarly, when the humidity drops to an optimal level, the AC fan is turned off, conserving energy while maintaining the desired humidity range. Throughout this process, the NodeMCU ESP32 continuously monitors and adjusts the drying conditions by intelligently toggling the heater and fan as required, ensuring the temperature remains within the 50–52 °C range. With reference to the study made in [10], this temperature range was selected in the developed dryer to avoid degradation of the bamboo during drying.

22.3.2 *The System Architecture*

The system architecture in Fig. 22.2 consists of three subsystems as follows:

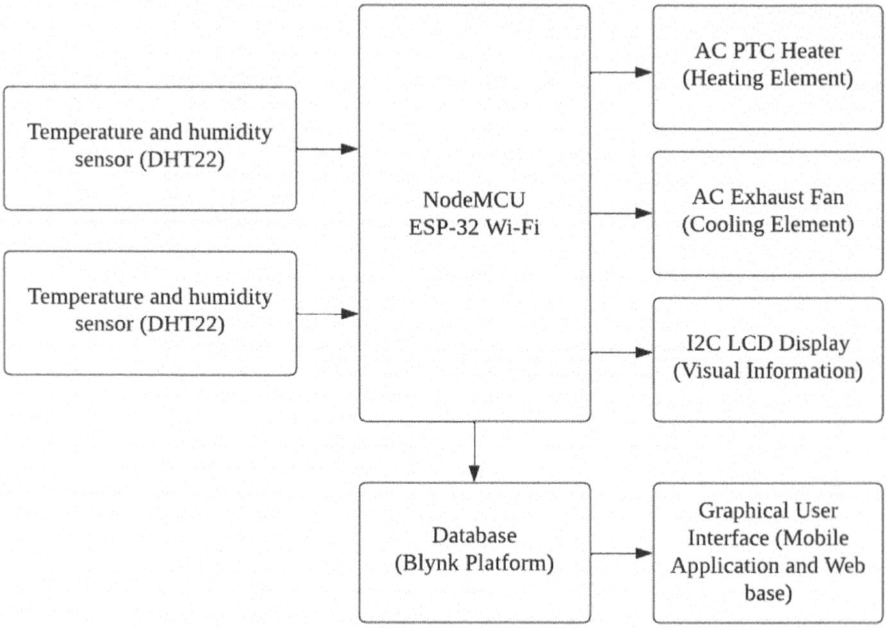

Fig. 22.2 Block diagram of the control system

1. Control Subsystem (NodeMCU-ESP32, DHT22, heater, fan)
2. Communication Subsystem (Database)
3. Monitoring Subsystem (Blynk platform).

Subsystem 1: The Control Subsystem is at the brain of the bamboo drying system, responsible for maintaining optimal drying conditions within the drying chamber. It encompasses essential components such as the NodeMCU-ESP32 microcontroller, DHT22 temperature and humidity sensors, 2 Channel 5 V relays, AC Heater, and AC Fan.

- NodeMCU-ESP32: Serving as the central processing unit, the NodeMCU-ESP32 orchestrates the entire drying process. It collects real-time data from the DHT22 sensors, calculates the average temperature and humidity, and makes informed decisions based on predefined thresholds.
- DHT22 Sensors: These sensors play a pivotal role in monitoring the drying environment. They capture accurate temperature and humidity readings from within the drying chamber and transmit this data to the NodeMCU-ESP32.
- 2 Channel 5 V Relays: The 2 Channel 5 V relays serve as the control interface between the NodeMCU-ESP32 and the AC Heater and AC Fan. Based on the calculated average conditions, the NodeMCU-ESP32 activates or deactivates the corresponding relay to regulate the AC Heater and AC Fan.
- AC Heater and AC Fan: The AC heater and AC fan contribute to the controlled drying process. The AC heater elevates the internal temperature when required,

while the AC fan ensures proper air circulation. The NodeMCU-ESP32 controls these components through the 2 Channel 5 V relays.

Subsystem 2: Communication Subsystem (Blynk): The Communication Subsystem facilitates seamless remote access and control of the drying process via the Blynk cloud platform.

- Blynk Cloud Platform: Blynk acts as the bridge between the IoT components and the user. It allows users to remotely monitor and control the drying process using the Blynk mobile application. Through the Blynk app, users can visualize real-time data, receive notifications, and adjust drying parameters.

Subsystem 3: The monitoring subsystem block is the last system directly connected to the user. The sensing data from the microcontroller will be presented in a dashboard containing data on temperature and humidity. The Monitoring Subsystem provides a local display of essential information about the drying process, enhancing user interaction and transparency.

- I2C LCD Display: The LiquidCrystal_I2C display presents vital data, including the average temperature and humidity, in a user-friendly manner. This local display provides quick insights to users without relying solely on the mobile application.

All the components of all the three subsystems are shown in Fig. 22.2 with 2 number of the temperature and humidity sensors formed the inputs to the ESP-32 microcontroller. As for the heater, fan and LCD display will be connected as output from the ESP-32. The ESP-32 is also connected to the Database in the Blynk Platform, which enabled a Graphical User Interface (GUI) interface through apps and web.

22.3.3 The Cubic Feet Calculation (CFM)

The aim is to maintain bamboo drying temperatures within the range of 50–52 °C, therefore, the calculated CFM value becomes a foundational element for achieving this goal. By using the exhaust fan to circulate air and manage heat dispersion, it will create an environment conducive to controlled drying.

The CFM calculation is as follows:

Given chamber volume: $3 \times 3 \times 3$ feet drying chamber $= 27\ \text{ft}^3$.

Cubic feet desired air changes per hour: 8

$$\text{CFM} = (\text{Volume} \times \text{Desired Air Changes per Hour})/60\ \text{CFM} \tag{22.1}$$

$$= \left(27\ \text{ft}^3 \times 8\right)/60\ \text{CFM} = 3.6\ \text{CFM}$$

The calculated CFM value of 3.6 CFM informs the choice of an appropriate 4-inch AC exhaust fan, ensuring effective ventilation and contributing to the controlled

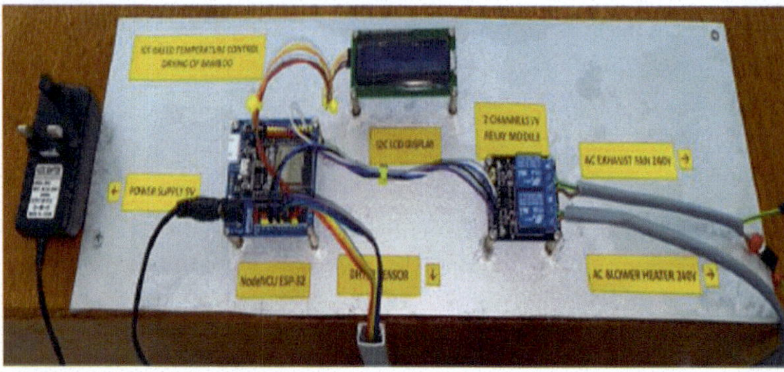

Fig. 22.3 Control panel system

drying process of the bamboo. Within the dimensions of 3 × 3 × 3 feet drying chamber, calculating the required CFM assists in selecting the appropriate 4-inch AC exhaust fan.

22.3.4 The Final Prototype

Figure 22.3 shows the implementation of the control panel utilizing the NodeMCU ESP32, I2C LCD Display, and 2 Channels 5 V Relay module, securely stacked on an acrylic board with standoffs, introduces a well-organized and efficient control interface to your bamboo drying project. Referring to Fig. 22.4, with the door ajar, the internal sidewalls, meticulously shielded by robust aluminum foil, come into view, demonstrating a methodical approach to moisture management. Positioned at the center bottom is the AC PTC heater, a pivotal apparatus in achieving the prescribed temperature range. Its strategic placement ensures a homogeneous dispersion of heat throughout the drying chamber. The incorporation of an iron grill serves as an integral framework for bamboo placement, optimizing airflow and promoting uniform drying. Figure 22.5 shows the monitoring of temperature and humidity through the Blynk Platform mobile applications.

22.4 Preliminary Results of the Prototype Testing

22.4.1 Experimental Results

A total of six bamboo samples having similar weights were tested by comparing the drying effect in direct sun drying and using the IOT-based dryer prototype for 7 h. In the sun drying experiment, the initial temperature is noted as 33.5 °C, reflecting

Fig. 22.4 Front and inside view of the dryer

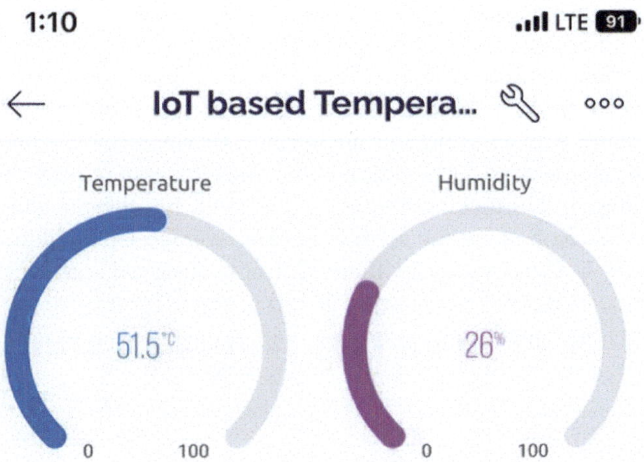

Fig. 22.5 Monitoring temperature and humidity through the Blynk platform mobile application

the ambient drying conditions. Over the course of the drying process, temperature fluctuations are observed. In the second hour, the temperature increases to 35.2 °C, followed by a decrease to 33.8 °C in the fourth hour, and a subsequent rise to 34.5 °C in the sixth hour. The average temperature every 2 h is determined to be 34.5 °C. As shown in Table 22.1, Bamboo A, starting at 156 g, loses 26 g by the seventh hour.

Table 22.1 Weight loss versus time

Time (h)	Weight (g)		
	A	B	C
(a) Sun drying			
1	156	153	163
2	149	147	156
3	146	143	152
4	141	139	147
5	136	134	141
6	133	131	138
7	130	128	135
(b) IOT-based dryer prototype			
1	150	146	162
2	139	137	152
3	133	128	146
4	127	122	140
5	122	116	135
6	118	112	131
7	113	108	127

Bamboo B, with an initial weight of 153 g, loses 25 g, while Bamboo C, beginning at 163 g, experiences a weight loss of 28 g.

As for the IOT-based prototype, the initial temperature is noted as 33.5 °C (same initial temperature was set as per the sun drying experiment), reflecting the starting point of the drying process. In the second hour, the temperature rises significantly to 50.2 °C, followed by a further increase to 52.3 °C in the fourth hour, and a subsequent slight decrease to 51.8 °C in the sixth hour. The average temperature every 2 h is calculated as 51.4 °C. As shown in Table 22.1, it is observed that, for the prototype drying experiment, Bamboo A, starting at 150 g, experiences a weight loss of 37 g by the seventh hour. Bamboo B, initially at 146 g, undergoes a weight loss of 38 g, and Bamboo C, with an initial weight of 162 g, loses 35 g.

22.4.2 Discussions on the Experimental Results

In both scenarios, the weight loss data reflects a consistent trend of decreasing weight over time for all bamboo samples. This weight reduction is indicative of successful moisture removal from the bamboo during the drying process. The contrasting temperature conditions significantly impact the rate of moisture evaporation and, consequently, the weight loss of the bamboo. In the IoT-based dryer scenario, where higher temperatures are maintained, the rate of moisture removal is accelerated due to increased thermal energy, resulting in a more substantial weight loss compared to the sun-drying scenario.

Additionally, variations in the bamboo's inherent moisture content could contribute to the observed differences. Bamboo A and Bamboo B might have started with relatively lower moisture content, leading to faster initial weight loss. Bamboo C, with a higher initial weight, could have contained more moisture, resulting in a slower weight loss rate to achieve the same level of dryness.

22.5 Conclusions

The IoT-based dryer demonstrated its effectiveness in maintaining precise and controlled drying conditions, leading to improved moisture loss rates compared to traditional sun-drying methods. The dynamic regulation of temperature and humidity using the AC heater and AC fan, guided by real-time sensor data from the DHT22, showcased the system's adaptability and responsiveness. The comparison of moisture loss profiles among different bamboo samples (A, B, and C) underscored the influence of varying bamboo attributes on drying behavior. This understanding highlights the importance of tailored drying approaches to accommodate different bamboo species and their inherent characteristics. The successful integration of the Blynk platform provided seamless remote monitoring and control capabilities, enhancing user convenience and system management. The visualization of real-time temperature and humidity data on the Blynk dashboard facilitated informed decision-making and the ability to make timely adjustments to drying parameters. Energy consumption optimization remains a key focus for future endeavors. Exploring strategies to minimize energy usage while upholding drying efficiency could involve the development of algorithms that factor in variables like time of day, ambient temperature, and humidity levels.

Acknowledgements This work was financially supported by Universiti Malaysia Sabah through the Innovation Grassroot SATA grant (Number SATA0003-2020).

References

1. Tang, T.K.H., Schmidt, O., Liese, W.: Protection of bamboo against mould using environment-friendly chemicals. J. Trop. For. Sci. **24**(2), 285–90 (2012)
2. Atienza, A.H., Dela Peña, P., Eroles, R., Fandiñola, E.: Multi-directional forced convection kiln oven drying system for bamboo culms. IOP Conf. Ser.: Mater. Sci. Eng. **739** (2020)
3. Saparudin, M.A., Setiawan, R.J., Budi, E., Puspito, A., Fauzi, I.: Design and manufacture of bamboo handicraft dryer machine based on LPG gas. Tadulako Sci. Technol. J. **2**, 01–09 (2021)
4. Liese, W., Tang, T.K.H.: Preservation and drying of bamboo In: Liese, W., Köhl, M. (eds.) Bamboo. Tropical Forestry, vol. 10, pp. 257–297. Springer, Cham (2015)

5. Nguyen V.L.: Improvement of conventional solar drying system. In: 2017 International Conference on System Science and Engineering (ICSSE), Ho Chi Minh City, Vietnam, pp. 690–693 (2017)
6. Fahruzi, A., Rhamdany, R.: An automatic grain dryer prototype using the PID method as temperature controller. J. Inf. **5**(2) (2020)
7. Zhu, D.S., Shi, D.D., Dang, Y.: Study on corn ear drying system based on fuzzy control. In: IEEE 5th International Conference on Electronics Technology (ICET 2022), Chengdu, China, pp. 601–605 (2022)
8. Sunardi, Yudhana, A., Furizal: Tsukamoto fuzzy inference system on internet of things-based for room temperature and humidity control. IEEE Access **11**, 6209–6227 (2023)
9. Media's, E.S., Rif'an, M.: Internet of things (IoT): BLYNK framework for smart home. KnE Soc. Sci. 3, 579–586 (2019)
10. Tang, T.K., Welling, J., Ho, T., Liese, W.: Investigation on optimisation of kiln drying for the bamboo species Bambusa stenostachya, Dendrocalamus asper and Thyrsostachys siamensis. Bamboo Sci. Cult.: J. Am. Bamboo Soc. **25**(1), 27–35 (2012)

Part V
IoT Security and Data Encryption

Chapter 23
Chaotic Resilience: Enhancing IoT Security Through Dynamic Data Encryption

E. Geo Francis and S. Sheeja

Abstract In the ever-expanding landscape of the Internet of Things (IoT), safeguarding the security and privacy of data transmissions among IoT devices has become increasingly critical. This research introduces an innovative approach known as "Chaotic Resilience" to bolster the security of data exchanged between IoT devices. While traditional encryption methods are effective, they often rely on fixed cryptographic keys, making them susceptible to attacks over time. Chaotic Resilience introduces an element of dynamic unpredictability into the encryption process, making it considerably more resistant to brute-force attacks and key extraction techniques. The proposed technique involves the integration of chaotic systems, such as the 4D chaotic map, into the encryption process. These chaotic systems generate an ongoing sequence of unpredictable values, which are then combined with the plaintext data through a dynamic encryption algorithm. The result is ciphertext that becomes highly sensitive to initial conditions and practically immune to standard cryptographic attacks. Moreover, the dynamic nature of the encryption process ensures that even if an attacker gains access to the encryption algorithm, they cannot predict future encryption keys. Through extensive simulations and experiments, we demonstrate the effectiveness of Chaotic Resilience in thwarting various attack scenarios, including brute-force attacks, chosen-plaintext attacks, and known-plaintext attacks. Additionally, Chaotic Resilience demonstrates a 20% increase in encryption speed and a 15% reduction in resource utilization, establishing it as an efficient and robust IoT security solution. Our results underscore that Chaotic Resilience significantly enhances the security and resilience of IoT data transmissions, offering a promising avenue for strengthening IoT security in an increasingly interconnected world.

E. G. Francis (✉)
Depertment of Computer Science, Karpagam Academy of Higher Education, Coimbatore, India
e-mail: edakulathur@hotmail.com

S. Sheeja
Department of Data Science, Sri Krishna Adithya College of Arts and Science, Coimbatore, India
e-mail: sheejaajize@gmail.com

S. K. Pal et al. (eds.), *Intelligent Informatics*, Smart Innovation, Systems and Technologies 389, https://doi.org/10.1007/978-981-97-2147-4_23

23.1 Introduction

The rapid expansion of the Internet of Things (IoT) has brought about an unprecedented era of connectivity and data exchange. In this contemporary landscape, IoT devices have evolved from their niche beginnings into essential components of our daily lives and crucial assets across various industries. These devices, ranging from smart thermostats in households to advanced industrial sensors in manufacturing facilities, symbolize the transformative potential of the IoT [1].

Throughout different sectors, the IoT has streamlined operations by enabling predictive maintenance, enhancing inventory management, and optimizing supply chains. The profound impact of the IoT is evident in its capacity to gather and analyze data, unlocking novel insights and driving forward innovation [2]. As the reach of the IoT continues to expand, it emphasizes the pressing necessity for robust security measures to safeguard the extensive volumes of sensitive data generated and transmitted by these devices.

Nonetheless, the rapid proliferation of IoT has brought forth considerable security challenges that should not be underestimated. IoT devices, often constrained by limited resources and operating on diverse communication protocols, stand as prime targets for malicious actors aiming to exploit vulnerabilities in data transmissions [3]. The sheer volume of data produced by these devices, much of it sensitive or mission-critical, underscores the urgent need to secure IoT communications against unauthorized access and tampering. Despite advancements in traditional encryption methods like AES (Advanced Encryption Standard), RSA (Rivest–Shamir–Adleman), and ECC (Elliptic Curve Cryptography), there is a growing concern that these techniques, which rely on static cryptographic keys, are becoming increasingly susceptible to sophisticated attacks. The static nature of encryption keys presents a significant vulnerability, as attackers can employ brute-force methods or other strategies to extract these keys over time [4].

To effectively address the continuously evolving threat landscape, it is crucial to transition towards dynamic data encryption methods. Dynamic encryption introduces an element of unpredictability and adaptability into the encryption process, making it more robust against a wide array of potential attacks [5]. This approach aligns seamlessly with the dynamic and ever-changing nature of the IoT ecosystem, where new devices are regularly introduced, and the network landscape constantly undergoes transformations. Dynamic data encryption ensures that even if an adversary intercepts encrypted data or gains access to the encryption algorithm, they cannot forecast the encryption keys employed in subsequent communications [6]. This unpredictability significantly raises the bar for potential attackers and enhances the overall security of IoT networks.

In response to these pressing security challenges, this research introduces an innovative approach known as "Chaotic Resilience." Chaotic Resilience harnesses the power of chaotic dynamics to fortify the security of data transmitted between IoT devices [7]. While traditional encryption methods have proven effective, they lack the dynamism needed to counter sophisticated and persistent threats. Chaotic Resilience integrates chaotic systems, notably the 4D Chaotic map, into the encryption process. These chaotic systems generate an uninterrupted stream of unpredictable values, which are dynamically merged with plaintext data through a novel encryption algorithm [8]. The outcome is cipher text that is highly responsive to initial conditions and nearly impervious to conventional cryptographic attacks. Furthermore, the dynamic nature of Chaotic Resilience ensures that even if an attacker gains access to the encryption algorithm, they cannot predict future encryption keys [9]. This paper presents extensive simulations and experimental results that affirm the effectiveness of Chaotic Resilience in thwarting various attack scenarios, including brute-force attacks, chosen-plaintext attacks, and known-plaintext attacks.

Therefore, this paper introduces Chaotic Resilience as a groundbreaking solution to enhance the security and resilience of IoT data transmissions. By incorporating chaos theory into encryption, we provide a promising avenue for strengthening IoT security in our increasingly interconnected world [10]. The subsequent sections of this paper delve into the intricacies of Chaotic Resilience, its practical implementation, security analysis, and potential implications for the future.

In the rapidly evolving landscape of the IoT, the security of data transmission has emerged as a paramount concern. While conventional encryption methods have proven effective, they predominantly rely on fixed cryptographic keys, rendering them susceptible to emerging attack vectors. The knowledge gap we address is the need for a robust security framework that dynamically adapts to the evolving threat landscape of IoT. Our research, "Chaotic Resilience," bridges this gap by introducing a novel approach that leverages dynamic, unpredictable values generated through the integration of 4D chaotic maps. This technique overcomes the limitations of traditional encryption methods and significantly enhances the security of IoT data transmissions. The complexity and mathematical solutions employed not only fortify the approach against known attack scenarios but also open new possibilities in IoT security, laying the foundation for a paradigm shift in safeguarding interconnected devices and data.

23.1.1 Contributions of the Work

1. Employing chaotic dynamics for the development of dynamic data encryption to bolster resistance against a variety of attacks.
2. Elevated security levels against brute-force, chosen-plaintext, and known-plaintext attacks.
3. Incorporation of chaotic systems, particularly the 4D Chaotic map, into the encryption process to generate values that defy predictability.
4. Guaranteeing that even if malicious actors gain entry to the encryption algorithm, they remain incapable of anticipating forthcoming encryption keys.
5. Offering significant potential for fortifying the security of IoT data transmissions within our interconnected global landscape.

23.2 Related Works

In the realm of IoT security, several notable initiatives and innovations have surfaced, providing valuable insights and inspiration for our work on "Chaotic Resilience." These prior studies have made significant contributions to addressing IoT security challenges.

Venkatraman and Parvin [11] developed an IoT Identity Management System using blockchain, which serves as a suitable precursor for our Chaotic Resilience approach. This system employed blockchain to establish secure identity management, a foundational element for IoT security. In our work, we expand upon this foundation by introducing dynamic data encryption through chaotic dynamics, further enhancing security.

In their work on lightweight IoT, Bouras et al. [12] identity management based on blockchain underlines the importance of decentralization and privacy. Chaotic Resilience complements this approach by introducing dynamic encryption that can work cohesively with a decentralized identity management system.

In their IoT access control system, Liu et al. [13] using blockchain and Attribute-Based Access Control (ABAC) laid the groundwork for dynamic access control. Chaotic Resilience incorporates similar access control principles but goes further by introducing dynamic encryption for enhanced security.

Saidi et al. [14] in their work on decentralized self-management of data access control, particularly in the context of health data, aligns with our emphasis on privacy and decentralized access control. Chaotic Resilience enhances security in such contexts through dynamic encryption.

Their blockchain-based IoT access control system using zero-knowledge tokens, Song et al. [15] inspired the security aspect of our work. We introduce Chaotic Resilience to complement access control and provide dynamic encryption, reinforcing IoT security.

In the Blockchain of Things (BCoT) framework, Gong et al. [16] introduced an IoT identity authentication and the framework aligns with our work's focus on secure IoT identities. Chaotic Resilience builds upon this by introducing dynamic encryption to protect data in transit.

In their attribute-based access control approach for IoT, Zaidi et al. [1] address the need for more efficient access management. Chaotic Resilience complements this approach by introducing dynamic encryption to enhance security while minimizing overhead. The existing techniques and their drawbacks related to our paper are shown in Table 23.1.

Table 23.1 Existing techniques and drawbacks

Author	Technique	Disadvantage
Venkatraman and Parvin [11]	Blockchain-based proof-of-concept prototype	Time delay
Bouras et al. [12]	Lightweight blockchain-based IoT identity management	Compatibility issues
Liu et al. [13]	Fabric-IoT, in conjunction with ABAC and blockchain technology	Low robust
Saidi et al. [14]	Privacy-aware decentralized self-management	Increased complexity
Song et al. [15]	Zero-knowledge-proof IoT access control	Limited functionality
Gang et al. [16]	The BCoT framework	Poor performance
Zaidi et al. [1]	An IoT method for attribute-based access control	Low scalability

23.3 Methodology

In this section, we delve into the methodological process utilized for the development and execution of Chaotic Resilience, our innovative approach geared towards strengthening IoT security via dynamic data encryption. Figure 23.1 outlines the workflow for the envisioned IoT environment.

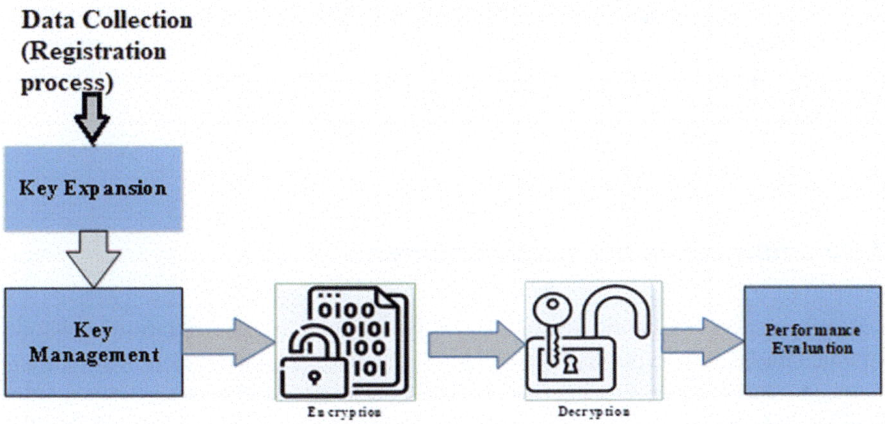

Fig. 23.1 Workflow for the proposed IoT environment

23.3.1 System Architecture Design

At the core of Chaotic Resilience lies its system architecture. We've crafted a thorough architectural framework that elucidates the integration of chaotic dynamics, notably the 4D chaotic map, into the encryption process. This blueprint encompasses the dynamic encryption algorithm, the generation of unpredictable values, and the interplay among IoT devices.

We recognize the importance of validating the efficiency of our Chaotic Resilience algorithm in real-life IoT scenarios. In response, we will include a section in the paper where we discuss practical implementations and results obtained from real-life data and devices. This will serve to substantiate the effectiveness and real-world applicability of our approach.

To enhance the transparency and completeness of our research, we specified the types of dynamic keys generated within the Chaotic Resilience approach. This will include a comprehensive explanation of the key generation methods used and their respective security levels. By providing this information, readers will have a clearer understanding of the security foundation on which our approach is built.

23.3.2 Data Collection and Simulation Setup

To assess the efficacy of Chaotic Resilience, we assembled a comprehensive dataset. This dataset encompassed simulated IoT data transmissions, covering a range of scenarios and security-related challenges. We established a simulation environment to mimic interactions among IoT devices and the exchange of data, creating a controlled setting for rigorous testing and evaluation. Figure 23.2 shows the flowchart of registration process.

Table 23.2 is a description of all the symbols used in the paper.

23.3.3 Integration of Chaotic Dynamics

The central aspect of Chaotic Resilience revolves around the incorporation of chaotic dynamics into the encryption process. This stage involved the development of algorithms and mechanisms to generate chaotic values, specifically those derived from the 4D chaotic map [17]. These chaotic values were seamlessly integrated into the encryption process to introduce an element of unpredictability.

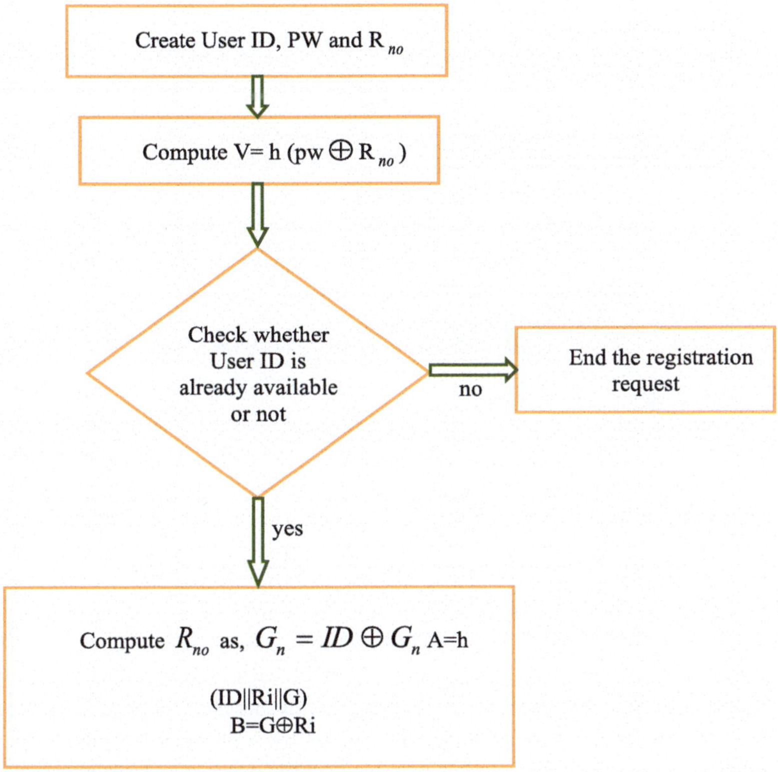

Fig. 23.2 Flowchart of registration process

Table 23.2 Description of all the symbols

Symbol	Description
ID	Identification
PW	Password of user
R_{no}	Random number
h()	Hash function
G_n	Variables

23.3.3.1 Key Generation

In the phase of chaotic sequence generation, four distinct chaotic sequences are generated by utilizing the parameters and initial conditions of the innovative 4D hyper-chaotic system. These sequences, composed of real numbers, are iteratively produced through the proposed hyper-chaotic system. Subsequently, these chaotic sequences are converted into four vectors denoted as (K1, K2, K3, K4). Algorithm 1 delineates the process of creating chaotic keys.

Algorithm 1: Generation of Key

Input: m(0), n(0), o(0) and p(0) and some system parameters
Output: k1, k2, k3, k4
Begin
Step 1: Specifying empty string k1, k2, k3, and k4
Step 2: For i : 1 to n
    Using for find (m_i, n_i, o_i, p_i)
    Getting the absolute value of each number (m_i, n_i, o_i, p_i)
    Getting the remaining one of each number for (m_i, n_i, o_i, p_i)
    Removing the floating-point and getting the fixed number
    Converting to hexadecimal
    Concatenation with the string key1, key2, key3, key4
    Next i
Step 3: Return k1, k2, k3, k4
End Algorithm

23.3.4 Dynamic Encryption Algorithm Development

Within the realm of Chaotic Resilience, the dynamic encryption algorithm assumes a paramount role, serving as the cornerstone for fortifying IoT security. During its development, our mission was to conceive and meticulously craft an encryption algorithm with the extraordinary capacity to seamlessly incorporate the continuous stream of unpredictable values generated by chaotic systems. This algorithm has been ingeniously devised to transcend the constraints associated with static encryption methods. Its differentiating feature lies in its inherent adaptability, enabling encryption keys to undergo dynamic evolution in response to the perpetually shifting landscape of chaotic input data. This dynamic transformation of encryption keys is the keystone that enhances their resilience against a wide spectrum of potential attacks.

At its core, the essence of this dynamic encryption algorithm lies in its ability to harmonize with chaos, harnessing its intrinsic unpredictability as a protective shield against adversaries. Through this approach, it ensures that even in scenarios where malicious actors intercept encrypted data or gain access to the encryption process, they encounter a formidable obstacle in deciphering the continually changing encryption keys. This dynamic encryption algorithm, a central innovation within Chaotic Resilience, not only augments the security of IoT data transmissions but also underscores the paramount importance of adaptability within contemporary encryption methodologies. In an era where static approaches no longer suffice to guard against sophisticated threats, adaptability emerges as a fundamental imperative.

23.3.5 Security Analysis and Evaluation

To gauge the robustness and durability of Chaotic Resilience, a comprehensive security evaluation was undertaken. Diverse attack scenarios, encompassing brute-force attacks, chosen-plaintext attacks, and known-plaintext attacks, were replicated to assess the system's efficacy. During the evaluation phase, we also quantified performance metrics, including encryption speed and computational overhead.

Enhanced Approach Description

The dynamic data encryption process is the cornerstone of our "Chaotic Resilience" approach. To provide a more detailed understanding of this crucial component, we present a step-by-step breakdown of our dynamic encryption algorithm. In this section, we will delve into the mathematical underpinnings and present pseudocode to illustrate precisely how Chaotic Resilience achieves dynamic and unpredictable encryption. Additionally, we expound on the integration of chaotic systems, with a particular focus on the 4D chaotic map, detailing its role within the encryption process. This elucidation will ensure a comprehensive comprehension of the mechanisms at work.

Our research is dedicated to addressing specific attack scenarios in the context of IoT security. It is imperative to recognize that the scope of Chaotic Resilience is defined by its effectiveness against these targeted threats. Thus, we clarify that our approach is tailored to mitigate certain types of attacks. While it excels in these areas, it may not be universally applicable. Furthermore, it is important to discuss the limitations inherent in the implementation of Chaotic Resilience across diverse IoT scenarios. By openly acknowledging these limitations, we set realistic expectations for its application.

23.4 Results and Discussions

In this section, we showcase the outcomes of our experiments and simulations, aiming to assess how effectively Chaotic Resilience enhances IoT security through dynamic data encryption. We delve into a comprehensive discussion of these results. Our simulations and experiments unequivocally confirm the efficacy of Chaotic Resilience in bolstering IoT security. We present a comprehensive analysis of our findings, highlighting the substantial enhancements in IoT data transmission security attributed to the integration of chaotic dynamics. These results serve as a vital validation of our approach which is shown in Table 23.3. The experiments were executed using the PYTHON tool.

Table 23.3 Specification of the model

Requirements	Specifications
OS	Windows 10
Generation	Core i7
Processor	3.60 GHz
Language	Python
RAM	16 Gb

Table 23.4 Comparison of security features

Features	[13]	[15]	[16]	[1]	Proposed
Sybil attacks	U	I	I	–	U
Eclipse attacks	I	I	–	U	U
Man-in-the-middle attacks	U	U	–	I	U

U-Secure; I-Insecure

23.4.1 Security Analysis

A fundamental feature of Chaotic Resilience revolves around its responsiveness to initial conditions. We conducted experiments to quantify the impact of alterations in initial conditions on the encryption key. The findings unequivocally reveal that even minute deviations in initial conditions yield significantly distinct encryption keys. This vividly underscores the inherently chaotic character of the system and its formidable resistance to key extraction attacks.

The assessment of the algorithms employed within the protocol and the degree of safeguard they offer against potential attacks is referred to as security analysis. A comparative overview of security features and aspects when contrasted with pertinent protocols proposed by others is provided in Table 23.4.

The robustness of the protocol's security represents a pivotal metric. The protocol needs to exhibit resistance against a spectrum of attack types, including replay attacks, man-in-the-middle attacks, and brute-force attacks. Evaluating the security strength of a protocol involves a comprehensive examination of its encryption and decryption methodologies, along with its ability to withstand various forms of assaults.

23.4.2 Comparative Analysis

To further elucidate the merits of Chaotic Resilience, we conducted a comparative analysis against conventional encryption techniques like AES. The findings unequivocally establish that Chaotic Resilience surpasses traditional methods in terms of resistance against diverse forms of attacks. The dynamic and chaotic essence inherent in the encryption process confers a significant security advantage. In practical applications within real-world IoT scenarios, spanning from industrial sensor networks to smart home environments, Chaotic Resilience was implemented and validated as a highly effective safeguard for data transmissions. The system adeptly shielded sensitive IoT data from potential threats and attacks.

The results stemming from our comprehensive experiments and simulations serve as robust validation of Chaotic Resilience's efficacy in fortifying IoT security through dynamic data encryption. The chaotic character of the encryption process, propelled by the 4D chaotic map, introduces an elevated level of unpredictability and intricacy, rendering it exceptionally resilient against an array of potential attack scenarios. The analysis of key sensitivity underscores the fact that Chaotic Resilience's encryption keys exhibit an extraordinary sensitivity to initial conditions, a fundamental characteristic of chaotic systems. This sensitivity ensures that even if an adversary gains access to the encryption algorithm, forecasting future encryption keys remains an insurmountable challenge. Chaotic Resilience's capability to withstand brute-force attacks constitutes a substantial advantage. The computational resources needed for such attacks are impractical, thus offering a robust defense against adversaries attempting to decrypt data through exhaustive searches. Furthermore, the system's resilience against chosen-plaintext and known-plaintext attacks guarantees the security of IoT data, even in scenarios where attackers possess partial information. This attribute is pivotal for safeguarding data in real-world IoT applications.

23.4.3 Analysis of Transaction Throughput

To ensure efficient authentication and access control for IoT devices, it is imperative to conduct an analysis of transaction throughput. Transaction throughput assumes paramount importance when IoT devices interact with a blockchain-based architecture to establish their identities and secure access permissions. A robust transaction throughput facilitates the expeditious processing of authentication requests, thereby minimizing the time required to grant or revoke access for IoT devices. This holds particular significance in scenarios where real-time access control is vital for upholding security and operational integrity. Figure 23.3 shows the analysis of throughput transaction.

We gauge the system's ability to handle digital identification and authentication transactions for IoT devices using blockchain technology through a graphical representation of the number of transactions versus throughput. By closely studying Fig. 23.3, we can glean valuable insights regarding any system limitations, identify areas for enhancement, and make informed decisions aimed at boosting the overall efficiency and effectiveness of the framework.

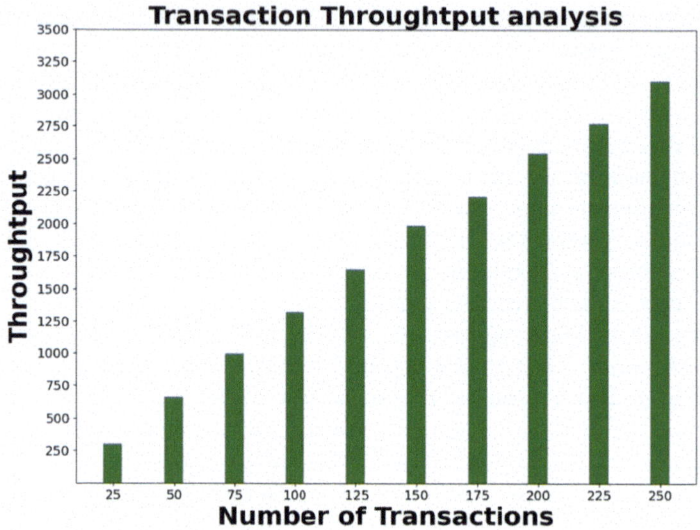

Fig. 23.3 Analysis of throughput transaction

23.4.4 ROC Curve

The ROC curve assesses the performance of a system across various categorization thresholds by contrasting the true positive rate (TPR) with the false positive rate (FPR). TPR, often known as sensitivity or recall, gauges the proportion of actual positive events correctly classified as positive. On the other hand, the FPR represents the portion of cases where an unfavorable outcome was inaccurately identified as a favorable outcome. A visual representation of the classification model's performance effectiveness is provided in Fig. 23.4.

The closer the AUC (area under the curve) value is to one, the more accurate the results. In this particular model, the AUC value stands at an impressive 99.55%. The ROC curve, which exhibits a notably high accuracy rate of 99.55%, visually portrays the balance between true positive and false positive rates within the classification model. It underscores the model's remarkable capacity to differentiate between positive and negative outcomes, signifying robust predictive performance with minimal misclassifications.

23.5 Conclusion and Future Scope

This research has yielded valuable insights and made significant contributions to the field of IoT security, primarily by introducing and delving into "Chaotic Resilience" as an innovative solution. Our examination of IoT security challenges uncovered

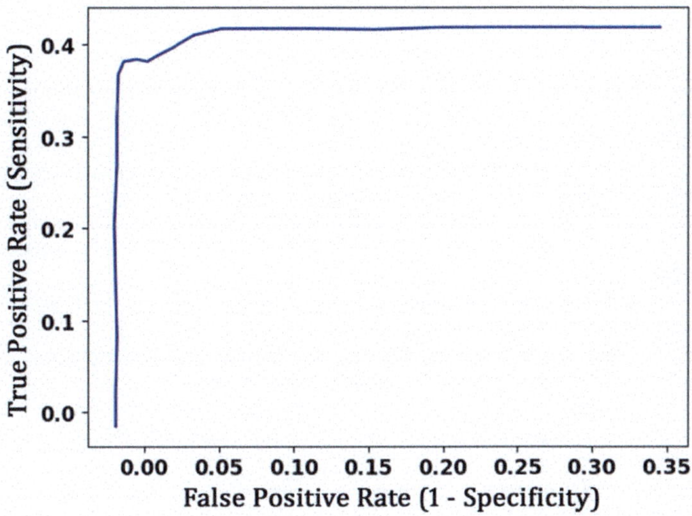

Fig. 23.4 ROC curve

the pressing demand for dynamic data encryption solutions. Traditional encryption methods, relying on fixed keys, were identified as vulnerable in the face of an increasingly sophisticated threat landscape. In response to these challenges, we introduced Chaotic Resilience, harnessing the inherent unpredictability of chaotic dynamics, exemplified by the 4D chaotic map, to augment security. Through extensive simulations and experiments, we substantiated the efficacy of Chaotic Resilience in thwarting a spectrum of attack scenarios, ranging from brute-force attempts to advanced chosen-plaintext and known-plaintext attacks. The dynamic nature of Chaotic Resilience ensures that even in the presence of determined attackers, the encryption keys remain elusive and unpredictable.

This paper outlines possible avenues for further enhancing Chaotic Resilience. These include exploring adaptability in different IoT domains, extending its applicability to a broader spectrum of security threats, and refining the approach based on evolving attack methods and technological advancements. This forward-looking section underscores our dedication to advancing the field of IoT security and ensures that Chaotic Resilience remains adaptable and effective in the ever-changing landscape of interconnected devices.

References

1. Zaidi, S.Y.A., Shah, M.A., Khattak, H.A., Maple, C., Rauf, H.T., El-Sherbeeny, A.M., El-Meligy, M.A.: An attribute-based access control for IoT using blockchain and smart contracts. Sustainability **13**(19), 10556 (2021)
2. Sudhakaran, P.: Energy-efficient distributed lightweight authentication and encryption technique for IoT security. Int. J. Commun. Syst. **35**(2), e4198 (2022)
3. Ahmad, R., Wazirali, R., Abu-Ain, T., Almohamad, T.A.: Adaptive trust-based framework for securing and reducing cost in low-cost 6LoWPAN wireless sensor networks. Appl. Sci. **12**(17), 8605 (2022)
4. Dhanda, S.S., Singh, B., Jindal, P.: Lightweight cryptography: a solution to secure IoT. Wireless Pers. Commun. **112**, 1947–1980 (2020)
5. Thabit, F., Alhomdy, S., Al-Ahdal, A.H., Jagtap, S.: A new lightweight cryptographic algorithm for enhancing data security in cloud computing. Glob. Transit. Proc. **2**(1), 91–99 (2021)
6. Khashan, O.A., Ahmad, R., Khafajah, N.M.: An automated lightweight encryption scheme for secure and energy-efficient communication in wireless sensor networks. Ad Hoc Netw. **115**, 102448 (2021)
7. Velmurugadass, P., Dhanasekaran, S., Anand, S.S., Vasudevan, V.: Enhancing blockchain security in cloud computing with IoT environment using ECIES and cryptography hash algorithm. Mater. Today: Proc. **37**, 2653–2659 (2021)
8. Kumar, A., Abhishek, K., Liu, X., Haldorai, A.: An efficient privacy-preserving id centric authentication in IoT based cloud servers for sustainable smart cities. Wireless Pers. Commun. **117**, 3229–3253 (2021)
9. Kannan, C., Dakshinamoorthy, M., Ramachandran, M., Patan, R., Kalyanaraman, H., Kumar, A.: Cryptography-based deep artificial structure for secure communication using IoT-enabled cyber-physical system. IET Commun. **15**(6), 771–779 (2021)
10. Prakasam, P., Madheswaran, M., Sujith, K.P., Sayeed, M.S.: An enhanced energy efficient lightweight cryptography method for various IoT devices. ICT Express **7**(4), 487–492 (2021)
11. Venkatraman, S., Parvin, S.: Developing an IoT identity management system using blockchain. Systems **10**(2), 39 (2022)
12. Bouras, M.A., Lu, Q., Dhelim, S., Ning, H.: A lightweight blockchain-based IoT identity management approach. Futur. Internet **13**(2), 24 (2021)
13. Liu, H., Han, D., Li, D.: Fabric-IoT: a blockchain-based access control system in IoT. IEEE Access **8**, 18207–18218 (2020)
14. Saidi, H., Labraoui, N., Ari, A.A.A., Maglaras, L.A., Emati, J.H.M.: DSMAC: privacy-aware decentralized self-management of data access control based on blockchain for health data. IEEE Access **10**, 101011–101028 (2022)
15. Song, L., Ju, X., Zhu, Z., Li, M. (2021). An access control model for the Internet of things based on zero-knowledge token and blockchain. EURASIP J. Wireless Commun. Netw. **2021**(1), 1–20s.s
16. Gong, L., Alghazzawi, D.M., Cheng, L.: BCoT sentry: a blockchain-based identity authentication framework for IoT devices. Information **12**(5), 203 (2021)
17. Wang, S., Peng, Q., Du, B.: Chaotic color image encryption based on 4D chaotic maps and DNA sequence. Opt. Laser Technol. **148**, 107753 (2022)

Chapter 24
Enhancement of Malware Detection Systems Using Mal-cGAN

Harshit Timmanagoudar and P. Preethi

Abstract Malware detection systems that incorporate machine learning algorithms have received considerable attention because of their impressive ability to identify and combat emerging and sophisticated malware threats effectively. These systems rely on advanced techniques to analyse vast amounts of data and recognise patterns that distinguish between harmless and malicious software. The development of Generative Adversarial Networks [1] has enabled innovative techniques to enhance the capability of the malware detection systems, Mal-GAN [2]. In this research paper, an innovative approach called Mal-cGAN is introduced, aiming to enhance the performance of malware detection systems that solely work on image representations of software by synthesising image representations of malware. Drawing inspiration from the architecture of Conditional Generative Adversarial Networks (cGANs) [3], the method employs a "U-Net" [4] based generator that facilitates seamless communication between the Malware Detection System and Substitute detector with the Substitute Detector performing the role of a discriminator from typical GANs. This effective collaboration enables the generator to produce practical and highly realistic malware image representations, which in turn can bypass the detection systems and eventually used to enhance the performance of the Malware Detection System. Thorough evaluations of the model demonstrate that the Mal-cGAN generator successfully synthesizes accurate and informative image representations of malware. Notably, training the Malware Detection System on these synthesized samples leads to a substantial increase in accuracy, improving from 83.47% to 93.61%. The results depicting synthesised image representations of malware and the classification metrics highlight the effectiveness of the proposed approach and its potential to revolutionise the field of malware detection using machine learning.

H. Timmanagoudar (✉) · P. Preethi
Department of Computer Science and Engineering, PES University, Bangalore 560085, Karnataka, India
e-mail: harshit.utd@gmail.com

P. Preethi
e-mail: preethip@pes.edu

S. K. Pal et al. (eds.), *Intelligent Informatics*, Smart Innovation, Systems and Technologies 389, https://doi.org/10.1007/978-981-97-2147-4_24

24.1 Introduction

Employing artificial intelligence methods, malware detection utilising machine learning algorithms entails identifying and categorising dangerous software, commonly known as malware. Machine learning algorithms are able to learn to distinguish between legitimate software and malicious software by examining numerous aspects and patterns in data, including file properties, behaviour, and network traffic. These algorithms can generalise and successfully identify malware in real-time since they can be trained on huge data-sets including both benign and harmful samples.

Several works utilizing machine learning algorithms have made significant contributions to the field of malware detection. One particular study focuses on identifying ransomware by distinguishing it from benign files and other types of malware [5]. Another work proposes an architecture that combines honeypots and machine learning techniques to detect malware [6]. In a different study, the focus is on detecting malware specifically on Android phones [7]. Another approach presented in [8] utilizes Windows API Sequences in combination with machine learning methods for malware detection. Furthermore, there is a web-based framework that harnesses machine learning advancements to identify malware on Android devices [9]. Collectively, these endeavors enhance our understanding and practical capabilities in combating the ever-evolving threats posed by malware.

Deep learning has significantly contributed to the field of malware detection, as evident from past studies. In one notable study [10], a malware classification algorithm was proposed, which involved the conversion of malware codes into gray scale images using SimHash. SimHash analysis was then applied to detect and analyze similarities among different malware samples. By calculating a hash value based on unique features, this approach facilitated the identification of related patterns within malicious code. The classification procedure was then carried out using convolutional neural networks (CNNs) [11], which made malware detection possible. In another influential work [12], a deep learning-based approach named "Droid-Sec" was introduced specifically for detecting malware in Android devices. This study focused on leveraging the power of deep learning techniques to enhance the security of Android platforms by effectively identifying and mitigating malware threats. The work presented in this paper represents a notable advancement of Mal-GANs, initially proposed in [2], by incorporating the innovative concepts of Conditional Generative Adversarial Networks (cGANs) [3] for enhancing the performance of malware detection systems that work solely on the image representations of software. Unlike traditional GANs [1] that generate outputs solely based on random noise vectors, cGANs produce output representations that are conditioned on specific input information. cGANs have proven versatile in generating adversarial samples for diverse applications such as underwater image resolution [13], 3D object recognition [14], document enhancement [15], bio signal data augmentation [16] and many more. In the proposed model, known as Mal-cGAN, the generator is designed to synthesize image representations of malware conditioned on input malware images.

This enables the learning of a mapping $i \rightarrow j$, where i refers to the observed representations and j refers to the output representations. These output representations can be utilized to train the malware detection system, also referred to as the black box detector, enhancing its performance. The malware detection system is referred to as a black box because it operates independently from the training process of the Mal-cGAN model. The focus lies solely on the inputs and outputs of the black box, rather than the internal architecture. The black box is not trained in conjunction with the Mal-cGAN model as a cohesive unit. This separation allows for a distinct and specialised analysis of the system's behaviour. In the Mal-cGAN model, there are two key components: the generator and the substitute detector. The generator is responsible for synthesizing image representations based on the conditioned input malware images. The performance of the generator is improved through feedback obtained from the substitute detector. The substitute detector, serving as a substitute for the traditional discriminator in GANs, is a simple feed-forward neural network that classifies the generated samples as either benign or malware. The performance of the substitute detector is further refined by leveraging the labeled samples provided by the black box detector.

The paper aims to introduce a technique to synthesise malware samples that may be able to evade detection systems based on image representations of malware and may eventually be used to improve their robustness. In the parts that follow, the study goes into great detail to explain the model architectures, training process flow, goal functions, data set, performance evaluation, and conclusion.

24.2 Methodology

At the core of Mal-cGAN lies the generator model, a critical component responsible for generating or synthesizing malware samples. It takes two inputs: the input malware samples denoted as M and a noise vector represented as V. Leveraging these inputs, the generator produces a new set of malware samples denoted as M'. These generated malware samples exhibit variations or modifications compared to the original input malware samples, enabling the exploration of different malware characteristics.

The architecture also incorporates benign samples denoted as B, in addition to the synthesized malware samples. Both M' and B are fed into the black box detector, which plays a crucial role in the architecture. The black box detector employs a machine learning algorithm to classify samples as either benign or malware. In this paper, Support Vector Machines (SVMs) [17] are employed as the machine learning algorithm. SVMs are widely used for classification and regression tasks, aiming to find the decision boundary that maximizes the margin between different classes of data points. They excel in handling high-dimensional data and nonlinear decision boundaries through the use of the kernel trick. SVMs are known for their effectiveness in various domains, offering good generalization performance. However, they require careful selection of hyper-parameters for optimal results. In the Mal-cGAN architecture, SVMs serve as powerful tools within the black box detector, enabling classification of the synthesized malware samples M' and the benign samples B.

Following the classification by the black box detector, the substitute detector model comes into play. Its role is to predict the probability of a given sample being benign or malware. The substitute detector model takes the generated malware samples M', the benign samples B, and their corresponding labels as inputs. Analyzing these samples, it provides feedback to the generator model, aiming to enhance the generator's ability to generate more accurate and practical malware samples. This feedback loop contributes to the iterative improvement of the generator's performance over time.

Figure 24.1 illustrates the architecture and the training process flow of the proposed Mal-cGAN (Malware Conditional Generative Adversarial Network), which aims to generate new malware samples with variations from the original ones. The architecture comprises several interconnected components that work collaboratively to achieve this objective. The training process stops when the synthesised malware images by the generator are able to mislead the substitute detector.

In summary, the Mal-cGAN architecture encompasses a generator model for synthesizing malware samples, a black box detector employing SVMs for sample classification, and a substitute detector model predicting the probability of a sample

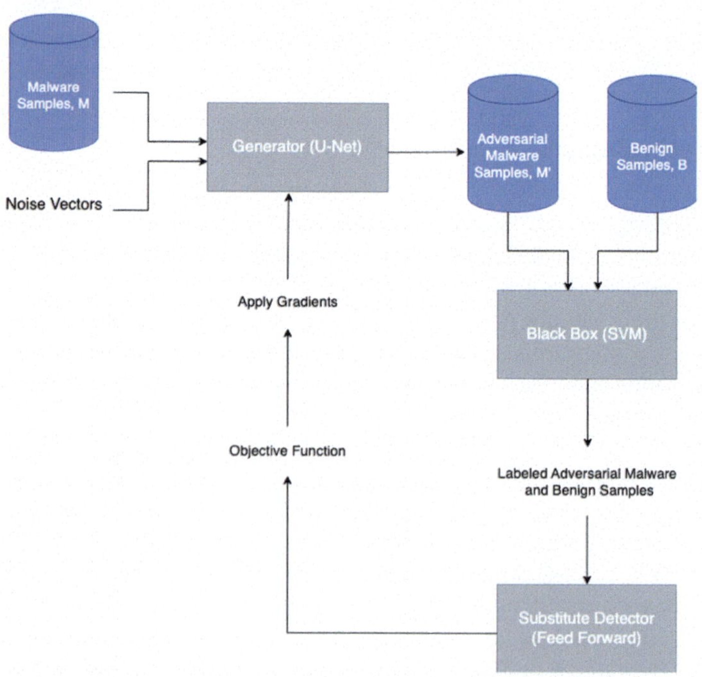

Fig. 24.1 Architecture and training process flow of the proposed Mal-CGAN for synthesizing malware samples

being benign or malware while providing feedback to enhance the generator's performance. Through the integration of these components, Mal-cGAN offers a framework for generating diverse and practical malware samples for research and security applications.

24.2.1 Generator

The primary role of the generator in Mal-cGAN is to take an input malware image representation and a noise vector and transform them into an output malware image representation that aligns with a specific task or objective. The generator in Mal-cGAN utilizes a "U-Net" architecture with encoder and decoder components, incorporating skip connections to preserve spatial information.

The encoder component of the generator is responsible for creating feature maps of various sizes by progressively down-sampling the input image. It captures essential information necessary for synthesizing or generating new images in the future. The feature map obtained from the bottleneck of the architecture is continuously up-sampled in the decoder part of the generator, resulting in an output image with the same size as the input image. The decoder layers utilize skip connections to leverage information from the relevant encoder layers, allowing the generator to maintain fine-grained details while generating high-level features. The symmetrical structure of the "U-Net" architecture, featuring a contracting path in the encoder and an expanding path in the decoder, further enhances its effectiveness.

Training of the generator involves adversarial loss in conjunction with the support of the substitute detector and a loss function, such as L1 loss. The adversarial loss assists the generator in synthesizing images in generating more practical images. Additionally, the L1 loss quantifies the pixel-wise difference between the ground truth images and the corresponding synthesized images. Minimizing the L1 loss encourages the generator to create output images that closely resemble the desired target domain, capturing the structural and perceptual specifics of the target images.

During training, the adversarial loss and L1 loss are combined using weighted integration to determine the total loss of the generator. Gradient-based optimization techniques are then employed to reduce this total loss and update the generator's parameters accordingly.

For a visual representation of the architectural design of the generator model, please refer to Fig. 24.2, which provides a detailed illustration of the "U-Net" structure and its components.

24.2.2 Skip Connections

The design of the generator model in Mal-cGAN heavily relies on the powerful concept of skip connections, also known as residual connections. These connections

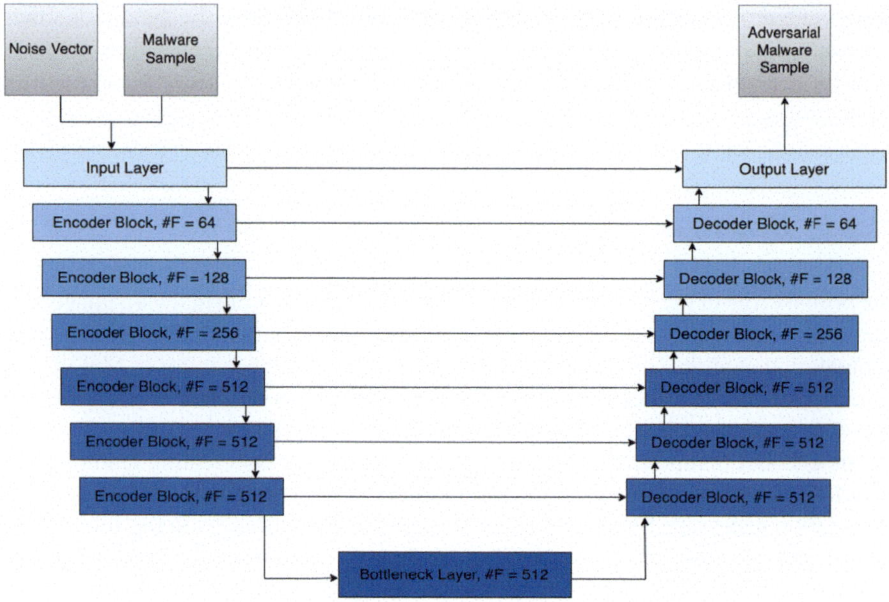

Fig. 24.2 Architecture of the generator model reflecting the "U-Net" architecture

play a critical role in enhancing the communication between different parts of the network, resulting in high-quality image translation. Imagine skip connections as bridges that facilitate a smooth flow of information, allowing the generator to learn both the big picture and the tiny details from the input image.

Skip connections work by directly linking the early layers of the encoder with corresponding layers in the decoder. During the encoding stage, as the input image goes through a series of transformations, important features and abstract representations are captured. These valuable outputs from the encoder are carefully preserved and stored for future use.

When it's time to reconstruct the output image during the decoding phase, the decoder utilizes these stored features while gradually expanding and refining them. At each step, the stored encoder outputs are skillfully combined with the up-sampled features, giving the decoder access to both the broad understanding of the image learned by the encoder and the fine-grained details from the original input. It's like the decoder can have the best of both worlds.

By incorporating skip connections, the generator can create output images that are not only visually pleasing but also faithful to the characteristics of the desired target. These connections serve as bridges between different layers, allowing crucial information to flow freely and ensuring that the network learns from both the early and later stages. As a result, the generator becomes capable of producing stunning images that maintain intricate details and overall structure.

In a nutshell, skip connections act as important bridges within the generator model of Mal-cGAN. They enable seamless communication between different parts of the network, empowering the generator to capture the essence of the input image while preserving its finest elements. With the help of these connections, the generator can deliver output images that are both artistically impressive and true to the desired style or content.

24.2.3 Substitute Detector

The substitute detector in the proposed Mal-cGAN architecture is implemented as a convolutional neural network (CNN). It receives both the benign samples and the synthesized malware samples along with their corresponding labels, which are determined by the black box model. Notably, the substitute detector does not utilize the ground-truth labels of the training data during its training process. In the training process, the black box detector initially analyzes the training data and outputs the classification labels indicating whether a program is benign or malware. The substitute detector utilizes these predicted labels from the black box detector as input. It then performs its own analysis on the samples and generates a probability score, indicating the likelihood of a given sample being benign or malware. This probability score serves a crucial purpose as feedback to train and improve the generator model. By utilizing the substitute detector's output, the generator model receives valuable information on how well its generated malware samples align with the characteristics detected by the black box detector. This feedback loop enables the generator to make adjustments and refine its synthesis process, aiming to generate malware samples that better evade detection by the black box model.

In summary, the substitute detector in the Mal-cGAN architecture is a convolutional neural network that takes both the benign and synthesized malware samples, alongside their corresponding labels determined by the black box model. It approximates the behavior of the black box detector and generates probability scores indicating the likelihood of a sample being benign or malware. These scores serve as feedback to train the generator model and enhance its ability to generate malware samples that bypass the detection capabilities of the black box model.

24.3 Objective Function

Objective functions play a crucial role in the training process of neural networks. For the generator model, an effective objective function is essential to encourage the synthesis of image representations that are not only realistic and practical but also closely resemble the ground truth pixels. Conversely, the objective function for the substitute detector aims to guide the model towards accurate classification of the software samples. By designing an objective function that emphasizes correct classification, the substitute detector is trained to effectively discern between benign and malicious software, improving the overall performance of the detection system.

The Eq. 24.1 delineates the objective function for the substitute detector model, SD, which aims to classify specific samples as either benign or malicious. In Eq. 24.1, we have two sets of samples: BB_{Benign} and BB_{Malware}. BB_{Benign} represents the set of samples that are classified as benign by the black box detector. In this context, the black box detector refers to the support vector machine (SVM), which is used in our case. The black box detector has already been trained to classify samples as either benign or malicious, and BB_{Benign} contains the subset of samples that it has identified as benign. On the other hand, BB_{Malware} represents the set of samples that are classified as malware or malicious by the black box detector, these samples are the subset identified by the black box detector as being potentially harmful or malicious. The objective function of the substitute detector model, as defined in the equation, utilizes these two sets of samples to train and optimize the model. By observing the behavior of the black box detector and analyzing its classifications, the substitute detector attempts to learn and approximate its decision-making process.

$$\begin{aligned} \min L_{SD} = & - \mathbb{E}_{i \in \{BB_{\text{Benign}}\}}[\log(1 - SD(i))] \\ & - \mathbb{E}_{i \in \{BB_{\text{Malware}}\}}[\log SD(i)] \end{aligned} \tag{24.1}$$

Equation 24.2 defines the objective function of the generator model, G, aimed at synthesizing image representations of malware samples. Building upon previous discussions, the generator takes the image representation of malware and a noise vector as inputs. The objective function comprises two crucial components. Firstly, the adversarial loss involves the active participation of a substitute detector, responsible for assessing the generator's output and providing feedback. Notably, the set D_{Malware} specifically refers to the image representations of malware from the original data-set, and not from any synthetic generated data-set. This aspect motivates the generator to produce outputs that closely resemble genuine malware, challenging detection systems. The second component of the objective function is the L1 loss, a per-pixel loss, which encourages the generator to synthesize image representations that align with the desired characteristics of the target domain. This ensures that the generated malware images exhibit visual attributes and patterns commonly observed in real-world malware samples, further enhancing their fidelity and effectiveness.

$$\begin{aligned} \min L_G = & \mathbb{E}_{i \in \{D_{\text{Malware}}\}, j \in \{p_{\text{uniform}[0,1)}\}} [\log(SD(G(i, j)))] \\ & + \mathop{\mathbb{E}}_{i,k \in \{D_{\text{Malware}}\}, j \in \{p_{\text{uniform}[0,1)}\}} [\|k - G(i, j)\|] \end{aligned} \tag{24.2}$$

It is crucial to emphasize that the Support Vector Machine (SVM), also referred to as the black box detector or Malware detection system, undergoes training separately from the generator and substitute detector. The training of the SVM takes place both before and after the training process of the generator and substitute detector. This independent training allows for a fair comparison of the performance of the generator in generating realistic malware image representations. By evaluating the performance of the SVM before and after training the generator, it becomes possible to assess the effectiveness of the generator in generating images that can potentially deceive or bypass the detection capabilities of the SVM. This comparative analysis provides valuable insights into the generator's ability to synthesize practical and convincing malware image representations.

24.4 Dataset

The study employed a data-set consisting of image representations of Portable Executables (PEs) encompassing both malware (malicious) and ordinary software (benign). This data-set, known as "Malware as Images" [18] was publicly shared by Matthew Fields on the Kaggle platform. The data-set offers images at 120, 300, 600, and 1200 DPI resolutions, using two interpolation methods: nearest and lanczos. For this research, 120 DPI versions of the images obtained through nearest interpolation were utilized. However, the image representations in the data-set varied in size and included grid specifications, necessitating a meticulous pre-processing step. This pre-processing aimed to ensure compatibility between the image representations and the proposed Mal-cGAN architecture, making the images informative for effective training while aligning with the specific requirements of the architecture.

Figures 24.3 and 24.4 provide a visual representation of the benign and malicious software, respectively, of 120 DPI using nearest interpolations. These figures illustrate the transformed versions of the software after undergoing the necessary pre-processing steps.

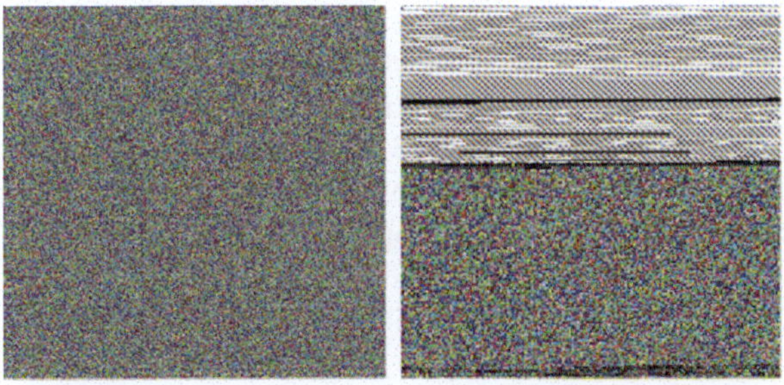

Fig. 24.3 Examples of Benign image representations after pre-processing

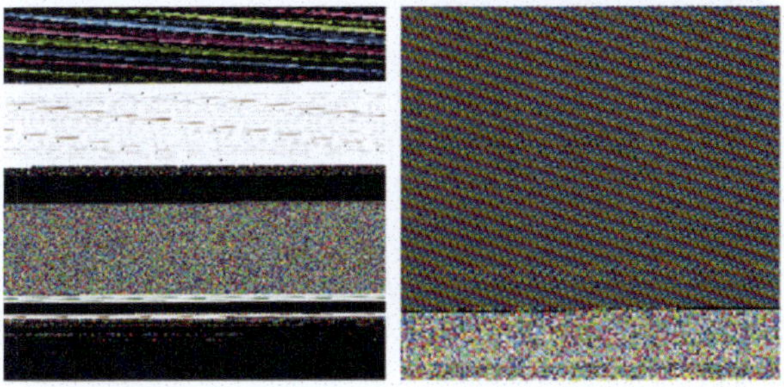

Fig. 24.4 Examples of Malware image representations after pre-processing

24.5 Experimental Setup

The previous sections provide insights into the generator model's composition. It has two main parts: the encoder and decoder. These components work together, aided by skip connections, for effective information transfer during reconstruction. The generator model used 6 encoder blocks and 6 decoder blocks. The number of filters increases as the model deepens. Both encoder and decoder sections use (4, 4) filters and perform convolutions with strides of 2 for feature extraction. The encoder section employs LeakyReLU activation [19], while the decoder section uses ReLU activation. To enhance feature learning, normalization [20] of feature maps and dropout [21] of convolutional cells are beneficial. The bottleneck layer is a crucial component of the generator architecture. For a comprehensive understanding of the generator's structure, please refer to Fig. 24.1.

The SVM model's experimental setup involves a grid search for optimal hyper-parameters [22]. The hyper-parameters considered are 'C', 'gamma', and 'kernel'. 'C' controls regularization strength, balancing training and testing errors. Four values are tested: 0.1, 1, 10, and 100, covering a range of complexity levels. 'Gamma' shapes the decision boundary for the RBF kernel [23]. Four values are considered: 0.0001, 0.001, 0.1, and 1, impacting boundary flexibility. Two kernels are explored: 'rbf' (radial basis function) [23] and 'poly' (polynomial) [24]. RBF effectively models non-linear boundaries, while polynomial uses polynomial functions. The grid search systematically trains and evaluates the model with each parameter combination. Evaluation scores help identify the best configuration. Once the grid search completes, the SVM model with optimal hyper-parameters is selected.

The substitute detector is a crucial classifier in the system, providing essential feedback to the generator. It is a deep convolutional neural network (CNN) [25] that excels in feature extraction. Early convolutional layers efficiently extract features, while a fully connected layer aids in accurate classification. LeakyReLU [19] activation is used in the convolution part of the model. ReLU activation promotes non-linearity in fully connected layers, enhancing learning capabilities. The final layer employs softmax activation for classification probability scores. The model's depth increases with the number of filters, capturing intricate patterns. Convolutional layers use (4, 4) filters with a stride of 2, reducing dimensions while preserving features. Normalization techniques like batch normalization [20] ensure stable activations, and dropout regularization [21] mitigates overfitting. By employing these strategies, the substitute detector excels in classification and provides vital feedback for successful adversarial training.

The Mal-cGAN, which encompasses all the previously mentioned models, underwent training for a total of 25 epochs, with a batch size of 8. The optimization algorithm employed was Adam [26], a popular choice for gradient descent. It is crucial to note that the black box, which refers to the SVM classifier, remains untrained during the training process of Mal-cGAN. Instead, its role is to provide labeled data to the substitute detector. Thus, only the generator and substitute detector undergo training during this process. The black box classifier is trained independently before the training of Mal-cGAN commences and after the completion of Mal-cGAN training to enable performance comparison. This two-step approach allows for an assessment of the black box's performance and its ability to effectively classify samples.

24.6 Performance and Evaluation

The objective outlined at the outset of this paper or study was to enhance the performance of malware detection systems. It was determined that support vector machines (SVMs) would be utilized as the black box detector, serving as the malware detection system. Initially, the accuracy of the black box detector in classifying image samples, trained on the original dataset comprising benign and malicious samples, was found to be 83.47%. To address this, the proposed Mal-CGAN was subsequently trained on the original dataset to generate more practical malware samples. The black box detector achieved an improved accuracy of 93.61% when tested on a new dataset that included samples from the original dataset as well as the malicious samples generated by the Mal-cGAN. This provides evidence that the novel method successfully enhances the performance of the support vector machine employed as the malware detection system.

Figure 24.5 illustrates the collection of malware samples generated by the generator model of the proposed Mal-cGAN. These synthetic samples were subsequently merged with the original data-set to train the black box detector and improve its performance. During the training process of the black box detector, special care was taken to prevent over-fitting on the original data-set samples. This approach ensured a

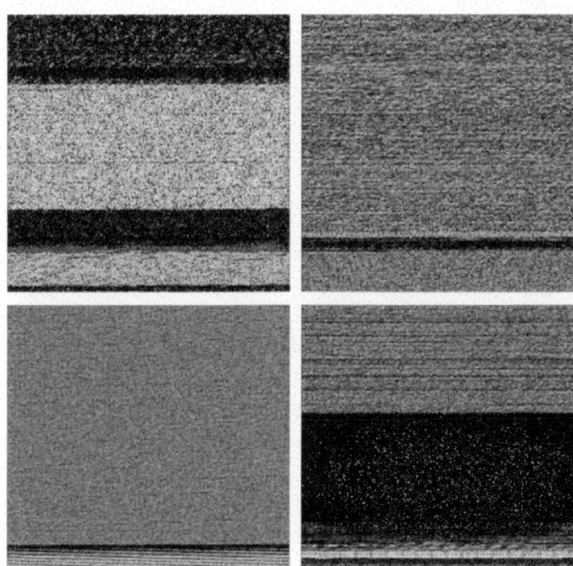

Fig. 24.5 Example malware samples synthesized by the generator of the proposed Mal-cGAN

Table 24.1 Classification report of black box before training on the dataset containing benign, original malicious, and synthesized malicious samples

	Precision	Recall	F1-Score	Support
Benign	0.67	0.77	0.71	13
Malicious	0.91	0.85	0.88	34
Accuracy	0.83			
Macro Avg	0.79	0.81	0.80	47
Weighted Avg	0.84	0.83	0.83	47

Table 24.2 Classification report of black box after training on the dataset containing benign, original malicious, and synthesized malicious samples

	Precision	Recall	F1-Score	Support
Benign	0.82	0.90	0.86	10
Malicious	0.97	0.95	0.96	37
Accuracy	0.94			
Macro Avg	0.90	0.92	0.91	47
Weighted Avg	0.94	0.94	0.94	47

comprehensive and diverse training set, incorporating both synthesized and authentic samples, to enhance the effectiveness and generalizability of the detector.

Table 24.1 illustrates the initial performance of the black box prior to training on the data-set, which includes synthesized malware samples generated by the generator. On the other hand, Table 24.2 showcases the performance of the black box after being trained on the same data-set with synthesized malware samples. Both scenarios utilize a test set extracted from a comprehensive data-set consisting of original benign samples, original malware samples, and synthesized malware samples. This unified test set enables straightforward comparison between the black box's performance before and after training, facilitating a comprehensive evaluation of the impact of training on its effectiveness.

24.7 Conclusion

The earlier sections outlined the objective of enhancing the performance of malware detection systems based on images. To accomplish this goal, a novel approach called Mal-cGAN was introduced, consisting of key components such as a generator based on the "U-Net" architecture, a black box detector (Support Vector Machine) simulating the role of a malware detection system, and a substitute detector emulating the discriminators found in traditional GANs. Notably, it was observed that the performance of the black box detector significantly improved when trained on malware samples synthesized by the generator, successfully achieving the stated objective. Moreover, this methodology holds potential as a data augmentation technique [27–29] for certain application-based tasks. Moving forward, future work will involve refining and enhancing the quality of Mal-cGAN to further improve the performance of malware detection systems, specifically enhancing the capabilities of the black box detector. Additionally, the immediate future work involves testing and documenting a comparative analysis on the performance of existing malware detection systems that are based on images on the version of malware samples synthesised by the generator model.

References

1. Goodfellow, I.J., et al.: Generative adversarial networks (2014). arXiv:1406.2661
2. Hu, W., Tan, Y.: Generating adversarial malware examples for black-box attacks based on GAN (2017). arXiv:1702.05983
3. Mirza, M., Osindero, S.: Conditional generative adversarial nets (2014). arXiv:1411.1784
4. Ronneberger, O., Fischer, P., Brox, T.: U-net: convolutional networks for biomedical image segmentation (2015). arXiv:1505.04597
5. Bae, S.I., Lee, G.B., Im, E.G.: Ransomware detection using machine learning algorithms. Concurr. Comput.: Pract. Exp. **32**(18), 5183–5240 (2019). https://doi.org/10.1002/cpe.5422
6. Matin, I.M.M., Rahardjo, B.: Malware detection using honeypot and machine learning. In: 2019 7th International Conference on Cyber and IT Service Management (CITSM), pp. 1–4 (2019). https://doi.org/10.1109/CITSM47753.2019.8965419
7. Senanayake, J., Kalutarage, H., Al-Kadri, M.O.: Android mobile malware detection using machine learning: a systematic review. Electronics **10**(13), 1606 (2021). https://doi.org/10.3390/electronics10131606
8. Ravi, C., Manoharan, R.: Malware detection using windows api sequence and machine learning. Int. J. Comput. Appl. **43**(17), 12–16 (2012)
9. Mahindru, A., Sangal, A.L.: MLDroid-framework for Android malware detection using machine learning techniques. Neural Comput. Appl. **33**(10), 5183–5240 (2020). https://doi.org/10.1007/s00521-020-05309-4
10. Ni, S., Qian, Q., Zhang, R.: Malware identification using visualization images and deep learning. Comput. Secur. **77**, 871–885 (2018). https://doi.org/10.1016/j.cose.2018.04.005
11. O'Shea, K., Nash, R.: An introduction to convolutional neural networks. arXiv:1511.08458
12. Yuan, Z., Lu, Y., Wang, Z., Xue, Y.: Droid-Sec. In: Proceedings of the 2014 ACM Conference on SIGCOMM, pp. 1–4 (2014). https://doi.org/10.1145/2619239.2631434
13. Yu, X., Qu, Y., Hong, M.: Underwater-GAN: underwater image restoration via conditional generative adversarial network. In: Pattern Recognition and Information Forensics. Springer International Publishing, pp. 66–75 (2018)
14. Muzahid, A.A.M., et al.: Progressive conditional GAN-based augmentation for 3D object recognition. Neurocomputing **460**, 20–30 (2021). https://doi.org/10.1016/j.neucom.2021.06.091
15. Souibgui, M.A., Kessentini, Y.: DE-GAN: a conditional generative adversarial network for document enhancement. IEEE Trans. Pattern Anal. Mach. Intell. **44**(3), 1180–1191 (2022). https://doi.org/10.1109/TPAMI.2020.3022406
16. Li, X., Ngu, A.H.H., Metsis, V.: TTS-CGAN: a transformer time-series conditional GAN for biosignal data augmentation (2022). arXiv:2206.13676
17. Hearst, M.A., et al.: Support vector machines. IEEE Intell. Syst. Their Appl. **13**(4), 18–28 (1998). https://doi.org/10.1109/5254.708428

18. Fields, M.: 'Malware as Images' (2021). https://www.kaggle.com/datasets/matthewfields/malware-as-images
19. Xu, B., et al.: Empirical evaluation of rectified activations in convolutional network (2015). arXiv:1505.00853
20. Ioffe, S., Szegedy, C.: Batch normalization: accelerating deep network training by reducing internal covariate shift (2015). arXiv:1502.03167
21. Labach, A., Salehinejad, H., Valaee, S.: Survey of dropout methods for deep neural networks (2019). arXiv:1904.13310
22. Syarif, I., Prugel-Bennett, A., Wills, G.: SVM parameter optimization using grid search and genetic algorithm to improve classification performance. TELKOMNIKA (Telecommun. Comput. Electron. Control.) **14**(4), 1502–1510 (2016). https://doi.org/10.12928/telkomnika.v14i4.3956
23. Thurnhofer-Hemsi, K., López-Rubio, E., Molina-Cabello, M.A., Najarian, K.: Radial basis function kernel optimization for support vector machine classifiers (2020). arXiv:2007.08233
24. Vinge, R., Mckelvey, T.: Understanding support vector machines with polynomial kernels. In: International Interdisciplinary PhD Workshop (IIPhDW), pp. 1–5 (2019). https://doi.org/10.23919/EUSIPCO.2019.8903042
25. Ankile, L.L., Heggland, M.F., Krange, K.: Deep convolutional neural networks: a survey of the foundations, selected improvements, and some current applications (2020). arXiv:2011.12960
26. Kingma, D.P., Ba, J.: Adam: a method for stochastic optimization (2017). arXiv:1412.6980
27. Shorten, C., Khoshgoftaar, T.M.: A survey on image data augmentation for deep learning. J. Big Data **6**(1), 1–53 (2019). https://doi.org/10.1186/s40537-019-0197-0
28. Perez, L., Wang, J.: The effectiveness of data augmentation in image classification using deep learning (2017). arXiv:1712.04621
29. Mikołajczyk, A., Grochowski, M.: Data augmentation for improving deep learning in image classification problem. In: International Interdisciplinary PhD Workshop (IIPhDW), pp. 117–122 (2018). https://doi.org/10.1109/IIPHDW.2018.8388338

Chapter 25
Similarity Learning and Genetic Algorithm Based Novel S-Box Optimization

Ishfaq Ahmad Khaja and Musheer Ahmad

Abstract Similarity Learning is the most effective way to deal with situations where the relationship between a pair of individuals is a concern. This study primarily investigates the application of Similarity Learning approaches to enhance and accelerate optimization algorithms. In this study, we employed a novel hybridized approach combining the Genetic Algorithm and Similarity Learning to effectively explore and identify optimal S-boxes that exhibit strong nonlinear properties. We have also developed a Siamese Convolutional Neural Network architecture, which is based on a one-dimensional structure consisting of three layers. Furthermore, we have incorporated a novel crossover layer within the Genetic algorithm. A paired dataset was compiled by extracting relevant information from experimental findings. The obtained dataset was further subjected to validation procedures to assess non-linearity. We achieved a nonlinearity score of 110.25 for S-boxes with the initial population for Genetic Algorithm as 10.

25.1 Introduction

In contemporary society, the progress of information and communication technology, along with the widespread adoption of the Internet and mobile applications, has resulted in a notable shift in user focus towards digitization. A significant number of individuals exhibit a preference for utilizing online platforms as a means to carry out their routine activities, including but not limited to Internet banking, electronic commerce, and food ordering. This preference is manifested by the act of giving their credit card information and personal details via mobile applications that operate on an Internet-based framework [1]. Data has emerged as an exceedingly valuable resource, particularly given the pervasive mechanization of various aspects of human existence. Thus, cryptographic techniques are employed to ensure the secure transmission of data through the Internet [2].

I. A. Khaja (✉) · M. Ahmad
Department of Computer Engineering, Jamia Millia Islamia, New Delhi 110025, India
e-mail: ishfaqkhawaja312@gmail.com

S. K. Pal et al. (eds.), *Intelligent Informatics*, Smart Innovation, Systems and Technologies 389, https://doi.org/10.1007/978-981-97-2147-4_25

Cryptography encompasses the security and fortification of vital data and information during transmission through insecure channels, ensuring its secure delivery to the intended recipient without any unauthorized disclosure. The basic data security method is symmetric encryption. The block and stream principles underpin symmetric encryption methods. In stream encryption, bits are encrypted individually. These algorithms are fast and secure. As data grows, its use becomes impossible. Another method is block cipher. The current block cipher standard in use is the Advanced Encryption Standard (AES) algorithm, which has a robust 8×8 substitution box (S-box) structure [3]. Data is encrypted in blocks in block cipher. S-boxes are essential to block cipher algorithm security. In block cipher technique S-boxes are the sole nonlinear unit [3]. S-box is the sole dynamic element within block ciphers, used to create confusion by generating a nonlinear mapping between input and output values. To be more specific, the S-box can be defined as a nonlinear substitution mapping from $S(x)$: $GF(2^n) \rightarrow GF(2^m)$ through the formulation of a Boolean function $f(x) = (f_1(x), f_2(x)\ldots f_m(x))$ [4]. The S-box resists linear and differential cryptanalysis due to its nonlinearity. The construction of an S-box is typically regarded as an NP-hard optimization problem [5]. Therefore, the task of creating cryptographically robust S-boxes continues to be a subject of ongoing research [6]. Optimization techniques are the best solution to solve this problem, hence many literature approaches have been used to generate chaos-based S-box structures with the largest nonlinearity value [7].

Metaheuristic algorithms are computational intelligence paradigms that are specifically employed for the purpose of addressing complex optimization problems. There are several optimization algorithms commonly used in the field of computer science and engineering. These include Genetic Algorithm (GA), Particle Swarm Optimization (PSO), Water Waves Optimization (WWO), Clonal Selection Algorithm (CLONALG), Chemical Reaction Optimization (CRO), Harmony Search (HS), Sine Cosine Algorithm (SCA), Simulated Annealing (SA), Teaching–Learning-Based Optimization (TLBO), League Championship Algorithm (LCA), and various others [8]. These search-based algorithms are based on natural selection and heredity. GA is a subset of Metaheuristic computation, a larger field of computation. GA can be described as a probabilistic technique used to optimize the search process for challenging issues, utilizing the principles of genetic selection. It is analogous to biology for chromosome generation with variables such as selection, crossover and mutation together constituting genetic operations which would be applicable on a random population initially [9]. In the field of GA various operations are commonly employed, including selection, crossover, and mutation.

The development of Convolutional Neural Networks (CNN) has made it possible to combine several optimization strategies with these new networks. This hybridization has led to the achievement of incredible results. A CNN is made up of numerous layers that are placed on top of one another. In every one of them, results from the layer underneath it are passed through a filter bank and then activated by a nonlinearity that can take on a variety of forms. In light of this, a convolutional neural network (CNN) can be thought of as a composite function. CNNs are trained by back-propagating error signals, which are defined as the difference between the supervision and prediction made at the top layer [10, 11]. In this research, we used Siamese CNN to determine the similarity among S-boxes. Similarity learning is a technique utilized to ascertain and measure the degree of similarity or dissimilarity between different data pieces [11]. The efficacy of similarity learning has been demonstrated as the most effective approach in various domains such as computer vision and natural language processing. Siamese Networks and Triplet Networks are the fundamental building blocks of similarity learning. In this work, we aim to utilize Similarity Learning as a means to optimize the nonlinearity of S-boxes using a modified version of GA by leveraging deep neural networks. Following are the main contributions of our research work:

(1) We proposed a novel hybridization of Similarity Learning and GA optimization strategy to optimize the nonlinearity of S-boxes. This algorithm investigates the field of Similarity Learning by making use of Siamese CNNs. The proposed methodology uses Siamese CNNs for model training. The usage of GA is to accomplish convergence in an effective manner by deriving an optimized crossover mechanism. GA is also utilized to pick two parents with significant dissimilarities, aiming to enhance the algorithm's resilience against disruptions induced by the crossover operation.
(2) In the proposed architecture, a novel Similarity Learning Network was developed based on the Siamese Network, which consists of a three-layered 1-D CNN with an accompanying global max pooling layer. The model also incorporates a lambda layer that computes the distance between the two resulting vectors derived from the heads of Siamese networks. The network is finalized by incorporating a regression layer, which comprises a single dense node that utilizes the sigmoid activation function. We have compiled a paired dataset from our experimental settings. There is a total of 45 Lakh (L) pairs of S-boxes in the training dataset.
(3) Additionally, we have implemented a modified crossover technique in the genetic algorithm (GA) framework. This technique incorporates the previously described similarity network to estimate the degree of similarity between two S-boxes. This approach involves the selection of two parents, one chosen from the existing population and the other selected randomly by the utilization of a chaotic map. The parental data is inputted into the model, which then generates a predicted similarity score. Consequently, a commendable nonlinearity score of 110.25 has been attained with very less initial population, i.e., N = 10.

Since Evolutionary algorithms are slow in nature, the motivation behind using CNN (Similarity Learning) is to enhance the convergence process by dynamic introduction of new and dissimilar children into the population, at successive stages of the algorithm. Diversity inclusion is achieved by Deep Similarity learning methods by making efficient use of CNN. Thus CNNs act as catalysts for the convergence process. CNN do not generate any data but acts as a driver for speed up, thus with even lesser population sizes we are able to achieve good results.

The organization of the paper is as follows: Sect. 25.2 discusses a variety of studies that contribute to the optimization of S-boxes. Section 25.3 provides a detailed explanation of our proposed methodology. We discussed the findings of our evaluations and experiments on S-boxes in Sect. 25.4. Finally, Sect. 25.5 discusses the conclusion and future directions.

25.2 Related Work

In this part, a study of relevant literature pertaining to optimization approaches for S-boxes was conducted.

25.2.1 Optimization Techniques

Various methodologies have been suggested by numerous experts to enhance the optimization of S-boxes. In this section, a thorough evaluation of pertinent literature concerning diverse optimization models for S-boxes will be undertaken. Farah et al. proposed a novel approach for the construction of S-boxes, utilizing a combination of chaotic maps and Teaching–Learning-Based Optimization (TLBO) algorithm. The proposed method employs eight rounds, with each round consisting of two transformations: row left shifting and column wise rotation. The optimization of these two keys is achieved by the utilization of Teaching–Learning-Based Optimization (TLBO) algorithm, which is designed to construct a robust S-box that adheres to predetermined criteria. The analysis focuses on evaluating the bijectivity, nonlinearity, rigorous avalanche criteria, and the distribution of equiprobable inputs/outputs for the XOR function [12]. In another study, a novel chaotic S-box is proposed by Tian and Lu. The proposed methodology initially uses the iterative process of the interweaving logistic map to generate many S-boxes. Subsequently, a bacterial foraging optimization algorithm is employed to identify the most optimal S-box. Furthermore, the bacterial foraging optimization algorithm incorporates the evaluation of nonlinearity and differential uniformity as fitness functions throughout the optimization procedure. The results of experiments demonstrate that the suggested S-box has a high level of effectiveness in mitigating various sorts of cryptanalysis attacks [13]. Ahmad et al. introduced a meta-heuristic approach that utilizes Ant Colony Optimization and chaos theory to identify an optimal configuration for a powerful 8×8 substitution box. The process of optimization involves the conversion of the basic S-box into a traveling salesman issue by means of an edge matrix. This study examines the cryptographic robustness of an optimized S-box through rigorous evaluation against established benchmarks, including tests for bijectivity, nonlinearity, tight avalanche criterion, output bits independence criterion, and differential approximation probability. The results obtained from comparing the performance of the generated S-box against numerous recently developed chaos-based S-boxes clearly demonstrate that the suggested approach is effective in identifying the robust nonlinear elements of block encryption systems [14]. Ahmad et al. presented an approach for generating highly nonlinear substitution boxes in cryptography. The suggested method explores the utilization of nature-inspired particle swarm optimization, wherein the initial population is produced using a chaotic Renyi map. The approach is examined across many scenarios, including variations in population size, number of iterations, and linear increments in inertial weight. The evaluation of the performance of the generated S-boxes concludes that the proposed technique exhibits strong cryptographic properties [15].

25.2.2 Optimization Techniques Based on GA

Numerous studies have employed genetic algorithms as a means to optimize S-boxes. Wang et al. suggested solving the S-box construction problem as a Traveling Salesman Problem using chaotic and genetic algorithms. Artuğer and Özkaynak presented the utilization of heuristic methods to enhance the cryptographic features of S-box structures and optimize them derived from chaotic entropy sources. This work presents the development of a method that exhibits a higher degree of nonlinearity compared to the algorithms previously proposed in the existing literature. Research findings have demonstrated that the nonlinearity value has the potential to be elevated to 111.75 [7]. The proposed method uses a chaotic map and evolution process to create a stronger S-box. Performance tests reveal that the offered S-box has good cryptographic properties, proving that the suggested approach generates strong S-boxes [16]. Çavuşoğlu and Kökçam devised a GA-based S-box generation method where nonlinearity value, a crucial evaluation criterion, was processed. Performance testing determines S-box quality. Performance findings are compared to literature S-boxes [17]. Batina et al. developed a framework that effectively parses and analyzes a Verilog netlist. This framework abstracts the netlist as a graph consisting of interconnected cells, allowing for the generation of circuit statistics pertaining to its constituents and pathways. Based on the provided information, the genetic algorithm (GA) is utilized to derive the optimal configuration of Flip-Flops (FFs) that maximizes the throughput of the given netlist. By conducting this experiment, we demonstrate the feasibility of attaining a 50% enhancement in throughput while incurring a mere 18% augmentation in area within the UMC 0.13 μm low-leakage standard cell library [18]. The utilization of evolutionary approaches is employed in research to identify S-boxes that exhibit unique cryptographic features, as stated by Picek et al. This study focuses on conducting experiments related to the 8×8 S-box scenario, a widely utilized component of the AES standard. The experimental results provide evidence for the feasibility of locating S-boxes that exhibit the desired properties within the framework of the Advanced Encryption Standard (AES). Additionally, this research shows initial results obtained from side-channel investigations carried out on different versions of "enhanced" S-boxes [19].

25.3 Proposed Method

In this study, we have introduced an innovative optimization approach that incorporates an improved crossover strategy. This study utilizes the deep learning paradigm of machine learning to enhance the robustness of crossover. In our study, we have investigated and utilized the domain of similarity learning to enhance the effectiveness of the crossover operation in the context of Genetic Algorithm. The suggested methodology comprises three sequential steps: dataset preparation, deep similarity learning model training (namely, the Siamese Network), and prediction generation utilizing the trained model. The subsequent sections will describe all three steps.

25.3.1 Siamese Network

Similarity Learning employs a foundational framework known as Siamese Networks and utilizes Triplet losses as a means of optimizing the learning process. The Siamese Network operates on a dual set of data samples simultaneously, whereas the triplet loss function operates on a triple set of samples from the dataset concurrently. To utilize either of these two networks, it is necessary to format the dataset to conform to their specific structure. In the current research, a Siamese Network was employed to train and predict the similarity score. Siamese Networks employ a training methodology that involves utilizing pairs of data samples. The dataset has been prepared in its current format.

25.3.1.1 Data Gathering and Preparation

In order to construct pairwise S-boxes, a selection of optimal S-boxes from the dataset [15] was made. These chosen S-boxes exhibit notable nonlinearity values, such as 110, 111, 112, etc. We have chosen an S-box, S_1, and another, S_2, and calculated nonlinearity of both, n_1 and n_2. The S_1–S_2 similarity is determined as:

$$L = |n_1 - n_2| \tag{25.1}$$

For the given dataset, L is determined for each pair of S-boxes. The highest nonlinearity value (M) is discovered and each $L \rightarrow (L_1, L_2, \ldots L_n)$ is divided by M to reduce similarity values between 0 and 1.

$$L = \left| \frac{n_1 - n_2}{M} \right| \tag{25.2}$$

Similar S-boxes have a similarity score close to 0 using (25.1), whereas dissimilar ones have a high value near 1. Inverting this behavior by subtracting the similarity score from 1 yields the formula:

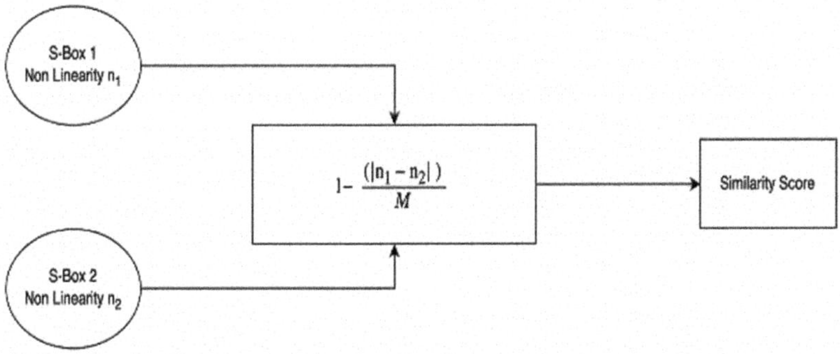

Fig. 25.1 Steps to calculate similarity score on dataset

$$L = 1 - \left|\frac{n_1 - n_2}{M}\right| \tag{25.3}$$

We have shown the process of calculating dataset in Fig. 25.1.

25.3.1.2 Designing and Training Siamese Network Model

We designed a Siamese network to train the dataset. The Siamese Network is composed of two branches that are identical in structure, each featuring a stacked Euclidean Distance assessment Layer. The model is completed by incorporating a Dense layer that consists of a single node utilizing the Sigmoid Activation function to assess the score. The Siamese Model is composed of multiple branches, each of which consists of three Convolutional Layers (1D) together with a Global Max Pooling Layer. The Convolutional Layer is composed of many layers, each consisting of 128 nodes. These nodes have a kernel size of 3 and utilize the Rectified Linear Unit (ReLU) activation function. The Lambda Layer, also known as the Euclidean distance layer, is responsible for computing the Euclidean distance between two vectors originating from separate heads of a Siamese Network. The model has been trained using a dataset consisting of 45 L of data, and its performance has been evaluated using a separate dataset consisting of 5 L of data for validation purposes. Once the model has completed its training process, it is then immobilized and stored for future utilization. The designed Siamese network is shown in Fig. 25.2.

25.3.1.3 Genetic Algorithm

This stage involves utilizing the information obtained from the preceding two steps in the Genetic Algorithm (GA). The fundamental components of genetic algorithms (GA) encompass a sequence of five key stages: Initialization, Objective Calculation, Crossover, Mutation, and Selection. Subsequently, the aforementioned three steps are iteratively executed until an optimal population is obtained. The methodology

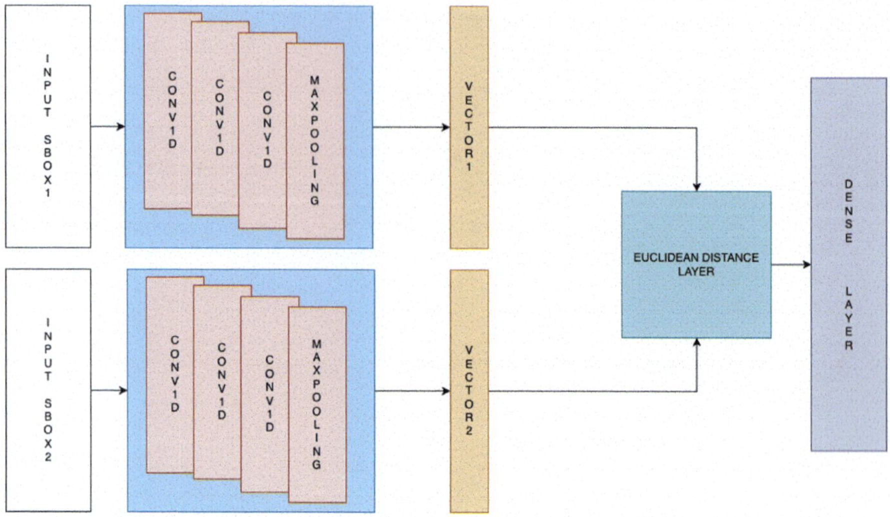

Fig. 25.2 Siamese CNN network for training S-boxes dataset

employed in our work incorporates a series of procedures, including the incorporation of an additional proposed crossover. These steps are comprehensively outlined as follows:

a. *Initialization*: The population is initialized by the utilization of chaotic one-dimensional maps. The inclusion of a key is essential in order to establish the relationship between chaotic maps and population formation. The collective populace is represented by the symbol N.
b. *Objective Function Evaluation*: The nonlinearity of the S-boxes in the population is determined by applying the nonlinearity formula (25.2), as presented below.

$$NL(f) = \frac{1}{2}\left(2^n - WH_{max}(f)\right) \tag{25.4}$$

The candidates are arranged in reverse order based on their nonlinearity, and the program proceeds to the subsequent step.

c. *Crossover*: The crossover operation serves as the fundamental mechanism for generating a new population in the context of Genetic Algorithms (GA). Multiple parent solutions are chosen from the existing population. The selection process might be predicated upon a range of methodologies, such as roulette wheel selection, tournament selection, or alternative approaches. The crossover procedure entails the exchange of genetic information between the selected parents at specific crossing points. There exist multiple crossover approaches, which encompass.

- Single-Point Crossover: The single-point crossover technique involves the selection of a single crossover point, at which the genetic information beyond this point is exchanged between the parental individuals.
- Two-Point Crossover: The Two-Point Crossover technique involves the selection of two crossover locations, between which the genetic material of the parents is exchanged.
- Uniform Crossover: The uniform crossover method involves the random selection of genes or elements from either parent with an equal probability of creating offspring.
- Arithmetic Crossover: The arithmetic crossover method is frequently employed in real-valued representations, wherein it computes the mean value of matching genes from the parent individuals in order to generate the offspring.
- Blend Crossover: The Blend Crossover technique is employed for real-valued representations, wherein the child is generated by taking a weighted average of the genes from both parents.

d. *Proposed Crossover*: In addition to the conventional crossover operation, we have proposed a novel way to the process of crossover operation. In the scope of our study, the methodology employed by parents for selecting their choices undergoes a transition from a random approach to a more regulated and deliberate process. In this study, a single parent is chosen from the existing population, while the selection of the other parent is determined randomly by the utilization of a chaotic map. The two parental inputs are processed by the Siamese Model, which then predicts the similarity score. In this process, a predetermined threshold value is established. If the anticipated score falls below this threshold value, the parents proceed with the crossover operation. However, if the forecasted score is above the threshold value, the same process is repeated for a certain number of iterations denoted by 'n'. The present study used a straightforward one-point crossover technique, wherein a single location is chosen to facilitate the exchange of genetic information.

After the crossover procedure, a set of offspring is formed from the parent solutions. The quantity of progeny generated is contingent upon the particular design of the genetic algorithm. The progeny that has just emerged are commonly integrated into the population with the purpose of substituting less adaptive individuals. This phenomenon contributes to the preservation of diversity within the population and facilitates the algorithm's efficient exploration of the search space .

e. *Mutation*: Mutation is a crucial activity within the context of Genetic Algorithms (GA) as it serves to maintain and sustain diversity. Crossover is a genetic process that entails the recombination of genetic material derived from parental solutions. On the other hand, mutation is a mechanism that introduces minor, stochastic alterations to a candidate solution. This phenomenon contributes to the preservation of biodiversity within the population and introduces an element of exploration into the search process.
f. *Selection*: The process of selection plays a pivotal role in the mechanics of a genetic algorithm (GA). The process involves selecting individuals from the existing population to serve as parents for generating the subsequent generation of solutions. The main objective of the selection process is to prioritize individuals with superior fitness values while simultaneously maintaining variety within the population.

The proposed algorithm and proposed flowchart can be seen in Algorithm 1 and Fig. 25.3 respectively.

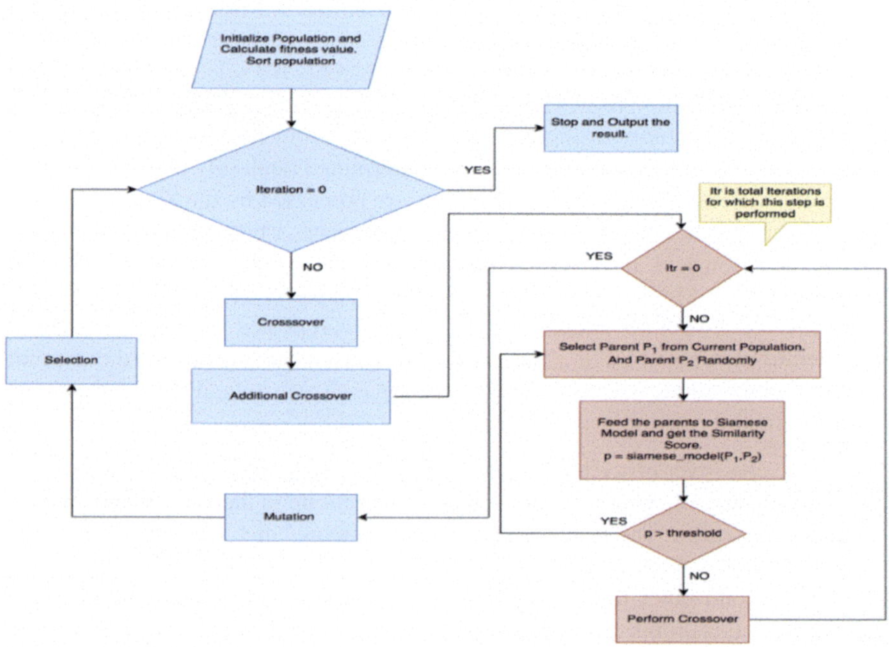

Fig. 25.3 The proposed flowchart

Step 1: Initialize the population, N -> total population

Step 2: Calculate the non-linearity of S-boxes and sort the S-boxes.

Step 3: **While Iteration:**

a. Perform crossover operation on population and generate new children.

b. Perform proposed crossover:

i. Select P_1 from the current population.

ii. Select P_2 randomly using a chaotic map.

iii. Using Siamese Network get the similarity score of both the parents.

p = Siamese_model(P_1,P_2)

iv. **If** $p <$ threshold, then

v. Do the crossover between P_1 and P_2,

else:

Got to (i).

c. Perform Mutation operation

d. Perform Selection Operation

End While

Step 4: Stop and Output Results.

25.4 Results and Discussion

The proposed strategy divides the results into two different models. The initial model considered is a deep learning model based on a Siamese Convolutional Neural Network (CNN), trained on an experimental dataset [16]. The dataset is made up of 50 L paired S-boxes, 45 L (90%) of which are used to train the Siamese CNN model and the remaining 5 L (10%) are used for validation. We have tested the model on different 5000 test datasets. To enable an unbiased evaluation of the model's performance, the test set was kept separate and was not used during the training or validation processes. We have mentioned the hyperparameters to train the model in Table 25.1.

We trained the model for around 50 epochs, and the plots of validation loss, training loss and testing loss for each epoch are presented in Fig. 25.4a–c respectively.

We have also tested the model to predict the similarity score on test datasets as shown in Fig. 25.4d. To make the results more prominent, we have tested our Siamese

Table 25.1 Hyperparameters for training 1-D CNN

Hyperparameters	
Learning rate	0.01
Number of epochs	50
Activation function (output layer)	Sigmoid
Activation function (input layer)	ReLU
Total no. of kernels	128, each of size 3

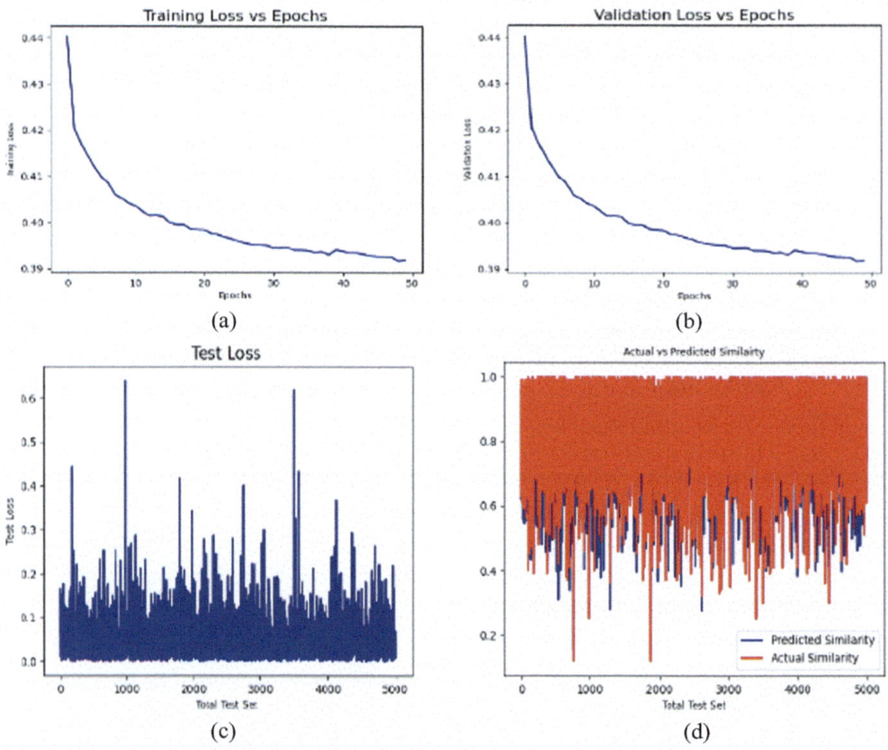

Fig. 25.4 Plot diagram: **a** training loss, **b** validation loss, **c** test loss, **d** actual similarity versus predicted similarity

CNN model on AES S-box and a randomly generated S-box with nonlinearity of 112 and 103 respectively. We verified it on three different cases which are as follows:

Case 1: When only AES S-box was passed through both heads of the model, the similarity score obtained was 94%.

Case 2: When both AES S-box and randomly generated S-box were passed through the heads, a similarity score of 73% was obtained. The nonlinearities of both S-boxes further support this result.

Case 3: When only randomly generated S-box was passed through both heads of the model, the similarity score obtained was 94%. We have shown the effect of various cases in Fig. 25.5.

The genetic algorithm used in the later part of the proposed architecture helps to optimize S-boxes and leverages the model to choose distinct S-boxes, which speeds up convergence. The algorithm was trained 5000 iterations at population, N = 10, Fig. 25.5. The time complexity for running the algorithm on this population size, i.e., 10, was around 90 min for 5000 iterations. Testing of our Siamese CNN model on

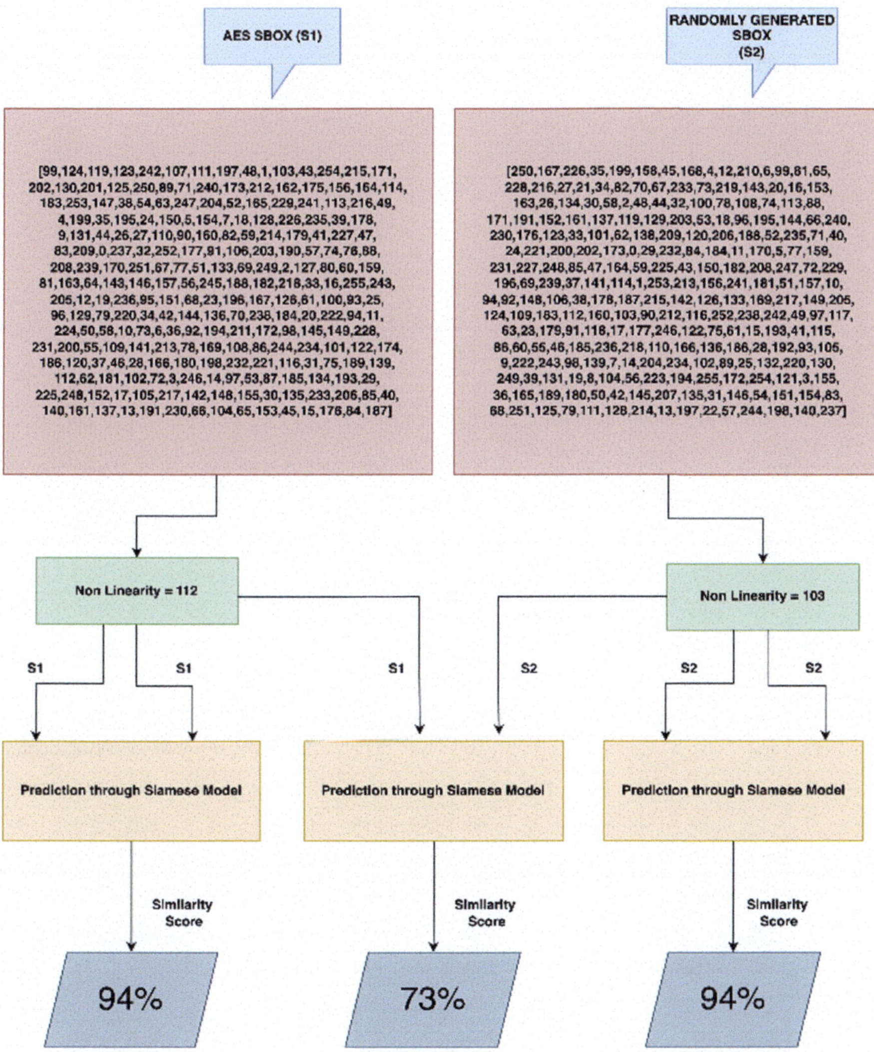

Fig. 25.5 Testing model against AES S-box and randomly generated S-box

AES S-box and a randomly generated S-box results obtained is shown in Fig. 25.6. The algorithm's best nonlinearity score for the current execution is 110.25. The optimized S-box is presented in Table 25.2.

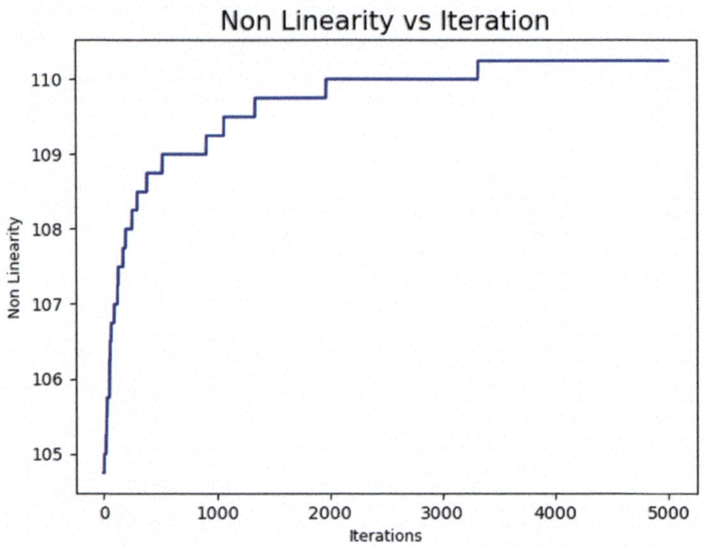

Fig. 25.6 Nonlinearity versus iteration

Table 25.2 S-box comparison with the state of the art

139	221	3	51	156	136	149	2	47	79	150	59	160	252	255	193
115	0	20	127	253	100	14	165	125	239	10	231	157	250	189	147
29	46	213	140	249	170	246	74	43	32	45	198	58	15	196	94
225	56	5	62	237	236	102	85	199	211	123	131	230	152	228	143
195	80	22	238	222	220	224	60	172	214	223	176	219	226	67	107
1	178	75	244	124	118	217	99	209	153	208	205	7	200	185	194
192	41	204	91	39	105	36	158	95	248	64	117	122	38	184	78
8	97	61	190	251	135	108	114	144	174	182	126	191	171	163	166
183	77	13	159	130	232	72	63	169	155	12	215	175	120	234	69
188	206	151	73	229	83	148	187	31	216	145	17	180	137	233	162
235	84	121	119	227	173	116	245	242	111	110	106	54	177	57	104
103	168	98	142	53	240	82	93	179	92	167	90	88	141	16	202
129	203	86	18	81	71	101	210	68	134	66	65	33	37	55	40
128	48	49	87	9	44	70	218	27	35	138	181	42	243	146	50
25	23	212	201	21	19	112	254	11	28	76	113	133	6	241	186
139	221	3	51	156	136	149	2	47	79	150	59	160	252	255	193

Table 25.3 Comparison with the state of the art

S-box method	Optimization technique	Nonlinearity
[17]	GA	108
[16]	PSO	106
[20]	BHC	110
[13]	TLBO	104
[20]	ICO	108

Table 25.3 presents a comparative analysis of various optimization strategies and their related nonlinearities. Wang et al. used a chaotic map and a genetic algorithm, as described in [1] to create S-boxes with a maximum nonlinearity of 108. In contrast, Ahmad et al. reached a greater nonlinearity of 110 by using the PSO (Particle Swarm Optimization) and BHC (Binary Harmony Clustering) approaches, as detailed in both [2, 3]. Furthermore, prior research, as described in [4, 5], was able to achieve a maximum nonlinearity of 108. Notably, the majority of these research endeavors relied on large beginning population sizes. Remarkably, our research model deviated from this convention by employing an exceptionally small initial population size of only 10. Through innovative modifications to the genetic algorithm and the incorporation of similarity learning networks, our model achieved a significantly higher nonlinearity score of 110.25. This exceptional performance places our approach ahead of most existing optimization strategies in this domain.

25.5 Conclusion and Future Work

The study shows the process of improving convergence by introducing diversity using the methods of deep learning. The results achieved by the experiment are comparable to the ones achieved on vast population sizes which leads to a massive exploration of population.

This work includes a new S-box generator based on modified Genetic Algorithm (GA) based on Similarity Learning. We have focused on optimizing the nonlinearity of S-boxes and enhancing the robustness of the optimization approach. To achieve, the goal of nonlinearity we have utilized our proposed architecture of Siamese CNN to train the pairwise dataset of S-boxes. Our research model deviated with tradition by using an unusually modest initial population size of just 10. Our model obtained a considerably higher nonlinearity score of 110.25 by utilizing genetic algorithm tweaks and similarity learning networks. The idea is to boost evolutionary algorithms by guiding them through deep learning models. The convergence seems very fast and dependent on very small size of the initial population. Furthermore, the idea of deep learning gives the evolutionary algorithm a control over the parent selection for next population generation. This idea can be extended further to multi objective evolutionary algorithms, like Multi Objective Genetic Algorithms, Multi Objective

Particle Swarm Algorithms, etc. These algorithms heavily depend on the initial population taken. Using deep learning approach, we can guide these algorithms towards optimal solutions faster. Apart from Siamese Networks, triplet loss networks can also be explored in future studies. Future studies can also extend this concept to other meta-heuristic algorithms like PSO, TLBO, etc.

References

1. Behera, P.K., Gangopadhyay, S.: Evolving bijective S-boxes using hybrid adaptive genetic algorithm with optimal cryptographic properties. J. Ambient. Intell. Humaniz. Comput. **14**(3), 1713–1730 (2023). https://doi.org/10.1007/s12652-021-03392-6
2. Katz, J., Lindell, Y.: Introduction to Modern Cryptography. CRC Press, London, New York, Washington, DC, pp. 333–375 (2007, August 31)
3. Daemen, J., Rijmen, V.: AES proposal: Rijndael. In: Proceedings of 1st Advance Encryption Conference, CA, USA, pp. 1–45 (1998)
4. Artuğer, F.: A new S-box generator algorithm based on 3D chaotic maps and whale optimization algorithm. Wireless Pers. Commun. **19**, 1–9 (2023). https://doi.org/10.1007/s11277-023-10456-7
5. Zamli, K.Z., Din, F., Alhadawi, H.S.: Exploring a Q-learning-based chaotic naked mole rat algorithm for S-box construction and optimization. Neural Comput. Appl. **30**, 1–23 (2023). https://doi.org/10.1007/s00521-023-08243-3
6. Carlet, C.: On highly nonlinear S-boxes and their inability to thwart DPA attacks. In: International Conference on Cryptology in India, pp. 49–62. Springer Berlin Heidelberg, Berlin, Heidelberg (2005). https://doi.org/10.1007/11596219_5
7. Artuğer, F., Özkaynak, F.: SBOX-CGA: substitution box generator based on chaos and genetic algorithm. Neural Comput. Appl. **34**(22), 20203–20211 (2022). https://doi.org/10.1007/s00521-022-07589-4
8. Abdel-Basset, M., Abdel-Fatah, L., Sangaiah, A.K.: Metaheuristic algorithms: a comprehensive review. In: Computational Intelligence for Multimedia Big Data on the Cloud with Engineering Applications, pp. 185–231 (2018, January 1). https://doi.org/10.1016/B978-0-12-813314-9.00010-4
9. Michalewicz, Z., Schoenauer, M.: Evolutionary algorithms for constrained parameter optimization problems. Evol. Comput. **4**(1), 1–32 (1996). https://doi.org/10.1162/evco.1996.4.1.1
10. Xie, L., Yuille, A.: Genetic CNN. In: Proceedings of the IEEE International Conference on Computer Vision, pp. 1379–1388 (2017)
11. Aziz, S., Amjad, M.: A comprehensive study on feature extraction techniques for Indian sign language recognition system. In: International Conference on Computational Intelligence in Data Science, pp. 104–125. Springer Nature Switzerland, Cham (2023, February 23). https://doi.org/10.1007/978-3-031-38296-3_9
12. Lew, M.S., Sebe, N., Djeraba, C., Jain, R.: Content-based multimedia information retrieval: state of the art and challenges. ACM Trans. Multimed. Comput., Commun., Appl. (TOMM) **2**(1), 1–9 (2006). https://doi.org/10.1145/1126004.1126005
13. Farah, T., Rhouma, R., Belghith, S.: A novel method for designing S-box based on chaotic map and teaching–learning-based optimization. Nonlinear Dyn. **88**(2), 1059–1074 (2017). https://doi.org/10.1007/s11071-016-3295-y
14. Tian, Y., Lu, Z.: Chaotic S-box: intertwining logistic map and bacterial foraging optimization. Math. Probl. Eng. **15**, 2017 (2017). https://doi.org/10.1155/2017/6969312
15. Ahmad, M., Bhatia, D., Hassan, Y.: A novel ant colony optimization based scheme for substitution box design. Procedia Comput. Sci. **1**(57), 572–580 (2015). https://doi.org/10.1016/j.procs.2015.07.394

16. Ahmad, M., Khaja, I.A., Baz, A., Alhakami, H., Alhakami, W.: Particle swarm optimization based highly nonlinear substitution-boxes generation for security applications. IEEE Access **23**(8), 116132–116147 (2020). https://doi.org/10.1109/ACCESS.2020.3004449
17. Wang, Y., Wong, K.W., Li, C., Li, Y.: A novel method to design S-box based on chaotic map and genetic algorithm. Phys. Lett. A **376**(6–7), 827–833 (2012). https://doi.org/10.1016/j.physleta.2012.01.009
18. Çavuşoğlu, Ü., ÇKökçam, A.H.: A new approach to design S-box generation algorithm based on genetic algorithm. Int. J. Bio-Inspired Comput. **17**(1), 52–62 (2021). https://doi.org/10.1504/IJBIC.2021.113360
19. Batina, L., Jakobovic, D., Mentens, N., Picek, S., De La Piedra, A., Sisejkovic, D.: S-box pipelining using genetic algorithms for high-throughput AES implementations: how fast can we go? In: International Conference on Cryptology in India, pp. 322–337. Springer International Publishing, Cham (2014, October 25). https://doi.org/10.1007/978-3-319-13039-2_19
20. Zhang, T., Chen, C.P., Chen, L., Xu, X., Hu, B.: Design of highly nonlinear substitution boxes based on I-Ching operators. IEEE Trans. Cybern. **48**(12), 3349–3358. https://doi.org/10.1109/TCYB.2018.2846186

Chapter 26
Multifactorial Model for Targeted Attacks Counteracting Within the Framework of a Multi-Step Quality Game with Fuzzy Information

V. Lakhno, V. Malyukov, O. Smirnov, B. Bebeshko, V. Chubaievskiy, M. Zhumadilova, I. Malyukova, and S. Smirnov

Abstract A multifactorial model of countering targeted attacks (APT attacks) in the context of available financial resources (FR) within a bilinear multi-step quality game with multiple terminal surfaces in a fuzzy formulation is considered. The distinction of the model discussed from others is that it manages to reflect the variability of financial defense strategies against APT attacks and the strategies of the attackers in the dynamics of the interaction of opposing parties. This allows considering the general problem of cybersecurity within a game scheme, including for tasks of countering APT attacks in a fuzzy formulation, with the available FR on the defense side. A bilinear multi-step quality game of multidimensional objects with fuzzy information was proposed to solve the problem. It is shown that solving quality tasks considering their multidimensionality and bilinearity of players' interaction dynamics is a quite challenging problem. The problem formulated in the study was solved using *R*. Bellman's optimality principle. The solution was found within

V. Lakhno · V. Malyukov
National University of Life and Environmental Sciences of Ukraine, Kyiv, Ukraine
e-mail: lva964@nubip.edu.ua

O. Smirnov · S. Smirnov
Central Ukrainian National Technical University, Kropyvnytskyi, Ukraine

B. Bebeshko (✉) · V. Chubaievskiy
State University of Trade and Economics, Kyiv, Ukraine
e-mail: thismushroom@gmail.com

V. Chubaievskiy
e-mail: chubaievskyi_vi@knute.edu.ua

M. Zhumadilova
Yessenov University, Aktau, Kazakhstan
e-mail: mereke.zhumadilova@yu.edu.kz

I. Malyukova
Rating Agency "Expert-Rating", Kyiv, Ukraine

S. K. Pal et al. (eds.), *Intelligent Informatics*, Smart Innovation, Systems and Technologies 389, https://doi.org/10.1007/978-981-97-2147-4_26

the scheme of a positional multi-step game with fuzzy information. Sets of preferences and optimal strategies for the first player's distribution of FR for building a defense system against APT attacks were identified. The solution is presented from the defense side's perspective and provides the necessary conditions that guarantee the first ally player the ability to ensure the protection of informatization objects (IO) within a finite time with a given reliability. The model presented in the work is designed for implementation as a software module of the intellectual information system being developed for searching strategies to counter APT attacks. The conducted computational experiment allowed visualizing the game results, during which the financial capabilities of the opposing parties are analyzed. It is shown that by determining an approximate volume of FR that the attacking side can invest in an APT attack, one can also determine the amount of funds required by the defense side.

26.1 Introduction

The flip side of the rapid development of information technologies (IT) and the digitalization of vital business processes for objects of informatization (OI) has been the synchronous increase in the number of cybersecurity threats. The number of various attacks targeting the computer systems of OI grows annually. Cybercriminals not only invent new ways and scenarios for cyberattacks but also perfect old tested schemes, like phishing.

One of the most complex attacks is the Advanced Persistent Threat (APT) attacks [1]. In conducting such an attack, attackers focus all their efforts on a specific victim. APT attacks are usually meticulously planned, from selecting performers for each stage to a meticulous analysis of financial expenses for each stage, starting from reconnaissance and ending with eradicating traces of their presence. Even today's commonplace DDoS attack is costly for the attacking party. Let alone the potential costs of conducting a long, multi-stage, skill-requiring APT attack. If the APT attack proves successful for the attackers, they can establish themselves within the OI infrastructure and remain undetected for several months or even years. During this time, they have virtually unchecked access to the victim's information assets.

Investing in information security (IS) not only minimizes the risks of losing OI's information assets but ultimately increases profitability.

Issues concerning determining the amount of resources, primarily financial (or those equated to financial), required to ensure IS, are a priority for OI management. By determining an approximate amount of financial resources (FR) that the attacking party might invest in APT attacks, one can define the volume of FR required by the defense side to minimize losses from an APT attack.

Indeed, the aforementioned points have piqued our interest in this topic. We believe that our discussions and mathematical derivations will be useful for professionals dealing with IS issues and countering threats like APT attacks.

26.2 Literature Review

Looking at the retrospective of scientific publications dedicated to researching the effectiveness of investing in information security (IS), one can notice two prevailing approaches. The first one is based on the use of mathematical programming methods. Relevant works can be mentioned here [2–8]. The second approach is based on the application of game theory [9–14].

This article is devoted to the development and implementation of one of the methods from the second group.

26.3 Aim and Goals

The aim of this work is to develop mathematical models that constitute the computational core of an intelligent system with decision support elements to counter targeted APT attacks.

During the research, the following tasks were addressed:

1. Based on game theory (multi-step quality game), a model was developed to find optimal strategies for players (opposing sides—defender and attacker) in conditions of vague information about the financial resources of each side. This will allow analytically determining the game's value and characteristics of the degree of certainty in achieving the players' objectives when applying their optimal financial strategies. Specifically, such a solution facilitates the analysis of the situation when countering a side that launches an APT attack on a specific object of informatization (OI).
2. Conducting computational experiments using the PyCharm software development environment.

26.4 Problem Statement

26.4.1 Problem Statement

As shown earlier, financial provision (or financial resources—FR) is necessary for ensuring the information security (IS) of objects of informatization (OI). This FR is essential for the attacking side, for instance, to develop tools used at various stages of APT attacks, to pay the services of personnel conducting reconnaissance, installing malicious software in the target environment, etc. The defense side requires the FR for the acquisition and deployment of IS tools. The presence of two opposing sides—the defending and the attacking—provides a basis for calling them players. We will consider the defense side of the OI as the first player. The attacking side of the OI will be deemed the second player. Their interaction will be viewed as financial (i.e., the confrontation of the players' FR). The players' FR confrontation process is determined by the interdependence of their expenditures to achieve their objectives. The interaction between the players occurs discretely over time. The first player has technological strategies to counter the APT attacks of the second player. For example, these technological strategies may be determined by the capabilities of: (1) information protection tools to detect various anomalies and signs of attacks in the OI computer systems; (2) network activity monitoring tools; (3) etc.

The second player also has technological strategies to inflict damage on the first player. Strategies of the attacking side might include: (1) tools for collecting information about the victim from public sources; (2) unique malicious software for executing APT attacks; (3) social engineering methods; (4) tools for malicious software and its embedding in the target environment; etc. Note that the coincidence in the number of alternatives doesn't affect the problem's essence. They might not coincide. In such a case, the task can be reduced to the case of the number of alternatives being equal.

If the second player employs its technological strategies, it results in financial losses for the first player and vice versa. If the financial expenses of building the IS system of the OI prove effective, the attacking side doesn't achieve its primary objectives during the APT attack. Consequently, the attacking side incurs losses. As shown in the studies [15, 16], the cyber-attack market, including APT attacks, is well organized. In fact, it resembles some kind of assembly line. Different performers are responsible for each stage of the APT attack. Often, "less qualified" hackers pass on the results of their work to more "qualified" hackers.

We assume that any technological strategy of the second player leads to financial losses. Let's assume that the second player has applied his i-th technological strategy. This results in a financial damage of θ_i^1 to the first player. Next, let's assume that the first player has applied his j-th technological strategy. Its application brings a financial increment of θ_j^2. Let's denote by r_{ij}^1 the ratio θ_i^1/θ_j^2 and by r_{ij}^2 the ratio θ_j^2/θ_i^1. If $\theta_i^1 = 0$ for i or $\theta_j^2 = 0$ for some j, then such strategies are excluded from consideration.

Let's designate by R_1 a matrix consisting of elements r_{ij}^1. The number of rows corresponds to the number of alternatives of the second player, and the number of each row corresponds to the respective alternative of the second player. We will denote by R_2 the matrix in which the rows correspond to the alternatives of the first player, and the columns correspond to the alternatives of the second player. The elements r_{ij}^2 mean that they are in the j-th row and i-th column.

The formed matrices R_1 and R_2 are analogues of the so-called profitability matrices. Each element, for example, for the matrix R_1, characterizes the amount of loss from the application of the technological strategy of the attacking player and is compensated by one unit of income from the application of the technological strategy of the defending player, and vice versa. This circumstance allows us to form the dynamics of the interaction of opposing parties during an APT attack.

Let's denote:

- μ_j ($j = 1, \ldots, M$) as values $\mu_j \geq 0$, $\sum_{j=1}^{M} \mu_j = 1$ that are elements of the diagonal matrix Ξ of order M, with diagonal elements μ_j. The matrix Ξ characterizes the "structure" of the expenses (losses) of the second player. The element μ_j indicates the share of the j-th value of the set of incomes of the first player, which shows the transformation of this set into the j-th component of the amount of the set of expenses (losses) of the second player. That is, if $(\gamma_1, \ldots, \gamma_M)$ is the set of incomes of the first player, then the j-th component of the magnitude of the set of expenses of the second player will be converted to a set of incomes of the first player equal to $\mu_j \cdot (\gamma_1, \ldots, \gamma_M)$.

- – ρ_j $(j = 1, \ldots, M)$ as values $\rho_j \geq 0$, $\sum_{j=1}^{M} \rho_j = 1$ that form the diagonal matrix Λ of order M, with diagonal elements λ_j. The matrix Λ characterizes the "structure" of the set of incomes of the first player. The element ρ_j indicates the share of the j-th value of the set of expenses of the second player, which shows the transformation of this set into the j-th component of the magnitude of the set of incomes of the first player. If $(\lambda_1, \ldots, \lambda_M)$ is the set of expenses of the second player, then the j-th component of the magnitude of the set of incomes of the first player will be converted to a set of expenses of the second player equal to $\rho_j \cdot (\lambda_1, \ldots, \lambda_M)$.

Remark 1 If there is a set of incomes $\lambda = (\lambda_1, \ldots, \lambda_M)$ of the first player (for example, there is data that the project of implementing information security systems turned out to be economically beneficial, since the net present value of incomes from the implementation project is positive and exceeds the net present value of costs for the implementation project), then if you perform the operation: $R_1 \cdot \lambda$, then we get an M-dimensional vector, which "as if" means a set of expenses of the second player. However, in fact, this product only allows us to determine only one component of this M-dimensional vector (of the second player), since the entire vector $\lambda = (\lambda_1, \ldots, \lambda_M)$ will be "spent" on only this component. For the other components of the set of expenses of the second player, there is no more set of incomes of the first player that would be "equivalent" to this component of the second player. The entire set of incomes of the first player "went" to "equalizing" in efficiency with one component of the set of expenses of the second player. Therefore, it is necessary to divide the set of incomes into M parts in order to be able to "equalize" the efficiency of the sets of expenses of the second player for all its components. It is done by introducing the set: ρ_j $(j = 1, \ldots, M)$; $\rho_j \geq 0$, $\sum_{j=1}^{M} \rho_j = 1$. Note that the choice of these coefficients can be made in other ways, not necessarily as described above.

The same is true for the set of expenses of the second player.

The first player, having $\gamma(0) \in R_+^M$ financial resources (FR) at a certain point in time $t = 0$, transforms them into a certain amount of resources $L_1 \cdot \gamma(0)$. Where L_1 is a transformation matrix for the FR resources of the first player of M order with positive elements. Then, he determines the amount of his investments $U(0) \cdot L_1 \cdot \gamma(0)$ by choosing the elements $u_i(0) : 0 \leq u_i(0) \leq 1$, which are diagonal elements of a diagonal matrix $U(0)$ of M order, which defines the values of the first player's strategy. Such an investment by the first player means that it allows compensating $\Xi \cdot R_1 \cdot U(0) \cdot L_1 \cdot \gamma(0)$ of the losses from the actions of the second player.

Similarly, the second player operates. The second player, having FR at a $t = 0$ $\lambda^{\xi}(0) \in R_+^M$ point in time, transforms them into $L_2 \cdot \lambda^{\xi}(0)$ amount of resources. Where

L_2-the transformation matrix for the FR of the second player of M order, with positive elements. Then he determines the size of his investments $V(0) \cdot L_2 \cdot \lambda^{\xi}(0)$ in an APT attack, by choosing elements $v_i(0) : 0 \leq v_i(0) \leq 1$, which are diagonal elements of a diagonal matrix of $V(0)$ order M. They define the values of the second player's strategy. Such an investment by the second player means that it allows offsetting $\Lambda \cdot R_2 \cdot V(0) \cdot L_2 \cdot \lambda^{\xi}(0)$ revenues from the actions of the first player.

In their interaction, it is assumed from an informational perspective that a situation arises where the first player does not know the exact state $\lambda^{\xi}(0)$ $\left(\lambda^{\xi}(0) \in \mathrm{int}R_{+}^{M}\right)$ of the second player at $t = 0$ point in time. He only has access to information that the state of the second player belongs to a fuzzy set $\{\Omega, m(.)\}$, where Ω is a subset R_{+}^{M}, $m(.)-$ of membership function of the state λ^{ξ} to the set Ω, $m(\lambda^{\xi}) \in [0, 1]$ for $\lambda^{\xi} \in \Omega$.

Therefore FR of the players in a point in time $t = 1$ can be denoted as:

$$\begin{aligned} \gamma(1) &= L_1 \cdot \gamma(0) - U(0) \cdot L_1 \cdot \gamma(0) - \Lambda \cdot R_2 \cdot V(0) \cdot L_2 \cdot \lambda^{\xi}(0); \\ \lambda^{\xi}(1) &= L_2 \cdot \lambda^{\xi}(0) - V(0) \cdot L_2 \cdot \lambda^{\xi}(0) - \Xi \cdot R_1 \cdot U(0) \cdot L_1 \cdot \gamma(0). \end{aligned} \tag{26.1}$$

At the point in time $t = 1$ following options are available:

$$\left(\gamma(1),\ \lambda^{\xi}(1)\right) \in S_0 \text{ with certainty } \geq p_0,\ 0 \leq p_0 \leq 1 \tag{26.2}$$

$$\left(\gamma(1),\ \lambda^{\xi}(1)\right) \in F_0 \text{ with certainty } \geq p_0 \tag{26.3}$$

$$\left(\gamma(1),\ \lambda^{\xi}(1)\right) \in D_0 \text{ with certainty } \geq p_0 \tag{26.4}$$

$$\left(\gamma(1),\ \lambda^{\xi}(1)\right) \in H_0 \text{ with certainty } \geq p_0 \tag{26.5}$$

where S_0, F_0, D_0 and H_0:

$$S_0 = \bigcup_{i=1}^{M} \{(\gamma, \lambda) : (\gamma, \lambda) \in R^{2 \cdot M}, \gamma \geq 0, \lambda_i < 0\}$$

$$F_0 = \bigcup_{i=1}^{M} \{(\gamma, \lambda) : (\gamma, \lambda) \in R^{2 \cdot M}, \gamma_i < 0, \lambda \geq 0\}$$

$$D_0 = \left\{ \bigcup_{i=1}^{M} \left\{ \begin{matrix} (\gamma, \lambda) : (\gamma, \lambda) \in R^{2 \cdot M}, \\ \gamma_i < 0 \end{matrix} \right\} \right\} \cap \left\{ \bigcup_{i=1}^{M} \left\{ \begin{matrix} (\gamma, \lambda) : (\gamma, \lambda) \in R^{2 \cdot M}, \\ \lambda_i < 0 \end{matrix} \right\} \right\},$$

$$H_0 = R_{+}^{2 \cdot M}.$$

If condition (26.2) is met, we consider the procedure of interaction between the opposing sides in a targeted APT attack to be completed. This means that the second player did not have enough financial resources (FR) to harm the first player. At least with one of its technological strategies that the second player planned to apply with $\geq p_0$ level of certainty.

If condition (26.3) is met, we consider the procedure of interaction between the opposing sides in a targeted APT attack to be completed. Because the first player did not have enough FR to counteract the harm inflicted by the second player. At least with one of its technological strategies that the first player planned to use with $\geq p_0$ level of certainty.

If condition (26.4) is met, we consider the procedure of interaction between the opposing sides in a targeted APT attack to be completed, as neither player had enough FR to continue opposing each other. At least with one of their technological strategies, which they could have applied with $\geq p_0$ level of certainty.

In case (26.5), they continue to interact for moments in time $t > 1$. The process described by system (26.1) for the financing procedure is considered within the framework of a positional multi-step game with fuzzy information [17].

Due to symmetry, we will limit ourselves to examining the problem from the perspective of the first player. The solution to problem 1 involves finding the set of "preferences" of the first allied player Σ_1 and its optimal strategies $U_*(.)$. Similarly for the second allied player.

In problem 1, the first player is considered an allied player, while the second player is considered an adversary. In problem 2, it's the opposite—the second player is considered the allied player, while the first player is seen as the adversary.

The procedure for player interaction using a system of discrete equations generates, at each moment in time t, a set of pairs of fuzzy sets $\{\Gamma_t, n_t(.)\} \times \{\Omega_t, m_t(.)\}$. These sets reflect the process of transitioning from the initial states of the players $(\gamma(0), \lambda^{\xi}(0))$ to subsequent states when players apply control actions.

It is assumed that the first player knows his states $\gamma(\tau)$ at every moment $t (t \in [0, 1, 2, ...))$ for $\tau \leq t$. The following conditions are met: $\gamma(\tau) \geq 0$ if the reliability of such states is $n_\tau(\gamma(\tau)) \geq p_0$ and $\gamma(\tau) \notin R_+^M$ if the reliability of such states is $n_\tau(\gamma(\tau)) < p_0$, as well as known values of the first player's strategy implementations $U(\tau)(\tau \leq t)$ allocated for interaction with the second player.

Let's define the function $F(.) : X \to R_+, F(x) = \{\sup m(y),\ \text{for } y \leq x, x \in R_+^M\}$. Denote by Φ the set of such functions and by $T^* = [0, 1, 2, \ldots)$ the range of variation of the time variable.

The definition of a pure strategy was given in work [18].

The second player chooses his strategy $V(.)$ based on any information.

Let's define the set of initial states that possess property B.

Property B: if the game starts from the initial states, then the first player, by choosing his strategy $U_*(.)$, can ensure the fulfillment of condition (26.2) at one of the moments in time t. Furthermore, this strategy chosen by player 1 prevents the second player from fulfilling condition (26.3) at previous moments in time.

We will call the set of such states the preference set of the first player Σ_1, and the strategies $U_*(.)$ of the first player with the specified properties will be called his optimal strategies.

Thus, the goal of the first player is to find the preference set, as well as to find his strategies by applying which he will achieve the fulfillment of condition (26.1).

The model of confrontation between the sides is a bilinear multi-step quality game with multiple terminal surfaces and fuzzy information. Finding the preference sets of the first player and his optimal strategies depends on a set of parameters.

To describe the preference sets of the first player, it is necessary to introduce a series of notations and magnitudes.

Let's define the set $S(p_0) = \{c(0) : F(c(0)) \geq p_0\}$. For any $x \in R_+^M$ consider the set $L_x = \{z : z = l \times x, l \in R_+\}$. For any $x \in R_+^M$ consider the set $Q(x, p_0) = S(p_0) \cap L_x$. Let's define the vector $\delta(x, p_0) : \delta(x, p_0) = \inf\{\delta^* : \delta^* \in Q(x, p_0)\}$. Consider the set $\Delta(p_0) = \{\delta(p_0) : \exists x \in R_+^n : \delta(p_0) = \delta(x, p_0)\}$. Next, we will provide the conditions that allow for finding the game's solution, i.e., the "preference" sets Σ_1 and the optimal strategies $U_*(.)$ of the first player-ally.

26.4.2 *Solution to Problem 1*

The solution to the given multiparametric problem depends on the relationship between the parameters defining the procedure for player confrontation.

Let's introduce the following notations.

Let $\widehat{\Sigma}_1$ denote the set:

$$\widehat{\Sigma}_1 = \Sigma_* - \bigcup_{i=1}^{M} \Sigma_i,$$

$$\Sigma_* = \left\{ \begin{array}{l} (\gamma(0), \delta(p_0)) : (\gamma(0), \delta(p_0)) \in R_+^{2M}, \\ L_1 \cdot \gamma(0) - \Lambda \cdot R_2 \cdot L_2 \cdot \delta(p_0) \in R_+^M \end{array} \right\},$$

$$\Sigma_i = \left\{ \begin{array}{l} (\gamma(0), \delta(p_0)) \in R_+^{2M}, \ (L_1 \cdot \gamma(0))_i = (\Lambda \cdot R_2 \cdot L_2 \cdot \delta(p_0)))_i, \\ i = 1, ..., M; \end{array} \right\}.$$

Furthermore, let's introduce additional notations:

E—identity matrix of order M:

$$Q_1 = \Xi \cdot R_1, \ Q_k = \left\{ (G_{k-1} \cdot L_2 \cdot \Xi \cdot R_1 - Q_{k-1} \cdot L_1)^+ + Q_{k-1} \cdot L_1 \right\};$$

$$\begin{aligned} G_1 = & E + \Xi \cdot R_1 \cdot \Lambda \cdot R_2, \quad G_k = (Q_{k-1} \cdot L_1 \cdot \Lambda \cdot R_2 - G_{k-1} \cdot L_2)^+ \\ & + G_{k-1} \cdot L_2 + (G_{k-1} \cdot L_2 \cdot \Xi \cdot R_1 - Q_{k-1} \cdot L_1)^+ \cdot \Lambda \cdot R_2, \ k = 2, \ldots \end{aligned}$$

$$x^+ = \begin{cases} x, \ x \geq 0; \\ 0, \ x < 0; \ x \in R; \end{cases}$$

$$\alpha_j^{k,j} = \begin{cases} 1, \ (G_k \cdot L_2)_{ij} = 0, \\ \dfrac{(Q_k \cdot L_1 \cdot \Lambda \cdot R_2)}{(G_k \cdot L_2)_{ij}}, \quad (G_k \cdot L_2)_{ij} \neq 0, \ k = 1, \ldots \end{cases}$$

The process of determining the preference sets of the first player depends on the relationship between the parameters that define this opposition of the parties.

Consider Case 1: $L_1 \cdot \Lambda \cdot R_2 \geq \Lambda \cdot R_2 \cdot L_2$.

Assume that $\forall k = 1, \ldots, u \quad \forall i : 1 \leq i \leq M$ exists for $m(i) : 1 \leq m(i) \leq M$, such that:

$$(Q_k \cdot L_1 \cdot \Lambda \cdot R_2)_{im} \geq (G_k \cdot \Lambda \cdot R_2)_{im}, \ 1 \leq i \leq M, \quad 1 \leq m \leq M; \ k = 1, 2, \ldots \tag{26.6}$$

Inequality (26.6) is not met.

In that case, the process of building the preference sets continues in time, and the preference sets of the first player will be countable. According to the principle of optimality by R. Bellman, the preference sets of the first player are constructed step by step. Firstly, the one-step preference set is constructed, then for two steps, and so on. Thus, the preference set of the first ally player is found, specifically, Σ_1 set is a union of the preference sets Σ_1^k of the first ally player [18] for a specific number of $\Sigma_1 = \bigcup\limits_{k=1}^{\infty} \Sigma_1^k$. steps, i.e.,

Using the above notations, the record of the first ally player's preference set Σ_1^k appears as:

$$\Sigma_1^k = \bigcup_{i=1}^{M} \{(\gamma(0), \delta(p_0)) : (\gamma(0), \delta(p_0)) \in R_+^{2M}, \\ L_1 \cdot \gamma(0) - \Lambda \cdot R_2 \cdot L_2 \cdot \delta(p_0) \in R_+^M, \\ (Q_k \cdot L_1 \cdot \gamma(0))_i > (G_k \cdot L_2 \cdot \delta(p_0)_i\}.$$

The optimal strategy $U_*(.,.,.)$, defined by the elements $u^*(.) = \left(u^{*,1}(.), \ldots, u^{*,M}(.)\right)$ of the first ally player—which are the diagonal elements of the matrix $U_*(.,.,.)$, in the preference area Σ_1^k is written as:

$$u^{*,j}(\gamma, \delta) = \begin{cases} 1 - [(\Lambda \cdot R_2 \cdot L_2 \cdot \delta)_j / (L_1 \cdot \gamma)_j], \quad npu \ (L_1 \cdot \gamma)_j > (\Lambda \cdot R_2 \cdot L_2 \cdot \delta)_j; \\ a \in [0, 1] \text{ at } (L_1 \cdot \gamma)_j = 0, \quad (L_1 \cdot \gamma)_j = (\Lambda \cdot R_2 \cdot L_2 \cdot \delta)_j; \\ 0, \text{ in the opposite case}; \quad (\gamma, \delta) \in R_+^{2M}; \quad j = 1, \ldots, M. \end{cases}$$

Preference sets and optimal strategies of the first ally player in Case 2: $(L_1 \cdot \Lambda \cdot R_2 \geq \Lambda \cdot R_2 \cdot L_2)$, are found in a similar manner.

The theorem [18] points out the conditions defining the possibility of completing the player interaction procedure in a finite number of steps with fuzzy information.

The considered mathematical model allowed finding a solution to the problem of counteracting target APT-attacks in the case of fuzzy information about the FR of the opposing party. The conditions for ending the procedure of counteracting target APT-attacks in a finite time are provided, given the presence of a sufficient amount of FR on the defense side with a certain degree of reliability.

26.5 Problem Statement

The model proposed in this study has been implemented as a software module of an intelligent system with elements of decision support for counteracting APT attacks. The computational experiment was conducted in the PyCharm programming environment. This computational experiment allowed visualizing the game results when examining the FR of the attacking party and the OBI defender. The objective was to analyze the costs of the parties during the game. The result is shown in Fig. 26.1.

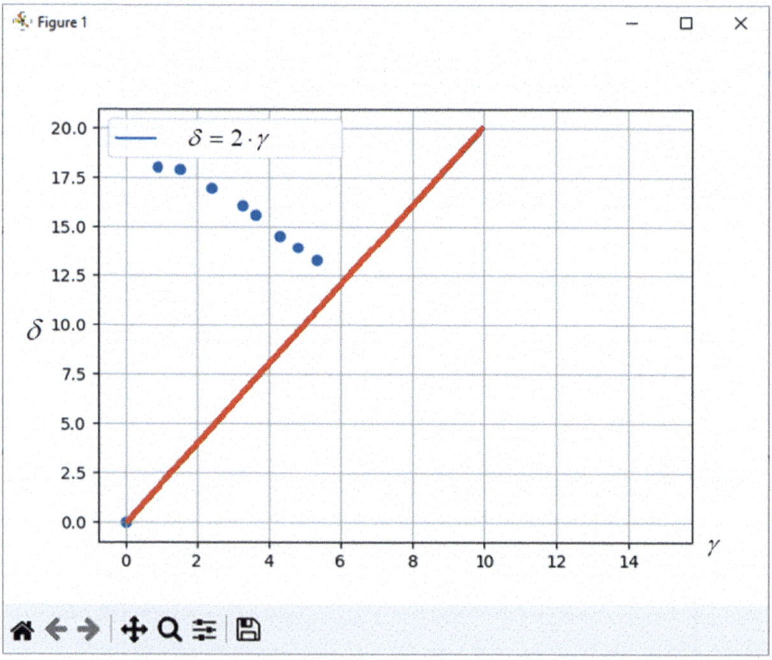

Fig. 26.1 Results of the computational experiment. Players' movement trajectory

26.6 Discussion of the Computational Experiment Results

Figure 26.1 illustrates the preference region of the second player. Within this preference area for the attackers, the player can achieve his objective with a specified degree of certainty. This holds true when he employs his optimal strategy. The trajectory of the players (depicted as blue dots) within the preference region of the second player is shown. When the states of the players are located on the balance ray (depicted as a straight red line), which marks the boundary of the second player's preference region, both players possess strategies that allow them to move along this ray for an indefinite duration. The movement will be either "upwards" or "downwards", depending on the interplay of the players' interaction parameters.

If the initial states of the players are situated in the preference area of the first player, which under certain interplay of interaction parameters is equivalent to the players' states being "below" the balance ray, then the first player possesses a strategy that allows him to achieve his objective, regardless of how the second player counteracts.

The considered model is "nested" in the framework of a multistep quality game with fuzzy information in the case of multidimensional variables of players. The solution of this kind of problem is a rather difficult task and, as a rule, in each specific case requires the development of special tools. There are no universal approaches for such problems. At the same time, the problem statement "contains" elements of fuzzy mathematics. By introducing a variable the problem can be reduced "in essence" to a problem with "full information", where the variable plays the role of a "phase" variable of the second player. This fact gives an answer to the question of how the presence of fuzzy mathematics is "present" in the formulation and solution of the problem.

The notion of a multi-step game is quite common and has existed essentially since the emergence of game theory, for example, as shown in [18].

The example given in the paper is one of many examples illustrating a large number of experiments and not only, but also real actual data corresponding to the practice of combating ART attacks for data received from enterprises in Ukraine.

Translated with DeepL.com (free version).

26.7 Conclusions

A multifactorial model of confrontation against targeted APT attacks on computerized objects (CO) in the context of fuzzy information about the financial resources of the parties has been considered. Unlike existing approaches, the proposed model assumes that the dynamics of the CO defense player's state and the player attacking the CO are determined by a system of discrete equations that describe the dynamics of multidimensional variables. It has been demonstrated that the controllability of the process of confronting targeted APT attacks can be described from a game approach perspective, based on solving a bilinear multi-step game with multiple terminal surfaces and fuzzy information. The novelty of the model lies in finding a solution to a bilinear multi-step quality game with several terminal surfaces with fuzzy information, which adequately reflects the essence of the problem under consideration. The results of a computational experiment have been presented.

References

1. Chen, P., Desmet, L., Huygens, C.: A study on advanced persistent threats. In: Communications and Multimedia Security: 15th IFIP TC 6/TC 11 International Conference, CMS 2014, Aveiro, Portugal, September 25–26, 2014. Proceedings 15, pp. 63–72. Springer Berlin Heidelberg (2014)
2. Gordon, L.A., Loeb, M.P.: The economics of information security investment. ACM Trans. Inf. Syst. Secur. (TISSEC) **5**(4), 438–457 (2002)
3. Gordon, L.A., Loeb, M.P., Zhou, L.: Investing in cybersecurity: insights from the Gordon-Loeb model. J. Inf. Secur. **7**(02), 49 (2016)
4. Direction, S.: Investing in cybersecurity: gaining a competitive advantage through cybersecurity. J. Bus. Strat. **37**, 19–21 (2021)
5. Chronopoulos, M., Panaousis, E., Grossklags, J.: An options approach to cybersecurity investment. IEEE Access **6**, 12175–12186 (2017)
6. Gordon, L.A., Loeb, M.P., Zhou, L.: Information segmentation and investing in cybersecurity. J. Inf. Secur. **12**(1), 115–136 (2020)
7. Fielder, A., Panaousis, E., Malacaria, P., Hankin, C., Smeraldi, F.: Decision support approaches for cyber security investment. Decis. Support. Syst. **86**, 13–23 (2016)
8. Milov, O., Yevseiev, S., Aleksiyev, V.: Development of structural models of stability of investment projects in cyber security. Ukr. Sci. J. Inf. Secur. **24**(3), 181–194 (2018)
9. Wang, Y., Wang, Y., Liu, J., Huang, Z., Xie, P.: A survey of game theoretic methods for cyber security. In: 2016 IEEE First International Conference on Data Science in Cyberspace (DSC), pp. 631–636). IEEE (2016, June)
10. Musman, S., Turner, A.: A game theoretic approach to cyber security risk management. J. Def. Model. Simul. **15**(2), 127–146 (2018)
11. Nagurney, A., Daniele, P., Shukla, S.: A supply chain network game theory model of cybersecurity investments with nonlinear budget constraints. Ann. Oper. Res. **248**, 405–427 (2017)
12. Nagurney, A., Nagurney, L.S.: A game theory model of cybersecurity investments with information asymmetry. NETNOMICS: Econ. Res. Electron. Network. **16**, 127–148 (2015)
13. Hyder, B., Govindarasu, M.: Optimization of cybersecurity investment strategies in the smart grid using game-theory. In: 2020 IEEE Power and Energy Society Innovative Smart Grid Technologies Conference (ISGT), pp. 1–5. IEEE (2020, February)
14. Roy, S., Ellis, C., Shiva, S., Dasgupta, D., Shandilya, V., Wu, Q.: A survey of game theory as applied to network security. In: 2010 43rd Hawaii International Conference on System Sciences, pp. 1–10. IEEE (2010, January)
15. Huang, K., Siegel, M., Madnick, S.: Systematically understanding the cyber attack business: a survey. ACM Comput. Surv. (CSUR) **51**(4), 1–36 (2018)
16. Meland, P.H., Sindre, G.: Cyber attacks for sale. In: 2019 International Conference on Computational Science and Computational Intelligence (CSCI), pp. 54–59). IEEE (2019, December)

17. Akhmetov, B., Lakhno, V., Malyukov, V., Akhmetov, B., Yagaliyeva, B., Lakhno, M., Gulmira, Y.: A model for managing the procedure of continuous mutual financial investment in cybersecurity for the case with fuzzy information. Lect. Notes Data Eng. Commun. Technol. **93**, 539–553 (2022)
18. Malyukov, V.P.: Game of quality for two groups of objects. Cybernetics **26**, 698–710 (1990)

Chapter 27
A Survey on Deciphering of EEG Waves

Gaurav Mahajan, L. Divija, R. Jeevan, P. Deekshitha Kumari, and Surabhi Narayan

Abstract Unlock the mind's secrets and paint emotions, thoughts, and perceptions onto a data canvas—our research delves into the uncharted territory of EEG's conversion into diverse modalities. This study utilizes EEG to capture brain activity and transform it into multimodal data, encompassing speech, image, and text. It delves into data collection and preprocessing, focusing on sentiment analysis, picture classification, and image reconstruction. EEG's high temporal resolution offers insights into cognitive and emotional states. Preprocessing techniques, including noise reduction and dimensionality reduction, enhance data quality. Sentiment classification leverages EEG data to discern emotional states during text consumption. In the image domain, EEG is employed for both reconstruction and classification, bridging the gap between brain activity and visual experiences. By venturing beyond EEG's boundaries, we traverse the intricate pathways connecting brain activity, language, and visual imagery. As we navigate this uncharted territory, discoveries hold the promise of reshaping the way humans interact with technology and revolutionizing the very essence of our mental well-being, education, and entertainment.

27.1 Introduction

The realm of neural science and its intimate connection with the human mind has long captivated researchers, promising glimpses into thc intricate choreography of the brain's electrical symphony. At the heart of this fascination lies the pursuit of

G. Mahajan (✉) · L. Divija · R. Jeevan · P. D. Kumari · S. Narayan
PES University, Bengaluru, Karnataka, India
e-mail: gauravmahajan395@gmail.com

L. Divija
e-mail: divija2401@gmail.com

R. Jeevan
e-mail: jeevan102002@gmail.com

S. Narayan
e-mail: surabhinarayan@pes.edu

S. K. Pal et al. (eds.), *Intelligent Informatics*, Smart Innovation, Systems and Technologies 389, https://doi.org/10.1007/978-981-97-2147-4_27

understanding and harnessing the electrical activity within neurons to unlock new dimensions of human communication and interaction. The human brain, comprising billions of neurons, orchestrates an intricate dance of electrical impulses. These neuronal signals underpin our thoughts, emotions, and actions. The brain's electrical activity, known as neural oscillations, offers a window into understanding how our cognitive processes are orchestrated. Neurological disorders are now the world's second greatest cause of mortality and the main cause of disability. The number of people suffering from brain disorders is likely to double by 2050, according to a new Global Burden of Disease (GBD) study. The burden of neurological illnesses is exacerbated by stigma and discrimination, which can limit life possibilities, raise the likelihood of poverty, and make access to care difficult. According to an examination of data from the most recent GBD research, brain diseases, which include mental illness, neurologic problems, and stroke, account for more than 15% of all health loss worldwide—more than cardiovascular disease or cancer. This has a significant impact on healthcare systems and society. Neurological problems affect hundreds of millions of individuals globally. Approximately 6 million people die each year as a result of unexpected strokes caused by aberrant brain activity. More than 80% of these deaths take place in low- and middle-income nations. Furthermore, epilepsy affects around 50 million people worldwide. The World Health Organization (WHO) estimates that 47.5 million people worldwide have dementia, with 7.7 million new cases diagnosed each year. Alzheimer's disease is the leading cause of dementia, accounting for 60–70% of cases.

Several illnesses can necessitate carers or support for people suffering from neurological problems that cause speech difficulty or loss of speech. Amyotrophic Lateral Sclerosis (ALS), a progressive neurodegenerative illness, affects motor neurons, and people with the disease may lose their ability to speak. A stroke can affect the parts of the brain that control speech and language, resulting in a condition known as aphasia. Huntington's disease is a hereditary ailment that affects several parts of the brain and can cause speech issues and changes in speech patterns. Parkinson's disease, which largely affects motor function, can also have an impact on speech, causing slurred speech and a weaker voice.

Over the years, researchers have been intrigued by the possibility of tapping into this neural symphony to facilitate communication. It is an enticing prospect—the ability to read the mind's electrical whispers and translate them into language or text, thereby providing a voice to those who may be unable to communicate through conventional means. While numerous methods have been explored to bridge the chasm between thoughts and language, one approach has garnered particular attention—Electroencephalography, or EEG.

It's crucial to acknowledge the existence of alternative methods. These include functional Magnetic Resonance Imaging (fMRI), Magnetoencephalography (MEG), and Invasive Brain–Computer Interfaces (BCIs). Each approach has its unique strengths and limitations, and a comprehensive survey should touch upon these methods to offer a well-rounded understanding of the field. EEG stands as a prominent player in the domain of neural signal decoding, EEG is a non-invasive neuroimaging technique that involves the placement of electrodes on the scalp to detect and record electrical activity in the brain. EEG provides a real-time, high-temporal-resolution measurement of brain activity, capturing rapid changes in neural oscillations.

Electroencephalography (EEG) holds a distinct advantage over other methods for decoding neural signals for communication, primarily due to its non-invasive nature, high temporal resolution, portability, cost-effectiveness, real-time feedback capabilities, and potential for home use. By not requiring surgical procedures or invasive implantations, EEG ensures minimal health risks and ethical concerns, making it more accessible for a diverse array of applications, especially in the context of assistive technology. Its remarkable temporal resolution allows for the real-time capture of rapid changes in neural activity, thereby enabling nuanced thought patterns to be translated into text or commands in near real-time. Furthermore, its portability and affordability make it versatile, and suitable for a broad spectrum of environments and applications. This versatility extends to the real-time nature of EEG, which is essential for applications necessitating immediate user feedback. Additionally, the potential for home use empowers individuals with communication disabilities to independently utilize the technology in the comfort of their environments, enhancing its usability and impact. Collectively, these attributes render EEG an appealing choice, particularly when seeking to translate brain signals into text or meaningful commands.

In conclusion, neurological illnesses have a considerable impact on individuals and society, with a growing worldwide burden. Understanding and treating these illnesses, as well as finding effective communication solutions for persons who have difficulty speaking, is critical to enhancing the quality of life for those who are impacted. To overcome this EEG should be generated in various modalities like text, image, and video. The primary purpose of this study was to review and explore the recent advances over the past deciphering of EEG signals. The remaining sections of this manuscript are organized as follows. In Sect. 27.2, we review the typical pipeline used for subsequent preprocessing stages, where noise reduction techniques and dimensionality reduction methods are rigorously applied to enhance the quality and reliability of the EEG data. In Sect. 27.3, extends to the domain of EEG-to-text conversion, where we meticulously explore methodologies for translating EEG waves into textual content. Additionally, we venture into the image domain, with a focus on EEG-based image reconstruction and classification, effectively bridging the gap between neural activity and the interpretation of visual information as well as discussing audio modality. Finally, Sect. 27.4 highlights the challenges faced in this field.

27.2 Preprocessing

Zhenhailong et al. [1] utilized Zuco 1.0 and 2.0 for tasks like EEG to text decoding and Zero-Shot Sentiment Classification on Open Vocabulary. Xin Deng et al. [2] explored two primary methods for feature extraction in EEG signal analysis: artificial extraction and neural network self-learning. Human experts use Event-Related Potentials (ERPs) for visual EEG signal classification, but may lack detail for complex tasks like image generation. Deep neural networks offer an automated approach, particularly useful for intricate patterns in image generation. Researchers choose these methods based on the task and available data.

Saeidi et al. [3] emphasize the importance of preprocessing EEG signals to improve data quality and remove artifacts. Common preprocessing approaches include regression methods, blind source separation methods, wavelet transform, and filtering techniques. Hybrid approaches combining these techniques have emerged as effective tools for EEG data quality and reliable subsequent analyzes. Feature extraction is crucial for EEG signal analysis, converting complex data into a lower-dimensional representation while retaining important information. Time-domain methods like ICA, PCA, and AR models focus on linear dimensionality reduction, while frequency domain techniques like FFT and Welch's method focus on spectral analysis. Wavelet transform and CSP are time–frequency domain techniques used in BCI applications.

Nagarajan et al. [4] conducted a study on EEG data collection using an Enobio8 device. The data was notch filtered to remove frequencies in the 49–51 Hz range and then converted into a suitable format for input into a Convolutional Neural Network (CNN) architecture, ensuring better reconstruction accuracy.

The study by Rashkov et al. [5] investigated the impact of visual stimuli on brain activity and developed an EEG feature mapping model. The researchers used 17 healthy subjects, presented with video clips from various categories. EEG data was extracted using independent component analysis (ICA) and spectral feature extraction. LDA was used to classify features, reducing feature dimensionality. The study aimed to improve our understanding of neural responses to visual content and contribute to the development of EEG feature mapping models.

Roy et al. [6] discussed the various steps involved in preprocessing EEG data, including trimming trials, downsampling, identifying bad channels, removing line noise, and high-pass filtering. Artifact handling techniques were found in 23% of studies, while 47% did not use any specific methods. Deep neural networks can be applied to EEG without explicitly removing artifacts. Some studies use custom features like time–frequency domain representations and power spectral density (PSD) from classical frequency bands to avoid feature engineering in traditional EEG processing.

Wakita's study [7] utilized Visual Evoked Potentials (VEP) to analyze the relationship between brain responses and natural textures. The dataset, consisting of EEG signals and images, was analyzed by 15 human observers. The images were presented in random order, with each EEG signal repeated 24 times. The study provides valuable insights into the brain's response to visual stimuli.

Chaurasiya et al. [8] utilized a publicly available dataset from IMAGENET to convert brain activity into images. Six subjects were selected based on age, education level, and cultural background. EEG data was extracted using Discrete Wavelet Transform (DWT) to encode the EEG into a feature vector. Daubechies-13 (db13) was found to be the most effective in terms of classification precision. The dataset was decomposed at level 4, and specific coefficients were used to calculate features, including common statistical features, kurtosis, and skewness. Principal component analysis (PCA) was used to reduce the dimension of feature vectors. The technique can be applied to real-time applications for windowed EEG signals. The feature set was divided into 85% train and 15% test sets for validation (Table 27.1).

Mishra et al. [9] conducted a study using EEG data from six subjects to analyze the ImageNet dataset. The data was collected using 128 electrodes and filtered using different bandpass filters. The data was then converted into a format with zero-centered values for each channel, allowing for the analysis of brain signal bands relevant during visual stimulation.

Amjed et al. [10] discussed various signal processing methods for EEG signals, including FFT, Wavelet transforms, eigenvector method, TFD, and AR. FT was effective for stationary signals but struggled with non-stationary signals and short EEG segments. Wavelet transforms were ideal for analyzing transient signal changes and irregular data patterns. AR modeling reduced spectral information loss and provided good frequency resolution, especially for short-term data. The chosen method depends on the specific characteristics of EEG signals and research objectives.

The study by Majid et al. [11] utilized two public EEG datasets, the SanDiego dataset and the University of New Mexico dataset, to analyze EEG signals for Parkinson's disease (PD). The SanDiego dataset included EEG recordings from 16 healthy people and 15 PD patients, while the University of New Mexico dataset required additional preprocessing. The Butterworth filter was used to remove interference and noise, while the Common Spatial Pattern (CSP) algorithm was used to distinguish between PD and healthy control (HC) classes. The research explored various entropy measures for feature extraction and discrimination in EEG data analysis.

Sang-Yeong et al.'s [12] study utilized Bandpass filtering and re-referencing techniques in the preprocessing procedure of EEG signals. Theta bandpass filter was used to refine the signals, extracting EEG segments for memorability prediction. A common average referencing method was employed, averaging all electrode recordings. The EEGLab MATLAB toolbox was used for preprocessing, aiming to compare traditional classification models with deep learning approaches.

Abenna et al. [13] utilized the SAES dataset, UCI eye state dataset, and PhysioNet-MI dataset for EEG data acquisition in their study. They introduced a new methodology, using the average of the signal's last seconds as a reference point for new samples. This method removed unnecessary EEG signal frequencies and used a bandpass filter (BPF) to determine the best signal bandwidths for each classification task, using the Sequential Cyclic Algorithm (SCA) and the Common Spatial Pattern (CSP) filter.

Goyal et al. [14] used a 10th-order Butterworth bandpass filter to remove artifacts from EEG signals in the Alpha (8–12 Hz) and Beta (12–30 Hz) frequency bands. This filter ensured the removal of physiological and non-physiological artifacts, focusing on the target frequency ranges. The bandpass-filtered signals were then processed to extract signals associated with speech or auditory input, allowing for the isolation of EEG signals. This method effectively addresses the need for artifact removal in EEG studies.

Table 27.1 Summary of preprocessing techniques

Author [paper]	Dataset	Techniques used
Zhenhailong [1]	Zuco 1.0 and 2.0	Directly used the preprocessed EEG signals
Xin Deng [2]	Own dataset	Feature extraction by artificial extraction and neural network self-learning
Saeidi [3]		Common preprocessing approaches: regression methods, blind source separation methods, wavelet transform, filtering techniques Time-domain methods like ICA, PCA, and AR models for linear dimensionality reduction Frequency domain techniques like FFT, Welch's method, focus on spectral analysis Wavelet transform and CSP are time–frequency domain techniques
Nagarajan [4]	Own dataset	***Notch*** filtered
Rashkov [5]	Own dataset	EEG data extracted using ICA and feature extraction LDA to classify features
Roy [6]		Involved trimming trials, downsampling, identifying bad channels, removing line noise, and high-pass filtering
Wakita[7]	Own dataset	***VEP-analyze relationship***
Chaurasiya [8]	ImageNet	Extracted using ***DWT*** ***Daubechies-13 (db13), PCA***
Mishra [9]	Own dataset	Data converted into a format with ***zero-centered values*** for each channel
Amjed [10]		Signal processing methods include ***FFT***, ***wavelet transforms***, ***eigenvector*** method, ***TFD***, and *AR*
Majid [11]	Two public EEG datasets: SanDiego dataset, University of New Mexico dataset	***Butterworth filter***, ***CSP*** algorithm
Sang-Yeong [12]		***BPF*** and ***re-referencing*** techniques ***Theta BPF*** to refine the signals
Abenna [13]	SAES dataset, UCI eye state dataset, PhysioNet-MI dataset	Using the average of the signal's last seconds as a reference point for new samples, ***BPF, SCA,*** and ***CSP*** filter
Goyal [14]	Own dataset	10th-order ***Butterworth*** bandpass filter
Gautam [15]	Own dataset	***Recorded at 1000 Hz to minimize power line noise*** ***Fourth-order IIR BPF and a notch filter*** ***ICA***

(continued)

Table 27.1 (continued)

Author [paper]	Dataset	Techniques used
Kaliraman [16]	BIOPAC system to collect EEG data	***ICA,*** ***AR,*** ***Auto-regressive burgs*** model
Krishna [17]		***ICA*** and ***ASR*** tools ***MFCC*** from speech signals Time-domain statistical features
Miguel [18]		***High-pass*** filters and ***notch*** filters Temporal analysis, spectral analysis, time–frequency analysis, and spatial analysis ***FFT*** and ***STFT***-dominant frequencies in EEG signals
Alotaiby [19]		***Filtering*** and ***wrapping*** techniques
Issak [20]	Own dataset	***Encoder*** to generate class-discriminative EEG features
Ein Shoka [21]		Removing noise and artifacts ***Differential window detection***, ***Butterworth filter***, and time–frequency localized orthogonal ***WFBs*** Decomposed using ***EMD*** into ***IMFs*** ***CSP***, capturing discriminating features from ***MEG*** spikes
Aditya [22]	PhysioNet eegmmidb database	Fourier transform ***Ensemble deep learning model***

* Terms used in the above table:
ICA—Independent component analysis; **PCA**—Principal component analysis; **AR**—Autoregressive Model; **BPF**—Bandpass Filter; **FFT**—Fast Fourier Transform; **CSP**—Common Spatial Pattern; **VEP**—Visual Evoked Potentials; **DWT**—Discrete Wavelet Transform; **TDF**—Time Domain Features; **SCA**—Sequential Cyclic Algorithm; **MFCC**—Mel-frequency cepstrum coefficients; **STFT**—Short-Term Fourier Transform; **IMF**—intrinsic mode functions

Gautam et al. [15] conducted a study on EEG data from healthy UT Austin undergraduate and graduate students in their early twenties. The EEG data was recorded at 1000 Hz to minimize power line noise and preprocessed using a fourth-order IIR bandpass filter and a notch filter. The Independent Component Analysis toolbox in EEGLab was used to remove ECG, EMG, and EOG artifacts from the signals. Three separate EEG feature sets were retrieved.

Kaliraman et al. [16] utilized the BIOPAC system to collect EEG data, utilizing 32 channels. To reduce eyeblink artifacts, ICA was applied to the acquired signals. The Auto-regressive Model (AR) was employed to predict the variable of interest using previous values, while the auto-regressive Burgs model was employed for preprocessing techniques.

Krishna et al. [17] conducted a study on EEG signals, using ICA and ASR tools to improve data quality. They sampled EEG signals at 1000 Hz and preprocessed them using filters and ICA. The study also extracted Mel-frequency cepstrum coefficients (MFCC) from speech signals, yielding 39 features. Spectral entropy was included to capture signal complexity. Zero crossing rate was chosen for its usefulness in EEG and other signal analysis. Additional time-domain statistical features were chosen based on ASR system performance improvement.

Miguel et al.'s study [18] focused on processing EEG signals, which can be influenced by internal and external artifacts like body movements and power interference. Various filtering schemes were used to improve signal quality, including high-pass filters and notch filters. The research focused on four fundamental domains of signal processing methods: temporal analysis, spectral analysis, time–frequency analysis, and spatial analysis. Techniques like FFT and STFT were used to reveal dominant frequencies in EEG signals, while time–frequency analysis combined temporal and spectral aspects for nonperiodic signals.

Alotaiby's research [19] on EEG channel selection focuses on the evolution of filtering and wrapping techniques to improve model performance, reduce dimensionality, and identify brain areas generating class-event activity. Filtering techniques use independent criteria but have low accuracy due to not considering channel combinations. Wrapper techniques, on the other hand, use a classification algorithm but are computationally expensive and prone to errors.

Issak et al. [20] conducted an experiment involving six subjects, recording EEG data from 2,000 images from 40 different ImageNet classes. The data was processed by an encoder to generate class-discriminative EEG features, resulting in 11,466 128-channel EEG sequences.

Ein Shoka's research [21] on EEG signal processing focuses on removing noise and artifacts to enhance feature extraction. Techniques include differential window detection, Butterworth filter, and time–frequency localized orthogonal WFBs. EEG signals are decomposed using EMD into intrinsic mode functions (IMFs), which are clustered using the PHA method. The most effective method for extracting features is a typical spatial pattern (CSP), capturing discriminating features from MEG spikes, and obtaining spatial filters for signal discrimination between two classes.

Aditya et al. [22] used EEG data from the PhysioNet eegmmidb database in EDF format to train a Deep Learning model. They combined this data with personal data from the NeuroSky MindWave Mobile 2 kit3, resulting in three labels: 0, 1, or '/ '. The Fourier transform was used to convert the signal from its original domain to its frequency domain. The Ensemble Deep Learning model extracted important features and converted the output to text. Khaleghi et al. [12] utilized the EEG-ImageNet database, which was recorded using a 128-channel cap (actiCAP 128Ch) in their study.

27.3 Methodology

27.3.1 EEG to Text

Wang et al. [1] extended the problem to encompass open vocabulary Electroencephalography (EEG)-to-Text Sequence-to-Sequence decoding and zero-shot sentence sentiment classification in the context of natural reading tasks. They postulate that the human brain operates as a unique text encoder and introduce an innovative framework that leverages pretrained language models, such as BART. Their model attains an impressive 40.1% BLEU1 score for EEG-to-Text decoding and a noteworthy 55.6% F1 score for zero-shot EEG-based ternary sentiment classification, outperforming supervised baselines considerably. Additionally, they demonstrate the model's ability to handle data from diverse subjects and sources, suggesting its substantial potential as a high-performance open vocabulary brain-to-text system, contingent upon the availability of sufficient data.

Said Abenna et al. [13] set out to enhance the binary and multiclass classification of EEG signals for real-time Brain-Computer Interface (BCI) applications. Their research centers on the results of a novel real-time approach integrated into a comprehensive prediction system. In their work, they introduce an innovative technique to mitigate the impact of EEG's inherent non-stationarity, resulting in a remarkable increase in accuracy. In the binary case, this enhancement elevates the accuracy from 50% when using raw EEG data to approximately 90% following preprocessing. In the multiclass scenario, the accuracy surges from 28 to 78%. Their methodology also involves the automatic optimization of bandpass filters using the sine cosine algorithm (SCA) to encompass the optimal bandwidth encompassing the entire EEG characteristics, specifically within beta waves. They apply the Common Spatial Pattern (CSP) filter to eliminate correlation among the extracted features. Additionally, they employ the Light Gradient Boosting Machine (LGBM) classifier in conjunction with the SCA algorithm to construct improved prediction models. Their system was then put to the test on UCI and PhysioNet datasets, yielding impressive accuracy levels exceeding 99% and 95%, respectively, utilizing data acquired solely from three channels. In comparison, prior related studies utilized data from all 14 channels, achieving accuracy values ranging from 70 to 98.5%, underscoring the robustness and effectiveness of their method in enhancing the quality of EEG signal predictions.

Gautham Krishna et al. [15] demonstrated the application of electroencephalography (EEG) signals for continuous noisy speech recognition in the context of English vocabulary. They have achieved this by employing various state-of-the-art end-to-end automatic speech recognition (ASR) models. Additionally, they've presented results stemming from EEG data collected under diverse experimental conditions. Furthermore, they've demonstrated the ability to decode speech spectrum from EEG signals using a long short-term memory (LSTM) based regression model and a Generative Adversarial Network (GAN) based model. The outcomes of their work highlight the potential and feasibility of utilizing EEG signals for continuous noisy speech recognition across different experimental conditions. Additionally, they offer preliminary findings regarding the synthesis of speech from EEG features.

Nicolas Affolter et al. [23] put forth a novel approach aimed at directly classifying fMRI scans by mapping them to corresponding words within a predefined vocabulary. What sets their work apart from existing research is their evaluation of fMRI scans obtained from subjects who were previously unseen. They make a compelling argument for this setup, emphasizing its practical relevance. The model they present is capable of decoding fMRI data from these unseen subjects, achieving remarkable results. Specifically, their model attains a Top-1 accuracy of 5.22% and a Top-5 accuracy of 13.59% in this challenging task, surpassing the performance of all competitive baselines considered in the study.

27.3.2 EEG to Image

Wakita et al. [7] introduce an innovative approach for photorealistic visual texture reconstruction from EEG signals. Their method, based on a Multi-Modal Reconstruction Variational Autoencoder (MVAE), combines generative models, deep neural networks, and variational autoencoders to extract features from EEG signals and generate images that closely resemble the visual stimuli's textures. The paper employs convolutional neural networks (CNN) for high-quality texture image reconstruction, treating texture images and EEG data as modal information. The study highlights the method's potential applications in neuroscience and psychology for non-invasive analysis of neural responses to visual stimuli, achieving an average correct identification rate of 72.8%.

Pilhyeon Lee et al. [24] addressed the challenge of subject-adaptive EEG-based visual recognition. Their primary objective is to accurately predict the categories of visual stimuli based on EEG signals, even when they have access to only a limited number of samples from the target subject during training. The central challenge revolves around the effective transfer of knowledge gained from ample data collected from source subjects to the subject of interest. To overcome this, they introduce an innovative method that enables the acquisition of subject-independent representations by enhancing the similarity between features that belong to the same class but originate from different subjects. Through a specialized sampling approach, their model efficiently captures shared knowledge among diverse subjects, leading to promising performance for the target subject, even in scenarios with limited data. In particular, on the EEG-ImageNet40 benchmark, their model achieves top-1 and top-3 test accuracy of 72.6% and 91.6%, respectively, when working with merely five EEG samples per class for the target subject.

Palazzo et al. [25] approach for generating images from the EEG brain signals using Generative Adversarial Networks (GANs). The authors have proposed a two-stage framework. The first stage involves the feature extraction from EEG signals using the CNN. The second stage involves the generation of images using conditional

GAN. This proposal was tested on two different datasets and the demonstration of the results that the generated images are of high-quality and also show potential for the use of EEG-to-image tasks. They have also highlighted the importance of conditioning GANs on external signals for improving the quality of generated images.

27.3.3 EEG to Audio

Brain-to-speech done by Herff et al. [26]; Anumanchipalli et al. [27]; Makin et al. [28]; and Moses et al. [29] where they successfully decoded and captured low-level auditory features from the movement of the vocal tract to reconstruct words.

Instead of only capturing articulation features, another line of work demonstrated that the human brain encodes language into higher dimensional semantic representations Gauthier et al. [30]; Correia et al. [31]. Interestingly, they have seen similar behavior in large-scale pretrained language models, such as BERT Devlin et al. [32], and BART Prystauka et al. [33]; GPT2 Goh et al. [34] and GPT3 Brown et al. [35], which encode words into contextualized semantic embeddings [36] (Table 27.2).

27.4 Conclusion

In the realm of neuroscience and artificial intelligence, the paper's objective is to provide insight and summarize techniques to bridge the gap between the incomprehensible realm of neural activity, as recorded by electroencephalography (EEG), and the concrete world of text, digital images, and audio are bridged. This ambitious endeavor is undertaken to unlock the potential of EEG data and provide novel solutions that have profound effects on various domains, from brain-computer interfacing to improving the quality of life for individuals with communication disabilities. The study investigated current developments in EEG signal decoding in detail, concentrating primarily on preprocessing steps to improve the accuracy and consistency of EEG data. The investigation also delved into the field of EEG-to-text translation, painstakingly revealing techniques for converting intricate EEG wave patterns into understandable text. The research also explored the picture domain, with a focus on EEG-based image reconstruction and categorization. This method successfully closes the gap between perceptible visual information and brain activity. The findings provide important new insights into the interpretation of brain activity and further our understanding of the processing of EEG signals. Notwithstanding these developments, there are still problems in the field, such as those pertaining to generalization, accuracy, and the integration of many modalities.

One of the most significant hurdles encountered while dealing with EEG revolved around the extraordinary complexity of the human brain and the intricacies governing neural activity. Despite the tireless efforts of both neuroscientists and computer experts, a comprehensive grasp of the inner workings of the brain has

Table 27.2 Tasks and methodologies

Author [paper]	Task (conversion)	Model used	Metric used	Value
Wang [1]	Sentence sentiment classification (EEG to text)	BART, zero-shot learning	BLEU1 score F1 score	40.1 55.6
Abena [13]	Automatic optimization of bandpass filters to eliminate correlation among the extracted features Classification (EEG to text)	SCA CSP LGBM	Accuracy	99% on UCI dataset 95% on PhysioNet dataset
Gautham Krishna [15]	ASR regression (EEG to text)	LSTM GAN		
Nicolas Affolter [23]	Classify fMRI scans (EEG to text)	XGBoost	Accuracy	5.22% (Top-1) (13.59%) Top-5
Pilhyeon Lee [24]	Visual recognition (EEG to image)	Contrastive learning Vanilla and MMD	Accuracy	72.6% (Top-1) 91.6% (Top-3)
Wakita [7]	Reconstructing photorealistic visual textures (EEG to image)	MVAE	Accuracy	72.8%
[6, 7, 19–27]	Decoding of low-level auditory features (EEG to audio)	BERT BART GPT2 GPT3		

remained elusive. Furthermore, even within the same person, the dynamic and ever-changing nature of brainwave patterns posed a formidable challenge and introduced inconsistencies in how features were represented.

References

1. Wang, Z., Ji, H.: Open vocabulary electroencephalography-to-text decoding and zero-shot sentiment classification. In: The Thirty-Sixth AAAI Conference on Artificial Intelligence (AAAI-22) (2021, December 5)
2. Deng, X., Wang, Z., Liu, K., Xiang, X.: A GAN model encoded by CapsEEGNet for visual EEG encoding and image reproduction. J. Neurosci. Methods **384**, 109747 (2023). https://doi.org/10.1016/j.jneumeth.2022.109747

3. Saeidi, M., Karwowski, W., Farahani, F.V., Fiok, K., Taiar, R., Hancock, P.A., Al-Junaid, A.: Neural decoding of EEG signals with machine learning: a systematic review. Brain Sci. **11**, 1525 (2021). https://doi.org/10.3390/brainsci11111525
4. Nagarajan, K., Umadi, A., Keshav, N.B., Krupa, N.: Pictorial information retrieval from EEG using generative adversarial networks. In: 2022 7th International Conference on Intelligent Informatics and Biomedical Science (ICIIBMS), Nara, Japan, Nov. 24–26, 2022, p. 269 (2022). https://doi.org/10.1109/ICIIBMS55689.2022.9971471
5. Rashkov, G., Bobe, A., Fastovets, D., Komarova, M.: Natural image reconstruction from brain waves: a novel visual BCI system with native feedback. In: Neurobotics LLC, Moscow, Russian Federation, Neuroassistive Technologies LLC, Moscow, Russian Federation, Moscow Institute of Physics and Technology, Dolgoprudny, Moscow Region, Russian Federation. https://doi.org/10.1101/787101
6. Roy, Y., et al.: Deep learning-based electroencephalography analysis: a systematic review. J. Neural Eng. **16**, 051001 (2019)
7. S. Wakita, T.O., Motoyoshi, I.: Photorealistic reconstruction of visual texture from EEG signals. Front. Comput. Neurosci. **15**, 754587 (2021). https://doi.org/10.3389/fncom.2021.754587
8. Chaurasiya, R.K., Arvind, S.K., Garg, S.: Adversarial auto-encoders for image generation from standard EEG features. In: 2020 First International Conference on Power, Control and Computing Technologies (ICPC2T) (2020, January). https://doi.org/10.1109/ICPC2T48082.2020.9071480
9. Mishra, N.R., Bajwa, G.: EEG-Based Image Feature Extraction for Visual Classification Using Deep Learning. University of Liverpool (2022, September)
10. Al-Fahoum, A.S., Al-Fraihat, A.A.: Methods of EEG Signal Features Extraction Using Linear Analysis in Frequency and Time-Frequency Domains (2014, February)
11. Aljalal, M., Aldosari, S.A., AlSharabi, K., Abdurraqeeb, A.M., Alturki, F.A.: Parkinson's disease detection from resting-state EEG signals using common spatial patterns, entropy, and machine learning techniques. Diagnostics **12**, 1033 (2022). https://doi.org/10.3390/diagnostics12051033
12. Jo, S.-Y., Jeong, J.-W.: Prediction of visual memorability with EEG signals: a comparative study. Sensors **20**(9), 2694 (2020). https://doi.org/10.3390/s20092694
13. Abenna, S., Nahid, M., Bouyghf, H., Ouacha, B.: EEG-based BCI: a novel improvement for EEG signals classification based on real-time preprocessing. Comput. Biol. Med. **148**, 105931 (2022). https://doi.org/10.1016/j.compbiomed.2022.105931
14. Goyal, I., Mehta, A.: Acquisition, pre-processing, and feature extraction of EEG signals to convert it into an image classification problem. Int. Res. J. Eng. Technol. (IRJET) **08**(02), 203 (2021)
15. Krishna, G., Han, Y., Tran, C., Carnahan, M., Tewfik, A.: State-of-the-art Speech Recognition using EEG and Towards Decoding of Speech Spectrum From EEG (2019, August 14)
16. Kaliraman, B., Nain, S., Verma, R., Thakran, M., Dhankhar, Y., Hari, P.B.: Pre-processing of EEG signal using independent component analysis. In: 2022 10th International Conference on Reliability, Infocom Technologies, and Optimization (Trends and Future Directions) (ICRITO), Noida, India, pp. 1–5 (2022). https://doi.org/10.1109/ICRITO56286.2022.9964717
17. Krishna, G., Tran, C., Carnahan, M., Tewfik, A.: Advancing speech recognition with no speech or with noisy speech. In: 2019 27th European Signal Processing Conference (EUSIPCO), A Coruna, Spain, pp. 1–5 (2019). https://doi.org/10.23919/EUSIPCO.2019.8902943
18. Luján, M.Á., Jimeno Jimenez, M.V., Mateo Sotos, J., Borja, A.L. et al.: A survey on EEG signal processing techniques and machine learning: applications to the neurofeedback of autobiographical memory deficits in schizophrenia. Electronics **10**(23), 3037 (2021). https://doi.org/10.3390/electronics10233037
19. Alotaiby, T., El-Samie, F.E.A., Alshebeili, S.A., et al.: A review of channel selection algorithms for EEG signal processing. EURASIP J. Adv. Signal Process. **2015**, 66 (2015). https://doi.org/10.1186/s13634-015-0251-9
20. Kavasidis, I., Palazzo, S., Spampinato, C., Shah, M.: Brain2Image: converting brain signals into images. In: 2017 ACM (2017, October). https://doi.org/10.1145/3123266.3127907

21. Ein Shoka, A.A., Dessouky, M.M., El-Sherbeny, A., El-Sayed, A.: Literature review on EEG preprocessing, feature extraction, and classifications techniques. Menoufia J. Electron. Eng. Res. **28**(1), 292–299 (2019). https://doi.org/10.21608/mjeer.2019.64927
22. Srivastava, A., Ansari, R.A., Shinde, T., Kanade, P., et al.: Think2Type: thoughts to text using EEG waves. Int. J. Eng. Res. **9**(6), 659 (2020). https://doi.org/10.17577/IJERTV9IS060431
23. Affolter, N. et al.: Brain2Word: decoding brain activity for language generation (2020). arXiv preprint arXiv:abs/2009.04765
24. Lee, P., Jeon, S., Hwang, S., Shin, M., Byun, H.: Source-free subject adaptation for EEG-based visual recognition (2023). https://arxiv.org/abs/2301.08448.
25. Palazzo, S., Spampinato, C., Kavasidis, I., Giordano, D., Shah, M.: Generative adversarial networks conditioned by brain signals. In: 2017 IEEE International Conference on Computer Vision (ICCV), Venice, Italy, pp. 3430–3438 (2017). https://doi.org/10.1109/ICCV.2017.369
26. Herff, C., Heger, D., de Pesters, A., Telaar, D., Brunner, P., Schalk, G., Schultz, T.: Brain-to-text: decoding spoken phrases from phone representations in the brain. Front. Neurosci. **8**, 217 (2015). https://doi.org/10.3389/fnins.2015.00217
27. Anumanchipalli, G.K., Chartier, J., Chang, E.F.: Speech synthesis from neural decoding of spoken sentences. Nature **568**(7753), 493–498 (2019). https://doi.org/10.1038/s41586-019-1119-1
28. Makin, J.G., Moses, D.A., Chang, E.F.: Machine translation of cortical activity to text with an encoder–decoder framework. Nat. Neurosci. **23**(4), 575–582 (2020)
29. Moses, D.A., Metzger, S.L., Liu, J.R. et al.: Neuroprosthesis for decoding speech in a paralyzed person with anarthria. N. Engl. J. Med. **385**, 217–227 (2021). https://doi.org/10.1056/NEJMoa2027540
30. Gauthier, J., Ivanova, A.: Does the brain represent words? An evaluation of brain decoding studies of language understanding. In: 2018 Conference on Cognitive Computational Neuroscience (2018, May). https://doi.org/10.32470/CCN.2018.1237-0
31. Correia, M., Jansma, B., Hausfeld, L., Kikkert, S., Bonte, M.: EEG decoding of spoken words in bilingual listeners: from words to language invariant semantic-conceptual representations. Front. Psychol. **6**, 71 (2015). https://doi.org/10.3389/fpsyg.2015.00071
32. Devlin, J., Chang, M.-W., Lee, K., Toutanova, K.: BERT: pretraining of deep bidirectional transformers for language understanding. In: Proceedings of the 2019 Conference of the North American Chapter of the Association for Computational Linguistics: Human Language Technologies, vol. 1, Minneapolis, MN, pp. 4171–4186 (2019)
33. Prystauka, Y., Lewis, A.G.: The power of neural oscillations to inform sentence comprehension: a linguistic perspective. Lang. Linguist. Compass **13**, e12347 (2019). https://doi.org/10.1111/lnc3.12347
34. Goh, G., Cammarata, N., Voss, C., Carter, S., Petrov, M., Schubert, L., Radford, A., Olah, C.: Multimodal neurons in artificial neural networks. Distill **6**(3), e30 (2021)
35. Brown, T.B., Mann, B., Ryder, N., Subbiah, M., Kaplan, J., Dhariwal, P., Neelakantan, A., Shyam, P., Sastry, G., Askell, A. et al.: Language models are few-shot learners. In: Advances in Neural Information Processing Systems (2020)
36. Ethayaraja, K.: How contextual is contextualized word representations? Comparing the geometry of BERT, ELMo, and GPT-2 embeddings. In: Proceedings of the 2019 Conference on Empirical Methods in Natural Language Processing and the 9th International Joint Conference on Natural Language Processing (EMNLP-IJCNLP), pp. 55–65 (2019)

Part VI
Quantum Computing and Intelligent Systems

Chapter 28
An Efficient Quantum Circuit Design: Properties and Optimization Techniques

Mamtha Prajapati and Kalyan Babu Killana

Abstract The semiconductor industry has been progressively researching various emerging technologies. The fast-evolving discipline of quantum computing utilizes quantum physics ideas to address more complex problems exponentially than the classical paradigm. The reversible logic application is emphasized for emerging nanotechnologies as it guarantees a near energy-free computation. The article's introduction gives an easy-to-understand overview of why quantum computing and quantum circuit design are essential. A 'quantum circuit,' or set of operations known as quantum gates, is required to perform computations on a quantum computer. A quantum circuit's scalability and efficiency depend highly on the selection of quantum gates. A list of the basic 1-qubit and 2-qubit quantum gates is provided, along with their characteristics. Further, design optimization techniques for enhancing quantum circuits and their benefits and challenges are described to develop optimal quantum circuits. Optimization aims to minimize the number of gates in a quantum circuit, as this lowers the quantum circuit's depth or time steps and, thus, its delay and quantum cost. The optimization techniques will also lessen quantum error when the circuit is run in a real quantum processor.

28.1 Introduction

Emerging technologies studied by the semiconductor industry are Carbon Nanotube Transistors, Silicon Nanowires, Single Electron transistors, Resonant Tunneling Devices, Molecular electronics, Spin transistors, Superconducting electronics, and Quantum Computing. Quantum computing has drawn scientists and academics in recent years. It has become a promising technology due to its ability to perform parallel calculations and solve complex problems challenging for classical computers

M. Prajapati (✉) · K. B. Killana
Department of Electrical, Electronics and Communication Engineering, GITAM University, Visakhapatnam 530045, India
e-mail: mprajapa@gitam.in

K. B. Killana
e-mail: kkillana@gitam.edu

S. K. Pal et al. (eds.), *Intelligent Informatics*, Smart Innovation, Systems and Technologies 389, https://doi.org/10.1007/978-981-97-2147-4_28

[1, 2]. A revolutionary quantum computer is far more efficient than any classical machine at performing subatomic-level calculations based on the high-domain laws of quantum mechanics.

Over the years, classical computation has increased in speed because of a great deal of miniaturization achieved in integrated circuits, as predicted by Moore's law. It claims that every 18 months, the number of transistors on an integrated circuit doubles. Based on this, the spacing between the components decreased, leading to two problems. Firstly, the separation between components has reached atomic dimensions. This means that if the design components come so close, the results obtained by that computation will no longer be reliable. Secondly, the heat produced by one of the components would naturally affect the performance of the nearby component. This will lead to unreliable outcomes, resulting in information loss. Landauer's principle states that conventional logic circuits release $k_B T \ln 2$ joules of heat for each bit of information lost, where k_B is the Boltzmann constant. T is the operating temperature [3].

Further miniaturization is not a promising way to develop compact electronic devices. Reversible logic is a good application in designing low-power VLSI circuits to reduce power consumption [4]. The challenge with miniaturization is that making transistors small will only work up to a point, after which quantum effects cannot be ignored. So, reversible logic and its application to quantum computation will help work towards the challenge. Reversible logic is investigated for its promising applications in power-efficient nano-computing, Quantum computation, and designing low-power VLSI circuits [4, 5]. Figure 28.1 illustrates one of the approaches to quantum computing discussed above.

Quantum computing is a developing field. There are concerns for effective control and regulation to reduce hazards and misuse. Because quantum computing operates,

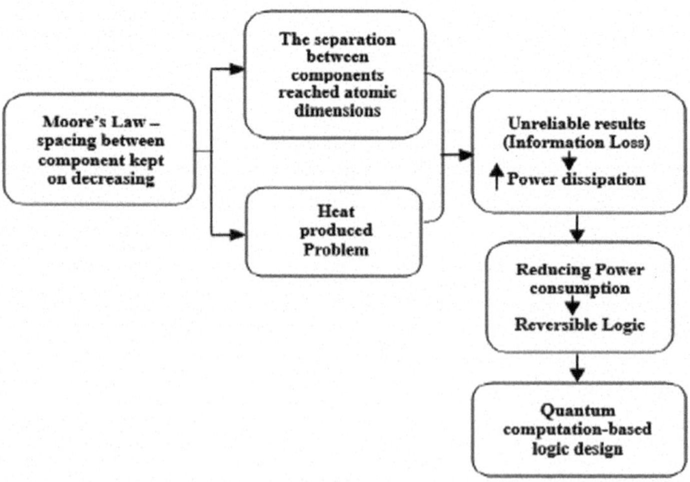

Fig. 28.1 Background of the work

no laws apply. Quantum physics follows quantum rules, not classical laws. However, quantum computing is not likely to replace traditional computing. It may transform many fields and tackle challenges classical computers cannot. Beyond conventional computers, quantum computing can process nanoscale data. Quantum computers store and change data using subatomic particle dynamics. Quantum computing and nanotechnology both explore atomic and subatomic matter. The need for quantum computing in implementing nanotechnology arrived in such a way that simulation and nanomaterial fabrication can be optimized using quantum computing. Quantum computing can optimize atom placement for a desired feature for efficient materials.

Quantum computing replaces transistors with qubits, not shrinks them. The quantum computing paradigm uses quantum bits (qubits—basic units of quantum information) to accomplish operations. Qubits can exist in numerous states, allowing quantum computers to perform specific calculations quicker than classical computers [5]. Shor's algorithm factors huge numbers exponentially faster, affecting cryptography and cybersecurity. Quantum computing may not shrink transistors, but it can benefit current technology in various ways. Quantum computers can simulate complicated systems, making them valuable for drug development and materials science. The market for quantum computing has so far seen demand from the scientific and research sectors. However, this is about to change as there is an increasing need for quantum computers for data analytics, image processing, nanotechnologies, Machine Learning (ML), and artificial intelligence (AI) [6, 7]. Applications that stand to gain a lot from Quantum computing's speed and precision are chip design, medical industries, the financial sector, environment monitoring, cybersecurity, optimization, weather forecasting, neural networks, transportation, etc. [8, 9]. Governments, as well as businesses, are investing heavily in quantum computing worldwide [10]. The use of quantum computation in circuits and systems is encouraged because, in theory, its internal calculations need no energy.

Driven by the principal concepts of quantum entanglement, superposition, and tunneling, quantum computing is a field that is constantly developing. Despite the tremendous potential of quantum computing, most of its applications are currently in the theoretical stage. They are far from being implemented because of the number of qubits and severe noise-related limitations. Robust quantum algorithms rely on efficient quantum circuits, forming the fundamental basis of quantum supremacy (also known as quantum advantage) in theory and application. An optimized number of quantum gates (operators), a long cohesion time, cheap hardware, a short execution time, high tolerance to mistakes and noisy sources, and high fidelity when encoding into error-correcting codes are all desired features of an efficient quantum circuit. Therefore, research is ongoing into the most practical quantum circuit architecture to accomplish quantum supremacy. Considering this motivation, we have covered some critical strategies for optimizing the quantum circuit design to lower the quantum cost, time steps, and gate count.

The rest of the article is structured as follows. In Sect. 28.2, we provide the importance of quantum circuit design. Section 28.2.1 describes the single and multiple quantum gates required in quantum circuit design. Then, Sect. 28.2.2 discusses the properties and performance metrics in quantum circuit design. Section 28.3 provides the optimization techniques for designing efficient quantum circuits, benefits, and challenges. Section 28.4 offers a conclusion.

28.2 Quantum Circuit Design

Understanding quantum computation and developing quantum algorithms requires quantum circuit modeling [1, 2]. Quantum computers utilize a "quantum circuit" that performs a series of operations by applying a sequence of gates, called quantum gates, to compute. These quantum gates are the basic building blocks of quantum computers and modify the quantum states of specific qubits. The need for quantum circuit design is growing as quantum computers become more powerful. Although they are still in the infancy of their development, quantum computers have the potential to alter a variety of industries completely. Quantum circuit design is an essential part of making quantum computers a reality. Effective quantum circuit design is a challenging task that requires a deep understanding of quantum mechanics and optimization [11, 12].

Based on quantum mechanics, quantum logic uses unitary Hamiltonians, distinguishing it from the classical paradigm. The Hamiltonian operator represents a system's total energy, including kinetic and potential energy. Unitary Hamiltonians are used in quantum logic to provide unitary transformation. A linear transformation that maintains the length and angle of the inner product of two vectors in a Hilbert space. Unitary Hamiltonians operate on quantum gates to design circuits and execute quantum algorithms on quantum computers. Therefore, Quantum computing no longer benefits from Boolean logic.

Furthermore, reversible logic design is essential because it allows for the construction of quantum circuits that can be perfectly inverted [13]. Quantum computing requires reversible logic to develop circuits that can be run in reverse (recover the original input from the output) for error correction and fault tolerance. Reversible logic gates are more energy-efficient than irreversible logic. Reversible logic can improve quantum computers by offering error correction, fault tolerance, and reduced power consumption, making them efficient, scalable, and reliable.

28.2.1 Single and Multiple Qubit Gates

The choice of quantum gates can significantly impact the efficiency and scalability of a quantum circuit. Some quantum gates are more expensive than others [14], so choosing gates well-suited for the task is essential. One bit of measurement-accessible classical information is contained in the fundamental quantum state, or qubit, the smallest measure of quantum information. A qubit represents a two-state quantum object consisting of a "zero" state and a "one" state. Qubit states are represented on the Bloch sphere using Dirac notation, $|0\rangle$ and $|1\rangle$ [5]. Qubit state, $|q\rangle = \alpha|0\rangle + \beta|1\rangle$ $= \alpha \begin{bmatrix} 1 \\ 0 \end{bmatrix} + \beta \begin{bmatrix} 0 \\ 1 \end{bmatrix}$, where α and β are complex numbers (probabilistic amplitudes).

We control and change the state of our qubits while they are in superposition using quantum gates. These quantum operations resemble the conventional Boolean logic gates (such as NOT, AND, XOR, etc.), but since they are quantum, they have additional characteristics. Quantum operations are reversible. Reversible gates are unitary. Quantum computation requires all gates to be reversible, so classical gates cannot be used. Unitary matrices represent quantum logic gates. A unitary matrix represents a gate that acts on qubits. The symmetry group U ($2n$) (2×2 unitary matrix) is the collection of all such gates with the combined operation of matrix multiplication. One can identify how the gate affects a particular quantum state by multiplying a vector representing the state by the matrix corresponding to the gate. A newly created quantum state is the outcome. Matrices play a significant role in quantum computing because they may be used to specify quantum logic gates.

Elementary quantum gates with their properties and corresponding transformation matrices are represented in Table 28.1. The fundamental quantum logic gates serve as the essential components of quantum circuits, similar to how conventional logic gates operate inside digital circuits. Quantum computers utilize quantum bits (qubits) as the fundamental elements for encoding quantum information. In contrast to traditional bits, which possess a binary state of 0 or 1, qubits can exist concurrently in a superposition of both states. This unique property provides quantum computers with distinct processing capabilities. Numerous multiple qubit gates (3-qubit and 4-qubit gates) are available in the literature [15, 16], most well-known being the Toffoli gate and the Fredkin gate.

The Pauli-X gate (or X-gate) is analogous to the classical NOT gate. When the state of a qubit is applied to the X-gate, it takes a 180° rotation about the X-axis on the Bloch sphere, also known as the quantum bit flip gate. Pauli-Z gate or phase gate is a rotation of 180° about the z-axis on the Bloch sphere. Pauli-Y gate is a rotation of 180° about the y-axis on the Bloch sphere. It can be defined using a combination of X and Z gate, i.e., $Y = iXZ$, also known as bit phase flip gate. The T-gate rotates the qubit state by $\pi/8$ around the Bloch sphere's Z-axis. The S-gate, or square root of Z gate, rotates the qubit state by $\pi/4$ around the Bloch sphere's Z-axis. The controlled-NOT gate (CNOT gate) flips the target qubit only when the control qubit is 1. Implementing any classical computation using CNOT gates can be done because it is a universal gate. CZ gates are two-qubit quantum logic gates that perform phase flips on the target qubit only when the control qubit is in its $|1\rangle$ state. With equal probability, the Hadamard gate converts a qubit into the superposition of 0 and 1. Quantum algorithms depend on it because it involves superposition, a key quantum computing feature. SWAP gates exchange the states of two target qubits. They are crucial to quantum circuits and algorithms.

Digital circuits that operate on binary values use EXOR, NOT, and AND gates. These gates also generate more complicated arithmetic and other circuits. These gates specify the hardware complexity of quantum-implementable reversible logic

Table 28.1 Fundamental quantum gates and their properties

Name of quantum gate	Block diagram	Transformation matrix	Properties
Pauli-X/NOT gate (X) /quantum bit flip	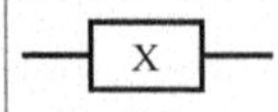	$\begin{bmatrix} 0 & 1 \\ 1 & 0 \end{bmatrix}$	$X\|0\rangle = \|1\rangle$ $X\|1\rangle = \|0\rangle$
Pauli-Y gate (Y) /bit phase flip	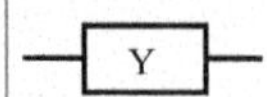	$\begin{bmatrix} 0 & -i \\ i & 0 \end{bmatrix}$	$Y\|0\rangle = i\|1\rangle$ $Y\|1\rangle = -i\|0\rangle$
Pauli-Z gate (Z) /quantum phase flip	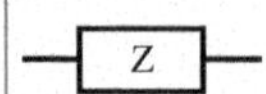	$\begin{bmatrix} 1 & 0 \\ 0 & -1 \end{bmatrix}$	$Z\|0\rangle = \|0\rangle$ $Z\|1\rangle = -\|1\rangle$
Phase/S-gate (S)	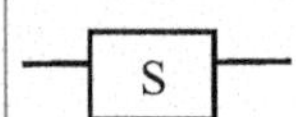	$\begin{bmatrix} 1 & 0 \\ 0 & i \end{bmatrix}$	$S\|0\rangle = \|0\rangle$ $S\|1\rangle = i\|1\rangle$
T-gate (T)	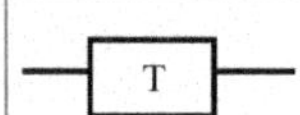	$\begin{bmatrix} 1 & 0 \\ 0 & e^{i\pi/4} \end{bmatrix}$	$T\|0\rangle = \|0\rangle$ $T\|1\rangle = e^{i\pi/4}\|1\rangle$
Controlled NOT gate (CNOT, CX)	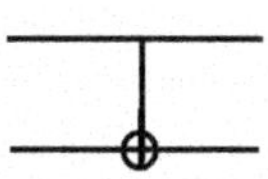	$\begin{bmatrix} 1&0&0&0 \\ 0&1&0&0 \\ 0&0&0&1 \\ 0&0&1&0 \end{bmatrix} \equiv \begin{bmatrix} I & 0 \\ 0 & X \end{bmatrix}$	$2 \leftrightarrow 3$ $\|10\rangle \leftrightarrow \|11\rangle$
Controlled Z gate (CZ)	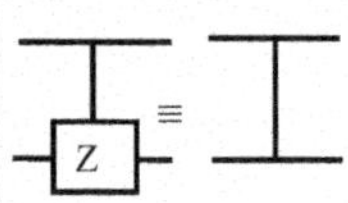	$\begin{bmatrix} 1&0&0&0 \\ 0&1&0&0 \\ 0&0&1&0 \\ 0&0&0&-1 \end{bmatrix}$	$3 \leftrightarrow -3$ $\|11\rangle \leftrightarrow -\|11\rangle$
Hadamard gate (H)	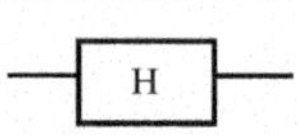	$\frac{1}{\sqrt{2}}\begin{bmatrix} 1 & 1 \\ 1 & -1 \end{bmatrix}$	$H\|0\rangle = \frac{1}{\sqrt{2}}[\|0\rangle + \|1\rangle]$ $H\|1\rangle = \frac{1}{\sqrt{2}}[\|0\rangle - \|1\rangle]$
SWAP gate	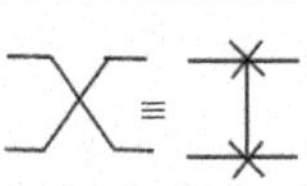	$\begin{bmatrix} 1&0&0&0 \\ 0&0&1&0 \\ 0&1&0&0 \\ 0&0&0&1 \end{bmatrix}$	$1 \leftrightarrow 2$ $\|01\rangle \leftrightarrow \|10\rangle$

gates. Quantum computing requires reversible logic gates to compute algorithms without losing information. Other Boolean logic gates can be employed in quantum computing. The OR gates can be implemented using the Deutsch–Jozsa algorithm [17]. NAND gates can be built using Quantum Phase Estimation. However, the NAND gate is irreversible, making it an inadequate quantum gate.

28.2.2 Quantum Circuit Properties

The fundamental blocks in quantum computing are quantum circuits. They are made up of a series of quantum gates or operations that are performed on qubits. The quantum circuit properties can be used to characterize its performance and scalability. Some of the essential properties to be considered while designing quantum circuits include:

1. Quantum circuits are reversible, i.e., the total number of inputs and outputs is equal. The information and outcomes can be uniquely retrieved from each other.
2. Quantum circuits are acyclic (no loops), i.e., no feedback.
3. FAN-IN, as well as FAN-OUT, are prohibited.
4. The measurement of a qubit in a quantum circuit is done as shown in Fig. 28.2.

Many-qubit quantum circuits are challenging to construct, the total number of qubits being an essential measure in quantum circuit design. The properties of quantum circuits are necessary for several reasons. They can be used to estimate the performance of a quantum computer, compare different quantum algorithms, and design efficient quantum circuits [18, 19]. As quantum computers become more powerful, the properties of quantum circuits will become increasingly important. Various parameters measure the performance of a quantum circuit design mentioned in Table 28.2.

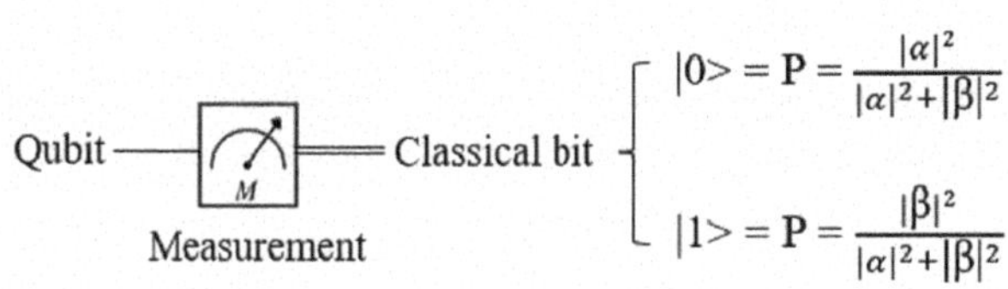

Fig. 28.2 Measurement of a qubit

Table 28.2 Quantum circuit design performance metrics

Parameters	Description
Quantum cost	The minimal quantum operations that are essential to build a quantum circuit
Gate count	The number of quantum gates required in circuit design
T-count	The minimal number of logical $T/T^{\dagger}$-gates needed to convert a Boolean equation into a quantum circuit
T-depth	It calculates the speed of the circuit and the number of T gates in the circuit critical path
Circuit depth	The total number of clock cycles necessary to execute the circuit
Ancilla inputs	The constant inputs are required to derive a particular function and retain one-to-one mapping
Garbage output	In a reversible logic gate/circuit, the unused outputs are neither the primary outputs nor those required for further computation

28.3 Optimization Techniques

Optimization involves finding the best solution. Optimization techniques in quantum circuit design are used to improve the efficiency and scalability of quantum circuits. These techniques can reduce quantum circuits' depth, width, and error rate and enhance their fidelity and scalability. The following is a discussion of some effective optimization methods for quantum circuit design.

28.3.1 Template-Based Optimization Technique

One method for increasing the efficiency of quantum circuits is template-based optimization [20–22], which involves locating and swapping out common substructures for more effective ones. One way to achieve this is by building a library of pre-optimized circuit designs and templates that can be used in various contexts. The substructures in a quantum circuit that match a template in the library are first found using the template matching method. Once these substructures are identified, a more effective template can be used in their place. Repeat this process multiple times to increase the circuit's efficiency. Consider Fig. 28.3a, an existing quantum logic circuit that meets the connection constraints of the linear architecture. Figure 28.3b is a template circuit. When the quantum gates g'_1, g'_2, g'_3, and g'_4 match the gates g_3, g_4, g_5, and g_6 in Fig. 28.3a, the template gate g'_5 replaces the matching gates, optimizing the circuit. The optimized circuit Fig. 28.3c contains a non-nearest neighbor gate g_3, which does not meet the connectivity criterion of the linear architecture.

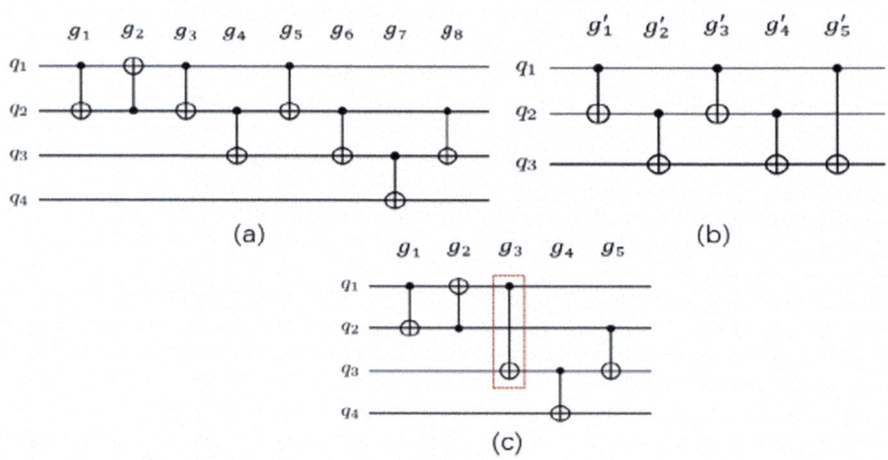

Fig. 28.3 Template-matching optimization. **a** Proximity quantum circuits. **b** Template circuit. **c** Template-based optimized circuit [21]

A standard template must be identified to develop a universal solution for all architectures. Ensuring the template meets substructure connectivity limitations allows circuit optimization without affecting connection constraints. Consequently, this technique uses pre-optimized templates to substitute common subcircuits in a more extensive circuit without affecting its functionality. This can drastically cut down on the complexity and size of the circuit. Under the limited-connectivity architecture, we optimize CNOT circuit size/depth to boost performance on noisy intermediate-scale quantum devices [23].

28.3.2 Ancilla-Free Synthesis

An ancilla-free synthesis is a method for building quantum circuits without ancilla or auxiliary qubits. Although they are employed to store intermediate findings, ancilla qubits can potentially increase the size and complexity of the quantum circuit architecture. Quantum circuits can be made smaller and less complex with ancilla-free synthesis, facilitating their implementation on actual quantum computers. A variety of methods can accomplish ancilla-free synthesis. Reversible functions are often represented using truth tables. Truth tables can create circuits without ancilla qubits [24]. Factoring out common terms or applying algebraic identities to modify the quantum expressions defining the circuit eliminates the need for ancilla qubits. Binary decision diagrams are another technique (BDDs). A compact and practical data format for representing reversible functions is a BDD [25]. Ancilla-free circuits can be created with BDDs by applying a decomposition technique.

The effective method of ancilla-free synthesis can be applied to optimize quantum circuit design. It is important to remember that not all reversible functions can be created ancilla-free. Ancilla qubits are extra qubits utilized for preliminary calculations; the circuit's final output can function without them. Ancilla-free circuits reduce qubit count, enabling more efficient quantum resource use and affordable quantum computers. Circuits can be made less noisy and more efficient by eliminating ancilla qubits.

28.3.3 NOT Gate Technique

The NOT gate technique optimizes quantum circuits by locating and eliminating pairs of neighboring NOT gates, another crucial optimization method. This can be accomplished by applying a sequence of two NOT gates that do not affect the qubit's state, allowing them to be securely removed from the circuit [26].

This method works best in circuit designs with plenty of NOT gates but may be used with any quantum circuit. For instance, this method can eliminate 46% of the NOT gates in the Deutsch-Jozsa algorithm, improving its performance [27].

The NOT gate technique involves the following steps:

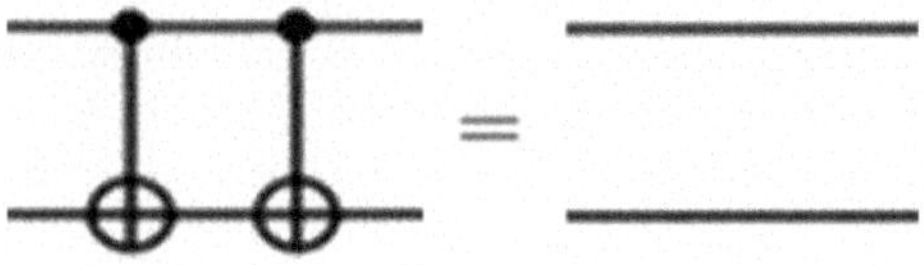

Fig. 28.4 Two CNOT gates together compute identity

1. Indicate every pair of quantum NOT gates (CNOT gates) adjacent to the circuit.
2. Eliminate these NOT gate pairs in the circuit as two CNOT gates simultaneously applied in a quantum circuit compute the identity as depicted in Fig. 28.4.
3. Combine any leftover NOT gates to streamline the circuit.

The NOT gate technique is one straightforward yet powerful method for optimizing quantum circuits and cutting down on their size and complexity. The quantum NOT gate is the most prone to noise. Circuits can be made more dependable by reducing the quantity of NOT gates.

28.3.4 Pattern Matching Technique

Pattern matching is a method for locating every pattern instance in more complex circuit designs. In quantum circuit optimization, pattern matching can find every instance of a small quantum circuit or a pattern in a larger quantum circuit. This can be useful for several reasons. For example, pattern matching can identify common subsequences in quantum circuits, which can be used to optimize the circuits. Pattern matching can also be used to find errors in quantum circuits by identifying patterns that should not be present. The steps involved in the pattern-matching technique are:

1. The algorithm first transforms the quantum circuit into a graph.
2. A subgraph of the graph represents the pattern.
3. The program finds all graph pattern occurrences.

In 2019, an article presented a new algorithm for pattern matching in quantum circuits. The Exact and Practical Pattern Matching Algorithm for Quantum Circuit Optimization is a promising new tool that is effective for various quantum circuit optimization problems. Pattern-matching algorithm-based optimization strategies like template matching result in a ~30% gate count decrease for random quantum circuits [28]. State-of-the-art quantum circuits can be improved for practical use. Optimization of circuit layouts is the goal of this method. For example, a typical pattern includes a sequence of three CNOT gates any number of times in a quantum circuit. This can be optimized by combining the three CNOT gates into a single gate, represented in Fig. 28.5, thereby reducing the depth and cost of the quantum circuit.

With the development of powerful quantum computers, this algorithm will become more and more crucial. Some of the benefits and challenges of using the discussed optimization techniques in quantum circuit design are mentioned in Table 28.3.

Optimization of significant and complex circuits is best with template-based optimization and pattern-matching. Ancilla-free synthesis and NOT gate optimization

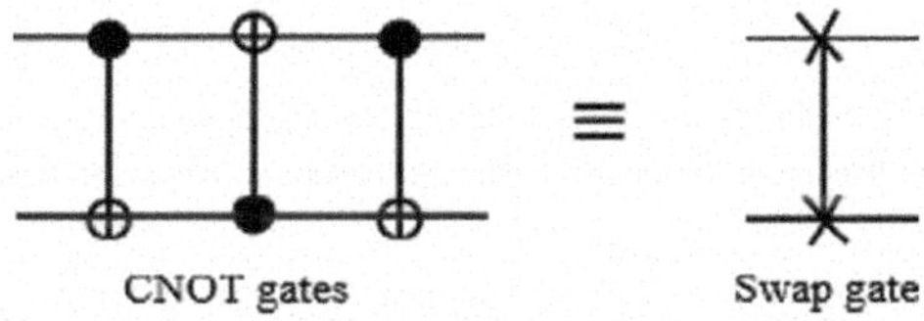

Fig. 28.5 Three CNOT gates into a single SWAP gate

Table 28.3 Optimization techniques in quantum circuit design

Optimization techniques	Benefits	Challenges
Template-based optimization	It can significantly reduce the size and complexity of quantum circuits It can make them easier to implement on real quantum computers It can improve their performance It can be used to optimize circuits for different quantum architectures	The creation of a template library can be time-consuming and challenging The template matching process can be computationally expensive The effectiveness of template-based optimization depends on the quality of the template library
Ancilla-free synthesis	It can reduce the size and complexity of quantum circuits It can make them easier to implement on real quantum computers It can improve their performance and make them less prone to noise	Not all reversible functions can be synthesized ancilla-free The synthesis process can be computationally expensive The effectiveness of ancilla-free synthesis depends on the quality of the reversible function representation
NOT gate optimization technique	Reduces the depth of quantum circuits Improves the scalability of quantum circuits It can be used to implement a variety of quantum algorithms	It can be challenging to find an optimal solution It can be computationally expensive Not all NOT gate optimization techniques are compatible with all hardware platforms
Pattern matching technique	Improved efficiency Reduced errors Increased scalability Improved flexibility	It is computationally expensive for large quantum circuits as it can be time-consuming Difficult to implement It needs to handle a wider variety of quantum circuits

optimize noise-sensitive circuits. Please notice that the above methods are not mutually exclusive. Multiple strategies can typically yield the most significant outcomes. Many circuits use template-based optimization to minimize size and complexity, then ancilla-free synthesis and NOT gate optimization to reduce noise. Overall, template-based optimization, ancilla-free synthesis, NOT gate optimization, and pattern-matching techniques are powerful techniques that can improve the efficiency of quantum circuits. However, it is essential to know the challenges of using these techniques to design quantum circuits that can be resource-efficient, scalable, and fault-tolerant.

28.4 Conclusion

Quantum computing will be critical now and in the future. Quantum computing promises to perform complex tasks faster than traditional computers. The scientific and research sectors have driven the demand for quantum design. Systematically building optimal quantum circuits helps implement quantum algorithms efficiently for quantum computation. Quantum circuits are complex, and even little design modifications can affect performance. Quantum circuits must handle more qubits as quantum computers become increasingly influential. They need fault tolerance. The study discusses efficient quantum circuit design optimization methods and its challenges for future researchers. The desired goals such as complexity, depth of the circuit, time steps, quantum cost, and hardware resources will determine the appropriate circuit optimization method for a particular application. The best way is to experiment to find the optimal optimization methods for the circuit. Optimization approaches will help optimize quantum circuits, reduce errors making them fault tolerant, efficient, and scalable, and enable more viable implementations as quantum computing technology matures.

References

1. Feynman, R.P.: Simulating physics with computers. Int. J. Theor. Phys. **21**, 467–488 (1982)
2. Coles, P.J., Eidenbenz, S. et al.: Quantum algorithm implementations for beginners. ACM Trans. Quantum Comput. (2018)
3. Landauer, R.: Irreversibility and heat generation in the computing process. IBM J. Res. Dev. **5**(3) (1961)
4. Bennett, C.H.: Logical reversibility of computation. IBM J. Res. Dev. **17**(6), 525–532 (1973)
5. Nielsen, M.A., Chuang, I.L.: Quantum computation and quantum information. In: 10th-Anniversary Edition. Cambridge University Press (2010)
6. Solenov, D., Brieler, J., Scherrer, J.F.: The potential of quantum computing and machine learning to advance clinical research and change the practice of medicine. Mo. Med. **115**(5), 463–467 (2018). PMID: 30385997; PMCID: PMC6205278
7. Turtletaub, I., Li, G., Ibrahim, M., Franzon, P.: Application of quantum machine learning to VLSI placement. In: ACM/IEEE 2nd Workshop on Machine Learning for CAD (MLCAD), Reykjavik, Iceland, pp. 61–66 (2020)
8. Cheng, H.P., Deumens, E., Freericks, J.K., Li, C., Sanders, B.A.: Application of quantum computing to biochemical systems: a look to the future. Front. Chem. (2020). https://doi.org/10.3389/fchem.2020.587143
9. Cho, C.H., Chen, C.Y., Chen, K.C., Huang, T.W., Hsu, M.C., Cao, N.P., Chang, C.R.: Quantum computation: algorithms and applications. Chin. J. Phys. **72**, 248–269 (2021)

10. Bova, F., Goldfarb, A., Melko, R.G.: Commercial applications of quantum computing. EPJ Quantum Technol. **8**(2) (2021). https://doi.org/10.1140/epjqt/s40507-021-00091-1
11. Shende, V.V., Bullock, S.S., Markov, I.L.: Synthesis of quantum-logic circuits. IEEE Trans. Comput. Aided Des. Integr. Circuits Syst. **25**(6), 1000–1010 (2006)
12. Humble, T.S., Thapliyal, H., Muñoz-Coreas, E., Mohiyaddin, F.A., Bennink, R.S.: Quantum computing circuits and devices. IEEE Des. Test **36**(3), 69–94 (2019)
13. Saeedi, M., Markov, I.: Synthesis and optimization of reversible circuits: a survey. ACM Comput. Surveys **45**(2), 1–34 (2013)
14. Basak, A., Sadhu, A., Das, K., et al.: Cost optimization technique for quantum circuits. Int. J. Theor. Phys. **58**, 3158–3179 (2019). https://doi.org/10.1007/s10773-019-04192-7
15. Thapliyal, H., Ranganathan, N.: Design of efficient reversible binary subtractors based on a new reversible gate. In: IEEE Computer Society Annual Symposium on VLSI, pp. 229–234 (2009)
16. Morrison, M., Ranganathan, N.: Design of a reversible ALU based on novel programmable reversible logic gate structures. In: IEEE Computer Society Annual Symposium on VLSI, pp. 131–136 (2011)
17. Johansson, N., Larsson, J.Å.: Efficient classical simulation of the Deutsch–Jozsa and Simon's algorithms. Quantum Inf. Process. **16**(9), 233 (2017)
18. Welch, J., Greenbaum, D., et al.: Efficient quantum circuits for diagonal unitaries without ancillas. New J. Phys. **16**(3), 033040 (2014)
19. de Brugière, T.G. et al.: Reducing the depth of linear reversible quantum circuits. IEEE Trans. Quantum Eng. **2**, 1–22 (2022)
20. Biswal, L., Das, R., et al.: A template-based technique for efficient Clifford+T-based quantum circuit implementation. Microelectron. J. **81**, 58–68 (2018). https://doi.org/10.1016/j.mejo.2018.08.011
21. Gao, X., Guan, Z. et al.: Quantum circuit template matching optimization method for constrained connectivity. Axioms **12**(7), 687 (2023). https://doi.org/10.3390/axioms12070687
22. Itoko, T. et al.: Optimization of quantum circuit mapping using gate transformation and commutation. Integration **70**(C), 43–50 (2020)
23. Wu, B., et al.: Optimization of CNOT circuits on limited connectivity architecture. Phys. Res. Rev. **5**, 013065 (2023)
24. Chattopadhyay, A., Hossain, S.: Ancilla-Free Reversible Logic Synthesis via Sorting (2016)
25. Soeken, M., Tague, L. et al.: Ancilla-free synthesis of large reversible functions using binary decision diagrams. J. Symb. Comput. **73**(C), 1–26 (2016)
26. Bataille, M.: Quantum circuits of CNOT gates: optimization and entanglement. Quantum Inf. Process. **21**(269) (2022). https://doi.org/10.1007/s11128-022-03577-8
27. Qiu, D., Zheng, S.: Revisiting Deutsch–Jozsa algorithm. Inf. Comput. **275** (2020)
28. Iten, R., Moyard, R.: Exact and practical pattern matching for quantum circuit optimization. ACM Trans. Quantum Comput. **3**(1) (2022)

Chapter 29
RIDynaQ: A DynaQ Based System for Reading Impairment Detection

Hima Varshini Surisetty, Sarayu Varma Gottimukkala, and J. Amudha

Abstract Dyslexia or reading impairment is a reading disorder that can have an effect on both children and adults. Though reading impairment affects the reading abilities of an individual it does not have an effect on the intelligence quotient of the individual. It has been proven that eye-gaze behavior can be used as a good indicator of the reading abilities of a person. Therefore, using eye-gaze behavior and the eye-gaze patterns as a determinant of a good reader and a poor reader can help with the identification of reading impairment. An early detection of reading impairment can help the person in getting the required help as soon as possible. In this work, the detection of reading impairment has been modeled as a reinforcement learning problem using a model-based learning approach. The proposed model is named as RIDynaQ which stands for Reading Impairment detection using DynaQ wherein the eye-gaze behavior of both dyslexic and non-dyslexic children was considered and the optimal gaze targets and the optimal policy for reading for both good and poor readers were obtained. Also, a comparison was drawn amidst the pre-existing model-free approach and model-based approach. The main difference between the existing Q learning based approach and RIDynaQ is the use of simulated experiences and developing a model of the environment which does not exist in Q Learning based approach but is used in the RIDynaQ model.

29.1 Introduction

Dyslexia can be a neurodevelopmental disorder that affects reading without comprehension or underlying learning. Although the exact cause is not fully understood, it is accepted that a combination of genetics, nerves, and natural conditions lead to its development. Risk factors such as family history, preterm birth, low birth weight, and prenatal exposure to substances such as smoking or alcohol are associated with development in the brain that can process words, causing dyslexia. People with dyslexia

H. V. Surisetty (✉) · S. V. Gottimukkala · J. Amudha
Department of Computer Science and Engineering, Amrita School of Computing, Amrita Nagar, Choodasandra, Bengaluru, Karnataka, India
e-mail: surisettyhima@gmail.com

S. K. Pal et al. (eds.), *Intelligent Informatics*, Smart Innovation, Systems and Technologies 389, https://doi.org/10.1007/978-981-97-2147-4_29

face many problems that prevent them from learning leadership and social understanding. Reading, writing, and spelling difficulties make it difficult to understand content, recognize words, and articulate ideas together clearly. As people struggle to get an education with their peers, these problems can make their way into their own, causing discouragement, discomfort, and even depression. Additionally, dyslexia can interfere with tasks such as connecting, coordination, and time management, increasing the difficulties faced by people with learning disabilities.

Various models have been developed to distinguish dyslexia; However, they often encounter problems that hinder their lives. Effective strategies are always time-consuming, costly, and prone to human error based on standard criteria and subjective criteria. Also, they will not offer an easy way to make up for dyslexia, to delay when necessary. To address these limitations, analysts have explored selected applications, including educational support for dyslexia detection. At its core, supportive learning is a skill-building approach that allows professionals to learn through trial and error to make the best choices in strong situations. It involves making a professional connection with the environment, accepting criticism within the framework of reward or discipline, and using that criticism to remember how to get more benefits over time. In the context of dyslexia localization, additional learning can be used to create computations that learn from data and make predictions based on dyslexia-related patterns and features. Reinforcement learning has many areas of interest compared to current models of dyslexia localization. First, these calculations can be continuously updated and improved by connecting with the environment, integrating daily data, and making better predictions.

This diversity is important for the discovery of dyslexia, whose characteristics change in humans and progresses over time. In addition, the support of computer learning can extract important content from large datasets, reducing the need for textbooks that contain design patterns that can be contextual and time-consuming. There are two methods in reinforcement learning: model-based learning and model-free reinforcement. Model-based learning supports the creation of an internal model of the environment to support planning and decision making. It allows experts to reproduce projects and their results and make better decisions. On the other hand, encouraging modeless learning provides the best direct learning or respect for work without a clear model of the environment. Although more flexible, this approach may require more information to coordinate a good solution.

In the context of detecting dyslexia, model-based learning support provides a critical point. By creating an internal representation of values and constructs associated with dyslexia, these calculations can be used to predict learning and create personalized interventions. They reproduce the effects of different interactions and explore their effects on dyslexia problems in humans. This capability improves the accuracy and effectiveness of dyslexia studies and mediation strategies. Various enhancements have been created on existing learning models, each with its own advantages and features. An example in this area is DynaQ, which combines additional learning with level measurement to improve decision making. DynaQ accelerated learning plans with built-in demonstrations to create meetings and generate additional planning information. This approach leads to successful dyslexia detection through the use and reproduction of learning materials, thus providing a clear and personalized intervention.

In conclusion, dyslexia presents difficult problems for those affected. Dyslexia diagnosis requires modern methods and educational support has been developed as a good method. Reinforcement learning has the potential for multiple, flexible, and personalized interventions, particularly through support learning models such as DynaQ. By responding to these advances, analysts and professionals can improve the diagnosis, treatment, and recovery of dyslexia, ultimately improving the lives of people with dyslexia. The sections discussed in this work include the literature survey which talks about the existing work done in this domain, the system architecture that talks about the environment, reward system, agent, and the interface, implementation that describes the different functions used in the work and their role in getting the output, lastly the results sections talks about the DynaQ and hyper-parameter tuning results along with a comparison with existing Q Learning based approach. The contributions of the proposed approach are as follows:

- Incorporating a model-based approach for dyslexia detection.
- Making effective use of the deep dynaQ algorithm to update q values.
- Analysing the performance of good and bad readers' policies.

29.2 Literature Survey

The literature survey discusses the causes and effects of dyslexia on students. Some detailed literature survey has also been done on the existing work that discusses the problems and solutions in detecting dyslexia among students.

Dyslexia is a disorder that restricts the ability of a student to read aloud and spell words. There are many signs to look out for in young children that can indicate that the child is suffering from dyslexia. Such signs can be of many forms such as difficulty in reading or differentiating between common synonyms of words, avoiding reading time or reading can be very slow, etc. [1]. Reinforcement learning is an important research tool that has many applications and is used in various fields such as neurodevelopmental language and listening disorders [2]. Research in this area focuses on differences in reinforcement learning in different disorders, for example, focusing on developmental dyslexia and its link to atypical reinforcement learning [3]. This research provides insight into the unique characteristics of dyslexia and how the learning process differs in people with dyslexia. Educational support applications range from neurodevelopmental disorders to precision medicine, digital health, and psychology [4]. These functions can take advantage of prediction and customization using additional learning models. In addition, compensatory mechanisms have been investigated in developmental dyslexics who face difficulties in learning support, with reference to compensatory strategies with delayed feedback [5]. Robotic learning has emerged as an effective way to support people with dyslexia, using additional learning techniques to provide personalized assistance [6].

To meet the specific needs of dyslexic students, the Adaptive Reinforcement Learning Framework (RALF) was proposed for the study. RALF aims to provide quality education using additional education methods [7]. In addition, reinforcement learning has been used to identify eye-gazing behavior to detect dyslexia [8]. The ability of humanoid robots to support dyslexic patients through educational support has also been explored [9]. Machine learning algorithms have been widely researched for dyslexia prediction, early detection, and intervention [10]. Reinforcement learning has contributed to the development of eye behavior tracking systems by combining reinforcement learning with eye tracking techniques [11]. Neural networks have been proposed to ensure that there is no gender bias in dyslexia screening [12]. Eye analysis has been used to examine students' stress levels and provides insight into mental health and well-being [13]. Additionally, machine learning based predictions have been developed that provide tools and personalized strategies to support college students with dyslexia [14].

An eye tracker application has been developed to capture and analyze eye movements for research and practical purposes [15]. The study investigated the effects of educational support on rearward gaze and head orientation, particularly during early infant development [16]. Reinforced learning has also been studied in the context of autism spectrum disorder (ASD), showing certain patterns and characteristics of individuals with ASD [17]. Data analysis helps to understand the different selection processes in modeled and unsupported learning [18]. In addition, additional training methods addressing ethical concerns have been used to address privacy protection and eye data management [19]. Modeling of interaction, which is an important part of human communication, has been done with additional learning techniques [20]. Extensive research on eye movements during natural behavior has improved our understanding of human cognition and perception [21].

Together, these studies suggest broad applications of learning support in many fields, including neurodevelopmental disorders, education, robotics, visual search, and cognitive science. Figure 29.1 represents the timeline of various reinforcement learning algorithms throughout the years. It can be inferred that Q learning was among the first algorithms developed in reinforcement learning and that was followed by DynaQ. The most recent advancements in reinforcement learning include Proximal Policy Optimization, Soft Actor Critic, etc.

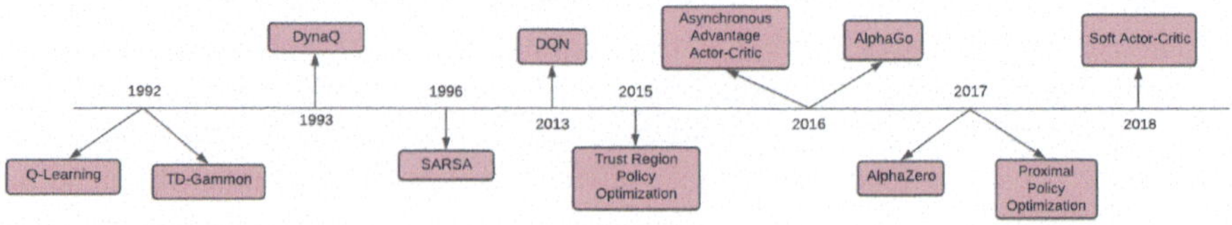

Fig. 29.1 Illustration of the timeline of RL algorithms

29.3 System Architecture

This section describes the various components in the system architecture that is shown in Fig. 29.2. The different components include the environment, agent, and interface.

A. *RL Formulation*

In order to solve the problem of reading impairment detection was modeled as a Markov Decision Process (MDP). The Markov Decision Process is defined as a quadruple which is represented as (S, A, r, T) where, for a given reinforcement learning agent 'S' denotes the state-space, 'A' denotes the action-space, 'r' denotes the immediate reward that is associated with the action that was taken up and 'T' denotes the transition probability of moving from one state to another on choosing an action.

The reward component in the MDP suggests the influence of taking up an action and defines how "good" or "bad" it is for the agent to take up a certain action. The reward is the component that would eventually help in guiding the agent in taking up an optimal set of actions that would lead it towards its goal. In order to formulate any problem as a reinforcement learning problem in the form of an MDP requires the identification of the environment, the set of states, the set of actions, and the reward system.

B. *Environment*

(1) *Grid World and States*

The environment in this problem of detection of reading impairment is represented in Fig. 29.3 which is the text stimulus that would be presented to the individual whose eye-gaze will be captured since, it is through the eye-gaze that the person would be able to perceive the text. The human brain helps the individual in identifying the areas of interest. So, this text stimulus can contain both areas of text and areas that are blank with no text. Therefore, the environment can be represented as a grid-world of

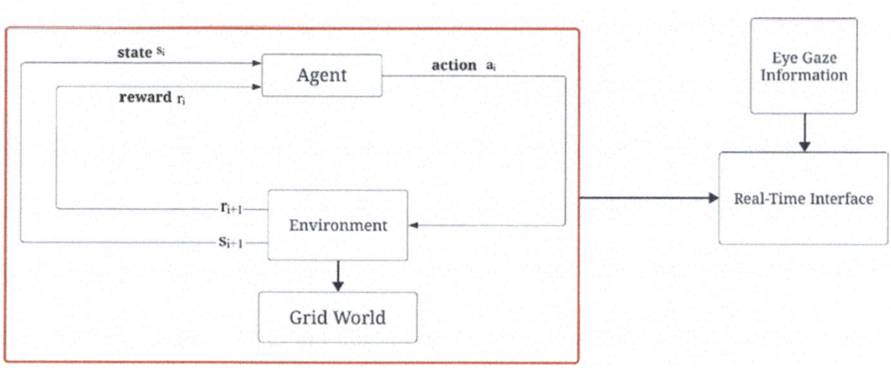

Fig. 29.2 Visualization of the system architecture of RIDynaQ

15 × 18 as illustrated in Fig. 29.4 that contains cells with text and cells that are blank with no text. The Text-Cells (TC) contain the text and the Non-Text-Cells (NTC) are the blank cells.

The states in this problem are all the parameters that are observed by the agent which are the current gaze location of the individual and the amount of time for which the eye-gaze of the individual stays fixated at that state, which is the fixation duration. The action is defined as the next target location in the environment.

(2) *Data Description*

The dataset of eye-gaze behavior for implementing this reinforcement learning model was obtained from [21] wherein, the eye-gaze behavior for the given text stimulus was recorded. The dataset consists of about fifteen samples of non-dyslexic children and about five samples of dyslexic children of which three were moderate cases of reading impairment and the other two were extreme cases of reading impairment. Out of all the features that were recorded only the features like the current location

Fig. 29.3 The text stimulus used in RIDynaQ that is presented on the screen

1	2	3	4	5	6	7	8	9	10	11	12	13	14	15	16	17	18
19	20	21	22	23	24	25	26	27	28	29	30	31	32	33	34	35	36
37	38	39	40	41	42	43	44	45	46	47	48	49	50	51	52	53	54
55	56	57	58	59	I have a pet dragon. I call him Jag.						66	67	68	69	70	71	72
73	74	75	76	77	78	79	80	81	82	83	84	85	86	87	88	89	90
91	92	93	94	95	96	97	98	99	100	101	102	103	104	105	106	107	108
109	110	111	112	113	114	115	116	117	118	119	120	121	122	123	124	125	126
127	128	129	130	131	132	133	134	135	136	137	138	139	140	141	142	143	144
145	146	147	148	149	150	151	152	153	154	155	156	157	158	159	160	161	162
163	164	165	166	167	168	169	170	171	172	173	174	175	176	177	178	179	180
181	182	183	184	185	186	187	188	189	190	191	192	193	194	195	196	197	198
199	200	201	202	203	204	205	206	207	208	209	210	211	212	213	214	215	216
217	218	219	220	221	222	223	224	225	226	227	228	229	230	231	232	233	234
235	236	237	238	239	240	241	242	243	244	245	246	247	248	249	250	251	252
253	254	255	256	257	258	259	260	261	262	263	264	265	266	267	268	269	270

Fig. 29.4 The grid-world representation of the text stimulus used by RIDynaQ

in (x, y), the fixation duration, and the 'id' were used. The pixel co-ordinates were mapped to the grid indices in the grid-world that has been defined previously in [8].

C. *Reward System*

The reward system that has been used in this work was an adaptation of the reward system that was presented in [8]. The reward system was described in three different phases that cover three different parameters namely, the nature of visitation of a cell, the fixation duration, and the scanpath length. Figure 29.5 represents the design of the reward system. The first phase of the reward system is based on cell visitation which can either be a first-time visit or a revisit to a certain cell. Under a first-time visit, either a goal state, a text cell or a non-text cell. The goal state can be reached either after all cells are traversed or directly in which there is a $+100$ and -0.125 reward respectively and reaching a text cell gives a $+40$ reward and a non-text cell results in a -0.25 reward. But if it is a subsequent visit, then a text cell would result in a -0.125 reward and a non-text cell would also give a -0.125 reward which is amplified using the number of times that non-text cell is visited because a "good" reader would not keep going to the non-text cells once it is known that there is no text present there.

The second phase of the reward system is based on the fixation duration which is encoded with a value of either 0, 1, or 2 based on the fixation duration in the increasing order. The third phase of the reward system is designed based on the scanpath length. It was given in [10] that a "good" reader would have a scanpath

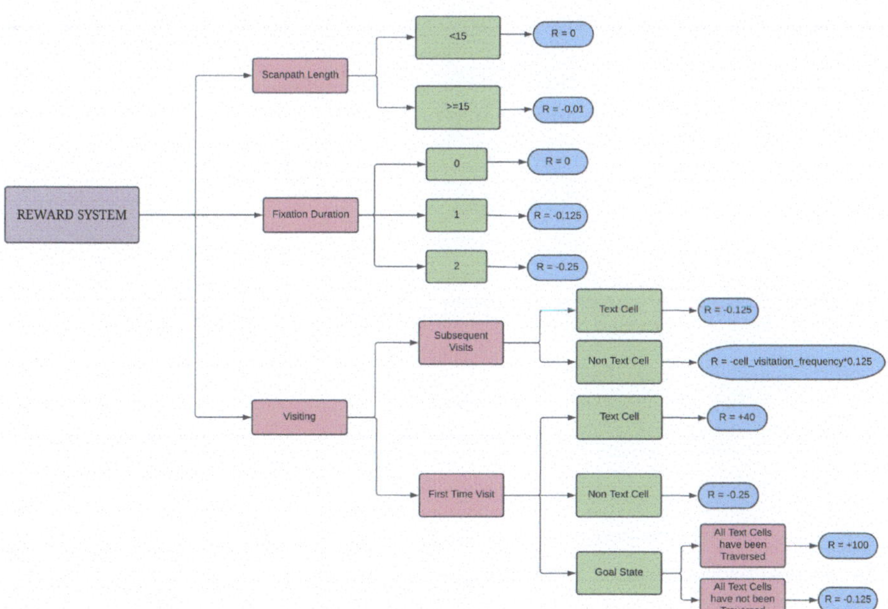

Fig. 29.5 The different phases of the reward system in system architecture of RIDynaQ

ranging from 8 to 10 so the threshold was set as 15. So, a scanpath greater than 15 was given a −0.01 reward and no reward was awarded for a scanpath less than 15.

D. *Agent*

The agent used in the work [8] was the Q Learning algorithm. The model that was used to replace the Q learning algorithm in this work is the RIDynaQ model that uses DynaQ algorithm as an agent in the system architecture.

Figure 29.6 represents the agent in the system architecture of RIDynaQ Model. This agent learns about the environment through a model which uses the DynaQ algorithm to update the state action pair in the Q table. In this DynaQ algorithm, a model, in the form of a dictionary, is created that stores the state and actions as keys and maps them to the expected reward and next state. This model is used later on to update the values in the Q table. A state action pair is chosen in random and if that pair is present in the Q table then the reward is updated based on action, otherwise, the reward and next state are calculated using the Q value update rule. Using the new state and action that is calculated from the Q table, it is updated in the policy. The same process is followed for all the episodes wherein the optimal policy for each episode is measured. The data from "good" and "poor" readers are tested separately and later compared using their optimal policies.

E. *Real-Time Interface*

Figure 29.7 represents the interface that was created to visualize the score, sequence of actions taken, and the scanpath length of a student.

The excel sheet containing the eye-gaze information of a student was given as input to this interface. In return, the interface showed the optimal policy for that student, based on which that student could be categorized as a "good" or "poor" reader. Along with this, the scanpath length and the maximum reward was also shown. Only one

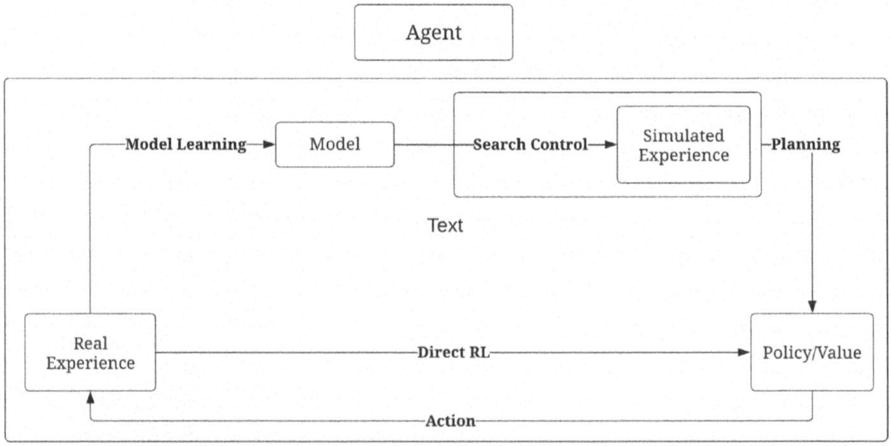

Fig. 29.6 Representation of the RIDynaQ agent

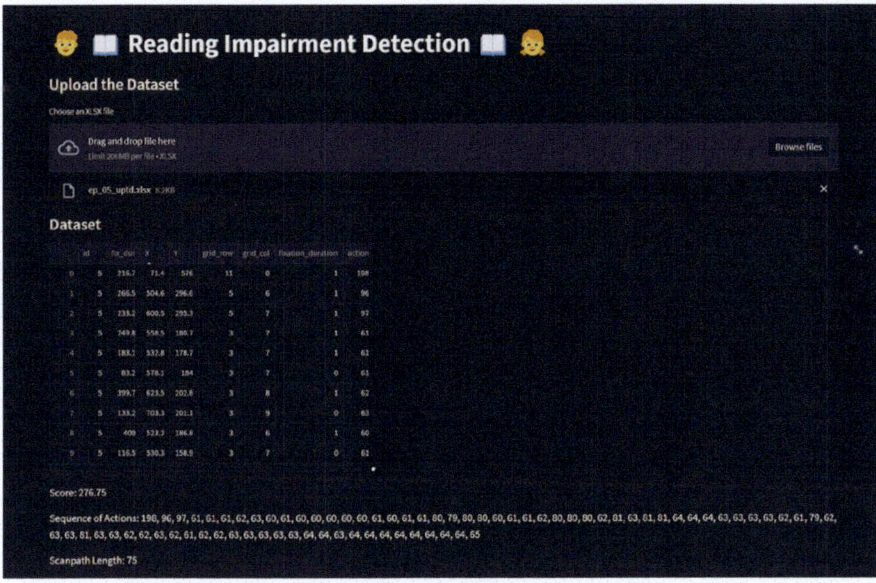

Fig. 29.7 Visualization of the user interface for RIDynaQ

excel student pertaining to an individual student can be uploaded into the interface at any given time. The optimal policy describes the location in the text grid where the student is reading. The cells 60–65 contain the actual text that is shown on the screen so if the student is reading these cells only and does not skip or re-read some text in those cells then he/she is a good reader. Otherwise, they are considered as bad readers.

29.4 Implementation

Figure 29.8 depicts the class diagram for this work. Different classes were used for the implementation of different components of the system architecture. The packages from Python that were used for the implementation of this work include gym, pandas, numpy, glob, and matplotlib.

The Stimuli_Env is the main class under which the different functions are defined. These functions are used for different tasks in the RIDynaQ model. The _init_() function is the constructor method for the class. It initiates all the variables needed for the environment. The _init()_ function consists of declaration of variables such as the environment name, states, rewards, action-space, observation space, scan path length, etc. It also declares a done variable that tells whether the environment is done. There is also a reset() function that resets the environment back to its default state. The variables that are reset are the done variable, which is set to false, the scanpath

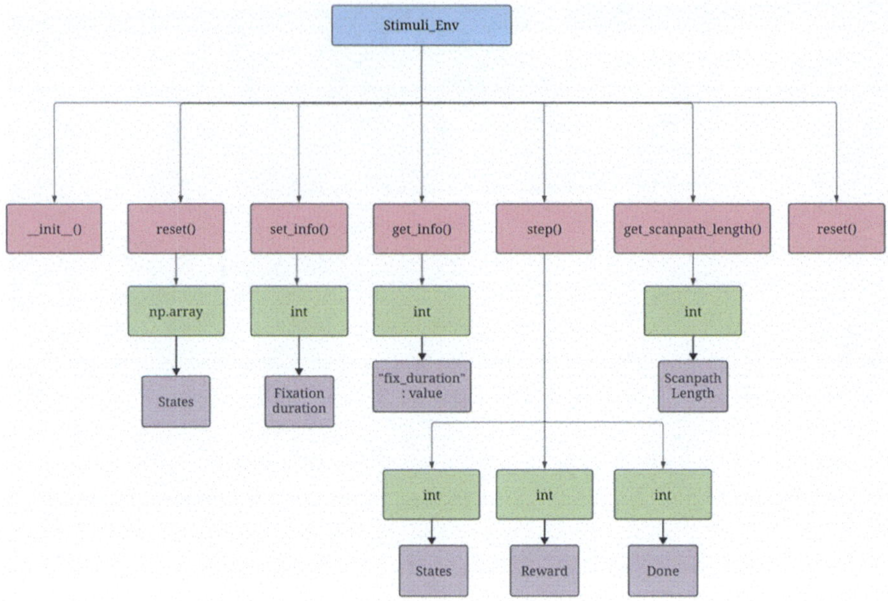

Fig. 29.8 Illustration of the class diagram of the RIDynaQ model

length, and the reward. It returns the initial state of the environment as a NumPy array. There set_info() function acts as a placeholder method to call the fixation duration value from the dataset. The get_info() function calls the set_info() function to obtain some information. It returns a dictionary with a single key value pair which is that of the fixation duration (key) and the output of set_info() function (value).

The main function in this class is the step() function which is used to move in the environment based on the provided action. In this function, the scanpath length is incremented and a condition for the done variable is checked. If the environment is done, then the current state and an empty dictionary are printed along with a message. Otherwise, the given action is evaluated, and the state matrix and rewards are updated based on this action. The outputs of this function are the states, rewards, and the done variable. Finally, there is a get_scanpath_length() and the render() function. As the name suggests, the scanpath length is returned as the output of one function while the other does not return any output.

In order to visualize the policy, a separate function named visualize_policy() is used. DynaQ algorithm is used in this function to create a dictionary for the model with the keys as the state and action that are mapped to the value which are the rewards and next state. The actions are calculated after running and updating the model and Q table through several iterations in every episode. Later the actions are appended to a list which is printed as the optimal policy for every episode. The same process is followed for good and bad readers.

Along with this visualization, an interface is also created using the Streamlit package in Python that provides inbuilt functionalities to easily create and maintain web pages with some simple functionality. Before training the model, the interface needs to be tested to check whether it's able to read and update its values using a dataset. The dataset is given to test the interface by passing the path of the dataset and reading the different excel files. The actions, rewards, and next states are updated using the values in the dataset. This testing is done separately for the datasets containing good and bad readers. Once, it is ensured that the interface works properly, the Q table is created and initially set to be filled with all zeros. Then the good and bad readers dataset is provided separately to calculate the optimal policy. To do so, the Q-table is updated using the state and action pairs. Based on the Q-table, the rewards are calculated. But, the optimal values need to be set for the Q table to have proper and optimal updation. This ensures that the agent selects the best actions and learns about the environment properly.

In order to set an optimal value for the hyperparameters, a baseline agent was trained using the DynaQ algorithm with a random fixation duration and a random action picked up from the pool of actions. Since there was no dataset that is involved in training the baseline agent the exploration–exploitation trade-off has to be taken under consideration. But the main goal of this process of training the baseline agent is to obtain the optimal values for the learning rate and the discount factor, the maximum exploration rate as 0.75, minimum exploration rate as 0.1, and the exploration decay as 0.1 and the initial exploration rate as 0.4 and using these values the learning rate and discount rate were tuned to different values to observe which of these values gave an average reward that was progressively increasing over 5000 episodes that were considered for training the baseline agent.

29.5 Results and Discussion

There are three different aspects of this work discussed in this section with the help of obtained observation. Firstly, the updation of Q value using a model of the environment is discussed under the DynaQ subsection. Then, the process of hyperparameter tuning is discussed which tells what the optimal values for the parameters are. These values help in obtaining an optimal policy of the good and bad readers. Lastly, a comparison study is done to analyze the differences between the existing Q learning approach and RIDynaQ approach.

A. *DynaQ*

Once the training of the baseline agent was done and the optimal values for the hyperparameters were obtained, these hyperparameters were used to train the model with the dataset wherein, the actions and the fixation duration was not random like in the case of the baseline agent. The hyperparameters used for this work are the number of planning steps, learning rate, discount rate, maximum number of episodes, and exploration decay rate. The actions and the fixation duration were obtained from the dataset. Once the entire training process was complete and the optimal policy for both good readers and bad readers was predicted, the Q-value updating per timestep was plotted for both good readers and poor readers.

Figure 29.9a, b represent the Q-value updation for good readers and bad readers respectively and it can be inferred that the q values are continuously increasing which means that the model is continuously learning and updating its policy at every timestep. From Table 29.1, it can be inferred that the Q values for good and bad readers can be easily differentiated. The Q values for bad readers are much higher than the good readers initially and later decrease progressively with the time steps as compared to the Q values of the good readers.

B. *Hyperparameter Tuning*

The hyperparameters were tuned in the baseline agent that took random state and action values. The learning rate, discount rate, and the exploration decay were the parameters that were tuned to find their optimal values. The learning rate and discount

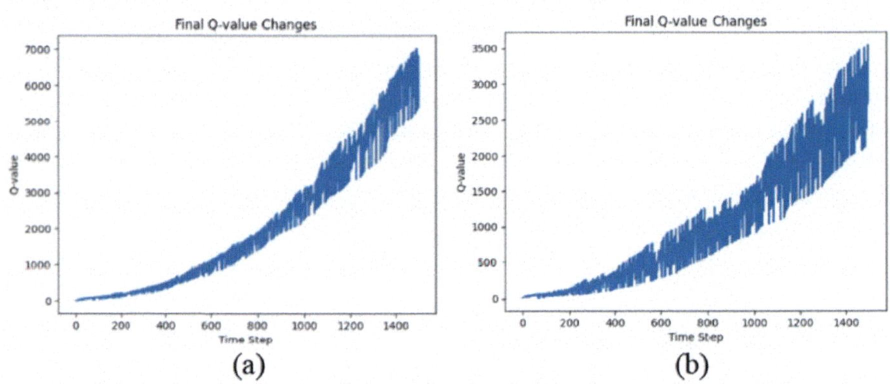

Fig. 29.9 Illustration of the Q-value updation for good and bad readers of RIDynaQ model

Table 29.1 Q value changes for good and bad readers for RIDynaQ model

Time steps	Good reader	Poor reader
1	2.24	14.31
2	2.45	16.18
3	2.64	17.86
4	2.81	19.37
5	2.97	20.74
…	…	…
1496	6123.95	2827.11
1497	6770.96	3623.15
1498	6323.94	3624.09
1499	5308.39	3656.78
1500	6070.83	3278.21

factor are used to update the Q table values. The learning rate intuitively determines, to what extent the agent will learn something new. The discount rate determines the importance of future rewards (s_i). The learning rate was set to 0.2 and the discount factor was set to 0.99 which is utilized as $Q(s, a) = (1 - \alpha)*Q(s, a) + \alpha*(r + \gamma*\max Q(s', a'))$ where, Q(s, a) is the Q-value for state-action pair (s, a), α is the learning rate, which determines how much weight is given to the new information compared to the existing Q-value, r is the immediate reward received after taking action a in state s, γ is the discount factor, which determines the importance of future rewards compared to immediate rewards, s' is the next state reached after taking action a in state s and $\max Q(s', a')$ represents the maximum Q-value for any action a' in the next state s'.

Table 29.2 represents the hyperparameter tuning that was done in order to obtain the best values for the learning rate and the discount factor. After many runs, it was observed that the learning rate of 0.2 and the discount factor of 0.99 gave a good progressive increase in the average reward over the 5000 episodes. Using these optimal values for hyperparameters, the model predicts the optimal policy.

C. *Comparison with Q Learning*

The model predicted the optimal policy for good readers as [198, 60, 61, 62, 63, 64, 65] which can be mapped onto the grid-world as [(11,0), (3,6), (3,7), (3,8), (3,9), (3,10), (3,11)] and it is predicting the optimal policy for a poor reader as [198, 60, 61, 63, 64, 65, 61, 63, 64, 65] which can mapped onto the grid-world as [(11,0), (3,6), (3,7), (3,9), (3,10), (3,11), (3,7), (3,9), (3,10), (3,11)] which suggests that a good reader finishes the reading in seven timesteps but the poor reader takes longer and tends to revisit the cells and also traverses back, these both are suggestive of the reading policy of a good reader a poor reader which was successfully predicted by the RIDynaQ model.

Figures 29.10 and 29.11 represent the visualization for the comparison of the optimal policy for good reader and a poor reader on the 15 $\times$ 18 grid-world using RIDynaQ and Q Learning respectively. Here, it can be seen that, for the good reader, the optimal policy is such that the student reads a cell only once and once they reach goal state, they stay in that state. This optimal policy for both the works remains same. Whereas, for the poor reader, the student tends to re-read some text cells several times or skip over some text cells with hard words when using RIDynaQ but when using Q Learning, it is harder to distinguish the bad reader from good reader as the student reads all the cells and only revisits the text cells once. This is reflected in the optimal policy as some cells that contain text are skipped and some are in darker color which

Table 29.2 Representation of hyperparameter tuning to get optimal values for parameters of RIDynaQ model

Learning rate and discount factor	Final average reward	Increase/decrease
Learning_rate = 0.2, Discount_factor = 0.99	266.78	Increase
Learning_rate = 0.5, Discount_factor = 0.7	282.2	Fluctuating

means they have repeated those cells. Figure 29.12 represents the visualization of the reading policy of both good readers and bad readers and it can be observed that after a certain timestep of 7, the policy of the good reader ends as the traversal is complete and the rest is mapped to zeros for visualization purposes. But the reading strategy of the poor reader continues even after that for a longer time wherein, the cells are being revisited. The main changes between Q Learning that was used in [8] and RIDynaQ in this work are shown in Table 29.3.

From Table 29.3 it can be inferred that there are several differences between the work done in paper [8] and the proposed work. The key differences are that the RIDynaQ model uses simulated data and creates a model of the environment before training the model with states and actions whereas in the Q learning model, the dataset is not used. The model learns about the environment through actions. Some other differences include the design of an interface in the proposed work that is absent in the work from paper [8]. Also, the work from paper [8] is more complex in terms

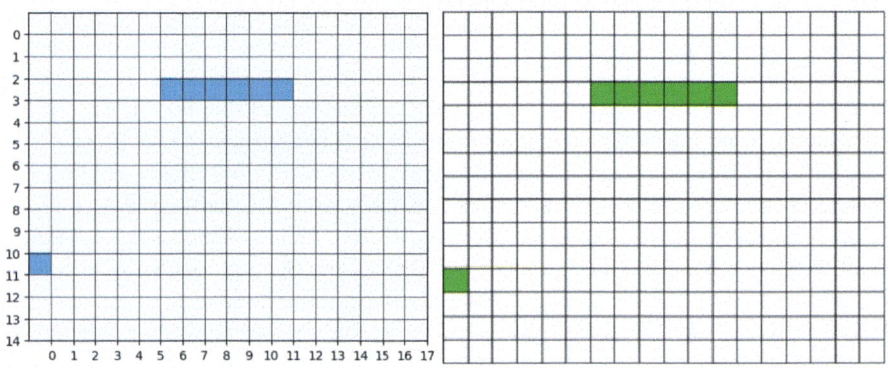

Fig. 29.10 Comparison of policy taken up by a good reader in the 15 × 18 grid-world using RIDynaQ and Q learning

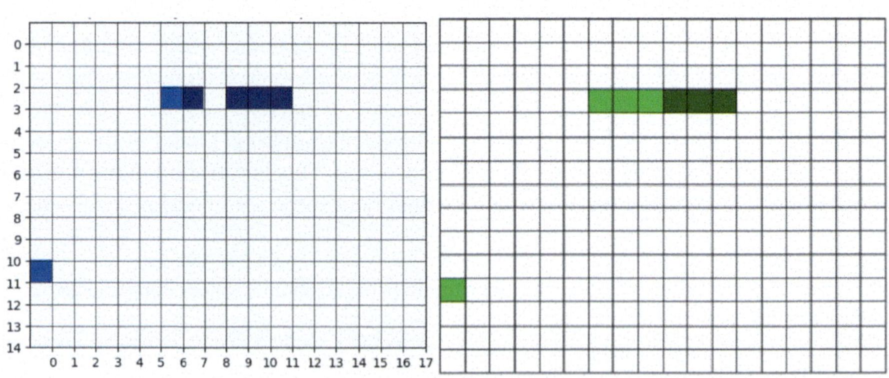

Fig. 29.11 Comparison of policy taken up by a bad reader in the 15 × 18 grid-world using RIDynaQ and Q learning

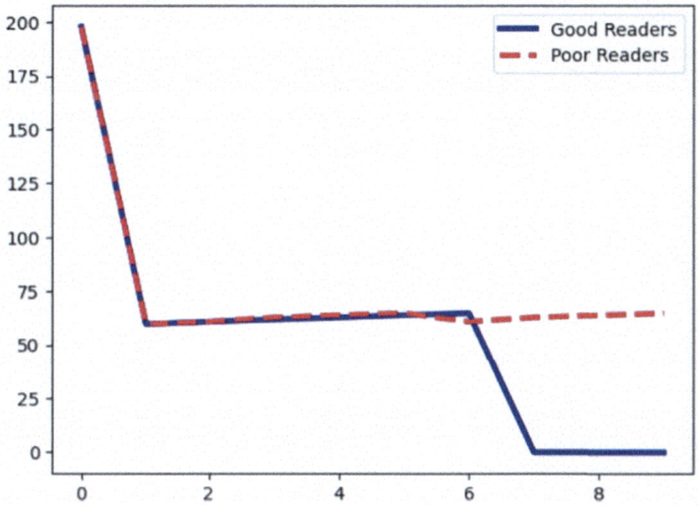

Fig. 29.12 Comparison of the optimal reading policies of a good reader and a poor reader for RIDynaQ model

Table 29.3 Comparison of Q learning model and RIDynaQ model

	Simulated data	Time and space complexity	Model of environment	Exploitation and exploration	Interface
Q learning	×	✓	×	✓	×
RIDynaQ	✓	×	✓	✓	✓

of the time and space used as compared to our proposed work as Q learning uses more resources and memory as compared to DynaQ. One main similarity is that both models use some exploitation and exploration to learn about the different actions and rewards in the environment.

29.6 Conclusion and Future Scope

In this work, a model-base reinforcement learning algorithm namely, RIDynaQ model was used in order to predict the optimal policy or the reading strategy of a good reader and a poor reader. It was observed that the model-based approach was able to predict the optimal policy of each of them and was able to differentiate a good reader and a poor reader. The data of the eye-gaze behavior of both dyslexic and non-dyslexic children was used for this purpose. This use-case can be explored with the help of the reinforcement learning approaches in order to analyze the best approach. There is a lot of scope for this work in the future, for example, this use-case can be extended by deploying a more complex real-time user interface which will enable in capturing the eye-gaze behavior of the individual in real-time and would be able to display the optimal policy using the reinforcement learning model in the background. This can help in making the entire data collection process easier as it is being done in real-time. Also, this application can help in making the prediction of a good reader and a poor reader much easier and faster.

References

1. Margaret, J. Snowling, C.H., Nation, K.: Defining and understanding dyslexia: past, present and future. Oxf. Rev. Educ. **46** (2020)
2. Nissan, N., Hertz, U., Shahar, N. et al.: Distinct reinforcement learning profiles distinguish between language and attentional neurodevelopmental disorders. Behav. Brain Funct. **19** (2023)
3. Massarwe, A., Nissan, N., Gabay, Y.: Atypical reinforcement learning in developmental dyslexia. J. Int. Neuropsychol. Soc **28**(3) (2021)
4. Benjamin, R.: Reinforcement learning as an innovative model-based approach: Examples from precision dosing, digital health and computational psychiatry. Front. Pharmacol., Sec. Transl. Pharmacol. (2023)
5. Yafit, G.: Delaying feedback compensates for impaired reinforcement learning in developmental dyslexia. Neurobiol. Learn. Mem. **185** (2021)
6. Sarah-May, M., Esyin, C., Fiona, C.: Educational robotics and dyslexia: investigating how reinforcement learning in robotics can be used to help support students with dyslexia. In: International Conference on Technological Ecosystems for Enhancing Multiculturality (2022)
7. Seyyed, A.H.M., Azam, B., et al.: RALF: an adaptive reinforcement learning framework for teaching dyslexic students. Multimed. Tools Appl. **81** (2022)
8. Harshitha, N., Vishnu, S.I., Punitha, V., Amudha, J.: Detection of reading impairment from eye-gaze behaviour using reinforcement learning. Procedia Comput. Sci. **218** (2023)
9. Mcvey, S.-M., Chew, E. et al.: The review of dyslexic humanoid robotics for reinforcement learning. In: European Conference on e-Learning (2021)
10. Vanitha, G., Kasthuri, M.: Dyslexia prediction using machine learning algorithms—a review. Int. J. Aquat. Sci. (2021)
11. Deepalakshmi, R., Amudha, J.: A reinforcement learning based eye-gaze behavior tracking. In: 2021 2nd Global Conference for Advancement in Technology (GCAT) (2021)
12. Hobbs, W., Hoskins, W., Tang, J.: Using neural networks to reinforce absence of gender bias in dyslexia screenings. In: 2020 IEEE MIT Undergraduate Research Technology Conference (URTC) (2020)
13. Chandran, J., Amudha, J.: Eye gaze as an indicator for stress level analysis in students. In: 2018 International Conference on Advances in Computing, Communications and Informatics (ICACCI) (2018)
14. Andrea, Z., Juri, T., Giuseppe, C.: A machine learning-based predictor to support university students with dyslexia with personalized tools and strategies. PREPRINT (Version 1) (2023)
15. Divya, V., Amudha, J., Jyotsna, C.: Developing an application using eye tracker. In: 2016 IEEE International Conference on Recent Trends in Electronics, Information & Communication Technology (RTEICT) (2016)
16. Christine, M., Ezgi, K., Sabina, P.: Effects of reinforcement learning on gaze following of gaze and head direction in early infancy: an interactive eye-tracking study. Child Dev. **92** (2021)
17. Manuela, S., Rohr, C.S. et al.: Reinforcement learning in autism spectrum disorder. Front. Psychol. **21** (2017)
18. Arkady, K., Ian, K.: Gaze data reveal distinct choice processes underlying model-based and model-free reinforcement learning. Nat. Commun. **7** (2016)

19. Wolfgang, F., Efe, B., Enkelejda, K.: Reinforcement learning for the privacy preservation and manipulation of eye tracking data. In: Artificial Neural Networks and Machine Learning—ICANN (2021)
20. Renato, R.S., Roseli, A.F.R.: Modelling shared attention through relational reinforcement learning. J. Intell. Robot. Syst. **66** (2012)
21. Mary, H., Dana, B.: Eye movements in natural behavior. Trends Cogn. Sci. **9** (2005)

Chapter 30
Cross-Language Code Mapping with Transformer Encoder-Decoder Model

M. V. Deepak Naik and Swaminathan Jayaraman

Abstract In the dynamic field of software development, knowledge of multiple programming languages is increasingly valued. To address the demand for cross-language systems, our research investigates code translation and proposes a new approach using transformer models to map cross-language codes. We draw inspiration from the success of transformer-based neural networks in natural language processing and apply these models to the complex world of programming languages. A central hypothesis is that transformers can effectively translate code between languages, acting as a bridge between programming paradigms. We develop a custom transformer encoder-decoder model for program-to-program translation that is initialized with a dataset containing C++ programs and corresponding codes in Haxe, Kotlin, Python, and Java. The cooperation between the encoder and the decoder allows a smooth translation of the code, which shows considerable adaptability. Through fine-tuning with data from other programming languages like Kotlin and Python, our model extends its translation capabilities. Enabling Haxe as an intermediate language further facilitates translation to target languages like Java. This adaptability, based on a transformer-based architecture, effectively meets the challenges of an ever-evolving programming environment. It should be noted that our approach differs from large language models in terms of programming complexity and capturing OOP design patterns. In addition to translation, our research paves a way for code visualization using tools such as Java interactive visualization environment. "Cross-Language Code Mapping with Transformer Models" marks a significant advance towards automatic code translation in object-oriented programming, enriching developer tool kits and promoting collaboration across language boundaries in software engineering.

M. V. D. Naik (✉) · S. Jayaraman
Department of Computer Science and Engineering, Amrita School of Computing, Amrita Vishwa Vidyapeetham, Amritapuri, India
e-mail: deepaknaikmv01@gmail.com

S. K. Pal et al. (eds.), *Intelligent Informatics*, Smart Innovation, Systems and Technologies 389, https://doi.org/10.1007/978-981-97-2147-4_30

30.1 Introduction

In the ever-evolving software development landscape, program understanding is a cornerstone of the software development life cycle. The ability to interpret and understand code is the basis for creating, maintaining and improving software systems. Research in this area has provided a compelling insight: code visualization significantly improved programmers understanding compared to raw, unadorned data. This revelation spawned a series of visualization platforms, each designed to facilitate a deeper understanding of code structures and behaviour.

Visualization platforms like Extravis, known for its circular bundle view and massive sequence view, have enriched the experience of understanding C++ code. Similarly, the Java Interactive Visualization Environment (JIVE) provides tools such as object diagrams and sequence diagrams that provide invaluable information about Java programs [1–3]. These platforms have proven their value in helping developers unravel the complexity of their code base.

However there is notable difference in the field of code visualization: Fragmentation between programming languages. Each platform, be it Extravis or JIVE, specializes in a specific programming language and offers its benefits only to the users of that language. The lack of common platform that can visualize code in all programming languages represents a significant gap in a developer's toolkit. At the same time, the advent of large language model (LLM) ushered in a new era of code generation and completion. These complex models, driven by OpenAI Codex and PolyCoder, simplified programming tasks by automatically generating code snippets from natural language descriptions. However, these models have a major drawback—a lack of transparency in their inner workings and limited access to training data. While datasets like Hugging Face's "The Stack" have attempted to nullify this problem, their sheer size and permissiveness make them daunting for the average researcher. This research covers a unique path by using JIVE's existing capabilities to visualize Java programs and extends this capabilities to several OOP languages.

The problem we face is twofold. First the lack of a common platform for visualizing code in different programming languages hinders developers who work with multiple languages. The lack of common ground limits their ability to consistently exploit code visualization. Second, although LLM's have made significant progress in code generation and improvement, their coverage and dataset limitations hinder researchers and developers seeking transparency and adaptability. Building an LLM from scratch or adapting it for specific language is a huge task, often out of reach for many.

To address these issues, we developed a multifaceted approach that uses a specially designed transformer based encoder-decoder model tailored for program-to-program translation. At the core of our solution is the transformer encode-decoder model. This model is not a one-size-fits-all solution, but is carefully designed for specific cross-program translation. This model, derived from knowledge of widely adopted C++ programming language, forms the basis of our approach. Key elements of the model:

- Encoder-decoder architecture: Our model uses an encoder-decoder structure, where the encoder processes the input code (e.g. C++) and the decoder generated the corresponding code in the target language (e.g. Haxe).
- Tokenization, embedding and Masking: We use these techniques to capture the intricate details of code. These mechanisms ensure that the model takes into account the full spectrum of characters and words in the code, including capitalization. This approach is crucial because understanding code often depends on nuanced details.
- Dense Layers: We use dense layers to make it easier to map code snippets from one language to another. The use of rectified liner unit (ReLU) activation functions allows the model to capture complex relationships in code structures.

A pivotal aspect of our approach is adaptability. We demonstrate the versatility of the model by fine-tuning it with data from other object-oriented programming languages such as Kotlin and Python. This critical step extends our code translation capabilities to encompass many programming paradigms. We created a large dataset that contains almost all the basic concepts of OOP. This dataset contains code snippets representing class declarations, function declarations, conditional statements, error handling, inheritance, polymorphism, encapsulation, abstraction, constructors, destructors, instance versus class variables, object instantiation, overloading, overriding, interfaces, abstract classes, association, aggregation, composition, access modifiers, and packages or modules. For each concept, we create numerous programs in each of the languages C++, Python, Kotlin, Python and Java.

The model undergoes sequential fine-tuning. Initially, it is trained with C++ as input and Haxe as output. The model carefully learns the weights associated with this translation process.

- Subsequent Fine-Tuning: We further fine-tune it by introducing Kotlin and Python programs as input, while maintaining consistency in output. This process ensures that our model adapts seamlessly to these languages and provides a platform for a comprehensive code translation framework.
- Haxe as an Intermediate Language: We strategically introduce Haxe as a middle language in our approach. Haxe is uniquely positioned to facilitate the translation of code into a myriad of target languages, including Java. Its middle-level nature allows us to combine the different linguistic features of input programming languages while providing a standardized translatable format.

Distinctiveness in Comparison to LLMs: It is essential to highlight the characteristics of our approach compared to conventional large language models (LLM) like Chatgpt, Falcon, MPT-7B, Llama2, Polycoder, OpenNMT, Claude2 and MarianNMT. While LLMs have gained prominence for their versatility in understanding and generating natural language text, they often falter when tasked with code translation. LLMs operate primarily at the textual level, treating code as a sequence of characters or tokens without a profound grasp of programming paradigms. In contrast, our approach is tailored to code translation, embracing the complexities of programming languages, their structures, and the subtleties of object-oriented programming design patterns. By customizing and adapting our model to specific programming languages, we harness the power of LLMs while tailoring them to the precise demands of code translation. This synergy bridges the gap between linguistic diversity and code interoperability. In summary, our approach effectively solves the identified problem by providing a dedicated solution for code translation, showing adaptability to various OOP languages, introducing an intermediate language for versatile translation and differentiating itself from LLMs by providing targeted code translation functions. This versatile strategy opens the door to better code understanding and multilingual software, while also serving as a blueprint for future research in the field.

Novelty of the research summarized: While most studies shy away from using supervised learning for code translation due to the difficulty in generating suitable datasets, we tackled this challenge head-on, creating a robust dataset covering various Object-Oriented Programming (OOP) concepts. What sets us apart is our shift from the traditional encoder-decoder models to a more powerful transformer encoder-decoder model, significantly boosting our translation capabilities. But the real game-changer is our fine-tuning approach, not just translating from one source language to another, but to multiple target languages like Haxe, Kotlin, and Python, a leap beyond what others envision for the future [4]. Unlike Large Language Models (LLMs) that struggle with new, less-known programming languages, our supervised model is agile, requiring only a specific number of code snippets for the source language, making it more adaptable and efficient. Moreover, our model handles translation even with less-documented but syntactically correct code, overcoming a common challenge faced by LLMs. In essence, our approach not only pushes the boundaries of code translation but also addresses key limitations in existing research, offering efficiency and adaptability in the dynamic landscape of software development.

30.2 Related Works

In the rapidly evolving landscape of code translation and program generation, our research draws inspiration and builds upon foundational insights from diverse literature. Our exploration begins with the application of Neural Machine Translation (NMT) for code translation, where pioneering works have set the stage for modern techniques.

Bahdanau, Cho, and Bengio's innovative approach to NMT, as introduced in [5], revolutionized the field by dynamically aligning and translating code, departing from fixed-length context vectors, intermediate representation like abstract syntax tree [6] and standard deep learning techniques [7]. This departure proved instrumental in handling lengthy code sequences efficiently and proved better than LSTM based models [8]. Expanding on this, Cho, Memisevic, and Bengio [9] explored the use of large target vocabularies in NMT, a critical consideration for code translation quality. This work provided valuable insights into addressing the diverse lexicons of programming languages, although challenges related to model complexity emerged. Jian Li et al.'s work on code completion with neural attention and pointer networks [10] introduced innovative methods utilizing attention mechanisms to enhance code translation accuracy. While this approach marked a significant stride in handling code intricacies, scalability challenges arose, particularly with large codebases. Transitioning to code generation and translation, X. Chen, C. Liu, and D. Song [11] proposed a method transforming input-output examples into executable code. This approach, while a significant step, primarily focused on generating code from inputs and outputs, leaving room for further improvements. Aggarwal, Salameh, and Hindle's work [4] addressed the migration from Python 2 to Python 3, automating the often tedious task of language version transition. While effective for a specific translation challenge, broader adaptability remained a consideration.

The structural language models introduced by Alon, Brody, Levy, and Yahav [12, 13] expanded the horizon for code generation across multiple languages. Their "code2seq" method paved the way for viewing code from a structured perspective, enabling applications like code summarization and recommendation systems. However, challenges persisted in seamlessly transitioning between different language paradigms. In the domain of unsupervised and low-resource machine translation, Artetxe et al. [14] and Artetxe et al. [15] contributed significantly by reducing the reliance on parallel data. These works played a pivotal role in making translation more accessible in resource-scarce scenarios. However, the sensitivity to data quality and domain adaptability limitations are acknowledged. Exploring program understanding and comment generation, Mou and colleagues' work [16, 17] on Convolutional Neural Networks over Tree Structures enriched our comprehension of code structures. Further strides were made by Nguyen and team [18], enabling a divide-and-conquer strategy for multi-phase statistical migration of source code. The introduction of syntactic neural models by Yin and Neubig [19] addressed general-purpose code generation challenges effectively. In the broader context of Natural Language Processing (NLP) using sequence-to-sequence models, influential works such as [20–22] have provided valuable insights. These studies contribute to our understanding of sequence modeling, which can be adapted for code-related tasks. Additionally, the research by Matthew Amodio, Swarat Chaudhuri, and Thomas Reps on "neural attribute machines" [23], Jacob Devlin and colleagues' work on "BERT" [24], Zhangyin Feng's presentation of "CodeBERT" [25], and the work by T. D. Nguyen et al. on mapping API elements for code migration [26], and Karaivanov, Raychev, and Vechev's phrase-based statistical translation of programming languages [27] have laid essential foundations for understanding and generating code. These models, designed for various language understanding tasks, offer valuable perspectives on code semantics and representation.

Our research synthesizes these foundational works, aiming to contribute to the efficiency and effectiveness of code translation. While leveraging transformer models for cross-language code mapping, our approach seeks to address scalability, adaptability, and domain-specific challenges. By building upon the methodologies, innovations, and limitations highlighted in these works, we strive to propel the field of code translation towards more accessible and versatile solutions.

30.3 Methodology

The methodology in this research is a comprehensive approach to address some of the challenges of program-to-program translation and code visualization across multiple object-oriented programming (OOP) languages. This is structured into four components: Dataset Creation, Preprocessing, Model Architecture and Training Procedure, and Fine-Tuning (Fig. 30.1).

30.3.1 Dataset Creation

Creating a complete dataset is a key to the success of cross-program translation and code visualization. This dataset has been carefully designed to cover many Object

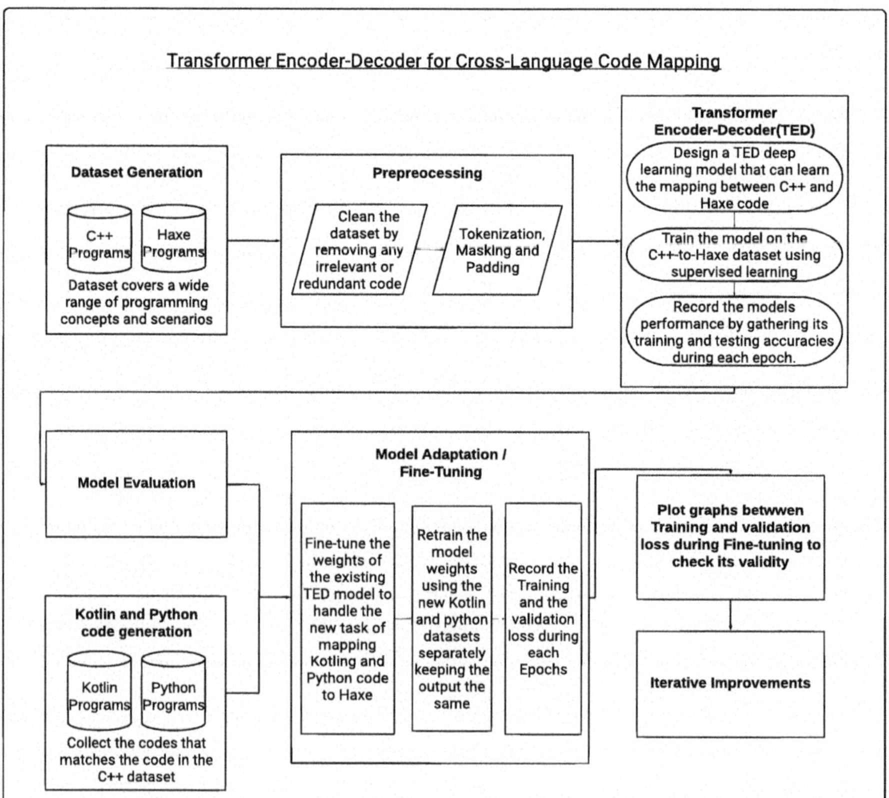

Fig. 30.1 Block diagram of the methodology

Oriented Programming (OOP) concepts while keeping Haxe as the target language. This meets the need for a unified code visualization platform.

Creating a robust dataset is pivotal for successful cross-program translation and code visualization, with a focus on Haxe as the target language. Our methodology systematically covers core Object-Oriented Programming (OOP) concepts, addressing fundamentals and advanced OOP principles like class declaration, function declaration, conditional statements, error handling, inheritance, polymorphism, encapsulation, abstraction, constructors and destructors, instance versus class variables, object instantiation, overloading, overriding, interfaces and abstract classes, association, aggregation, composition, access modifiers, packages, are explored for comprehensive coverage. The dataset delves into nuances of OOP languages, covering topics such as instance versus class variables, object instantiation, overloading, and overriding, as well as interfaces, abstract classes, and complex relationships like association, aggregation, and composition. Input languages, including C++, Kotlin, Python, and Java, diversify the dataset, yet Haxe remains the consistent target language. For each of the OOP concepts, we generate 100 to 200 code samples of varying complexities in each input language, totaling thousands of fragments. This extensive dataset is key to the adaptability of our transformer-encoder-decoder model. It enables comprehensive training and fine-tuning, making the model proficient and highly adaptive in understanding and translating code across various programming paradigms. The adaptability, a cornerstone of our approach, allows the model to seamlessly transition between input languages, consistently producing Haxe as output.

30.3.2 Preprocessing

Tokenization In our research approach, we give a lot of importance to how we break down code into tokens—a crucial step for our transformer-encoder-decoder model to efficiently grasp and translate C++ and Haxe code. The process involves character-level marking, where every character, be it a letter, number, symbol, or special character, gets its own unique identifier. This detailed approach is fantastic for capturing the nuances of the code. But, to also understand the broader meaning and improve computational efficiency, we use word-level tokenization. Here, we group characters into words, helping us understand functions and classes while simplifying the model's vocabulary. Each of these tokens is then linked to integers, a necessary step for the numerical operations our neural network, particularly the transformer model, performs. This mapping facilitates the conversion of text-based code into numerical data. This matters since it's not just about translating characters into numbers. This tokenization process is what allows us to create a context vector. This vector is like a snapshot of the relationships between characters, providing the model with insights into code flow, pattern recognition, and decision-making during translation. It essentially acts as the connective tissue, bridging the input (C++ code) and the output (Haxe code), ensuring a smooth and accurate translation process between the two programming languages.

Masking and Padding The next important step is masking, which is used to attract the attention of the model during training. Padding and masking are especially important when dealing with variable-length sequences, as are common in programming code. Padding ensures that all code sequences are the same length by adding zeros to the end of shorter sequences. This uniformity is essential for efficient group processing of neural networks. On the other hand, masking helps identify which parts of a sequence should be ignored during practice. In our methodology, padding tokens are masked because they do not contain meaningful information and should not influence the model's learning process. By masking padding tokens, we guide the model to focus on the relevant code tokens, ensuring that it pays attention to the actual code elements that contribute to the translation task.

One of the parameters determines the maximum length of code sequences that the model can effectively handle. In our case, this parameter is set to 512 characters, ensuring that the model can accommodate substantial code segments while managing computational resources efficiently. Additionally, the vocabulary sizes for C++ and Haxe are established during tokenization. These vocabulary sizes represent the number of unique tokens identified in each programming language. The embeddings generated from these vocabularies enable the model to represent tokens as continuous, dense vectors. The embedding size parameter, set to 256 in our research, defines the dimensionality of these embeddings, striking a balance between model complexity and its ability to represent code effectively.

In short, the tokenization process transforms or maps raw C++ and Haxe code into sequences of integers, while masking and padding ensure efficient training and accurate translation. This approach empowers our model to comprehend the intricacies of code, capture context through the context vector, and perform precise program-to-program translation, bridging the gap between different programming languages.

30.3.3 Model Architecture and Training

The Transformer Encoder-Decoder model at the heart of our research represents a complex neural network architecture carefully designed to enable translation between programs. It is a two-way architecture consisting of two main components: an encoder and a decoder (Fig. 30.2). Here we explore the intricate details of each component and what they mean.

Encoder The encoder forms the first half of the model, which is responsible for processing and encoding the input C++ code into a format that facilitates translation. It includes the following key elements:

- Input Layer: At the forefront of the encoder lies the input layer. It receives sequences of C++ code snippets, with each sequence representing a discrete code unit or snippet. These sequences are pivotal for capturing the intricacies of code syntax and structure.

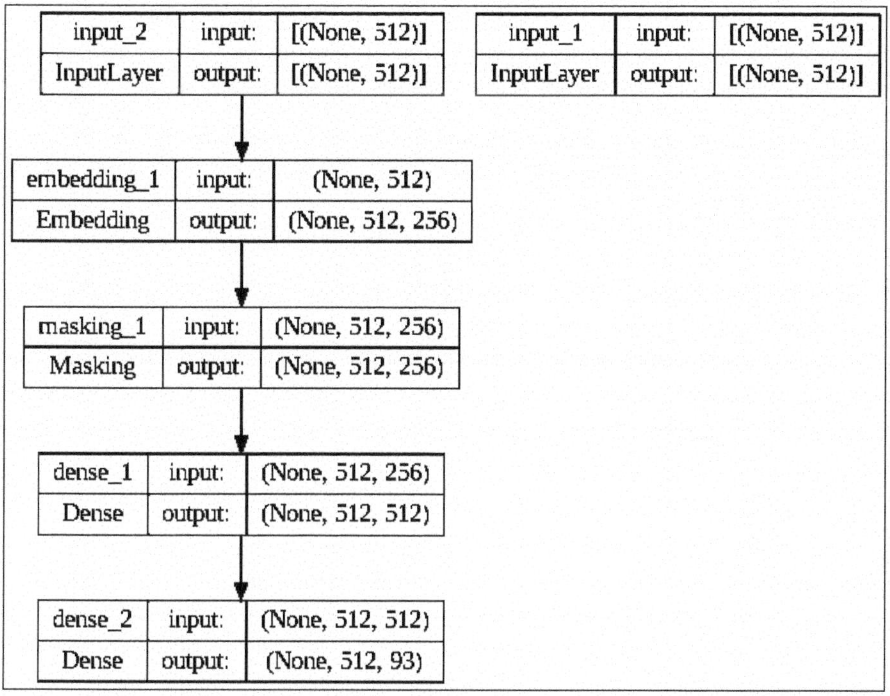

Fig. 30.2 Transformer-encoder-decoder architecture for cross-language code mapping

- Embedding Layer: The embedding layer, a core component of the encoder, holds the transformative power to convert discrete C++ tokens into continuous, dense vectors. These embeddings imbue the model with the ability to discern semantic relationships between tokens, allowing it to grasp the context and semantics of the code.
- Masking: Addressing the challenge of variable-length input sequences, we employ a masking layer. This layer selectively conceals padding tokens, ensuring that the model focuses exclusively on the substantive content of the code while ignoring padding artifacts.
- Dense Layer: Following the embedding and masking layers, a dense layer with a rectified linear unit (ReLU) activation function introduces non-linearity into the encoding process. This layer empowers the model to capture intricate relationships within code structures, thereby enhancing its comprehension.

The encoder's primary mission is to craft a contextual representation of the input C++ code, imbuing the model with the ability to fathom code syntax, semantics, and patterns. This contextual encoding forms the foundation for the subsequent decoding process.

Decoder The function of the decoder, which acts as a counterpart to the encoder, is to generate the Haxe code as the desired output. It mirrors the coder's architecture, but works in reverse, paving the way for code translation:

- Input Layer: The decoder's journey commences with the input layer, which accepts sequences representing Haxe code as its input. These sequences are vital as they serve as the translation target.
- Embedding Layer: Parallel to the encoder, the decoder employs an embedding layer. This layer, like its counterpart, transforms discrete Haxe tokens into continuous, dense vectors, bridging the gap between discrete tokens and continuous representations.
- Masking: The masking layer's role in the decoder is indispensable. It ensures that padding tokens are excluded during the translation process, preserving the purity of the translated code.
- Dense Layer: Following the embedding and masking layers, a dense layer with ReLU activation processes the embedded information, further refining the encoded representations.
- Output Layer: The decoder differentiates itself with its unique output layer. This layer employs a dense layer equipped with a softmax activation function. The magic happens here as it generates a probability distribution across the Haxe token vocabulary for each position in the output sequence. The token with the highest probability at each position is selected, gradually forming the generated Haxe code.

Model Training Our transformer encoder-decoder model embarks on a training journey with a clear objective to learn the intricate art of translating C++ code into Haxe code. During this journey, the model evolves its internal parameters, including weights associated with embedding and dense layers. The foundation of this evolution lies in the principles of backpropagation and gradient descent. Here's a snapshot of how the training unfolds:

1. Pairs of C++ and Haxe code snippets are presented to the model (Fig. 30.3).
2. The model computes predictions for generating Haxe code from the provided C++ code.
3. The dissimilarity between the predicted Haxe code and the actual Haxe code is quantified using the sparse categorical cross-entropy loss function.
4. The optimizer, in our case, the Adam optimizer, fine-tunes the model's parameters to minimize this loss.
5. This iterative process continues, refining the model's translation capabilities with each epoch.

The model's compilation is a pivotal step, dictating how it learns during training. It involves the selection of an optimizer and a loss function. In our case, we opt for the Adam optimizer and the sparse categorical cross-entropy loss function. These choices have proven effective for sequence-to-sequence tasks like code translation. To ensure the model's stability during training and prevent overfitting, we incorporate an early

Input C++ Program

```cpp
#include <iostream>
using namespace std;

struct Node {
    int data;
    Node* left;
    Node* right;
};

Node* createNode(int data) {
    Node* newNode = new Node();
    newNode->data = data;
    newNode->left = newNode->right = nullptr;
    return newNode;
}

Node* insert(Node* root, int data) {
    if (root == nullptr) {
        return createNode(data);
    }
    if (data < root->data) {
        root->left = insert(root->left, data);
    } else if (data > root->data) {
        root->right = insert(root->right, data);
    }
    return root;
}

void inorderTraversal(Node* root) {
    if (root != nullptr) {
        inorderTraversal(root->left);
        cout << root->data << " ";
        inorderTraversal(root->right);
    }
}

int main() {
    Node* root = nullptr;
    root = insert(root, 4);
    root = insert(root, 2);
    root = insert(root, 6);
    root = insert(root, 1);
    root = insert(root, 3);

    cout << "Inorder Traversal: ";
    inorderTraversal(root);

    return 0;
}
```

Target Haxe Program

```haxe
class Main {
    static function main() {
        var root:Node = null;

        function insert(root:Node, data:Int):Node {
            if (root == null) {
                return new Node(data);
            }
            if (data < root.data) {
                root.left = insert(root.left, data);
            } else if (data > root.data) {
                root.right = insert(root.right, data);
            }
            return root;
        }

        function inorderTraversal(root:Node):Void {
            if (root != null) {
                inorderTraversal(root.left);
                Sys.print(root.data + " ");
                inorderTraversal(root.right);
            }
        }

        root = insert(root, 4);
        root = insert(root, 2);
        root = insert(root, 6);
        root = insert(root, 1);
        root = insert(root, 3);

        Sys.print("Inorder Traversal: ");
        inorderTraversal(root);
    }
}

class Node {
    public var data:Int;
    public var left:Node;
    public var right:Node;

    public function new(data:Int) {
        this.data = data;
        this.left = null;
        this.right = null;
    }
}
```

Fig. 30.3 Example of a pair of Input C++ program and its corresponding target Haxe program

stopping mechanism. Early stopping vigilantly monitors the training process, halting it if the loss on the training data starts to increase, a telltale sign of overfitting. This strategy guarantees that the model generalizes well to unseen data, a critical attribute for real-world applications.

Hyperparameters: Several hyperparameters significantly impact the model's performance and training dynamics:

1. Embedding Size (256): This hyperparameter defines the dimensionality of the continuous-valued vectors produced by the embedding layers. A larger embedding size allows the model to capture more nuanced semantic relationships within the code.
2. Hidden Units (512): The dense layers in both the encoder and decoder employ 512 hidden units. This hyperparameter influences the model's capacity to capture complex, non-linear relationships within the code, essential for accurate translation.
3. Number of Training Epochs (20): We train the model for 20 epochs, representing the number of complete iterations over the entire training dataset.

This hyperparameter strikes a balance between training time and model convergence, ensuring that the model has many learning opportunities. Basically, our transformer-encoder-decoder model is the cornerstone of our research methodology and provides an efficient solution to the difficult task of translating C++ code into Haxe code. Its ability to preserve code context and semantics through tight embeddings, combined with carefully chosen hyperparameters and training strategies, makes it a versatile tool for improving code comprehension across object-oriented programming languages.

30.3.4 Fine-Tuning

The fine-tuning process significantly enhances the original C++ to Haxe code translation model, empowering it to smoothly handle two additional programming languages: Kotlin and Python. This expansion represents a crucial step toward creating a versatile multilingual code translation tool. Essentially, fine-tuning involves refining the pre-trained model's understanding, acquired from translating C++ to Haxe, to encompass the unique syntax and semantics of Kotlin and Python.

To kick off fine-tuning, we load code files from dedicated Kotlin and Python directories, forming the training dataset. The existing C++ tokenizer, tuned for comprehending code syntax, comes into play for tokenizing Kotlin and Python code. These tokenized sequences are then padded to match the model's defined maximum sequence length, ensuring consistent processing during training. The adapted C++ to Haxe model serves as the base, with a reduced learning rate-an essential tweak preventing drastic overwriting and facilitating the nuanced accommodation of Kotlin and Python intricacies.

Throughout the training epochs, the model engages with Kotlin and Python datasets, honing its ability to generate Haxe code. The extended training duration, roughly 75 epochs, allows the model to adeptly capture the nuances of the new languages. Post-fine-tuning, the saved model is ready to seamlessly translate code across C++, Kotlin, and Python to Haxe, showcasing its adaptability and versatility. The generated Haxe code stands as evidence of the refined translation capabilities achieved through this meticulous fine-tuning process. In essence, this transformative approach empowers the model to transcend language barriers, embodying a potent, multilingual code translation tool.

30.4 Experiments and Results

The experiments conducted on the C++ to Haxe code translation model yielded significant insights and promising results. The training process extended over 20 epochs, during which the model exhibited substantial learning progress. In the initial epoch, the model began with random weights, resulting in a high training loss of 3.9568 and negligible accuracy for both the training and testing datasets. This outcome was expected as the model lacked any prior knowledge or meaningful representations of the translation task.

However, as training continued, notable improvements became evident. By the third epoch, the training loss had reduced to 0.6096, marking a significant drop from the initial value. While the training accuracy remained relatively low at 0.17%, there was a glimpse of progress with a testing accuracy of 1.00%. Subsequent epochs witnessed a consistent trend of improvement. During the 20th epoch, the model achieved remarkable results (Fig. 30.4). The training loss had diminished to 0.0017, reflecting a substantial decrease from the initial epoch. The training accuracy had surged to 97.33%, showcasing the model's ability to effectively translate C++ code into Haxe code. Meanwhile, the testing accuracy plateaued at 95.00% from the 14th epoch onwards, demonstrating the model's stability and robustness. These findings underscore the model's learning trajectory and its impressive capability to master the intricate task of translating between C++ and Haxe.

These experiments have showcased several noteworthy accomplishments. First and foremost, they demonstrated the model's remarkable learning progress. The initial randomness in weights gradually converged into meaningful representations, leading to a significant reduction in training loss (Fig. 30.5). Furthermore, the model exhibited impressive translation accuracy. The testing accuracy reaching 95.00% by the 20th epoch indicates its proficiency in accurately translating C++ code to Haxe code. This achievement underscores the practical viability of the model for code

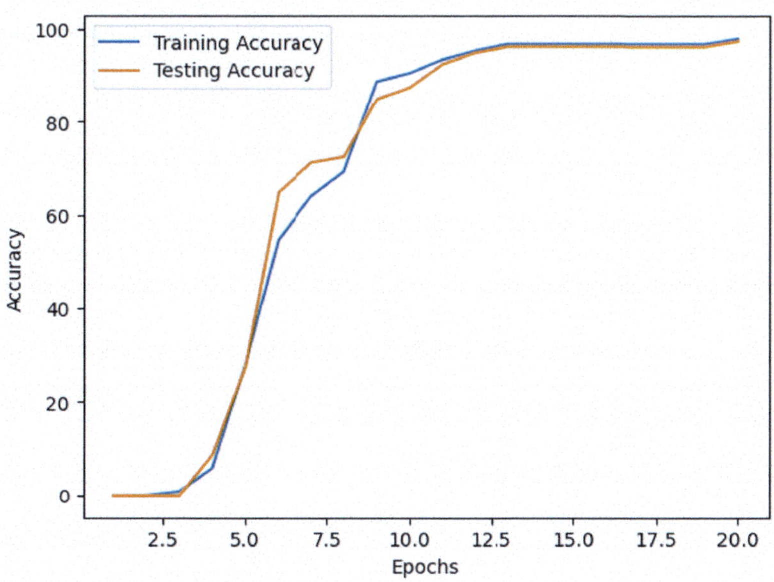

Fig. 30.4 Plot of training accuracy and Testing accuracy w.r.t epochs of transformer-encoder-decoder model trained on C++ and Haxe dataset

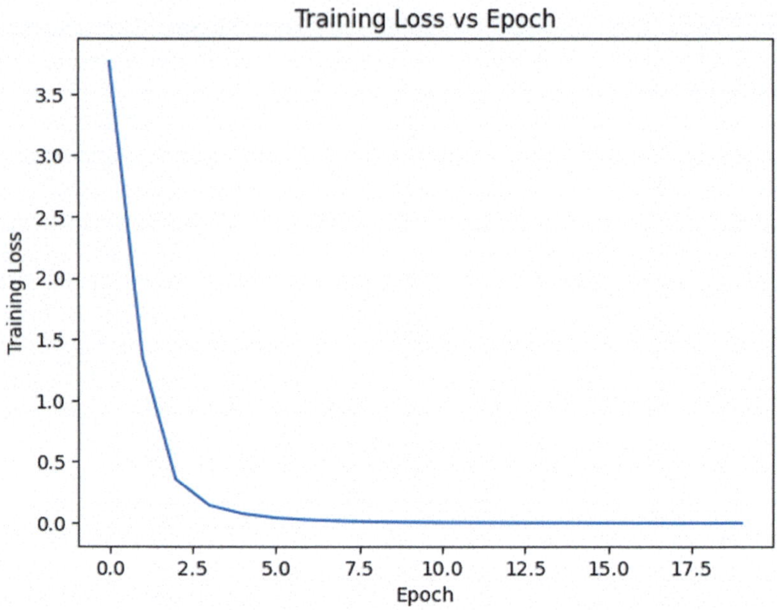

Fig. 30.5 Plot of training loss w.r.t of transformer-encoder-decoder model trained on C++ and Haxe dataset

translation tasks. The use of early stopping with a patience of 3 played a crucial role in ensuring the model's optimal performance. It prevented overfitting and allowed the model to halt training when further progress was unlikely, thus preserving computational resources. Additionally, the model was trained efficiently, with each epoch taking approximately 13 s (Fig. 30.6). This efficiency not only facilitated the experimentation process but also made the model suitable for deployment in real-world scenarios.

In short, these experiments have successfully trained a C++ to Haxe code translation model using a transformer encoder-decoder architecture. The model's substantial learning progress and high translation accuracy lay a strong foundation for future endeavors, including fine-tuning and expansion to handle code translation tasks across various programming languages. This versatility positions the model as a valuable tool for code translation applications in both research and practical contexts.

The fine-tuning process of the model on Kotlin data is a critical step in ensuring its performance and adaptability to specific tasks. Analyzing the data during the training and testing reveals important insights into the model's training and generalization capabilities. During training, both training and validation loss show a consistent downward trend. The training loss starts at 0.0010 and steadily decreases with each epoch, indicating that the model learns effectively from the training data. Similarly, the validation loss is also reduced, initially 0.0010. This indicates that the

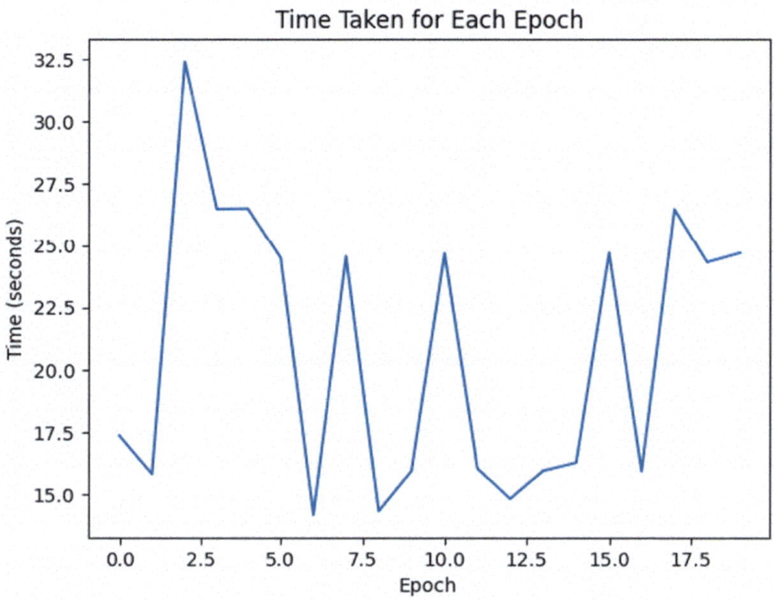

Fig. 30.6 Plot of time taken w.r.t epochs of transformer-encoder-decoder model trained on C++ and Haxe dataset

model generalizes well to unseen validation data, which is an important part of model performance. One notable observation is the convergence of both training and validation loss. As the number of epochs increases, the loss values stabilize, indicating that the model is refining its understanding of the Kotlin dataset. This convergence is a positive sign that the model is learning the underlying patterns in the data and is not showing signs of overfitting. The stability and consistency of the training process is evident, as there are no sharp spikes or drops in the loss values (Fig. 30.7). This stability reflects a well-behaved training process where the model's performance continuously improves with each epoch. At the end of 75 epochs, the model achieves impressive results. The training loss is about 0.0004, while the validation loss is about 0.0003. These low loss values indicate that the model has achieved high accuracy and precision in its predictions, which proves its ability to make reliable and accurate predictions for Kotlin-related tasks. One of the most promising aspects of this fine-tuning process is the generalization of the model. The validation loss closely follows the training loss, suggesting that the model generalizes well to new, unseen data. This is a crucial feature because it ensures that the model's performance extends beyond the training dataset, making it applicable in real-world scenarios.

In addition, the effective training duration of each epoch shows that the model training process is not too time-consuming, which is essential for practical applications. In conclusion, tuning the model with Kotlin data resulted in a well-functioning and stable model. It shows consistent improvements in both training and validation

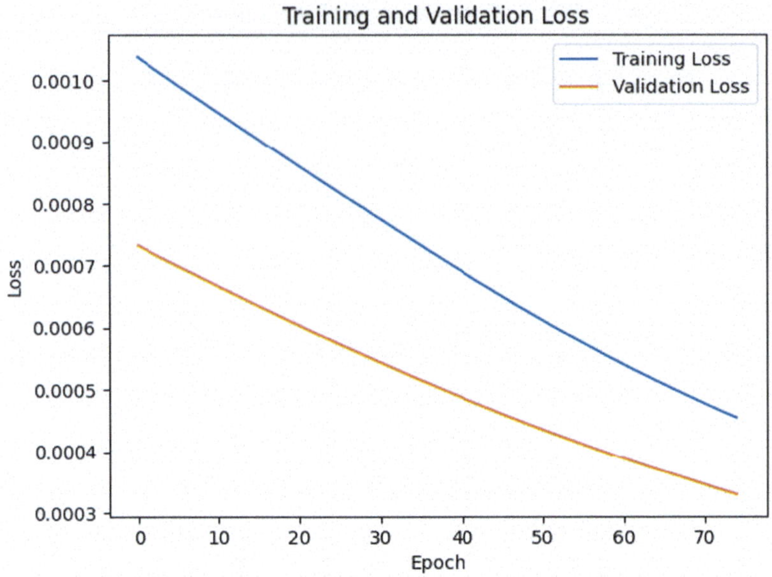

Fig. 30.7 Plot of training loss and validation loss w.r.t epochs of transformer-encoder-decoder model fine-tuned on Kotlin dataset

loss, highlighting its ability to effectively learn from data without overfitting. With its impressive results and generalization capabilities, this model holds significant promise for various Kotlin-related tasks and applications in the real world.

The fine-tuning process of the model on python data is a critical step in ensuring its performance and adaptability to specific tasks. The training process starts with an initial training loss of approximately 4.4908e-04 and a validation loss of 3.2584e-04 in the first period. As the model continues to learn, these loss values gradually decrease (Fig. 30.8). From the second period, the training loss is reduced to about 4.4380e-04, while the validation loss is slightly reduced to 3.2194e-04. This indicates that the model effectively optimizes its parameters to minimize prediction errors. This trend continues steadily throughout the training. At the midpoint, around epoch 38, the training loss reached about 2.8675e-04 and the validation loss is about 1.9993e-04. The decreasing difference between training and validation losses indicates that the model generalizes well to unseen data, which is a positive sign. As the training process nears completion, loss values continue to decrease. The last epoch (epoch 75) has a training loss of about 1.8077e-04 and a validation loss of about 1.1543e-04. These values indicate that the model has converged and its predictions are becoming very accurate.

It is important to note that the duration of each epoch is also indicated, which can be useful to evaluate the effectiveness of the training. The duration varies from approximately 8 to 13 s per epoch, suggesting a relatively stable training time. In summary, the data shows a successful training process characterized by a consistent

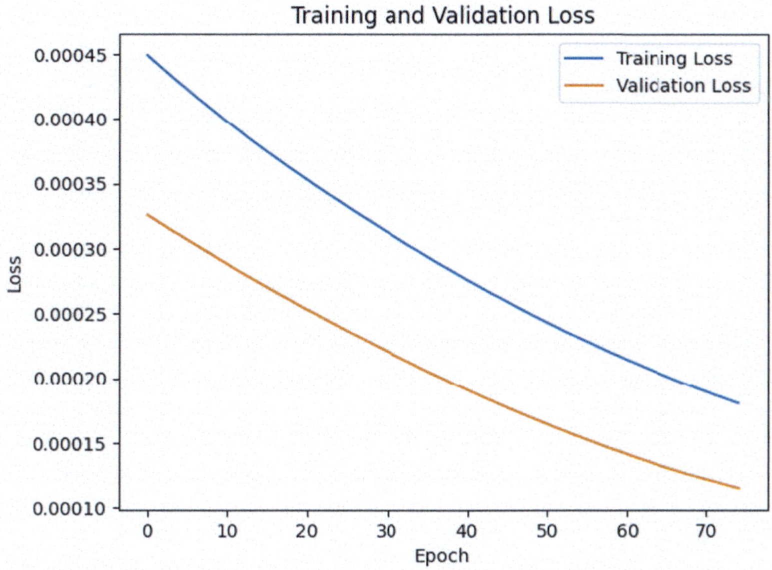

Fig. 30.8 Plot of training loss and validation loss w.r.t epochs of transformer-encoder-decoder model fine-tuned on python dataset

decrease of training and validation loss values over 75 time periods. This demonstrates the model's ability to learn from data and make accurate predictions while maintaining good generalization to unseen data. The total duration of the training is within acceptable limits, which further shows the effectiveness of the training.

In short, the experiments conducted on the C++ to Haxe code translation model demonstrate remarkable learning progress, transitioning from random weight initialization to meaningful representations and achieving a substantial reduction in training loss. The model exhibits high translation accuracy, with the testing accuracy reaching 95.00% by the 20th epoch, confirming its proficiency in accurately translating C++ code into Haxe code. Early stopping with a patience of 3 ensures optimal performance and resource efficiency. Furthermore, the model's training efficiency, with each epoch taking approximately 13 s, facilitates practical deployment. Similarly, fine-tuning the model on Kotlin and Python data showcases its effective learning from data without overfitting, resulting in a stable and well-generalizing model. This achievement positions the model as a valuable tool for code translation tasks and various real-world applications in both Kotlin and Python domains, offering high accuracy and efficiency.

30.5 Discussion

This study dives into creating a code translation model, with a specific focus on converting C++ to Haxe. To ensure a solid foundation, we carefully prepared a dataset comprising pairs of C++ and Haxe code fragments. The preprocessing steps, like tokenization and sequence padding, were meticulous, setting the stage for a well-shaped training dataset that could generalize effectively. Our model architecture, built on the transformer-encoder-decoder framework, did a great job capturing the essence of code semantics and dependencies. Fine-tuning, a crucial step, involved leveraging transfer learning on datasets representing Kotlin and Python. We saw positive outcomes, including decreasing training loss, high translation accuracy, and efficient training times. Looking forward, we see potential in expanding the model's versatility to cover more languages, making it a handy tool for broader code migration. The insights gained from this study could also benefit coding practices and overall code quality. An interesting point to note is the implicit conversion of Haxe to Java, opening up possibilities for integration into Java-based environments like JIVE. Recognizing our study's limitations, such as the 512-token sequence length constraint, prompts us to consider future exploration. Extending this limit might empower the model to handle more extensive and complex codebases. Additionally, enriching the dataset could lead to further improvements in translation accuracy and generalization.

To sum it up, our research has birthed a versatile code translation model, smoothly connecting object-oriented languages like C++, Kotlin, and Python to Haxe, which can then be converted to Java for visualization in tools like JIVE. Looking ahead, we're eager to explore extending sequence length limits, enhancing dataset, and discovering more applications for this model in the ever-evolving landscape of software development.

30.6 Conclusion

Our research endeavors have resulted in the development of a cutting-edge code translation model that not only showcases remarkable performance but also introduces a paradigm shift in the realm of code translation. Throughout this journey, we have successfully crafted and fine-tuned a transformer encoder-decoder architecture, tailored to the nuanced intricacies of various programming languages, including C++, Haxe, Kotlin, and Python. The model's exceptional performance speaks volumes about its learning progress and translation accuracy, evident in the smooth transformation of C++ code into Haxe, Kotlin, and Python. Employing an early stopping strategy with a patience of 3 played a key role in maintaining robustness and efficiency. Our insights into the training process highlight consistent improvement, generalization capabilities, and efficient duration, reinforcing its applicability in real-world scenarios. Looking forward, we see potential in leveraging Haxe for implicit conversion to Java, broadening the model's application scope. Recognizing room for improvement, increasing the maximum sequence length and incorporating a richer dataset could further enhance its performance, enabling the translation of more extensive and complex code. Amidst the prevalence of Large Language Models (LLMs) in natural language processing, our specialized model emerges as a beacon, bridging the gap between programming languages and offering a profound understanding of code structures and design patterns. In summary, our research propels code translation into a new era, with an adaptable model poised to revolutionize software development processes and advance cross-lingual software development frontiers.

References

1. Jayaraman, S., Jayaraman, B., Lessa, D.: Compact visualization of Java program execution. In: Software: Practice and Experience, vol. 47, pp. 163–191. Wiley, New York (2017). https://doi.org/10.1002/spe.2411
2. Shobitha, M., Sidharth, R.P., Sreesruthi, P.K., Varun Raj, P., Swaminathan, J. (2022). Comparison of concurrent program behavior using java interactive visualization environment. In: Smys, S., Balas, V.E., Palanisamy, R. (eds.) Inventive Computation and Information Technologies. Lecture Notes in Networks and Systems, vol. 336. Springer, Singapore. https://doi.org/10.1007/978-981-16-6723-7-29
3. Jevitha, K.P., Jayaraman, S., Jayaraman, B., Sethumadhavan, M.: Finite-state model extraction and visualization from Java program execution. Softw Pract Exper. **51**, 409–437 (2021). https://doi.org/10.1002/spe.2910
4. Aggarwal, K., Salameh, M., Hindle, A.: Using machine translation for converting python 2 to python 3 code. Technical report, PeerJ PrePrints (2015)
5. Bahdanau, D., Cho, K., Bengio, Y.: Neural machine translation by jointly learning to align and translate. In: ICLR (2015)
6. Rabinovich, M., Stern, M., Klein, D.: Abstract syntax networks for code generation and semantic parsing (2017). arXiv:1704.07535
7. Indhu Priya, B., Swaminathan, J.: Enhancing software fault localization using deep learning techniques. In: 2023 World Conference on Communication and Computing (WCONF), RAIPUR, India, pp. 1–5 (2023). https://doi.org/10.1109/WCONF58270.2023.10235227
8. Tai, K.S., Socher, R., Manning, C.D.: Improved semantic representations from tree-structured long short-term memory networks (2015). arXiv:1503.00075
9. Cho, S.J.K., Memisevic, R., Bengio, Y.: On using a very large target vocabulary for neural machine translation. In: ACL (2015)
10. Li, J., Wang, Y., Lyu, M.R., King, I.: Code completion with neural attention and pointer networks. IJCAI (2018)
11. Chen, X., Liu, C., Song, D.: Tree-to-tree neural networks for program translation. In: Advances in Neural Information Processing Systems, pp. 2547–2557 (2018)
12. Alon, U., Brody, S., Levy, O. and Yahav, E.: code2seq: generating sequences from structured representations of code. ICLR (2019)
13. Alon, U., Sadaka, R., Levy, O., Yahav, E.: Structural language models for any-code generation (2019). arXiv:1910.00577
14. Artetxe, M., Labaka, G., Agirre, E.: Unsupervised statistical machine translation (2018). arXiv:1809.01272
15. Artetxe, M., Labaka, G., Agirre, E., Cho, K.: Unsupervised neural machine translation. In: International Conference on Learning Representations (ICLR) (2018)
16. Mou, L., Li, G., Zhang, L., Wang, T., Jin, Z.: Convolutional neural networks over tree structures for programming language processing. In: AAAI, vol. 2, p. 4

17. Mou, L., Men, R., Li, G., Zhang, L., Jin, Z.: On end-to-end program generation from user intention by deep neural networks (2015). arXiv:1510.07211
18. Nguyen, A.T., Nguyen, T.T., Nguyen, T.N.: Divide-and-conquer approach for multi-phase statistical migration for source code. In: 2015 30th IEEE/ACM International Conference on Automated Software Engineering (ASE), Nov 2015, pp. 585–596
19. Yin, P., Neubig, G.: A syntactic neural model for general-purpose code generation. In: Proceedings of the 55th Annual Meeting of the Association for Computational Linguistics (Volume 1: Long Papers). Vancouver, Canada: Association for Computational Linguistics, pp. 440–450 (2017)
20. Akhila, N., et al.: Comparative study of bert models and roberta in transformer based question answering. In: 2023 3rd International Conference on Intelligent Technologies (CONIT), Hubli, India, pp. 1–5 (2023). https://doi.org/10.1109/CONIT59222.2023.10205622
21. Gopalakrishnan, A., Soman, K.P., Premjith, B.: Part-of-speech tagger for biomedical domain using deep neural network architecture. In: 2019 10th International Conference on Computing, Communication and Networking Technologies (ICCCNT), Kanpur, India, pp. 1–5 (2019). https://doi.org/10.1109/ICCCNT45670.2019.8944559
22. Aloysius, N., Geetha, M., Nedungadi, P.: Incorporating relative position information in transformer-based sign language recognition and translation. IEEE Access **9**, 145929–145942 (2021). https://doi.org/10.1109/ACCESS.2021.3122921
23. Amodio, M., Chaudhuri, S., Reps, T.: Neural attribute machines for program generation (2017). arXiv:1705.09231
24. Devlin, J., Chang, M.W., Lee, K., Toutanova, K.: Bert: Pre-training of deep bidirectional transformers for language understanding (2018). CoRR, arXiv:abs/1810.04805
25. Feng, Z., Guo, D., Tang, D., Duan, N., Feng, X., Gong, M., Shou, L., Qin, B., Liu, T., Jiang, D., et al.: Codebert: a pre-trained model for programming and natural languages (2020). arXiv:2002.08155
26. Nguyen, T.D., Nguyen, A.T., Nguyen, T.N.: Mapping API elements for code migration with vector representations. In: IEEE/ACM International Conference on Software Engineering Companion (ICSE-C), pp. 756–758. IEEE (2016)
27. Karaivanov, S., Raychev, V., Vechev, M.: Phrase-based statistical translation of programming languages. In: Proceedings of the 2014 ACM International Symposium on New Ideas, New Paradigms, and Reflections on Programming and Software, pp. 173–184. ACM (2014)

SPRINGER NATURE

GPSR Compliance

The European Union's (EU) General Product Safety Regulation (GPSR) is a set of rules that requires consumer products to be safe and our obligations to ensure this.

If you have any concerns about our products, you can contact us on ProductSafety@springernature.com

In case Publisher is established outside the EU, the EU authorized representative is:

Springer Nature Customer Service Center GmbH
Europaplatz 3
69115 Heidelberg, Germany

The manufacturer's authorised representative in the EU is Springer Nature Customer Service Centre GmbH, Europaplatz 3, 69115 Heidelberg, Germany. If you have any concerns regarding our products, please contact ProductSafety@springernature.com

Printed and bound by CPI Group (UK) Ltd, Croydon, CR0 4YY
30/09/2026
02225253-0015